Soft and Biological Matter

For further volumes:
http://www.springer.com/series/10783

"Soft and Biological Matter" is a series of authoritative books covering established and emergent areas in the realm of soft matter science, including biological systems spanning from the molecular to the mesoscale. It aims to serve a broad interdisciplinary community of students and researchers in physics, chemistry, biophysics and materials science.

Pure research monographs in the series as well as those of more pedagogical nature, will emphasize topics in fundamental physics, synthesis and design, characterization and new prospective applications of soft and biological matter systems. The series will encompass experimental, theoretical and computational approaches.

Both authored and edited volumes will be considered.

Diethelm Johannsmann

The Quartz Crystal Microbalance in Soft Matter Research

Fundamentals and Modeling

Springer

Diethelm Johannsmann
Institute of Physical Chemistry
Clausthal University of Technology
Clausthal-Zellerfeld
Germany

ISSN 2213-1736 ISSN 2213-1744 (electronic)
ISBN 978-3-319-07835-9 ISBN 978-3-319-07836-6 (eBook)
DOI 10.1007/978-3-319-07836-6
Springer Cham Heidelberg New York Dordrecht London

Library of Congress Control Number: 2014941727

Printed on acid-free paper

Springer is part of Springer Science+Business Media (www.springer.com)

Dedicated to my family

Preface

If we date the birth of the quartz crystal microbalance (QCM) to Günter Sauerbrey's paper from 1959 [1] the QCM is now more than 50-years old. Its use in vacuum is routine to the extent that the fact is not usually mentioned in the scientific literature. The QCM flew to Mars in 1997 [2, 3] and there are least four books covering it in detail. Why one more book?

Starting in the 1980s, the QCM got its second wind after it was immersed into liquids and after it was realized that the QCM is a surface-analytical tool with capabilities much beyond gravimetry. The advanced QCMs supply the bandwidth of the resonance in addition its frequency (where the "dissipation factor" of the QCM-D is equivalent to bandwidth), and they do so on a number of different overtones. This added information has opened the door to non-gravimetric applications of the QCM. "Non-gravimetric" sounds as if it was not known what—exactly—was measured. That may have been true at some instances in the beginning, but it is not true today. The QCM is a surface-acoustic-wave-based analytical instrument and the behavior of surface acoustic waves in complex media is not a mystery.

The widespread use of the QCM in liquids since 1990 was mainly driven by the novel instrumentation, but there also was a widespread interest in soft matter at interfaces. Typical samples were polymer brushes, supported lipid bilayers, polyelectrolyte multilayers, adsorbed proteins, cell cultures, and functional polymer films in a wider sense. The evolution of the QCM was paralleled by similar instrumental progress in optics (surface plasmon resonance spectroscopy, advanced ellipsometry, to name examples), by the long and multifaceted evolution of scanning force microscopy, and by improved electrochemical instrumentation. These instruments and the research done with them form the context of the QCM and the book therefore occasionally makes comparisons and indicates, where combinations can be of use.

The QCM is simple, basically, but to set up one's own instrument requires somewhat of a physics background. One needs a basic understanding of electrical circuits and one needs to pay attention to the construction of the crystal holder. If one is without such skills (or without the time to revive them), one can embark on QCM-based research with one of the commercial instruments. One may plug, play, and see what happens. This progress in the hardware means, among other things, that ever more users apply this technique to ever more diverse samples. As always,

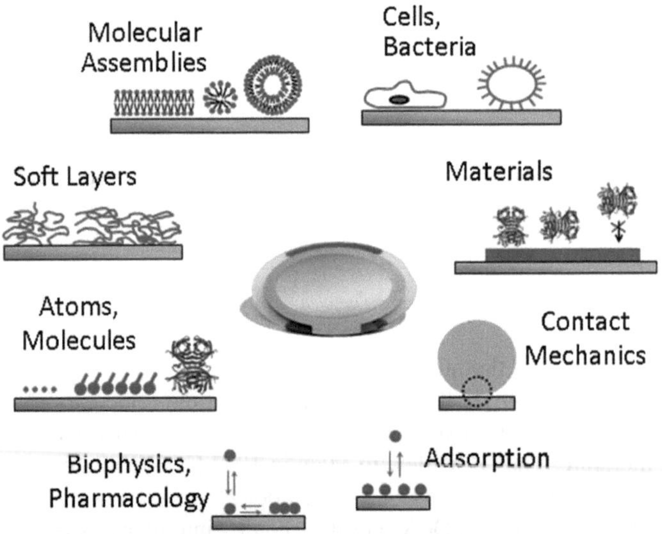

Fig. 1 Various applications of the QCM

plugging and playing is one thing, not having to worry about the instrument is another. The advanced QCMs open up new research opportunities, but there is a danger of not sufficiently appreciating the different sources of artifacts.

This book provides a detailed description of the background behind the technique and the data analysis methods, in a format that can be understood and appreciated by users with backgrounds ranging from biology to engineering and at a sufficient level of detail that it is educational. The analysis of acoustic shear waves builds on a number of fundamental concepts of physics, which many users of the technique do not usually come across. Short discussions of background material (viscoelasticity and contact mechanics, in particular) have therefore been included.

The author's thinking about the QCM is guided by what he calls the "small load approximation" (SLA). The SLA states that the frequency shift is proportional to the amplitude of the periodic stress at the resonator surface. A recent appraisal of the SLA can be found in Ref. [4]. The SLA is as old as acoustic sensing (meaning: older than the QCM) and it was not called small load approximation previously. The approximation needs a name and "small load approximation" is the best one the author could think of. The SLA was made the guiding principle for modeling. Of course choosing such a unique perspective has its pros and cons. It gives the book a clear structure. It (hopefully) simplifies the presentation of the more advanced topics. At the same time, it makes some of the familiar and time-honored results look unfamiliar. The Sauerbrey equation has suffered this fate. It is derived in the middle of the book (Sect. 8.1). Had the Sauerbrey relation been the main topic of the book, the formal machinery leading to the SLA and from there to the Sauerbrey result would have been too much of a good thing. With the SLA at hand, the discussion can move on from the Sauerbrey case to viscoelastic films, bubbles, slip, multilayers, proteins, and contact mechanics (Fig. 1). With the machinery established, it all falls in place.

The mathematics was kept at such a level that undergraduate students in any of the science disciplines should be able to follow every single step. The physics of the QCM in many respects makes for an interesting bit of science education. Numerous aspects have counterparts in optical spectroscopy, electrical engineering, quantum mechanics, rheology, or mechanics. Familiarizing oneself with the QCM, one renews old acquaintances and that in itself can be an enjoyable exercise.

The book is self-contained. At least this was the intention. As far as the QCM itself is concerned, all proofs are outlined step-by-step. Some derivations, where the mathematical machinery serves as proof only and contains little physical insight, have been moved to the appendices. The appendices also contain background information. The book walks on a somewhat narrow line with regard to anisotropy. Anisotropy is essential. Quartz would not be piezoelectric if it was isotropic. Yet, if anisotropy had been covered in its full beauty, this would have hidden some of the simpler concepts. In order to avoid the discussion of anisotropy, the parallel-plate model is often pushed to the foreground. Within the parallel-plate model, all tensor properties (just about all properties of a crystal other than density) reduce to complex numbers, which are called the "relevant tensor component." The "relevant component" is meant to be evaluated after the respective tensor has been transformed to the coordinate system of AT-cut quartz. The algebra leading to these tensor components is not needed and it is therefore not reproduced. Good books on the matter are Refs. [5] and [6]. The algebra does not hold any unsolved problems; anisotropy is well understood. Not coincidentally, the early masterpiece on anisotropy, the "Lehrbuch der Kristallphysik" by W. Voigt (first published in 1910), contains a thorough discussion of piezoelectricity. Voigt coined the term "tensor." Tensor algebra is not a black art, it is merely a matter of not letting oneself be intimidated by large equation systems.

With regard to the fundamentals, the book has to start somewhere. For example, while the reader is reminded of what complex numbers are, there is no detailed introduction to their properties. Readers not familiar with complex numbers will need to exercise the basic operations. Similarly, Taylor-expansions are often applied. The respective formula is then cited but not proven. Furthermore, resonances and waves are introduced only briefly. Resonances and waves are explained with much didactic zeal in the textbooks of physics; there is no point in trying to do better here.

The number of papers *making use* of the QCM is so large that it is futile trying to reference even a significant fraction. The book is brief on applications for a second simple reason, which is limited expertise. The QCM has become a standard tool in diverse areas of soft matter research and the life sciences. The book touches on selected applications only. Along similar lines: The QCM is a member of a wider class of acoustic sensors. Other acoustic sensors are mentioned, but the focus is on the QCM.

Much knowledge has been gained already and giving due credit to all previous workers is a difficult task. This shall not be an excuse not to try. However, the following remarks should not be understood as a historical account. While writing the book, the author discovered work, which he had not been remotely aware of. It is fair to assume that there is more work waiting for rediscovery. The author covers his own work in more depth than the work of others because this is what he knows best. Please do not misunderstand this as a lack of appreciation for the achievements of others.

Below is a list of datable accomplishments deserving a special mention:

- Starting with the early days of acoustic sensing, the book by Mason from 1948 certainly is a hallmark [7]. It contains many of the modern concepts. That includes the electromechanical analogy, equivalent circuits, the small load approximation, and an acute emphasis on the fact that the viscoelastic parameters of soft materials depend on frequency.
- What is called the tensor form of the small load approximation here (Sect. 6.1), was derived by Pechhold in 1959 and published under the title *Zur Behandlung von Anregungs- und Störungsproblemen bei akustischen Resonatoren* [8]. As the title says, Pechhold chose to publish in German (but so did Sauerbrey). The paper is not cited well; it is not even listed in the Web of Science. Its outcome is of fundamental relevance to the physics of all resonant acoustic sensors.
- Central to acoustic sensing of course is Günter Sauerbrey's paper from 1959 [1]. Sauerbrey is open about the fact that the influence of added mass onto the resonance frequency was known to many in the field. People used to tune resonators by marking the resonator surface with a pencil. There actually is a second brief paper from 1959 proposing the use of piezoelectric crystals for microweighing [9]. Sauerbrey put the idea into a concise form.
- As far as instrumentation is concerned, the QCM-D from Q-Sense has certainly made a remarkable contribution to the field. It has made the technique available to a broad community. It should be remembered, though, that the "QCM-D principle" is equivalent to impedance analysis. Knowing this context is worthwhile.

A recent book on the QCM and related piezoelectric sensors, edited by Janshoff and Steinem, appeared in 2007 [10]. The book has a rather wide scope; it covers diverse applications in-depth. An earlier book, edited by Arnau, covers similar content, but has more of an emphasis on acoustics, instrumentation, and modeling [11]. Still earlier, the books by Thompson and Stone [12], and by Balantine et al. [13] as well as volume 107 of the *Faraday Discussions* [14] collect knowledge about the advanced QCM and related techniques at a time, when this methodology was picking up momentum.

A book often quoted with regard to the classical QCM (operated in air or vacuum) is the one edited by Lu and Czanderna [15]. Still further back in time, the book chapter by Stockbridge gives an overview of the QCM applied in vacuum [16]. Quartz crystals in general are treated in-depth by Bottom [17] and by Salt [18]. Bottom is interested in resonators more than in sensors. He approaches the subject from the perspective of the practitioner, but still covers physical background (including anisotropy). The book by Rosenbaum [19] complements Bottom's book. Rosenbaum is more detailed with regard to theory and modeling than Bottom. The tutorial by Vig covers quartz resonators in time and frequency control [20]. There are a number of excellent reviews covering specific applications and topics, referenced further below in the respective context. For the newest developments, the literature surveys by Cooper and co-workers are a good source [21–23]. Cooper et al. are brief on technology and modeling; they are mostly interested in the growing range of applications.

A few remarks on notation:

- The formal machinery heavily builds on complex numbers and complex amplitudes. Complex numbers and amplitudes of complex time-harmonic functions were given a tilde (~) and a hat (^), respectively (example: $\tilde{u}(t) = \hat{u}\exp(\mathrm{i}\omega t)$). If one and the same letter is used with and without a tilde, the letter without tilde denotes the real part. The imaginary part then is small and was neglected. The prime (′) and the double prime (″) denote the real and the imaginary part of a complex quantity. What is often called the "modulus" of a complex quantity, $|\tilde{z}|$, is called "absolute value" here. This is intended to avoid confusion with the shear modulus.
- Bold letters denote vectors.
- The components of piezoelectric tensors relevant to the AT-cut carry the index "26". The text does not elaborate on tensor algebra. With regard to the index 26, we just follow the practice in other sources [19].
- In cases where variables have different meanings in the literature, they were mostly given an index, rather than risking confusion. For example, G is the shear modulus in rheology, and is the conductance in electricity. The electrical conductance is called G_{el} here, even though this makes the equations look clumsy. A few variables *do* have more than one meaning (D, n, ε). In these cases the context hopefully clarifies the variable's meaning.
- When inserting numbers into equations, the chosen values mostly pertain to the crystals used in the author's laboratory. These are AT-cut crystals ($Z_q = 8.8 \times 10^6$ kg m^{-2} s^{-1}) with a fundamental frequency of $f_0 = 5$ MHz. Less important than f_0 is the size. The author mostly uses 1-inch crystals. These have an active area around 30 mm^2.

References

1. Sauerbrey, G.: Verwendung von Schwingquarzen zur Wägung dünner Schichten und zur Mikrowägung. Zeitschrift für Physik **155**(2), 206–222 (1959)
2. http://qcmresearch.com/Images/sojourner.gif. Accessed 11 Feb 2013
3. http://en.wikipedia.org/wiki/Sojourner_(rover). Accessed 11 Feb 2013

4. Reviakine, I., Johannsmann, D., Richter, R.P.: Hearing what you cannot see and visualizing what you hear: interpreting quartz crystal microbalance data from solvated interfaces. Anal. Chem. **83**(23), 8838–8848 (2011)
5. Auld, B.A.: Acoustic Fields and Waves in Solids. Krieger Publishing Company, Florida (1990)
6. Royer, D., Dieulesaint, E.: Elastic Waves in Solids, Volumes I and II. Springer, New York (1999)
7. Mason, W.P.: Piezoelectric Crystals and Their Applications to Ultrasonics. Van Nostrand, Princeton (1948)
8. Pechhold, W.: Zur Behandlung von Anregungs- und Störungsproblemen bei akustischen Resonatoren. Acustica **9**, 48–56 (1959)
9. Oberg, P., Lingensjo, J.: Crystal film thickness monitor. Rev. Sci. Instrum. **30**(11), 1053–1053 (1959)
10. Steinem, C., Janshoff, A.: Piezoeletric Sensors. Springer, Heidelberg (2007)
11. Arnau, A.: Piezoelectric Transducers and Applications. Springer, Heidelberg (2004)
12. Thompson, M., Stone, D.C.: Surface-Launched Acoustic Wave Sensors: Chemical Sensing and Thin Film Characterization. Wiley-Interscience, New York (1997)
13. Ballantine, D.S. et al.: Acoustic Wave Sensors: Theory, Design and Physico Chemical Applications. Academic Press, San Diego (1997)
14. Faraday Discussions 107: Interaction of Acoustic Waves with Thin Films and Interfaces, The Royal Society of Chemistry (1997)
15. Lu, C., Czanderna, A.W.: Applications of Piezoelectric Quartz Crystal Microbalances. Elsevier, Amsterdam (1984)
16. Stockbridge, C.D.: Vacuum Microbalance Techniques. In: Behrndt, K.H. (ed.) vol. 5, 4th edn. Plenum Press, New York (1966)
17. Bottom, V.E.: Introduction to Quartz Crystal Unit Design. Van Nostrand Reinhold, New York (1982)
18. Salt, D.: Hy-Q Handbook of Quartz Crystal Devices. Van Nostrand Reinhold, New York (1986)
19. Rosenbaum, J.F.: Bulk Acoustic Wave Theory and Devices. Artech House Inc., Norwood (1988)
20. http://www.am1.us/Local_Papers/U11625%20VIG-TUTORIAL.pdf. Accessed 18 June 2014
21. Cooper, M.A., Singleton, V.T.: A survey of the 2001 to 2005 quartz crystal microbalance biosensor literature: applications of acoustic physics to the analysis of biomolecular interactions. J. Mol. Recognit. **20**(3), 154–184 (2007)
22. Becker, B., Cooper, M.A.: A survey of the 2006–2009 quartz crystal microbalance biosensor literature. J. Mol. Recognit. **24**(5), 754–787 (2011)
23. Speight, R.E., Cooper, M.A.: A Survey of the 2010 Quartz Crystal Microbalance Literature. J. Mol. Recognit. **25**(9), 451–473 (2012)

Acknowledgments

The author is looking back to more than 20 years of work with the QCM. All of this work was done together with others. A first thank you goes to the current and past members of the group and to those who have collaborated with us.

I would like to specifically mention Ilya Reviakine and Arne Langhoff. Ilya Reviakine has worked in Clausthal as a Humboldt fellow a few years ago and has returned numerous times (in part funded by the Humboldt foundation). He has strongly influenced my thinking about soft matter at interfaces and made numerous detailed contributions to the book. Arne Langhoff runs the instrumentation at Clausthal, co-supervises the students working with the QCM, and has been part of this effort in many other ways.

When I started my Ph.D. in 1988, I visited F. Eggers and T. Funk in Göttingen. These people put me on the right track early on, thank you.

I am happy to acknowledge specific contributions to the book by Jörg Adams, Ingo Alig, Anne Finger, Holger Fritze, Sigrid Goertz, Sylvia Hanke, Adriana Ispas, Rebekka König, Astrid Peschel, and Judith Petri.

Jean-Michel Friedt read the close-to-final version of the book line-by-line and made countless suggestions to contents, didactics, and style.

Contents

Chapter 1
Introduction

Abstract The QCM is an amazingly simple device. It consists of a disk of crystalline quartz. The acoustic resonances of this plate can be excited electrically because crystalline quartz is piezoelectric. The main application of quartz resonators is in time and frequency control. However, the resonance frequency and the resonance bandwidth depend on the resonator's environment and the plate can therefore be used as a frequency-based sensor. The chapter gives a brief tour through the modeling process, mostly building on the parallel plate and emphasizing the small load approximation (SLA). Models beyond the parallel plate as well as refinements of the SLA are also discussed. The chapter concludes with an overview of applications.

1.1 Quartz Crystal Resonators

The center piece of a quartz crystal microbalance (QCM) is a quartz crystal resonator (QCR). A QCR essentially is a disk of crystalline quartz as shown in Fig. 1.1. A typical thickness of the resonators used for sensing is 330 μm, a typical diameter is 1.5 cm. The plate carries electrodes on both sides. When a voltage is applied to the electrodes, the inverse piezo-effect induces a shear deformation. It is a shear deformation (rather than an elongation) because the crystal cut was chosen accordingly. If the frequency of the exciting AC-voltage matches one of the acoustic resonance frequencies of the plate, the amplitude of oscillation becomes large and the electric current drawn by the electrode becomes large at the same time. Measuring the current into the electrodes, one can easily find the resonance frequencies.

This sounds all very well, but it does not explain why the market for QCRs in their various forms amounted to about 4.5 billion US dollars in 2010 [1]. About 2 billion units are manufactured annually [2]. According to the tutorial by Vig, 3,000 tons of α–quartz were grown in 1997 [3]. Of the man-grown single crystals,

D. Johannsmann, *The Quartz Crystal Microbalance in Soft Matter Research*,
Soft and Biological Matter, DOI: 10.1007/978-3-319-07836-6_1,

quartz was second only to silicon in quantity grown. The amount of silicon was 3–4 times larger.

A first explanation for the large market volume certainly is the large demand for clocks. (Somewhat loosely, any element delivering a precise frequency is called a "clock" here. More exact would be "oscillators and filters".) Historically inclined readers may consult a book by Dava Sobel to find that this demand is not new [4]. Sobel reports that the classical pendulum clock cannot be used on a boat and why people were desperately looking for a substitute 300 years ago. The problem was not genuinely a problem of knowing the time, but rather one of navigation: An accurate clock is needed to calculate the position of a ship from the position of the sun. The academics (in Sobel's words: "the leading lights of [the] day") pushed hard for a solution based on astronomy. A accurate determination of the position of the moon and an equally accurate calculation thereof can yield the time, and this did work, in principle. However, the solution eventually adopted was contrived by John Harrison, a self-educated carpenter, who improved spring-driven clocks to the extent that they were accurate enough for navigation, robust enough to work on a boat, and simple enough to be used by the captain. Even today, navigation requires good clocks. The clocks carried by the GPS satellites are among the best money can buy. And even today, these clocks are better than the astronomical frequency standards, although the latter (the millisecond pulsars) also are pretty good [5].

It is argued in Sect. 15.8 that smaller is not always better, but smaller was better when the spring-driven clock replaced the classical pendulum clock and smaller was also better when the quartz clock replaced the spring-driven clock. Other advantages of QCRs are *simplicity*, *precision*, and *stability*.

Simplicity mostly is the consequence of the piezo-effect. Reading the resonance frequency of the crystal is easy because the mechanical vibration creates a voltage. Piezoelectricity is very interesting, but it is actually not quite as important for understanding QCRs as one might think. For a start, one may just remember that for reasons deep-down in solid-state physics, a quartz plate generates a voltage when squeezed. By exploiting the piezoelectric effect, a QCR can be electrically controlled. Of course someone has to design the electronic equipment. Arguably, the electronics is not simple, but it is to the author because he just buys it.

Precision is the ability to distinguish between two slightly different values. Precision differs from accuracy in that both these readings are allowed to deviate from the true values for a precise instrument, while no such deviation is allowed for an accurate instrument. Accuracy is not needed for QCM-based sensing because the data analysis entirely relies on frequency *shifts*. If the resonator's resonance frequency is unequal to the frequency specified on delivery, this will not affect the shift of frequency upon film deposition. (More exactly: This effect will be negligible. The Sauerbrey equation does contain the absolute frequency, but the potential error is minute.) Sensing here differs from timing, where a clock must be calibrated against a master clock in order to be of any use.

The precision of QCRs is excellent because the resonances are outstandingly narrow. They are much sharper than anything achievable with electrical circuits (at room temperature). The sharpness of a resonance is quantified by its quality factor, Q. Q is defined as $Q = f_r/(2\Gamma)$ with f_r the resonance frequency and Γ the half-bandwidth (Fig. 1.1c, Sect. 4.1.4). Quartz crystal resonators easily achieve quality factors of more than 10^5. (Q-factors exceeding 10^9 have been reported at cryogenic temperatures [6], but such instruments are outside the scope of this book.) For a resonance frequency of 5 MHz, the half-bandwidth can be less than 25 Hz. Because the bandwidth is so extraordinarily small, it is easy to determine the resonance frequency with good precision. Oscillators employing quartz crystals typically have a short-time reproducibility in frequency better than 10^{-6} (talking of cheap clocks).

The *stability* of QCR-based clocks and sensors is good, but it is not as good as the repeatability of frequency achieved with atomic clocks. Quartz crystals have a bit of a character: They age, they respond to stress and temperature, and they remember their history. There is a complex interplay between crystal defects and their migration, on the one hand, and stress and temperature, on the other. Quartz resonators respond slowly and hysteretically to fluctuations of temperature and stress. There are several ways to deal with this problem, but if long-time stability is the prime concern, microwave clocks are the instruments of choice. One might think that aging (at the level of 10^{-7}/year) should not be an issue in sensing because the short-time repeatability achieved in sensing is around 10^{-7}. However, the sources of drifts are also much more pronounced in sensing than in frequency control. Mounting a crystal in a liquid cell creates a static stress, which slowly relaxes over time. A typical drift is 1 Hz/h; a typical precision is 0.1 Hz. Frequency drifts are about as much of a challenge in sensing as precision.

When building a clock from the quartz crystal, one is always well advised to *not* expose the crystal to the environment. One encapsulates it in a tight box, keeps it stress-free, makes sure that it does not see contaminations, and maintains constant temperature. One separates the crystal from the environment as much as possible because the crystal is an excellent sensor. It responds to stress, to changes in temperature, and to contaminants. Before turning to sensing, we spend one section on the application which drove the materials development, which is time and frequency control.

1.2 Piezoelectric Resonators in Time and Frequency Control

The first piezoelectric oscillator was patented in 1918 by Nicholson from the Bell Telephone Laboratories [7]. This happened 38 years after the discovery of piezoelectricity by the Curie brothers in 1880. Nicholson employed Rochelle salt, which was widely studied at the time. Rochelle salt has the problem of being water soluble. Nevertheless the material was used commercially for quite some time (although not for oscillators [8]). The first quartz clock (the "crystal clock", as it was called [9]) was built in 1928 by Marrison and Horton [10]. Marrison and Horton employed a torsional rod, operating at 50 kHz. This clock still suffered from temperature-frequency coupling. Koga from the Tokyo Institute of Technology based his oscillator on a disk employing the thickness-shear mode [11]. Koga wrote down Eq. (1.5.2). Later he found the temperature-compensated cuts. The oscillator circuits (which essentially contain an amplifier with a resonator in the feed-back loop) were developed at this time. For instance, Pierce designed what today is called the Pierce circuit [12]. Oscillator circuits turn resonators into clocks. In the 1930s NIST (then the "National Bureau of Standards") defined its time standard based on quartz clocks. Quartz clocks remained in that function until they were replaced by the atomic clocks in the 1960s [13].

During World War II, quartz resonators were a key component of radar technology. The military was the technology driver. A letter to the editor of the LIFE Magazine, published in on August 23, 1943 says: *Sirs: In proportion to size these little glass like quartz wafers are perhaps the most remarkable of all the tools science has given to war. When the story of the almost incredible progress in research and manufacture of radio crystals can be told, it will prove to be a tale of one of the war's greatest achievements. No less significant will be the fruit of these advancements to a new world at peace where crystals will be the vibrating hearts of most telecommunication equipment.*

Gerald James Holton, Harvard University, Cambridge, Mass [14].

Holton was right: After Word War II quartz oscillators turned into a mass product [15]. In the process, methods were found to synthesize quartz rather than harvesting it from natural resources [16]. In recent years, much effort has been spent on making the resonators smaller, so that they fit into portable devices more easily. A certain drawback of the technology is that quartz crystals cannot be monolithically integrated with silicon chips. Monolithic integration is possible for the thin-film bulk acoustic wave resonator (FBAR) [17]. FBARs are mostly made of aluminum nitride (AlN). The fabrication process is compatible with CMOS technology. The FBAR functions in essentially the same way as the QCR. FBARs have been evaluated for sensing purposes [18–21], but their importance in sensing there is nowhere near their role in frequency control.

At the high end, quartz clocks compete with the microwave standards. The best oscillators made out of quartz (employing the BVA design [22, 23]) age at a rate below 4×10^{-9}/year. They cost a few thousand US dollars. The low-end rubidium clocks are comparable in price and in performance. At the time of writing, the NIST-F1 (a cesium fountain clock) reaches a frequency uncertainty of 3×10^{-16}; optical clocks are said to be even better. These are stable down to a level below 10^{-18} [24]. Progress continues and the future is always difficult to foresee. The author still anticipates that both the quartz crystal resonator and the quartz crystal microbalance will be around for a while. Both are workhorses in their fields and it is worthwhile to spend time understanding them.

1.3 The Quartz Crystal Microbalance

Since a precise measurement of a QCRs resonance frequency is easy and since the resonance frequency is affected by the resonators environment in a number of different ways, sensing applications abound. Following the tradition in much of literature on the QCM, the advanced QCM is called a "sensor" here. It senses the properties of its environment. The textbooks on chemical sensors usually state that a sensor must be cheap and simple. Unfortunately, the advanced QCMs are not what would be called cheap in the sensors community. But that might change. Is the advanced QCM simple? It can be, if put to work properly.

This book is limited to effects, which are caused by *a sample in contact with the resonator surface* in one way or another. QCRs can sense temperature, hydrostatic pressure [25], and static stress [26], but these devices are outside of the scope here. To the author, fluctuations of temperature, hydrostatic pressure, and static stress are disturbances to be avoided if ever possible.

The name "microbalance" goes back to Günter Sauerbrey [27]. The name suggests that the QCM is a quartz-based device for measuring small ("micro") masses. Actually it measures an areal mass density in the units of mass per unit area, rather than a mass. The two quantities should not be confused. The name QCM is convenient and it stuck, despite the fact that the material employed is not always quartz, that the QCM is today used for purposes other than weighing, and that even the prefix "micro" might replaced by "nano" because the sensitivity is well in the range of ng/cm^2. (The modern MEMS- and NEMS-based devices might be called "zeptobalances" [28]. Even a "yoctobalance" based on the resonances of carbon nanotubes was proposed [29].) As you can tell from the title of the book, the term "quartz crystal microbalance" in the author's opinion is a good choice, not only because of age. In particular, the name "balance" is appropriate even for the advanced QCMs—if slightly reinterpreted. On a fundamental level, the QCM measures periodic forces exerted at its surface (Sect. 4.6). It is a "force balance". Coincidentally, the forces turn out to be in the microNewton range, so that the prefix "micro" can stay. Depending on the exact configuration, one might want to call the QCM a stress balance, not a force balance. This distinction parallels the distinction between mass and mass per unit area.

Some further remarks on terminology:

- The abbreviations "QMB" and "QCMB" (both for "quartz crystal microbalance") also are around. To the author all three acronyms are equally legitimate and he has followed what he perceives to be the majority opinion.
- "PM" for "piezoelectric microgravimetry" denotes all piezoelectric sensing schemes targeting a mass. Most mass-sensitive detectors are plates; all of them are resonators or acoustic waveguides. Some mass-sensitive MEMS-resonators do not rely on piezoelectricity for actuation.
- In the context of acoustic resonators in their various forms, the QCM is called "TSM resonator" (for "thickness-shear mode") and "BAW resonator" (for "bulk acoustic wave"). The former property is much important than the latter.

The vibration of the QCM being of the thickness-shear type is critical for its use in liquids. Resonators employing the thickness-expansion mode (such as the standard FBARs, Sect. 15.6) cannot be used because these emit compressional waves. They interact with liquid samples much differently from the QCM. Unfortunately, the QCM is not truly a TSM resonator. Because of energy trapping, the mode of vibration has flexural admixtures (Sect. 7.6). To what extent a given mode couples to compressional waves requires a closer look.

The distinction between bulk acoustic wave (BAW) resonators and other configurations (such as the SAW devices, Sect. 15.4) is a different matter, because the frame of data analysis may actually be the same. Surface acoustic waves (if of the shear-horizontal type) interact with the sample in about the same way as waves permeating the entire resonator (bulk waves). The principles underlying the interpretation of QCM-data also apply to SH–SAW devices (shear-horizontal surface acoustic wave devices). All acoustic resonators making use of shear waves measure the shear-wave reflectivity at the interface with the sample. In this regard, the QCM is just one member of a family.

This is a good time to insert a brief explanation of why the resonance frequency of a quartz disk is sensitive to deposited mass. The fact itself is easy to understand. Figure 1.2 sketches the principle of operation. On resonance, the deformation pattern is a standing plane wave with antinodes at the surface. The wavelength of sound is twice the thickness of the plate. If a thin film is deposited on one surface and if this film has the exact same acoustic properties as the resonator, all the film does, is to increase the acoustic thickness of the plate. The wavelength of sound, λ, increases and the resonance frequency, f_r, decreases accordingly. One has

$$f_r = \frac{c_q}{\lambda} = \frac{c_q}{2(d_q + d_f)} \tag{1.3.1}$$

c_q is the speed of sound, d_q is the thickness of the resonator plate, and d_f is the film thickness.

For such samples, the Sauerbrey result (Eq. 8.1.2) follows from Fig. 1.2 and Eq. (1.3.1) with very little mathematics in-between. Equation (1.3.1) suggests that the frequency shift was mainly proportional to the film thickness. However, the film might differ in its acoustic properties from the plate itself. The more detailed analysis shows that the QCM measures the film's *mass per unit area* (not the film thickness), regardless of what the film's physical properties are. Actually, the sample may or may not be a homogeneous film. As long as the sample is rigid, the QCM measures the sample's area-averaged mass per unit area. Again: This applies as long as the sample is so thin, that it appears as rigid to the QCM. If the sample is not thin in this sense, this is by no means the end of a meaningful QCM experiment. The sample might be a soft film, a liquid, a polymer brush, or a biological cell. Those samples are the realm of the *non-gravimetric QCM* and those samples are, what this book mostly is about.

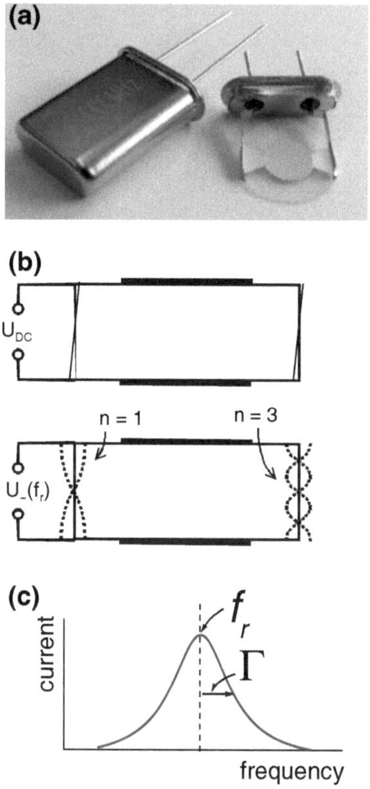

Fig. 1.1 **a** A quartz crystal in its encasement (*left*) and with the cap removed (*right*). Such crystals are employed in frequency control. The electrodes have keyhole shape to confine the oscillations to the center of the crystal (Chap. 7). Modern crystals often are smaller than the one shown here. **b** When a static voltage is applied to the electrodes, the crystal responds with a thickness-shear deformation. However, the static deformation is minute. The lateral displacement at the crystal surface is 3.1 pm/V. When the crystal is excited with an AC-voltage and when the frequency matches the resonance frequency of the thickness-shear mode, the displacement pattern is a sine wave with antinodes at the resonator surface. The number of nodal planes parallel to the resonator surface is equal to the overtone order, n. The *dotted lines* shows the fundamental ($n = 1$) and the third overtone ($n = 3$). **c** When the frequency of excitation is swept across the resonance, the amplitude of oscillation and the current drawn by the electrodes become large. From an electrical measurement of the current-voltage ratio, one infers the resonance parameters. Most important are the resonance frequency, f_r, and the half-bandwidth, Γ

Employing quartz resonators as sensors rather than clocks necessitates a few changes in the hardware. For instance, the holder must allow to remove the crystal for cleaning. Roughly speaking, there are two types of holders. The first type is similar to what is shown in Fig. 1.1. The crystal is lightly held by two wires. Contrasting to frequency-control applications, the contacts are not soldered, but just established by a soft spring (and possibly a small dot of a conductive glue). This holder exerts little stress. This is a benefit because stress makes the resonance frequency drift. The

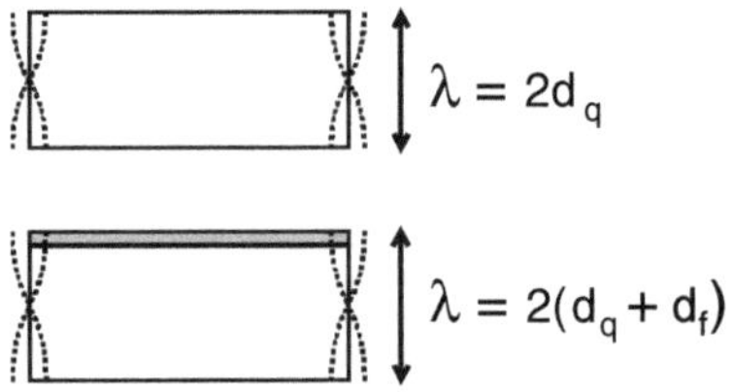

Fig. 1.2 Principle of operation of the quartz crystal microbalance. By depositing a film, one increases the thickness of the plate. Since the standing shear wave has antinodes at the resonator surface, a film increases the wavelength and decreases the resonance frequency. d_q: thickness of the crystal, d_f: film thickness, λ: wavelength of shear sound

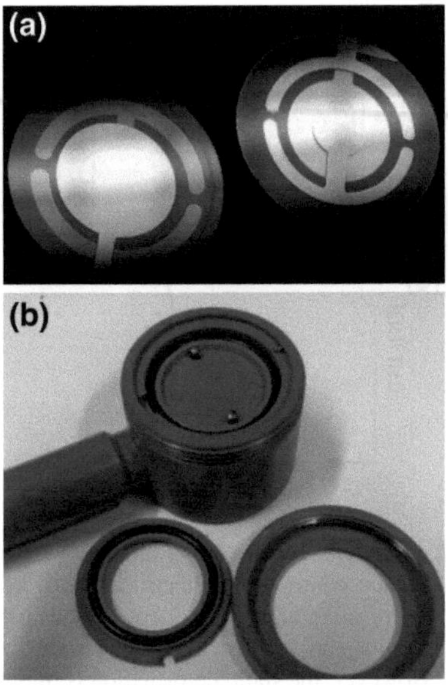

Fig. 1.3 **a** Images of crystals used for sensing in liquids taken from the front (*left*) and from the back (*right*). The front electrode (exposed to the sample) usually is larger than the back electrode. This ensures that the entire active area is homogeneously covered by the front electrode. A second purpose is to protect the sample from electric fringe fields (Sect. 14.2). Often, the front electrode is covered by a layer specifically interacting with the sample (a receptor layer). **b** A typical holder. Both electrodes are contacted with pins from the back. For sensing in liquids the crystal is usually mounted such that the back stays dry. This holder is suitable for large amounts of liquids. Small liquid cells right above the resonator are in use, as well

drawback is poor temperature control. Heat cannot be easily supplied or removed across the two wires. The second type of holder is shown in Fig. 1.3b. The crystal is held by O-rings around its edge. This mount establishes a better thermal control than the two-wire configuration. Also, it hides the back of the crystal from the sample. The front electrode wraps around the edge of the crystal, so that both electrodes can be contacted with soft elastic springs from the back. The stripes connecting to the pins are usually given the shape of anchors (Fig. 1.3a), so that the user does not have to pay much attention to the orientation of the crystal. The user should still make sure that the front electrode connects to ground and the back electrode connects to signal. The front electrode should be grounded in order to keep electric fringe fields away from the sample (Sect. 14.2). For the same reason the (grounded) front electrode is larger than the back electrode.

Advanced applications of the QCM introduce three more constraints. Firstly, it is essential that the crystal indeed is a parallel plate. Wedge distortions are detrimental. Secondly, the crystal surface must be optically polished. Rough surfaces in contact with a liquid will have an effect on the frequency shift, which is usually avoided. Overtone crystals happen to be polished anyway, because polishing improves the Q-factor on the overtones. A third requirement for advanced sensing applications is good crystal quality. In crystal growth, there is a trade-off between cost and the defect density [30]. High-quality crystals are needed for liquid sensing. At the time of writing, the price per crystal is 10 US dollars and above. Quartz crystals are not disposable items in the narrow sense of the word. The cheap crystals produced for consumer electronics (these cost a few cents) have been evaluated for liquid sensing [31], but are not widely used.

The electrode material can be adapted to the application. Gold is widely used because it is chemically inert. There usually is a thin adhesion layer of chromium or titanium underneath the gold. Resonators with coated electrodes (silica, titania, self-assembled monolayers of various kinds) are available from a number of different suppliers.

1.4 Operation in the Liquid Phase, Surface Specificity

In the 1980s, piezoelectric resonators were *again* immersed into liquids. They had been there before [32–35]. The extensive work by Mason, McSkimin, and their co-workers starting in the 1930s all concerned complex fluids. In the early days of acoustic sensing the resonators mostly were torsional rods. A profound understanding of the underlying physical principles was achieved. The book by Mason from 1948 collects the knowledge of the time [36]. It heavily builds on equivalent circuits and the electromechanical analogy. It also contains the small load approximation, although Mason quantifies dissipative processes by motional resistance rather than bandwidth.

Nomura [37, 38] and Bruckenstein [39] (two researchers often quoted for making the QCM wet) had an electrochemical application in mind. The goal was to determine the mass transfer in an electrochemical process. This worked and it still does. The instrument today is called electrochemical QCM (EQCM). Early reviews were published by Buttry and Ward [40] and by Schumacher [41]. Once the QCM worked reliably in liquids, it was also applied in the life sciences. Reviews can be found in Refs. [42–45], and [46].

Bruckenstein's and Nomura's work (and also the work be Konash and Bastiaans [47]) triggered the widespread use of the QCM in liquids, but attempts along this line had been made before. The review by Alder and Mc Callum [48] mentions two articles from 1969, where a QCM had been used as the cathode in an electrochemical process [49, 50]. Apparently, these experiments were perceived as being difficult; no one followed. With regard to non-electrochemical applications of the QCM in liquids, there are same scattered reports, notably some from Russia [51, 52]. Pechhold immersed the QCM into polymer melts [53]. As far as technology is concerned, it is hard to see, why the QCM in liquids only became widespread after 1985. Had more people taken an interest in running a QCM in liquids earlier, it would have happened earlier. Of course it *did* happen earlier, but not at the large scale seen today.

When QCMs were first used in liquids, it was a problem to find a driver, which could to handle the large viscous damping. One can either use a suitably adapted oscillator circuit or resort to a passive method called "impedance analysis" in this book. Ring-down as a routine technique of interrogation was introduced later [54, 55]. Arguably, ring-down was *re*introduced in the 1990s [56], but the same can be said about many other developments of that time. Once passive interrogation is in place, one realizes that resonance frequency and resonance *bandwidth* are both available for analysis and that, further, these parameters can be acquired *on a few different overtones*. It is this added information, which opens up the sensing opportunities beyond gravimetry [57].

Today there are a number of advanced QCMs on the market. "Advanced" here is supposed to mean that the instrument analyzes frequency as well as bandwidth (in one way or another). Some of the advanced QCMs do so on different overtones. Among the suppliers of advanced QCMs are 3T analytik (the "qcell T"), AWS (the "AWS 10" and the "AWS 20"), Gamry (an EQCM, the "eQCM 10 M"), initium (the "affinix series"), the Institute of Physical Chemistry of the Polish Academy of Sciences (IPC PAS, the "EQCM 5710"), Maxtek (now Inficon, the "RQCM"), Q-Sense (the "QCM-D"), Sierra Sensors (the "QCMA-1"), and SRS (the "QCM200"). The list intentionally leaves out QCMs, which do not measure the bandwidth, but still deserve to be called "advanced" with regard to liquid handling, automated data analysis, data acquisition on many channels, or combination with electrochemistry (EQCMs). Evidently, this definition of what should be called advanced is guided by the author's own interests.

Given that Mason and McSkimin spent their careers on running resonators in liquids, one wonders why they did not employ AT-cut quartz plates. They *did* think of this. In 1949 they published an article reporting on how they had used AT-cut quartz plates to extend the frequency range of their instrumentation to a few MHz [34]. However, they did not immerse the crystals themselves into the liquid but rather glued the crystals to the ends of cylindrical rods of fused quartz. The crystals were excited by a MHz RF-pulse, thereby launching a transverse acoustic wave in the rod. The wave travelled down the rod, was reflected at the rear end (which was immersed into the liquid), and returned to the crystal, which also served as the detector. The liquid's viscosity was inferred from the reflected amplitude. Mason et al. comment in the introduction on why they did not immerse the AT-cut crystals themselves into the liquid: *Consideration was given to the use*

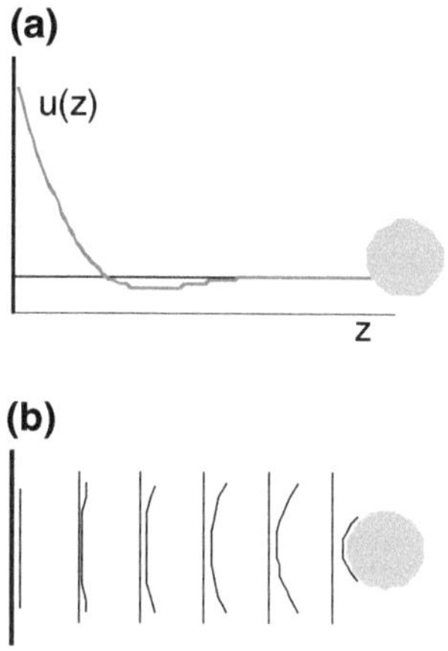

Fig. 1.4 Shear waves rapidly decay in liquids and in gases (**a**). The penetration depth at 5 MHz in water is 250 nm. Objects located outside this horizon are not seen by sensor. This contrasts to compressional waves, which may return to the sensor surface after having been scattered from objects in the bulk of the liquid (**b**)

of a thickness vibrating shear crystal of the AT or BT type, but is was found that the shear motion was too closely coupled to other modes of motion, such as flexure modes, to give reliable results. Hence another method had to be used [34].

Before answering the question of why this argument is correct in a way, but still does not prevent the use of piezoelectric plates in liquids, the crucial importance of the resonator's vibration being of the thickness-shear type needs to be addressed. Central to the success of the QCM in liquids is the fact that the motion of the resonator surface occurs into the tangential direction. Unfortunately, this is not rigorously true, but if it were grossly wrong, the QCM would work much less well in liquids than it actually does. A surface moving tangentially emits an acoustic *shear* wave. In liquids, shear waves (transverse waves) and compressional waves (longitudinal waves) are much different types of excitation. Soft matter is much more compliant to shear stress than to hydrostatic pressure. The latter changes the material's volume. Soft materials easily change shape, but not volume. For this reason, the wavelength of shear waves is much smaller than the wavelength of compressional waves. Even further: Shear waves decay rapidly (Fig. 1.4a) because of viscous damping. In water at 5 MHz, the depth of penetration is about 250 nm. Given this difference, researchers from the field of acoustics often do not even count shear waves as "acoustic waves". They view shear waves as boundary layers, and they have a point. The fact that shear waves are boundary layers is of immense practical importance. Because shear waves do not propagate in liquids, the QCM is "surface-specific". The QCM only sees objects within a distance of about a micron from the resonator surface. If the QCM was not a surface-specific device, it would see all objects in the cell, which give an acoustic contrast (Fig. 1.4b). That would much complicate the interpretation.

Going back to the paper by Mason et al. from 1949 [34], Mason and co-authors basically are right: AT-cut plates are not pure thickness-shear mode (TSM) resonators. There are flexural contributions to the mode of vibration (Sect. 7.6). The resonator surface moves both tangentially and normally. In air, the normal component can be larger than the tangential component. Given the flexural contributions, shifts of the resonance frequency caused by compressional waves must be expected. There are two complications, which Mason and co-workers did not sufficiently appreciate at the time. Both work to the advantage of sensing in liquids. Firstly, compressional waves are emitted in the wings of the amplitude distribution (close to the edge of the electrodes), where the sensitivity of the resonator is low. Secondly and more importantly, the liquid itself reduces the amplitude of the compressional waves. With regard to flexural deformation of the plate and compression of the liquid, the crystal actually is softer than the liquid. The crystal bends much less in water than it bends in air (Sect. 6.1.3). In hindsight, luck was of help when people started using the QCM in liquids. There were excellent reasons to expect that those experiments would fail. Experiment proved these worries to be unjustified.

When crystals were immersed into liquids, a problem on the modeling side was to distinguish effects of viscosity from effects of deposited mass. One hopes that the two effects are additive, as formulated on a heuristic basics by Martin et al. [58]. The treatment of viscoelastic multilayers (Sect. 10.2) shows that this is indeed the case in the thin-film limit. If effects of mass and of viscosity are additive, one still needs to understand the effects of viscosity. In this regard we deal with a classical rediscovery. Gordon and Kanazawa are widely quoted for the respective equation [59], but the relation occurs at various places earlier in the literature [51, 52, 60, 61] and follows rather straight-forwardly from Mason's work [36]. Mason did not employ AT-cut crystals and his equations therefore do not contain the parameter Z_q (the acoustic impedance of AT-cut quartz), but apart from that, Eq. 14.19 in Mason's book is the Gordon-Kanazawa result (Eq. 9.1.1). The result should be renamed, and "Gordon-Kanazawa-Mason result" in the author's opinion is the best option. This being said: The determination of viscosity is not a prime application of the advanced QCMs. For various reasons (Sect. 9.1) a QCM-based measurement of a viscosity is more difficult than the Gordon-Kanazawa-Mason result suggests. Reliable determination of a complex fluid's viscosity covering frequencies from quasi-static to the MHz range still is a challenge.

1.5 A Quick Tour Through Modeling

1.5.1 Resonances, Resonators, and the Small Load Approximation

Figure 1.1a shows the simplest possible representation of a resonator, which is the spring pendulum. A mass is attached to a rigid wall across a spring. The resonance frequency is given by

$$f_r = \frac{1}{2\pi}\sqrt{\frac{\kappa_R}{M_R}} \quad (1.5.1)$$

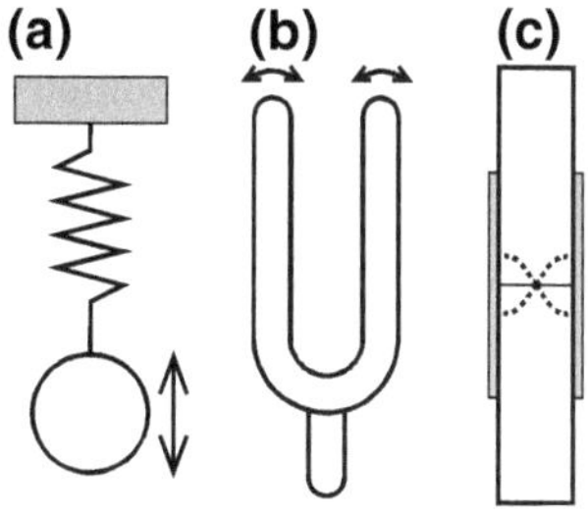

Fig. 1.5 Representations of resonators. **a** Idealized spring pendulum, composed of a mass-less spring and a rigid mass. **b** Tuning fork. **c** Piezoelectrically excited thickness-shear mode (TSM) resonator

κ_R is the spring constant and M_R is the mass. It is easy to see how this device can be turned into a balance. If one increases the mass, the frequency decreases. Inserting numbers, one finds that an increase of the mass by 1 % decreases the resonance frequency by about 0.5 %. (For the mathematically inclined readers: The factor of 1/2 between the change in mass and the change in frequency is a consequence of the Taylor expansion of the square root, which is $(1 + \varepsilon)^{-1/2} \approx 1 - \varepsilon/2$.)

Figure 1.5a is not only simple, it is too simple for our purposes. For instance, it does not indicate how the resonance should be excited, how the resonance frequency should be measured, and whether or not excitation and read-out have an influence on the resonance frequency. These problems were a major engineering challenge in the early time of pendulum clocks. The pendulum clock is controlled by a mechanism called "escapement" [62]. Its design is critical. A further problem with Fig. 1.5a is that the resonator is connected to a rigid wall at the top. There is no such thing as a rigid wall. The holder might have an influence on f_r. Most importantly, Fig. 1.5a assumes a mass-less spring and a rigid mass. Real spring-based resonators contain light-weight springs and rather stiff bodies as the masses. Still, when making the idealizations, one inevitably introduces errors (albeit possibly small ones).

Figure 1.5b shows a much improved representation of a resonator, which is the tuning fork. The tuning fork is an elastic body. Elasticity and inertia enter the force balance equation everywhere. On resonance, there is a time-periodic deformation pattern. The calculation of the deformation pattern is nontrivial, but the outcome qualitatively is known from experience (double arrows). On resonance, there is little deformation at the center. The tuning fork can be held by hand with little disturbance to the resonance. There is no need for a rigid wall. Excitation occurs by striking the tuning fork against a surface. Detection occurs acoustically. Both excitation and acoustic damping have a small influence on how the tune fork sounds, but the influence is small. The tuning fork was designed to make this influence small.

The representation in Fig. 1.5c comes as close to our needs, as a graphical representation can come. The resonator again is an elastic body. It does not contain mass-less springs or rigid masses. Although not indicated in the figure, there is an elegant mechanism for excitation and read-out, which is piezo-electricity.

What distinguishes this resonator from the tuning fork, is that the simple geometry (the parallel-plate geometry) easily allows to calculate the deformation pattern of the resonant mode. Of course the deformation pattern of the tuning fork can be calculated as well [63], but the result is not nearly as simple as the sine wave found for the parallel plate. The sine wave is indicated with dotted lines in Fig. 1.5c. Actually, the deformation patterns of the thickness-shear resonators employed in practice slightly differ from the sine wave. It is a bit more complicated. The plates are a little thicker in the center than at the edge; they form an acoustic cavity. In consequence, the amplitude of oscillation decays towards the edge and one can hold the crystal at the edge without overdamping the resonance. This of crucial importance (Chap. 7).

Once the deformation pattern is known, one can calculate the resonance frequency from the thickness of the plate and the speed of sound. The deformation pattern is a standing plane wave with a wavelength, λ, equal to twice the thickness of the plate, d_q. The wavelength, in turn, is equal to the speed of sound, c_q, divided by the resonance frequency, f_r. One has

$$2d_q = \lambda = \frac{c_q}{f_r} \tag{1.5.2}$$

The device can be turned into a balance by depositing a film on one side. Let the film thickness be 1 % of the plate's thickness and let the acoustic properties of the film be the same as the acoustic properties of the plate. From Eq. (1.5.2) one sees that a 1 % increase in mass leads to a 1 % decrease in frequency. This contrasts to Eq. (1.5.1), where a 1 % increase in mass reduced the frequency by 0.5 %. Clearly, only one of these two predictions is correct (better: is applicable to the problem at hand). It is the second one. Equation (1.5.2) is much better adapted to our problem than equation Eq. (1.5.1).

From Fig. 1.5c, one is readily convinced that the plate has more than one mode of vibration. The figure shows the mode with one single nodal plane in the center of the plate, but there might be more nodal planes. These higher modes have shorter wavelengths and correspondingly higher frequencies (Of course the tuning fork also displays higher modes. The idealized resonator in Fig. 1.5a does not.).

Given that the vibration pattern needs to be known in order to understand the resonator's behavior, one might think that the natural way of calculating a frequency shift would amount to calculating the vibration pattern twice: once with and once without the sample. That would be the rigorous way. However, if the changes to the resonance pattern induced by the sample are small enough (if the small load approximation applies), then there is an approximate way to calculate frequency shift, which avoids the second full calculation of the resonant mode. All that needs to be known is the periodic stress at the interface between the resonator and the sample. Calculating the traction at the resonator surface is less laborious than finding the vibration pattern a second time. If the small load approximation applies, there is an explicit relation leading directly from the periodic traction at the interface to the frequency shift. Of course one can calculate the resonance frequency shift without making use of the small load approximation. Within the frame of equivalent circuits (Sect. 4.5.2), this is not even difficult. One numerically finds those frequencies, where the electrical impedance across the electrode becomes zero. One does this twice, once with a sample present and once for the reference state. The drawback of this approach is that the more qualitative features of the resonator's response to the sample are hidden behind the numerical calculation.

Chapter 4 explains the parallel-plate model in three different ways. Arguable, one would have been enough. These three representations of the parallel plate have their strengths and weaknesses; they will appeal to readers with different backgrounds to different extent. All three are rooted in the literature on the QCM. Last not least, convincing oneself that they are all equivalent is a worthwhile exercise beyond the immediate subject of the QCM. In the following, the three formulations are briefly previewed.

1.5.2 Wave Equation and Continuity at Interfaces

The mathematically-inspired approach (Sect. 4.3) is elegant because it rests on two assumptions only. The first is the wave equation, which holds inside all media (including the resonator itself), and the second is continuity of stress and displacement at the interfaces. Everything else follows from the mathematical machinery. There is no need for further concepts or guiding principles. There is not even a need for physical interpretation. Once the equations are written down, one might as well forget what the variables mean. Put differently, this exact same formalism applies in other branches of physics. "$u(z, t)$" is a transverse acoustic wave here, but if it were an optical wave, the mathematics would still be the same. Of course the speed of sound and the density would then also acquire different meanings.

In order to really honor the beauty of the formalism, one would have to apply it in 3D and account for anisotropy, but a choice was made and anisotropy was left aside. For the sake of simplicity, the model was also kept one-dimensional. The displacement field is a plane wave. Its wave vector only has one component (called wavenumber) and all interfaces are perpendicular to the wave vector. These choices were made, hoping that undergraduates can then follow every single step of the derivation. There is one instance, though, where those readers, who have not been sitting through an introduction to linear algebra, might get lost. A homogeneous linear equation system only has a nontrivial solution if the determinant of the coefficient matrix is zero. Readers, who have never heard of this statement, need to either catch up on linear algebra or skip Sect. 4.3. Skipping Sect. 4.3 definitely is an option. The later chapters do not build on the vanishing determinant.

1.5.3 The QCM as an Acoustic Reflectometer

Section 4.4 takes a more physical approach and introduces the concept of the impedance. Needless to say: The term impedance has more than one meaning. In Sect. 4.4 the reader will need to make peace with at least two of them, which are the "acoustic wave impedance" and the "acoustic impedance". These can take the same value under certain conditions. They have the same physical unit, but they are not the same. (Adding to the confusion: What is called "acoustic wave impedance" here, is widely called "acoustic impedance" in the literature.) The acoustic wave impedance determines the acoustic reflectivity at interfaces. "Reflectivity" is another essential term. Reflectivity here always is a complex number. The reflectivity quantifies the reflected amplitude, not the reflected power. The Cambridge Handbook of Physics Formulas (Ref. [64]) calls this quantity "amplitude reflection coefficient" (to be distinguished from the "power reflectance coefficient"). Section 4.4 explains why the QCM can be viewed as a shear-wave reflectometer and how the shear-wave reflectivity at the interface between the resonator and a coating relates to the coating's thickness.

1.5.4 Equivalent Circuits

Section 4.5 repeats the same content in the language of electrical engineering. The author believes that some readers are, on the one hand, not truly at ease with waves and reflectivity, but are, on the other hand, at ease with how to compute the total electrical impedance of arrangements of electrical elements (resistors, capacitors, inductors). The rules, by which such impedances are computed, are the Kirchhoff rules. The Kirchhoff rules are recapitulated in Sect. 4.7. In simple words, impedances are additive if the respective elements are placed in series, while inverse impedances are additive, if the elements are arranged in parallel. Some readers may say: "How can anyone not know the Kirchhoff rules?" Yes, they are taught in high school, but so is grammar and spelling. Other readers will say: "Please remind me of the Kirchhoff rules". Readers not familiar with electrical networks can jump from the discussion of reflectivity in Sect. 4.4 straight to the applications.

A first reason to introduce equivalent circuits is the tradition in the literature. Equivalent electrical circuits have guided the thinking of the scientists using acoustic resonators since the early days of piezoelectricity. These people where electrical engineers by training and the resonators were parts of electrical networks (they still are). There is a second reason: Piezoelectric stiffening is not covered in Sect. 4.4 because piezoelectric stiffening is not needed for many of the applications. If piezoelectric stiffening is to be included, the wave picture becomes more complicated. One has to deal with coupled equations (Chap. 5). If one wants to account for piezoelectric stiffening correctly and rigorously without much mathematics, the Mason equivalent circuit is the rope to hang on.

To apply the Mason circuit is one thing, to derive it is quite another. The derivation is tedious (Sect. 4.7.4). Its application, however, is easy: Just add all impedances properly and search for the (complex) frequency at which the total electric impedance between the electrodes vanishes. Equivalent circuits are not cartoons. Applied correctly, they quantitatively embody the essence of the parallel-plate model.

1.5.5 Influences of the Electrical Boundary Conditions

One can understand the parallel plate without going into the details of piezoelectricity and this approach is taken in Chap. 4. Chapter 5 addresses certain details of piezoelectricity and piezoelectric stiffening, in particular.

1.5.6 Tensor Form of the Small Load Approximation, Third-Order-Perturbation

The small load approximation as derived in Sect. 4.6.1 applies to the parallel plate. A more general version of the SLA and a more general proof are needed for resonators with other shapes. This proof is provided in Sect. 6.1. It leads to the tensor form of the small load approximation.

The small-load approximation can be viewed as the first-order result of a perturbation scheme, which can be further refined. The refinements are spelled out in Sect. 6.2. They are relevant for the thin film in air.

1.5.7 Beyond the Parallel-Plate Model

The quartz crystals employed in practice are not parallel plates and the deviations definitely are of relevance. Since the deviations are small, they can be treated within a formalism called "paraxial approximation" in optics. (The paraxial approximation is a good start, but it also has short-comings.) The plates are slightly thicker in the center than at the edge. One can think of them as acoustic cavities. The surfaces of the cavity are shaped such that the acoustic beam is focused to the center of the plate. The same strategy is employed in optical cavities (laser cavities). Because the amplitude of oscillation is small at the edge, the plate can be mounted at the edge with little disturbance to its movement. The mechanism carries the name "energy trapping". Energy trapping has a number of consequences, discussed in Chap. 7.

1.6 Related Instruments, Combined Instruments

The book mostly is about the QCM itself; it does not cover acoustic sensors in general. Still, there is a technical context, briefly introduced in Chaps. 15 and 16. The QCM is member of a family of acoustic sensors, which all exploit the reflectivity of acoustic waves (shear waves, if applied in liquids) at interfaces. Chapter 15 briefly describes some of these.

Because QCRs are so simple, they can be easily combined with other surface-analytical techniques. Various combined instruments are discussed in Chap. 16. Most advanced is the combination with electrochemical equipment, which carries the name EQCM. The combination with optical reflectometry is intriguing because the model underlying the analysis of optical reflectivity is strikingly similar to the model underlying the QCM. Still, the acoustic thickness often is larger than the apparent optical thickness, and there is a deeper reason for that.

1.7 Overview of Applications

The book is limited to sensing problems, where the shift in frequency is somehow caused by what is happening at the resonator surface. It does not cover temperature and stress as sensing problems. Six different cases can be distinguished:

(a) *Inertial loading with rigid samples*

For rigid samples, shape does not matter. The sample shifts the resonance frequency in proportion to its mass per unit area. The principal application is gravimetric sensing in the gas phase. Gravimetry is simple as far as the physics is concerned. The challenge is in finding receptor layers, which perform well when judged by the triple-S-criterion (Sensitivity, Specificity, Stability).

(b) *Semi-infinite media and stratified layers*

For samples with no in-plane structure, the deformation pattern is a plane wave. The *acoustic multilayer formalism* describes these well. Central is the concept of the acoustic reflectivity. These samples include Newtonian liquids, complex liquids, viscoelastic films in a gas phase, viscoelastic films in a liquid phase, viscoelastic multilayers, and layers with continuous viscoelastic profiles.

(c) *Point contacts*

For macroscopic external objects touching the resonator across a contact with a diameter much smaller than the wavelength of sound and also much smaller than the size of the object, the QCM determines the stiffness of the contact. Contact stiffness is a difficult topic, but there are theories predicting it and also other experimental techniques for its determination.

If the external objects are small (bacteria, colloidal particles), they give rise to coupled resonances. These are analogs of molecular vibrations. The QCM applied to such samples performs vibrational spectroscopy on surface-attached colloids.

(d) *Structured samples with no separation of scales*

If a sample has in-plane structure on a scale, which is neither much smaller nor much larger than λ, modeling has to be carried out numerically. Of course such models need assumptions on geometry and materials parameters. Because the computation occurs numerically, it is difficult to trace the influences, which the various assumptions have on Δf and $\Delta\Gamma$. The machine runs forward very well. To apply it in the reverse direction and make statements about the sample entirely based on experimental values of Δf and $\Delta\Gamma$ with no further input is impossible.

(e) *Nonlinear interactions*

At large amplitudes, the sample's stress-strain relations can be nonlinear, which leads to an amplitude-dependence of the resonance frequency. Conversely, the amplitude-dependence can be used to study nonlinearities. Nonlinear interactions can also lead to steady forces, to particle detachment, to steady streaming, to second-harmonic generation, and to third-harmonic generation.

(f) *Frequency-based measurement of a sample's electric impedance*

Usually, one grounds the front electrode well in order to avoid effects of piezoelectric stiffening. If the front electrode is not grounded, this causes an influence of the sample's electrical impedance on the resonance frequency. The electrical impedance can be inferred from $\tilde{f}_r$ by switching between grounded front electrode and grounded back electrode. The applications of piezoelectric stiffening are discussed in Chap. 14.

Glossary

Variable	Definition
c_q	Speed of shear sound in AT-cut quartz plates
d_f	Thickness of film
d_q	Thickness of resonator plate
f_r	Resonance frequency
λ	Wavelength of sound
κ_R	Effective spring constant of the resonator
M_R	Effective mass of the resonator

References

1. http://de.wikipedia.org/wiki/Schwingquarz, Accessed 6 Feb 2013. The number of 4.5 billion USD includes all piezoelectric resonators (including tuneforks)
2. http://en.wikipedia.org/wiki/Crystal_oscillator, Accessed 6 Feb 2013
3. http://www.am1.us/Local_Papers/U11625%20VIG-TUTORIAL.pdf, Accessed 18 June 2014
4. Dava Sobel: Longitude: The True Story of a Lone Genius who Solved the Greatest Scientific Problem of His Time. Penguin, New York (1996)
5. http://www.technologyreview.com/view/418326/where-is-the-best-clock-in-the-universe/, Accessed 9 Feb 2013
6. Galliou, S., Goryachev, M., Bourquin, R., Abbe, P., Aubry, J.P., Tobar, M.E.: Extremely low loss phonon-trapping cryogenic acoustic cavities for future physical experiments. Sci. Rep. **3**, 2132 (2013)
7. Nicholson, A.M.: Generating and transmitting electric currents U.S. Patent 2,212,845, filed Apr 10, 1918, granted Aug 27, 1940
8. http://en.wikipedia.org/wiki/Potassium_sodium_tartrate, Accessed 15 Feb 2013
9. Marrison, W.A.: The Crystal Clock. Nat. Acad. Sci. Proc. **16**, 496–507 (1930)
10. Marrison, W.A.: The evolution of the quartz crystal clock. Bell Sys. Tech. J. **27**, 510–588 (1948) (Reprint online)
11. Koga, I.: Thickness vibrations of piezoelectric oscillating crystals. Phys A J. Gen. App. Phys. **3**(1), 70–80 (1932)
12. http://en.wikipedia.org/wiki/Pierce_oscillator, Accessed 15 Feb 2013
13. http://en.wikipedia.org/wiki/History_of_timekeeping_devices, Accessed 15 Feb 2013
14. http://www.ieee-uffc.org/main/history.asp?file=bottom, Accessed 15 Feb 2013
15. http://www.piezo.com/tech4history.html, Accessed 15 Feb 2013
16. Iwasaki, F., Iwasaki, H.: Historical review of quartz crystal growth. J. Cryst. Growth **237**, 820–827 (2002)
17. For an overview see Piazza: G.; Felmetsger, V.; Muralt, P.; Olsson, R.H.; Ruby, R., Piezoelectric aluminum nitride thin films for microelectromechanical systems. MRS Bull. **37**(11), 1051–1061 (2012)
18. Chen, D., Wang, J.J., Xu, Y., Li, D.H., Zhang, L.Y., Li, Z.X.: Highly sensitive detection of organophosphorus pesticides by acetylcholinesterase-coated thin film bulk acoustic resonator mass-loading sensor. Biosens. Bioelectron. **41**, 163–167 (2013)
19. Nirschel, M.: Label-free Biosensors: Thin-film Bulk Acoustic Resonators: Theory and Application of FBARs for Biomolecular Interaction. Südwestdeutscher Verlag für Hochschulschriften (2012)
20. Wingqvist, G.: AlN-based sputter-deposited shear mode thin film bulk acoustic resonator (FBAR) for biosensor applications—A review. Surf. Coat. Technol. **205**(5), 1279–1286 (2010)
21. Wingqvist, G., Bjurstrom, J., Liljeholm, L., Yantchev, V., Katardjiev, I.: Shear mode AlN thin film electro-acoustic resonant sensor operation in viscous media. Sens. Actuators B Chem. **123**(1), 466–473 (2007)
22. http://tf.boulder.nist.gov/general/pdf/214.pdf, Accessed 15 Feb 2013
23. http://www.oscilloquartz.com/, Accessed 28 Mar 2013
24. Hinkley, N., Sherman, J. A., Phillips, N. B., Schioppo, M., Lemke, N. D., Beloy, K., Pizzocaro, M., Oates, C. W., Ludlow, A. D.: An Atomic Clock with 10–18 Instability. Science **341**, 1215–1218 (2013)
25. http://www.quartzdyne.com/
26. EerNisse, E.P., Wiggins, R.B.: Review of thickness-shear mode quartz resonator sensors for temperature and pressure. IEEE Sens. J. **1**(1), 79–87 (2001)
27. Sauerbrey, G.: Verwendung von Schwingquarzen zur Wägung dünner Schichten und zur Mikrowägung. Zeitschrift für Physik **155**(2), 206–222 (1959)

28. Yang, Y.T., Callegari, C., Feng, X.L., Ekinci, K.L., Roukes, M.L.: Zeptogram-scale nanomechanical mass sensing. Nano Lett. **6**(4), 583–586 (2006)
29. Chaste, J., Eichler, A., Moser, J., Ceballos, G., Rurali, R., Bachtold, A.: A nanomechanical mass sensor with yoctogram resolution. Nat. Nanotechnol. **7**(5), 300–303 (2012)
30. Brice, J.C.: Crystals for Quartz Resonators. Rev. Mod. Phys. **57**(1), 105–146 (1985)
31. Su, X.D., Ng, H.T., Dai, C.C., O'Shea, S.J., Li, S.F.Y.: Disposable, low cost, silver-coated, piezoelectric quartz crystal biosensor and electrode protection. Analyst **125**(12), 2268–2273 (2000)
32. Mason, W.P., Baker, W.O., McSkimin, H.J., Heiss, J.H.: Mechanical Properties of Long Chain Molecule Liquids at Ultrasonic Frequencies. Phys. Rev. **73**(9), 1074–1091 (1948)
33. McSkimin, H.J.: Measurement of Dynamic Shear Viscosity and Stiffness of Viscous Liquids by Means of Traveling Torsional Waves. J. Acoust. Soc. Am. **24**(4), 355–365 (1952)
34. Mason, W.P., Baker, W.O., McSkimin, H.J., Heiss, J.H.: Measurement of Shear Elasticity and Viscosity of Liquids at Ultrasonic Frequencies. Phys. Rev. **75**(6), 936–946 (1949)
35. McSkimin, H.J.: Measurement of the Shear Impedance of Viscous Liquids by Means of Traveling Torsional Waves. J. Acoust. Soc. Am. **24**(1), 117 (1952)
36. Mason, W.P.: Piezoelectric Crystals and Their Applications to Ultrasonics. Princeton, Van Nostrand (1948)
37. Nomura, T., Okuhara, M.: Frequency-shifts of piezoelectric quartz crystals immersed in organic liquids. Analytica Chimica Acta, **142**, 281–284 (1982)
38. Nomura, T., Hattori, O.: Determination of micromolar concentrations of cyanide in solution with a piezoelectric detector. Analytica Chimica Acta, **115**, 323–326 (1980)
39. Bruckenstein, S., Shay, M.: Experimental aspects of use of the quartz crystal microbalance in solution. Electrochim. Acta **30**(10), 1295–1300 (1985)
40. Buttry, D.A., Ward, M.D.: Measurement of interfacial processes at electrode surfaces with the electrochemical quartz crystal microbalance. Chem. Rev. **92**(6), 1355–1379 (1992)
41. Schumacher, R.: The quartz microbalance—a novel-approach to the insitu investigation of interfacial phenomena at the solid liquid junction. Angew. Chem. Int. Eng. **29**(4), 329–343 (1990)
42. Thompson, M., Kipling, A.L., Duncanhewitt, W.C., Rajakovic, L.V., Cavicvlasak, B.A.: Thickness-shear-mode acoustic-wave sensors in the liquid-phase—a review. Analyst **116**(9), 881–890 (1991)
43. Janshoff, A., Galla, H.J., Steinem, C.: Piezoelectric mass-sensing devices as biosensors—An alternative to optical biosensors? Angew. Chem. Int. Eng. **39**(22), 4004–4032 (2000)
44. Bunde, R.L., Jarvi, E.J., Rosentreter, J.J.: Piezoelectric quartz crystal biosensors. Talanta **46**(6), 1223–1236 (1998)
45. Marx, K.A.: Quartz crystal microbalance: a useful tool for studying thin polymer films and complex biomolecular systems at the solution-surface interface. Biomacromolecules **4**(5), 1099–1120 (2003)
46. Rickert, J., Brecht, A., Gopel, W.: Quartz crystal microbalances for quantitative biosensing and characterizing protein multilayers. Biosens. Bioelectron. **12**(7), 567–575 (1997)
47. Konash, P.L., Bastiaans, G.J.: Piezoelectric-crystals as detectors in liquid-chromatography. Anal. Chem. **52**(12), 1929–1931 (1980)
48. Alder, J.F., McCallum, J.J.: Piezoelectric-crystals for mass and chemical measurements—a review. Analyst **108**(1291), 1169–1189 (1983)
49. Mieure, J.P., Jones, J.L.: Electrogravimetric trace analysis on a piezoelectric detector. Talanta **16**(1), 149 (1969)
50. Jones, J.L., Mieure, J.P.: A piezoelectric transducer for determination of metals at micromolar level. Anal. Chem. **41**(3), 484 (1969)
51. Borovikov, A.P.: Measurement of viscosity of media by means of shear vibration of plane piezoresonators. Instrum. Exp. Tech. **19**(1), 223–224 (1976)
52. Tabidze, A.A., Kazakov, R.K.: High-frequency ultrasonic unit for measuring the complex shear modulus of liquids. Meas. Tech. USSR **26**(1), 24–27 (1983)

53. Pechhold, W.: Eine Methode zur Messung des Komplexen Schubmoduls im Frequenzbereich 1–100 kHz. Acustica **9**, 39 (1959)
54. Rodahl, M., Hook, F., Krozer, A., Brzezinski, P., Kasemo, B.: Quartz-crystal microbalance setup for frequency and q-factor measurements in gaseous and liquid environments. Rev. Sci. Instrum. **66**(7), 3924–3930 (1995)
55. Hirao, M., Ogi, H., Fukuoka, H.: Resonance emat system for acoustoelastic stress measurement in sheet metals. Rev. Sci. Instrum. **64**(11), 3198–3205 (1993)
56. Sittel, K., Rouse, P.E., Bailey, E.D.: Method for determining the viscoelastic properties of dilute polymer solutions at audio-frequencies. J. Appl. Phys. **25**(10), 1312–1320 (1954)
57. Lucklum, R., Hauptmann, P.: Acoustic microsensors-the challenge behind microgravimetry. Anal. Bioanal. Chem. **384**(3), 667–682 (2006)
58. Martin, S.J., Granstaff, V.E., Frye, G.C.: Characterization of a quartz crystal microbalance with simultaneous mass and liquid loading. Anal. Chem. **63**(20), 2272–2281 (1991)
59. Kanazawa, K.K., Gordon, J.G.: Frequency of a quartz microbalance in contact with liquid. Anal. Chem. **57**(8), 1770–1771 (1985)
60. Stockbridge, C.D.: In: Behrndt, K.H. (eds.) Vacuum Microbalance Techniques, 4 edn., Vol. 5 Plenum Press, New York (1966)
61. Glassford, A.P.M.: Response of a Quartz Crystal Microbalance to a Liquid Deposit. J. Vac. Sci. Tech. **15**(6), 1836–1843 (1978)
62. http://en.wikipedia.org/wiki/Escapement, Accessed 9 May 2013
63. Han, S.M., Benaroya, H., Wei, T.: Dynamics of transversely vibrating beams using four engineering theories. J. Sound Vib. **225**(5), 935–988 (1999)
64. Woan, G.: The Cambridge Handbook of Physics Formulas. Cambridge University Press, Cambridge (2000)

Chapter 2
Methods of Read-Out

Abstract Resonance frequency and resonance bandwidth can be interrogated in three different ways, which are based on oscillator circuits, impedance analysis, and ring down. The techniques are described and compared.

2.1 Oscillator Circuits

QCRs in time and frequency control are always driven by oscillator circuits. An oscillator consists of amplifier and a resonator, where the latter is part of the feed-back loop (Fig. 2.1). A self-sustained oscillation occurs if the so-called Barkhausen condition is fulfilled. The gain of the amplifier must compensate the losses in the feed-back circuit and the phase shift accumulated in the feed-back circuit must be a multiple of 2π (The Barkhausen condition is a necessary, but not a sufficient condition for stable oscillation). An overview of different types of oscillators is given in Chaps. 1 and 5 of Ref. [1]. Oscillator circuits in a wider sense are covered in Ref. [2].

In order to be useful for sensing, oscillator circuits must be combined with frequency counters. Rather than counting the MHz frequency directly, one sometimes down-mixes the signal from the oscillator with a second, stable reference and counts the difference, where the latter typically is in the kHz range.

For sensing in liquids, oscillators have some disadvantages:

- Usually, the circuit is optimized for one particular overtone. Switching between overtones requires additional elements.
- Employing simple oscillator circuits one only obtains the resonance frequency, not the bandwidth. The bandwidth can be determined from the gain of the amplifier needed to drive the resonance because the amplifier's gain compensates the system's losses. Such devices have been built. Other instruments measure the motional resistance, R_1. However, the circuits providing this

D. Johannsmann, *The Quartz Crystal Microbalance in Soft Matter Research*, Soft and Biological Matter, DOI: 10.1007/978-3-319-07836-6_2,

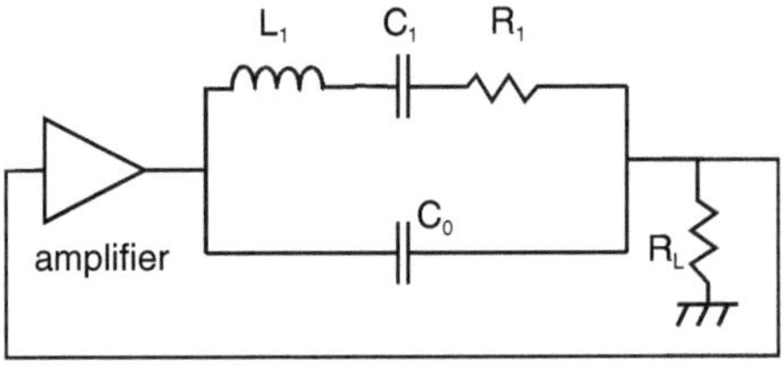

Fig. 2.1 In an oscillator circuit, the resonator and an amplifier together form an oscillator. Often the crystal is part of the feed-back circuit. The figure is much simplified. For a more detailed account of oscillator circuits see Ref. [1]. Adapted from Ref. [5]

information are more expensive. Also, the conversion from amplifier gain or motional resistance to resonance bandwidth is not trivial.

- A somewhat subtle problem follows from the fact that the oscillation frequency, f_{osc}, is not necessarily the same as the acoustic resonance frequency, f_r. In this book, the resonance frequency, f_r, is synonymous to the series resonance frequency, f_s (Note: This terminology deviates from other sources. In electronics, f_r often is what is named $f_{\varphi=0}$ in Fig. 2.2. For characteristic frequencies and their definitions see Fig. 9 in Ref. [3]). At the series resonance frequency both the amplitude of oscillation and the resonator's electrical conductance attain its maximum (Fig. 2.2, Sect. 4.5.3). Again, f_s is not strictly equal to the oscillation frequency. The Barkhausen condition is different from the electrical conductance being at its maximum. A difference between f_s and f_{osc} is not a problem as long as it remains constant, regardless of how the properties of the sample change. In sensing, one is only interested in frequency *shifts*. Unfortunately, $f_{osc} - f_s$ *does* depend on both the damping of the resonance and on the parallel capacitance (Sect. 4.5.4). These two parameters usually respond to the presence of a sample. For the parallel capacitance, there are compensation schemes [4], but the issue still requires attention.

These difficulties acknowledged, oscillator circuits may be not only cheaper than the passive schemes like impedance analysis and ring-down, but also more stable [5]. Of course this requires that the damping and the parallel capacitance are under control. Among the reasons is, that impedance analysis and ring-down turn the resonance on and off periodically, thereby introducing variable sources of heat and stress. This does not happen with oscillators. Also, the high-quality oscillators mostly employ the SC-cut, not the AT-cut. The SC-cut is not only temperature-compensated, but also insensitive to certain components of stress. However, the SC-cut is not of the thickness-shear type and can therefore not be used in liquids. More generally, the stability of oscillators has been optimized over many years [6]. Passive interrogation schemes cannot beat the frequency stability achieved in this long-lasting effort. Stability in this context can mean a number of different things, but the oscillator circuits out-perform the passive instruments, regardless of what definition is used.

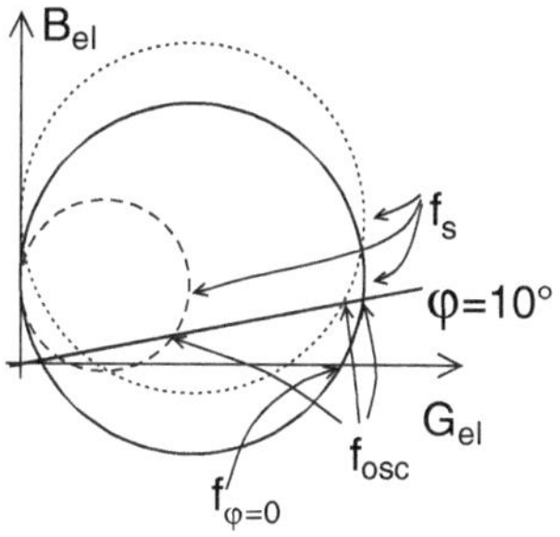

Fig. 2.2 Admittance diagram (see also Sect. 4.5.4). The frequency of oscillation of an oscillator, f_{osc}, is not usually the same as the series resonance frequency, f_s. It was drawn here as one of the two frequencies, where the phase of $\tilde{Y}_{el}$ is $\varphi = 10°$. The details depend on the resonator circuit. f_{osc} often is larger than f_s. The difference between the two frequencies *changes*, when the parallel capacitance increases (*dotted*), and also, when the motional resistance increases (*dashed*). These shifts usually amount to artifacts. The user is mostly interested in f_s

With regard to stability and precision, there is a fundamental difference between the study of complex samples with an advanced QCM, on the one hand, and gravimetric sensing on one single overtone, on the other. When studying complex samples, one fits a model with certain parameters to the shifts of frequency and bandwidth. Most of the time, the systematic errors are larger than the statistical errors and the accuracy of the results therefore does not improve much when the scatter on the individual frequency readings improves. Gravimetric sensing on one overtone is different. Here, the repeatability of this one frequency sets the limit of detection. If the damping and the parallel capacitance are under control, a good oscillator circuit then is the method of choice.

2.2 Impedance Analysis

Mapping out the electrical admittance of the resonator as a function of frequency is the most direct and transparent way of interrogation [8, 9]. An impedance analyzer sweeps the frequency across the resonance and measures the resonator's electrical admittance, $\tilde{Y}_{el}(\omega)$. There are different ways to connect the crystal to the analyzer. Reflectometry as shown in Fig. 2.3a is particularly easy. Reflectometry is efficient if the impedance of the device under test is in the range of 50 Ω. The resonator's impedance is calculated from the reflectivity and the impedance of the cable (often 50 Ω), making use of Eq. 4.2.11. In liquids, the impedance of resonator usually is much larger than 50 Ω. The series-through configuration (Fig. 2.3b) then yields more precise values of the device's impedance than reflectometry. Note: When wiring the resonator in the series-through mode, one needs to worry about grounding of the front electrode (see also Fig. 14.2b).

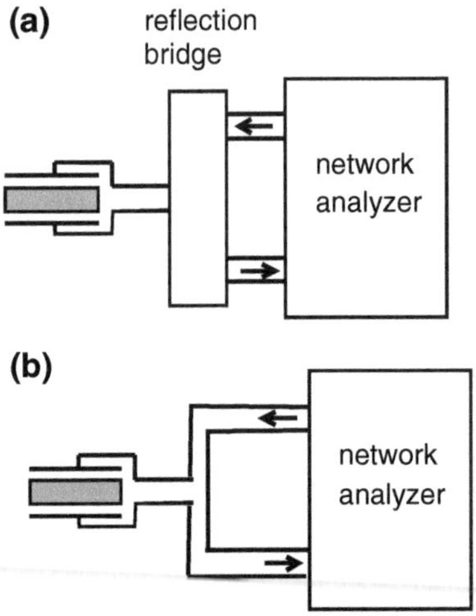

Fig. 2.3 In impedance analysis, a network analyzer (also: "impedance analyzer") determines the complex electrical admittance of the crystal. **a** The crystal is wired such that the measurement occurs in reflection. In that case, the resonator's electrical impedance is calculated from the reflected electrical amplitude by an equation analogous to Eq. 4.2.11. **b** A resonator wired in the series-through mode. Reference [7] discusses the comparative advantages of the different electrical configurations

At the resonance frequency, the real part of the electrical admittance displays a maximum. The corresponding frequency is the series resonance frequency, f_s. On resonance, the electrodes draw a large current in order to compensate the large piezoelectrically-induced polarization. There is a phase shift between current and voltage. The admittance, $\tilde{Y}_{el}(\omega) = G_{el}(\omega) + iB_{el}(\omega)$ with G_{el} the "conductance" and B_{el} the "susceptance" therefore is complex (Fig. 2.4). We come back to the exact dependence of G_{el} and B_{el} on frequency in the context of the 4-element circuit (Sect. 4.5.4).

Impedance analysis is particularly advantageous if the experimental configuration is complicated in one way or another. For instance, if there are long cables or if anharmonic sidebands interfere with the measurement, the user can recognize and diagnose potential problems based on the admittance curve. For an example see Fig. 7.3b. Oscillator circuits under such conditions often just stop to oscillate (or, worse, drift in frequency, although they should not) and the user cannot easily understand why this happens.

There are various ways to improve on impedance analysis, for instance by adding elements compensating for the parallel capacitance [10], by smart schemes of interrogation [11], and by detection schemes which are intermediate between oscillators and impedance analysis. There is also considerable activity in making impedance analyzers cheaper, possibly allowing for multichannel devices [12–14]. For a recent review see Ref. [15].

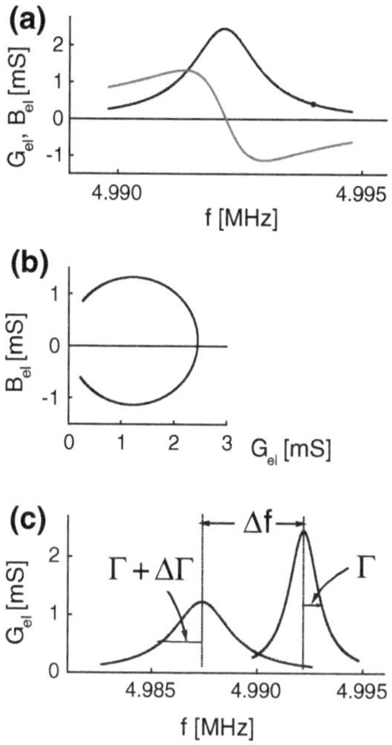

Fig. 2.4 Impedance analysis is based on the admittance curve as shown in panel (**a**). The electrical admittance, $\tilde{Y}_{el} = G_{el} + \mathrm{i}B_{el}$, is the inverse of the electrical impedance, $\tilde{Z}_{el}$, where both the impedance and the admittance are complex (Sect. 4.5.4). One typically analyzes the admittance curve because its real part (the conductance, $G_{el}(\omega)$) forms the well-known, symmetric resonance curve. $G_{el}(\omega)$ is at its maximum at the series resonance frequency. The polar diagram of the admittance curve shows a circle (**b**). Of primary interest in sensing are the shifts of resonance frequency Δf and half-bandwidth, $\Delta\Gamma$ (**c**). Panel **c** only shows the conductance curves for clarity. The imaginary part of the admittance curve (the susceptance, $B_{el}(\omega)$) displays the same shift and the same broadening

2.3 Ring-Down

Ring-down experiments are carried out in many physics undergraduate laboratories. The central piece of equipment is a pendulum, which the students set in motion by hand. They count the number of oscillation cycles per minute and measure the time which it takes, until the amplitude has decayed to half its original value. Sometimes, they also check whether the oscillation period depends on the excursion angle. It does (slightly) because the pendulum is a slightly nonlinear device. Nonlinear behavior is ignored in the following. Similar setups exist, where the conventional pendulum is replaced by spring-driven pendulum (often a torsion pendulum). From the resonance frequency and the known mass (or the known moment of inertia) the students infer the elastic stiffness of the spring. From the decay time they infer the loss modulus (Sect. 3.7).

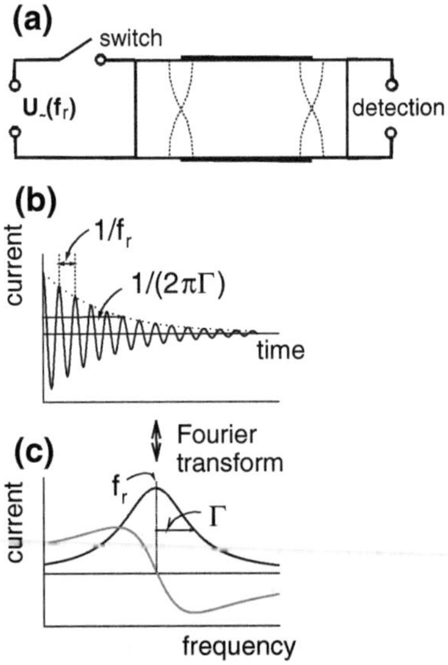

Fig. 2.5 In ring-down, the resonator is excited by an RF-pulse, the frequency of which is close to the expected resonance frequency [18]. In a second step, the excitation is shut off and the oscillation is allowed to decay (**a**). The current into electrodes is recorded. The decay time (the time when the envelope has decayed by a factor of e) is equal to $1/(2\pi\Gamma)$ (**b**). The decaying cosine observed in ring-down is the Fourier transform of the resonance curve as observed with impedance analysis (**c**)

Ring-down is conceptually simple and was employed in acoustic sensing as early as 1954 (Ref. [16], see also Ref. [17]). Q-Sense has commercialized an advanced QCM based on ring-down [18]. It is the QCM-D, where "-D" stands for "with Dissipation Monitoring". The term QCM-D is a trademark owned by Q-Sense. Excitation occurs with a radio-frequency-pulse (RF pulse) having a frequency matching the expected resonance frequency (Fig. 2.5a). Ring-down is called "impulse excitation" in Refs. [19] and [20]. Q-Sense sometimes calls the process "pinging". Once the excitation is turned off, the resonance decays freely. The current into the electrodes is recorded digitally (Fig. 2.5b). (The process of interrogation actually is slightly more complicated, but the details are unessential). The resonance frequency and the decay time follow from fitting the current-versus-time trace with a decaying cosine. Importantly, the information obtained from ring-down is equivalent to the information derived from impedance analysis. The dissipation factor, D, is equal to $2\Gamma/f_r$.

Ring-down can be faster than impedance analysis. The acquisition time per data point can be as low as a few times the inverse half-bandwidth, Γ^{-1}, while it takes about one second to sweep the frequency across the resonance in impedance analysis. Some of this speed is lost in an averaging process, but the user can trade precision against data acquisition rate. This trade-off is more difficult in impedance analysis.

Ring-down deserves a side remark because it is also widely practiced in nuclear magnetic resonance (NMR), where it is called "free induction decay". There is a difference though. In NMR, many resonances (many spins in different environments, having different Larmor frequencies) are excited at the same time by application of an RF-pulse, which is sufficiently broad in frequency to cover all resonances of interest. The data trace (the free induction decay) is a super-position of all decays originating from the different spins. The spectrum is retrieved from the decay by Fourier transformation. The Fourier transform of the decay contains many different lines. "Induction" here is magnetic induction, sometimes also called magnetic polarization. It is the equivalent of the current into the electrodes of a piezoelectric resonator. Ring-down as implemented in the QCM-D is slightly different from free-induction decay. In the QCM-D, excitation occurs with a narrow RF-pulse, selectively exciting the one resonance of interest. Of course an approximate a priori knowledge of the resonance frequency is needed to do this, but this is not a problem in practice because the resonance frequency changes slowly. The reason to not excite all resonances at the same time is that out of the very many resonances of the crystal only a few are of interest. These are the 01-modes (Sect. 7.2). It is more efficient to only excite the resonance of interest and to repeat this process for the different overtones, than it is to excite all resonances at the same time and to disentangle them by Fourier transformation. (Broadband excitation of all resonance has been recently reported by Resa et al. [21]; see the reference for details.)

Ring-down and impedance analysis are equivalent, in principle, but of course there are differences. Frequency-domain data (as acquired in impedance analysis) and time-domain data (as acquired in ring-down) are strictly equivalent to each other as long as the system obeys linear response. However, the QCM is a slightly nonlinear device and the consequences can be observed in both impedance analysis and ring-down if one looks closely enough. The consequences are not strictly the same in both types of interrogation. A second set of differences is linked to noise sources, errors, and error propagation. When stating that ring-down and impedance analysis were equivalent, we simplified the matter a little bit.

Glossary

Variable	**Definition (Comments)**
B_{el}	Electrical susceptance
C_0	Parallel capacitance (see Sect. 4.5.3)
C_1	Motional capacitance (see Sect. 4.5.3)
el	As a subscript: *el*ectrical
f	Frequency

f_{osc}	Oscillation frequency
f_r	Resonance frequency
f_s	Series resonance frequency (same as resonance frequency in this book)
G_{el}	Electrical conductance
L_1	Motional inductance (see Sect. 4.5.3)
R_1	Motional resistance (see Sect. 4.5.3)
R_L	Load resistance
t	Time
$U_\sim$	AC-voltage
$\tilde{Y}_{el}$	Electrical admittance ($\tilde{Y}_{el} = G_{el} + \mathrm{i}B_{el}$)
$\tilde{Z}_{el}$	Electrical impedance ($\tilde{Z}_{el} = 1/\tilde{Y}_{el}$)
Δ	As a prefix: a shift induced by the presence of a sample
Δf	Shift of resonance frequency (might have been called Δf_r; the index r was dropped for brevity)
$\Delta\Gamma$	Shift of the half-bandwidth (might have been called $\Delta\Gamma_r$; the index r was dropped for brevity)
Γ	Half-bandwidth ($2\pi\Gamma$: decay rate in a ring-down experiment)
ω	Angular frequency

References

1. Arnau, A.: Piezoelectric Transducers and Applications. Springer, Heidelberg (2004)
2. Gottlieb, I.M.: Practical Oscillator Handbook. Newnes, Boston (1997)
3. http://txccrystal.com/term.html. Accessed 10 Aug 2013
4. http://www.thinksrs.com/downloads/PDFs/Manuals/QCM200m.pdf. Accessed 14 Feb 2013
5. Montagut, Y.J., Garcia, J.V., Jimenez, Y., March, C., Montoya, A., Arnau, A.: Frequency-shift versus phase-shift characterization of in-liquid quartz crystal microbalance applications. Rev. Sci. Instrum. **82**(6), 064702 (2011)
6. Rubiola, E., Giordano, V.: On the 1/f frequency noise in ultra-stable quartz oscillators. IEEE Trans. Ultrason. Ferroelectr. Freq. Control **54**(1), 15–22 (2007)
7. http://cp.literature.agilent.com/litweb/pdf/5988-0728EN.pdf. Accessed 14 Feb 2013
8. Beck, R., Pittermann, U., Weil, K.G.: Impedance Analysis of Quartz Oscillators, Contacted on One Side with a Liquid. Ber. Bunsen Phys. Chem. Chem. Phys. **92**(11), 1363–1368 (1988)
9. Zimmermann, B., Lucklum, R., Hauptmann, P., Rabe, J., Buttgenbach, S.: Electrical characterisation of high-frequency thickness-shear-mode resonators by impedance analysis. Sens. Actuators B Chem. **76**(1–3), 47–57 (2001)

10. Calvo, E.J., Etchenique, R., Bartlett, P.N., Singhal, K., Santamaria, C.: Quartz crystal impedance studies at 10 MHz of viscoelastic liquids and films. Faraday Discuss. **107**, 141–157 (1997)
11. Kankare, J., Loilkas, K., Salomaki, M.: Method for measuring the losses and loading of a quartz crystal microbalance. Anal. Chem. **78**(6), 1875–1882 (2006)
12. Auge, J., Dierks, K., Eichelbaum, F., Hauptmann, P.: High-speed multi-parameter data acquisition and web-based remote access to resonant sensors and sensor arrays. Sens. Actuators B Chem. **95**(1–3), 32–38 (2003)
13. http://n2pk.com/. Accessed 29 Nov 2013, http://www.makarov.ca/vna.htm. Accessed 29 Nov 2013
14. http://sdr-kits.net/VNWA3_Description.html. Accessed 29 Nov 2013
15. Arnau, A.: A review of interface electronic systems for AT-cut quartz crystal microbalance applications in liquids. Sensors **8**(1), 370–411 (2008)
16. Sittel, K., Rouse, P.E., Bailey, E.D.: Method for determining the viscoelastic properties of dilute polymer solutions at audio-frequencies. J. Appl. Phys. **25**(10), 1312–1320 (1954)
17. Hirao, M., Ogi, H., Fukuoka, H.: Resonance emat system for acoustoelastic stress measurement in sheet metals. Rev. Sci. Instrum. **64**(11), 3198–3205 (1993)
18. Rodahl, M., Hook, F., Krozer, A., Brzezinski, P., Kasemo, B.: Quartz-crystal microbalance setup for frequency and q-factor measurements in gaseous and liquid environments. Rev. Sci. Instrum. **66**(7), 3924–3930 (1995)
19. http://en.wikipedia.org/wiki/Impulse_excitation_technique. Accessed 14 Apr 2013
20. Lucklum, R., Eichelbaum, F.: Interface circuits for QCM sensors. In: Steinem, C., Janshoff, A. (eds.) Piezoelectric Sensors, vol. 5, pp. 3–47. Springer, New York (2007)
21. Resa, P., Castro, P., Rodriguez-Lopez, J., Elvira, L.: Broadband spike excitation method for in-liquid QCM sensors. Sens. Actuators B Chem. **166**, 275–280 (2012)

Chapter 3
Essentials of Viscoelasticity

Abstract Viscoelastic dispersion is a rather essential element of the modeling process. The viscoelastic parameters of soft matter (such as shear modulus) depend on frequency because soft matter often relaxes on the time scale of the experiment. Mechanical relaxation can even be viewed as characteristic of soft condensed matter. The chapter discusses the basics of viscoelasticity and its relevance to QCM-based sensing.

3.1 General

Quantitative interpretation of QCM experiments often entails a thorough assessment of viscoelastic effects. Admittedly, viscoelasticity is not a popular subject in the life sciences, where many users of the advanced QCM are from. Bioengineers usually care more about the organization of molecules than about stress-strain relations. The organization of molecules and the variability thereof is intimately linked to biological function; the connection between MHz-viscoelasticity and biological function is less direct. Researchers from bioengineering will consequently tend to formulate models, which directly relate the set of complex frequency shifts, $\{\Delta\tilde{f}_n\}$, to molecular conformation. Such models need empirical input and their range of validity is limited to the types of samples covered by the empirical basis. An empirical basis and a limited scope are perfectly legitimate attributes of a meaningful model. Still, in the author's opinion it is helpful to insert viscoelasticity as an intermediate step. Understanding viscoelasticity entails an effort, but the effort is rewarded by a simplification and generalization of the models.

Stress-strain relations in their full generality can be a rather complicated subject [1, 2]. Below, we first formulate a set of simplifying assumptions and then focus on those aspects of viscoelasticity, which are relevant to the QCM. The central assumptions are linearity, homogeneity, isotropy, and incompressibility.

D. Johannsmann, *The Quartz Crystal Microbalance in Soft Matter Research*, Soft and Biological Matter, DOI: 10.1007/978-3-319-07836-6_3,

3.2 Linearity

Linearity is a fundamental concept, relevant in many different fields. It implies that a material's (more generally, a system's) response (such as magnetization, polarization, strain, or strain rate) is proportional the corresponding stimulus (the magnetic field, the electric field, or the stress). Most materials obey linear stress-strain relations as long as the stress is small enough. On a fundamental level, a system in a stable thermodynamic equilibrium behaves linearly as long the external excitations are comparable in magnitude to the thermal excitations. Soft matter usually obeys linear viscoelasticity at the levels of stress exerted by a quartz resonator. (The term "viscoelastic" itself usually is meant to imply linear behavior. Viscoelasticity always is linear viscoelasticity.)

While nonlinearities are the exception in QCM-based research, there are experiments, which can only be understood in terms of nonlinear response. An example is the study of point contacts at large oscillation amplitudes (Sect. 13.3). In this configuration, the area-averaged stress is small (as always in QCM experiments), but the local stress still is large. The stress-displacement relations in contact mechanics often are nonlinear. An example is the stick-to-slip transition [3]. If the tangential force transmitted through a contact drops sharply at the onset of sliding, then the force-displacement relation must be nonlinear. When analyzing QCM experiments on interfacial friction, nonlinearities must be accounted for.

3.3 Spatial Homogeneity

All materials are inhomogeneous on the molecular scale. In the following, a material is only considered as inhomogeneous, if it is structured on a scale comparable to the wavelength of sound. For the purpose of this estimate, the wavelength of sound may be replaced by the penetration depth of the shear wave, δ. δ is 250 nm at 5 MHz in water. δ increases with increasing viscosity (or more generally, with the stiffness of the medium) and it decreases with increasing frequency. Clearly, molecules do not constitute an inhomogeneity in this sense. For nanoparticles, there is room for debate. Gold nanoparticles have a diameter of around 10 nm. Under which conditions a dispersion of nanoparticles is recognized by the QCM as being a heterogeneous medium is discussed in Sect. 9.3.

For a homogeneous medium, the materials parameters (including the shear modulus) do not depend on geometry. Slightly heterogeneous materials are sometimes treated as "*effective media*" (Sect. 9.3). Within an effective medium theory one calculates effective materials parameters, which do depend on the small-scale structure. Parameters of influence might the size of embedded nanoparticles or their volume fraction. In a second step, one approximates the material as homogeneous on larger scales and hopes that all effects of small-scale heterogeneity are captured by the effective materials constants. These effective parameters can take values, which are impossible for unstructured matter.

For instance, the effective shear modulus can become negative in certain frequency bands [4]. When structured materials have been designed to have such peculiar properties, they are called metamaterials.

Chapter 10 is concerned with stratified samples, which possess structure *along the surface normal* (potentially with a scale comparable to λ), but not in the plane of the resonator. The standard acoustic multilayer formalism does apply to these sample; there is no need for an effective-medium formalism.

3.4 Isotropy

Viscoelasticity simplifies considerably if the material under study does not have a preferred direction. The viscoelastic behavior of isotropic materials is fully specified by two parameters, which are the shear modulus *G*, and the bulk modulus *K*. *K* is the inverse compressibility. In static elasticity, *G* and *K* have fixed values. For viscoelastic materials they are complex functions of frequency (Sect. 3.7). There are other choices of parameters, which can be converted to *G* and *K*, such as the elongational modulus (Young's modulus), *E*, and the Poisson ratio, ν.

Soft matter often is isotropic. A prominent counter-example are nematic liquid crystals. These are anisotropic, although they flow like a liquid. Nematics in contact with a QCM surface are discussed in Sect. 9.2.

3.5 Incompressibility

The defining property of soft condensed matter is a shear modulus, *G*, being much smaller than the bulk modulus, *K*. The bulk modulus (the inverse of the compressibility) is *not* small for condensed matter. Water flows easily, but it strongly resists any change in volume. Rubbers have a shear modulus of around a few MPa, which is about three decades below their bulk modulus. When $K \gg G$, compressional waves propagate much faster than shear waves. To be precise, the speed of longitudinal sound waves is governed by the longitudinal modulus, *M*, rather than the bulk modulus. One has $M = K + 4G/3$. *M* also is much larger than *G*. For many flow problems, one treats the material as incompressible, that is, one assumes $K \approx \infty$. This leaves *G* as the one viscoelastic parameter of relevance.

3.6 Time-Domain Experiments

A typical time-domain experiment is sketched in Fig. 3.1a. The material under test (for instance a layer of an adhesive) is attached to substrate, standing upright. A light-weight plate is glued to the other side of the layer. At $t = 0$, a weight is

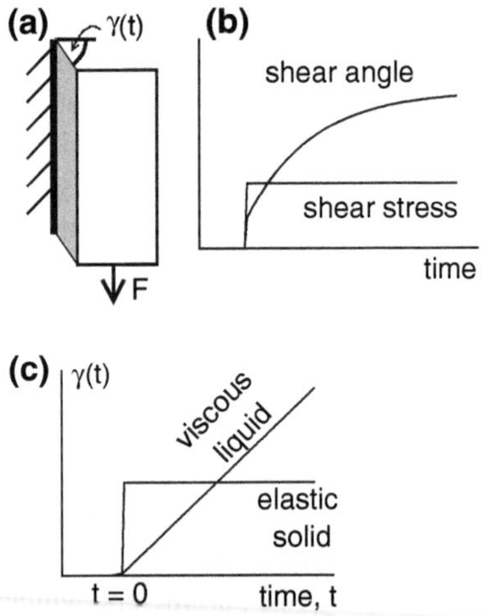

Fig. 3.1 **a** A creep experiment. At $t = 0$, a fixed shear stress $\sigma = F/A$ with A the area is applied. The sample responds with a shear deformation, $\gamma(t)$. **b** Shear stress and shear angle versus time. The shear stress is discontinuously set to a fixed value at $t = 0$ and stays constant thereafter. In this particular example, there is an instantaneous response of the shear angle, as well as delayed creep. **c** Limiting cases of elastic and viscous response

attached to the second plate, exerting a shear stress onto the sample. Initially, the layer responds to shear stress with a small, instantaneous shear strain, γ. A good adhesive will show no further creep, even over long durations of applied stress. This may change at high temperatures.

Figure 3.1b sketches a possible outcome of the creep experiment. There is an instantaneous as well as a delayed response. The shear angle assumes a finite value immediately after the weight has been attached and further increases with time. Panel C displays how a perfectly elastic and a perfectly viscous material would behave. For the former case, the elastic shear compliance, J, is given as $J = \gamma/\sigma$. For the viscous liquid one writes $\mathrm{d}\gamma/\mathrm{d}t = \sigma/\eta$ with $\mathrm{d}\gamma/\mathrm{d}t$ the shear rate and η the viscosity. In intermediate cases, the material's response is characterized by a function $J(t)$. $J(t)$ is the time-dependent shear compliance. Once $J(t)$ is known, one can predict the shear angle, $\gamma(t)$, for arbitrary stress histories as

$$\gamma(t) = \int_{-\infty}^{t} \left\{ J(t - t') \frac{\mathrm{d}\sigma(t')}{\mathrm{d}t'} \right\} \mathrm{d}t' \qquad (3.6.1)$$

Since $\gamma(t)$ depends on the stress at some time in the past, $J(t - t')$ is also called "memory kernel".

There is a corresponding experiment, where $\gamma(t)$ is controlled from the outside and the stress is measured (Fig. 3.2a). In terms of experimental realization, this requires some kind of feed-back loop. The strain must be constantly measured and the stress must be adjusted to keep the strain fixed. With fixed γ (instantaneously stepped up from zero to some constant value at $t = 0$) the external stress needed to maintain this exact level of strain, $\sigma(t)$, is characteristic of the material. Figure 3.2b shows elastic and viscous behavior as well as the mixed case. In analogy to the creep experiment one has

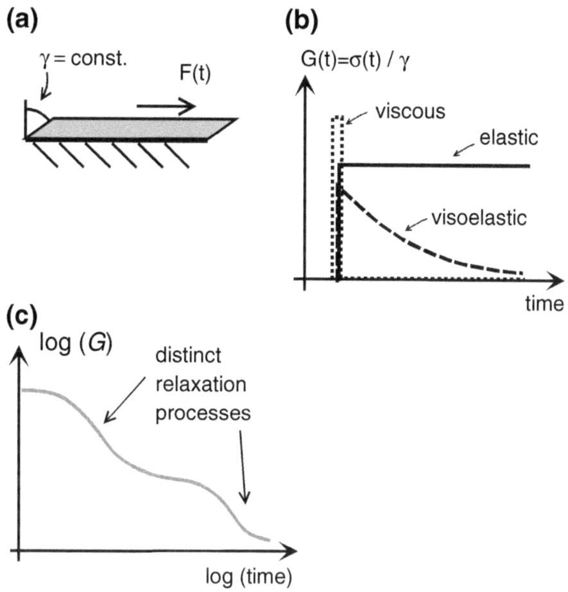

Fig. 3.2 **a** Geometry of the stress-relaxation experiment. **b** Evolution of $G(t)$ for a viscous material, an elastic material, and a viscoelastic material. **c** Ideally, one would wish to display the stress evolution on a log-log scale and identify different relaxations (with different underlying mechanisms) from the slope. The limited dynamic range of the instrument does not usually allow to cover such wide ranges in G and t

$$\sigma(t) = \int_{-\infty}^{t} \left\{ G(t - t') \frac{\mathrm{d}\gamma(t')}{\mathrm{d}t'} \right\} \mathrm{d}t' \tag{3.6.2}$$

$G(t - t')$ is the time-dependent shear modulus.

Following the tradition in polymer physics, the discussion of viscoelasticity is based on the shear modulus G. One might equally well put the compliance, J, into focus. $G(t)$ and $J(t)$ are related to each other, but the conversion is not trivial. (This is one reason for discussing viscoelasticity in frequency space. In frequency space, one has $\tilde{J}(\omega) = 1/\tilde{G}(\omega)$). On rather general grounds, one can make two statements on $G(t)$, which are that $G(t)$ can never be less than zero, and, further, that $G(t)$ must be a decreasing function of t. Both are a consequence of the fact that the material approaches its equilibrium state via relaxation processes only. Exceptions are the "meta-materials". These are structured internally, so that they behave unconventionally on the macroscopic scale. Ringing gels [5] also are structured in that sense, although they do not usually count as meta-materials.

What is shown in Fig. 3.2c is a stress relaxation history with more than one characteristic time. There are different mechanisms of stress relaxation, occurring on different time scales. Clarifying the underlying mechanisms gives clues to a material's functionality. Unfortunately, the experiment which would generate data as shown in Fig. 3.2c is very difficult. Both axes have logarithmic scales. In order to access such a wide time range, the step in shear angle would have to occur very fast and one would have to wait very long. In order to cover a few decades in shear modulus, the force sensor would need to have a large dynamic range. Rheological data covering a few decades in time or frequency *can* be produced making use of a phenomenon called time-temperature superposition (TTS). TTS is mostly applied to frequency-domain data and is discussed in this context, below.

3.7 Frequency-Domain Experiments, Viscoelastic Dispersion

For reasons covered in the textbooks of rheology, frequency-domain experiments are often better suited for rheological characterization than time-domain experiments. The information derived from these two types of experiments is equivalent as long as the material behaves linearly. In frequency-domain experiments, the externally applied stimulus (stress or strain) varies sinusoidally. The response (strain or stress) also is time-harmonic. In complex notation, stress and strain are complex amplitudes (as indicated by the hats, Sect. 4.1.1). The ratio of the two is the frequency-dependent shear modulus:

$$\tilde{G}(\omega) = G'(\omega) + \mathrm{i}G''(\omega) = \frac{\hat{\sigma}(\omega)}{\hat{\gamma}(\omega)} \tag{3.7.1}$$

$\tilde{G}(\omega) = G'(\omega) + \mathrm{i}G''(\omega)$ is complex because the stress has an in-phase component and an out-of-phase component relative to strain. G' is called storage modulus, G'' is called loss modulus.

$\tilde{G}(\omega)$ and $G(t)$ are related to each other by the Fourier-Laplace Transform [1]. The ratio G''/G' is the loss tangent, $\tan(\delta_L)$. δ_L is called δ in rheology. The index L was inserted to avoid confusion with the penetration depth of a shear wave. Using that the amplitude of the shear rate is equal to $\mathrm{i}\omega\hat{\gamma}$ for the oscillatory motion, the complex viscosity is found to be

$$\tilde{\eta}(\omega) = \frac{1}{\mathrm{i}\omega}\frac{\hat{\sigma}(\omega)}{\hat{\gamma}(\omega)} = \frac{\tilde{G}(\omega)}{\mathrm{i}\omega} \tag{3.7.2}$$

The shear compliance, $\tilde{J}(\omega)$, is defined as

$$\tilde{J}(\omega) = \frac{\hat{\gamma}(\omega)}{\hat{\sigma}(\omega)} \tag{3.7.3}$$

Clearly, one has $\tilde{J}(\omega) = 1/\tilde{G}(\omega)$. One writes $\tilde{J} = J' - \mathrm{i}J''$. There is a negative sign before J'' in order to ensure that J'' takes positive values (Sect. 4.1.2). Since $\tilde{J}$ and $\tilde{G}$ are complex, both G' and G'' have to be known in order to do the conversion. One has

$$\begin{aligned} J' &= \frac{G'}{G'^2 + G''^2}, \quad J'' = \frac{G''}{G'^2 + G''^2} \\ G' &= \frac{J'}{J'^2 + J''^2}, \quad G'' = \frac{J''}{J'^2 + J''^2} \end{aligned} \tag{3.7.4}$$

The conversion is simpler for the absolute values and the loss tangent. One has

$$\begin{aligned} |\tilde{J}| &= \frac{1}{|\tilde{G}|} \\ \tan(\delta_L) &= \frac{G''}{G'} = \frac{J''}{J''} \end{aligned} \tag{3.7.5}$$

Clearly, the value of the loss tangent does not depend on whether the discussion occurs in terms of modulus or compliance. The loss tangent is popular for that reason.

A material displays "viscoelastic dispersion" if its viscoelastic parameters depend on frequency. The name "dispersion" originates from optics, where the role of the compliance and modulus is taken by the refractive index. If the refractive index of a prism depends on frequency, the prism "disperses" the different colors, meaning that rays of different frequency exit the prism under different angles. A thorough understanding of QCM-experiments requires a thorough appreciation of viscoelastic dispersion.

Figure 3.3 shows a rheological spectrum taken on a long-chain linear polymer. There are two broad peaks in $G''(\omega)$, which correspond to two different relaxations. Relaxations here are to be distinguished from resonances. Both are of fundamental importance in understanding a material's dynamics and functionality. The absorption lines familiar from vibrational spectroscopy provide information about the motion of the molecule in question. There always is the problem of mode assignment. For any given absorption line, it is a priori not clear, to what type of motion this line corresponds. For relaxations there is a similar problem. Assigning the two peaks in Fig. 3.3 to two corresponding motions of the polymer is a classical problem of polymer physics. The fast relaxation occurs by rearrangements of individual segments, while the slow relaxation reflects the disentanglement of entire chains. These two separate relaxation mechanisms are characteristic of linear, long-chain polymers.

Relaxations differ from resonances in two respects. Firstly, there is no inertial term in the equation of motion of a polymer segment relative to its environment. The segment's inertia is negligible compared to the elastic and viscous forces exerted by its neighbors. If the segment's position is in an out-of-equilibrium state (for instance because of a suddenly imposed shear stress), the equilibrium will be approached exponentially or with a superposition of exponentials. There is no overshoot. The question of whether or not inertial forces can be neglected, resurfaces in the physics of the QCM at various points. For instance, inertia is not always negligible for nanoparticles immersed in a liquid and subjected to a MHz transverse acoustic wave. A second difference between resonances and relaxations

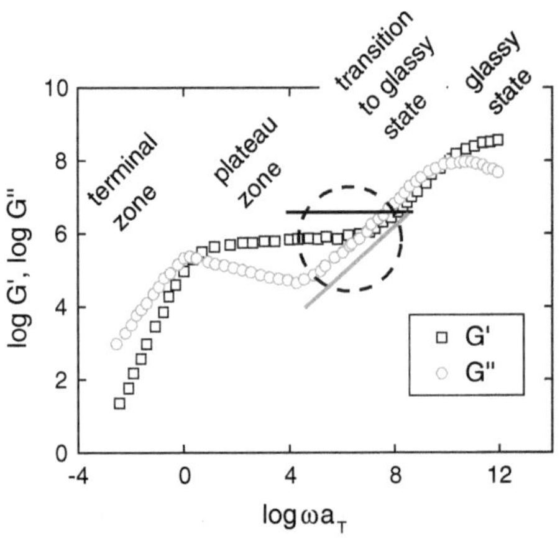

Fig. 3.3 The frequency-dependent variations in the viscoelastic properties typically span many orders of magnitude. Inside the limited frequency range of a QCM (slightly more than a decade) the dependence of G' and G'' on ω can be approximated by power laws. These correspond to straight lines in a double-logarithmic plot. The data were taken on a melt of linear chains of polyisoprene with a molecular weight of 130,000 g/mol. The reference temperature for the mastering process was $T_{ref} = 218$ K. The x-axis is a scaled frequency with a_T the "shift factor." The spectrum is "master curve", produced from spectra taken at different temperatures over smaller ranges of frequencies. The mastering process exploits the time-temperature superposition principle (see the main text at the end of this section). Adapted from Ref. [6]

is that the peaks in the response functions plotted versus frequency ($G''(\omega)$ in the case of viscoelasticity) are much broader for relaxations than for resonances. Take Fig. 3.3 as an example. The frequency axis covers more than 12 decades (note the logarithmic scale) and there are only two peaks. In infrared spectroscopy, for comparison, the so-called fingerprint region (often containing more than a dozen lines) stretches from about 500 and to about 2,000 cm^{-1}. The acquisition of spectra covering many decades in frequency is a challenge. The spectrum shown in Fig. 3.3 was not actually acquired with an instrument, which would cover such a wide range. The spectrum was stitched together from different spectra all taken over a narrower frequency range, making use of time-temperature superposition (TTS). The one technique, which routinely covers many decades in frequency is broad-band dielectric spectroscopy [7]. (Broad-band acoustic spectroscopy employing compressional waves can also cover many decades, but the effort is much larger [8].) Spectra of the dielectric constant versus log(ω) can evidence relaxations separated by a few decades in frequency, not making use of TTS. One would certainly wish to cover a similarly wide frequency range with shear-wave spectroscopy. In this way, one would be able to study samples, which do not obey TTS. Attempts in this direction are being made (Sect. 9.3), but the success is nowhere near to what dielectric spectroscopy can do.

Most text books on rheology use the shear modulus, $\tilde{G}$, to describe and compare materials. In the context of the QCM, the shear compliance, $\tilde{J}$, often is the more convenient variable. J' and J'' appear as coefficients in various Taylor expansions. Take Eq. 10.1.9 as an example, which relates the shifts of frequency and half-bandwidth to the viscoelastic properties of a film. The compliance is the pre-factor to the term containing the dependence on overtone order. J' and J'' are obtained from $\Delta f(n)$ and $\Delta\Gamma(n)$, respectively. Expressed in terms of the modulus, the real part of Eq. 10.1.9 would contain the term $G'/(G'^2$ and $G''^2)$, which would look more clumsy. More generally, QCM results deviate from the simple Sauerbrey picture if the sample is sufficiently *soft*. The trivial case (the Sauerbrey case) is obtained in the limit of infinite stiffness, that is, zero softness. The discussion of QCM results flows most smoothly if $\tilde{J}$ is made to quantify viscoelasticity because the trivial case corresponds to $\tilde{J} = 0$. (The argument is about language, not substance. "$J \approx 0$" sounds simpler than "$G \approx \infty$".)

The real and imaginary parts of response functions in the frequency domain (such as $J'(\omega)$ and $J''(\omega)$ or $G'(\omega)$ and $G''(\omega)$) are interrelated by the Kramers-Kronig relations [9]. Take the compliance as an example. $J'(\omega)$ and $J''(\omega)$ obey:

$$\begin{aligned} J'(\omega) &= \frac{2}{\pi} P \int_0^\infty \frac{\omega' J''(\omega')}{\omega^2 - \omega'^2} \mathrm{d}\omega' \\ J''(\omega) &= -\frac{2\omega}{\pi} P \int_0^\infty \frac{J'(\omega')}{\omega^2 - \omega'^2} \mathrm{d}\omega' \end{aligned} \tag{3.7.6}$$

P is the principal value in the Cauchy sense. (The integrand becomes infinite around the zero of the denominator, but the integral stays finite because the integrand goes to $+\infty$ on one side and to $-\infty$ on the other). ω' is a frequency to be integrated over (not the real part of a frequency). The Kramers-Kronig relations are the consequence of the fact that both $J'(\omega)$ and $J''(\omega)$ are calculated from the real function $J(t)$ by a Fourier-Laplace transform (Sect. 3.7). When deriving two functions of frequency from a one single function of time, the two functions of frequency cannot be independent from each other. The Kramers-Kronig relations express this interdependence. Analogous relations hold for $G'(\omega)$ and $G''(\omega)$. The Kramers-Kronig relations are not of immediate practical use because they require an integration over an infinite frequency range. One would have to know $J'(\omega)$ at *all* frequencies to compute $J''(\omega)$ and vice versa. While not immediately evident from Eq. 3.7.6, there is an important rule of thumb connected to the Kramers-Kronig relations: When $J''(\omega)$ is comparable in magnitude to $J'(\omega)$, then $J'(\omega)$ decreases with increasing ω. By analogy: When $G''(\omega)$ is comparable in magnitude to $G'(\omega)$, then $G'(\omega)$ increases with ω. For the quantitative details see Ref. [10]. A large imaginary component of the response function is indicative of relaxations with rates in the respective frequency range. Relaxations always imply that the real part of the response function depends on ω. Take the storage modulus as an example. If the sample relaxes at a rate comparable to the frequency of the experiment, the material's ability to store elastic energy is higher if excited at frequencies faster than the relaxation rate than if excited at lower frequencies. This is the very essence of a relaxation. Relaxation always entails to viscoelastic dispersion.

A side remark: Do not confuse a dependence of the viscoelastic constants on frequency with a dependence on shear rate (often labeled $\dot{\gamma}$). The steady-shear viscosity of polymer melts strongly depends on shear rate. Both the shear rate and the frequency have units of Hz. The dependence on shear rate is the consequence of nonlinear behavior, while the dependence on frequency goes back to relaxations. There can be correlations between the functions $\eta(\dot{\gamma})$ and $\eta'(\omega)$ [11], but these should not blur the distinction between the two effects.

Viscoelastic dispersion is nearly ubiquitous in QCM experiments, but only nearly so. There are cases, where it can indeed be neglected. Samples without viscoelastic dispersion are the ones, in which all relaxations have rates far away from the frequency window covered by the QCM. The two clear-cut cases are the elastic solid and the Newtonian liquid. Elastic solids have all relaxations at $\omega = 0$. Arguably, a relaxation at zero frequency is not a relaxation because it never actually takes place. For the sake of this argument, infinitely slow relaxations and very slow relaxations (for instance observed with glassy materials) are treated as a similar phenomenon. When all relaxations have rates much below the frequency of the experiment, the storage modulus has a finite value, independent of frequency. The loss modulus at the same time is much smaller than the storage modulus. The modulus is often called G_∞ because the frequency, at which it is measured, is much larger than the relaxation rate. The second case with trivial viscoelastic dispersion is the Newtonian liquid. The Newtonian liquid relaxes at a rate much faster than the frequency of the experiment. An example is water, which adapts its structure to external stress with in a few nanoseconds, at most [12]. For Newtonian liquids, the loss modulus is proportional to ω and the coefficient is called viscosity ($G'' = \omega\eta$ with η the viscosity). The storage modulus is much smaller than the loss modulus. Technically speaking, viscoelastic dispersion does not vanish for Newtonian liquids (G'' does depend on ω), but the common use of the term viscoelastic dispersion excludes the Newtonian liquid.

Less clear-cut cases, where the viscoelastic dispersion is noticeable but still small, are gels and long chain polymer melts with slow disentanglement. These have relaxations at small (or even zero) frequency *and* at frequencies much above the experimental frequency. (Actually, many elastic solids display some relaxation at high frequencies. This includes crystalline quartz, cf. Eq. 4.5.21.) In a certain frequency range between the slow and the fast relaxations, G'' is smaller than G'. $G''(\omega)$ shows a minimum somewhere close to the middle of this range. The range is called "rubber plateau" for long-chain linear polymers (Fig. 3.3). At frequencies higher than the frequency of minimum G'', G'' increases about linearly with ω. The example chosen for display in Fig. 3.3 contains such a frequency range, indicated by a dashed ellipse. $G'(\omega)$ is about constant in the region of $\log(\omega a_T) \approx 3$. In this same frequency range one also has $G''(\omega) \ll G'(\omega)$.

The case of $G' \sim const.$ and $G'' \sim \omega\eta$ is special. It carries some significance, because the assumption $G' \sim const.$ and $G'' \sim \omega\eta$ is used as the starting point for the so-called "Voigt-based model" (also: "Voigt model") [13], advocated by Q-Sense. The "Voigt-based model" can mean two different things. Initially, Voigt-based modeling implied, that the shear modulus should be expressed as $\mu + i\omega\eta$ with μ and η both being constant. μ is the storage modulus (called G' here) and $\omega\eta$ is the loss modulus (called G'' here). The parameter η is not a Newtonian viscosity. It is the ratio of G'' and ω, assumed to be independent of frequency. (A similar terminology is sometimes employed with regard to the quartz crystal, see Eq. 4.5.21.) Assuming $G' = const$ and G'' proportional to ω is not always wrong, but still a peculiar choice. Later, Q-Sense expanded the model and allowed μ and η to depend on frequency. With frequency-dependent μ and η, the model is much better. Apart from the fact that the variables should be renamed ($\mu \to G'$, $\eta \to G''/\omega$), the expanded model is perfectly appropriate. It is equivalent to the time-honored acoustic multilayer formalism. However, some researchers were left with the impression that frequency-dependent values for μ and η amounted to an advanced model, and that choosing μ and η as constant was a good start for simple systems. In the author's opinion, constant μ and η are rarely a good starting point.

The name "Voigt-based model" originates from a representation of the film as a spring and a dashpot arranged in parallel. This network carries the name Voigt model (also: "Voigt-Kelvin model") in rheology. It is a simple mechanical equivalent circuit of a viscoelastic solid in the low-frequency range. The Voigt model is only applicable at low ω because its loss modulus (equal to $\omega\eta$) goes to infinity at large ω, which is unphysical. Even at low frequencies, the Voigt model is mainly of didactic value. It brings across that a material may be able to store elastic energy and to also dissipate part of the energy at the same time. Since the spring and the dashpot are arranged in parallel, elastic behavior is found at zero frequency. This contrasts to the Maxwell model, which leads to viscous flow at $\omega = 0$. As a toy system, the Voigt model has its merits, but it should not be misunderstood as a realistic representation of a soft material. "Voigt-Kelvin materials" (which are supposed be described by the Voigt-Kelvin model) rarely are real materials, at least not in the MHz range.

Given that G' and G'' depend on frequency: How can QCM data be fitted at all? Should the problem not be underdetermined? The number of free parameters is larger than the number of experimental data. There are two experimental data points (Δf and $\Delta\Gamma$) from each overtone, but there also are two unknown viscoelastic parameters for each overtone. Further, there is the unknown thickness. In order to make fitting possible, one exploits the fact that viscoelastic spectra are smooth. They only show broad peaks, extending over decades on the logarithmic ω-scale. The QCM covers only a rather narrow frequency range of a bit more than a decade. The straight lines inside the dashed circle in Fig. 3.3 illustrate the matter. Over a range of a decade, the viscoelastic spectra can be approximated by power laws (Eq. 10.4.1). Power laws are straight lines in a log-log plot. When approximated by a power laws, $G'(\omega)$ and $G''(\omega)$ take the form of $G_{ref}'(\omega/\omega_{ref})^{\beta'}$

and $G_{ref}''(\omega/\omega_{ref})^{\beta''}$. β' and β'' are power law exponents, further discussed in Sect. 10.4. The index *ref* denotes a reference frequency, where $G' = G_{ref}'$ and $G'' = G_{ref}''$ (not to be confused with the reference state of the resonator). The number of free parameters in this way reduces to 5 (G_{ref}', β', G_{ref}'', β'', and the thickness). Fitting is possible if three or more overtones contribute to the experimental data set.

In Sect. 3.6 it was argued that it is difficult to identify different relaxations in time-domain experiments because the instrument would have to cover very fast and very slow processes at the same time. A similar problem exists in the frequency domain. In order to cover a wide range of relaxation rates, one would have to operate the apparatus at both very low and very high frequencies. Measurements at low frequency are time-consuming. Measurements at high frequencies are impractical because macroscopic machines have their own, poorly controlled resonances. Once the frequency of the measurement comes close to one of the resonances of the apparatus, the results cannot be trusted. One builds rheometers compact and rigid in order to push those resonance frequencies up, but a practical limit usually is in the range of a few kHz, at most.

A side remark: Conventional rheometers are non-resonant devices. For the measurements of viscoelastic constants, one can either exploit the resonances of the apparatus or stay away from all those resonances. The mixed situation must be avoided. Acoustic resonators exploit resonances (as the name says). The shear-wave reflectometer discussed in Sect. 15.1 works at MHz frequencies but still is a non-resonant device. With non-resonant devices, one is free to pick the frequency according to one's needs. However, one must be sure to indeed operate at frequencies far away from all resonances of the apparatus.

For a certain class of materials, the accessible frequency range can be extended by changing temperature. This phenomenon is referred to as time-temperature-superposition (TTS) [1]. Materials obeying TTS are called "thermorheologically simple". For such materials, rheological spectra acquired at low temperature are equivalent to spectra acquired at higher temperature and correspondingly higher frequency. If this equivalence holds, one can slow down the material's internal dynamics by cooling it, instead of making the instrument run faster [1]. The two types of measurements yield equivalent results. In a first step, one acquires rheological spectra at different temperatures, always using the limited frequency range of the instrument. In the second step, one shifts the individual spectra, $\tilde{G}(T,\omega)$, on the log(ω)-scale. The new frequency scale is the labeled "log(ωa_T)" with a_T the "shift factor". The value of the shift factor follows from the requirement that the shifted curves must lie above each other. One stitches these shifted spectra together and obtains one single "master curve" on a wider frequency scale. Again, the *x*-axis of the master curve is not strictly a frequency scale. The scale is log(ωa_T), not log(ω), but if time temperature-superposition-holds, log(ωa_T) is as good as log(ω). Figure 3.3 shows an example of such a master curve (adapted from Ref. [6]). (There are more advanced schemes of mastering than the one described above, see Ref. [14].)

Many interesting samples are *not* thermo-rheologically simple. Among them are the colloidal dispersions (Sect. 9.3). In this research area, acoustic resonators have acquired a special role. Using conventional rheometers in parallel to a set of acoustic resonators covering the kHz and the MHz range, one can measure the shear modulus over a frequency range of a few decades with no need to assume time-temperature-superposition [15]. The procedure is tedious, but still has been undertaken a few times. Ultimately, one would wish to acquire broad-band rheological spectra without resorting to TTS in the same way as this is done with dielectric spectroscopy [7]. Comparing shear moduli acquired with conventional rheometers (operating at low ω) and with acoustic resonators (operating at frequencies in the kHz and MHz range) is a step into this direction.

Since the frequency range of the QCM is so narrow, one wonders whether time-temperature-superposition would be a way to expand this range in the same way as this is done with conventional rheometers. This is possible [16] but it is not a routine approach. A major problem is the intrinsic temperature-dependence of the resonance frequency of a quartz resonator. This temperature dependence is weak for AT-cut crystals close to room temperature, but is significant, once the temperature is above 60 °C or below −20 °C. Temperature-frequency coupling is a serious experimental problem.

Glossary

Variable	Definition (Comments)
A	Area
a_T	Shift factor (to be used when producing a master curve, making use of time-temperature superposition)
E	Young's modulus ($E = G(1 + 2\nu)$)
F	Tangential force
$\tilde{J}$	Shear compliance ($\tilde{J} = 1/\tilde{G}$)
J'	Elastic compliance
J''	Viscous compliance
$\tilde{G}$	Shear modulus
G'	Storage modulus
G''	Loss modulus
K	Bulk modulus (an inverse compressibility)
M	Longitudinal modulus ($M = K + 4G/3$, governs the propagation of compressional waves, also called "plate modulus")
ref	As an index: *ref*erence frequency or *ref*erence temperature

t	Time
β', β''	Power law exponents (see Eq. 10.4.1)
γ	Shear angle
δ_L	Loss angle ($\tan(\delta_L) = G''/G' = J''/J'$, often called $\tan(\delta)$ in rheology)
$\{\Delta\tilde{f}_n\}$	As set of complex resonance frequencies acquired at the different overtone orders
$\tilde{\eta}$	Viscosity ($\tilde{\eta} = \tilde{G}/(i\omega)$)
η	"Viscosity" in "Voigt-based modeling" (equal to G''/ω)
ν	Poisson ratio
μ	Shear modulus as used in "Voigt-based modeling" (equal to G')
ω	Angular frequency

References

1. Shaw, M.T., MacKnight, W.J.: Introduction to Polymer Viscoelasticity. Wiley, New York (2005)
2. Larson, R.G.: The Structure and Rheology of Complex Fluids. Oxford University Press, Oxford (1998)
3. Persson, B.N.J.: Sliding Friction: Physical Principles and Applications. Springer, New York (2000)
4. Fang, N., Xi, D.J., Xu, J.Y., Ambati, M., Srituravanich, W., Sun, C., Zhang, X.: Ultrasonic metamaterials with negative modulus. Nat. Mater. **5**(6), 452–456 (2006)
5. Gradzielski, M., Bergmeier, M., Muller, M., Hoffmann, H.: Novel gel phase: a cubic phase of densely packed monodisperse, unilamellar vesicles. J. Phys. Chem. B **101**(10), 1719–1722 (1997)
6. http://www2.mpip-mainz.mpg.de/documents/akbu/pages/mechdiel.html, Accessed 4 May 2013
7. Kremer, F., Schönhals, A., Luck, W.: Broadband Dielectric Spectroscopy. Springer, New York (2002)
8. Kaatze, U., Hushcha, T.O., Eggers, F.: Ultrasonic broadband spectrometry of liquids: a research tool in pure and applied chemistry and chemical physics. J. Solution Chem. **29**(4), 299–368 (2000)
9. Arfken, G.B., Weber, H.J., Harris, F.E.: Mathematical Methods for Physicists: A Comprehensive Guide, 7th edn. Academic, Waltham (2012)
10. Mobley, J., Waters, K.R., Miller, J.G.: Causal determination of acoustic group velocity and frequency derivative of attenuation with finite-bandwidth Kramers-Kronig relations. Phys. Rev. E **72**(1), 016604 (2005)
11. Marrucci, G.: Dynamics of entanglements: a nonlinear model consistent with the Cox-Merz rule. J. Nonnewton. Fluid Mech. **62**(2–3), 279–289 (1996)
12. Nandi, N., Bhattacharyya, K., Bagchi, B.: Dielectric relaxation and solvation dynamics of water in complex chemical and biological systems. Chem. Rev. **100**(6), 2013–2045 (2000)
13. Voinova, M.V., Rodahl, M., Jonson, M., Kasemo, B.: Viscoelastic acoustic response of layered polymer films at fluid-solid interfaces: continuum mechanics approach. Phys. Scr. **59**(5), 391–396 (1999)

14. Ferry, J.D.: Viscoelastic Properties of Polymers. Wiley, New York (1980)
15. Fritz, G., Pechhold, W., Willenbacher, N., Wagner, N.J.: Characterizing complex fluids with high frequency rheology using torsional resonators at multiple frequencies. J. Rheol. **47**(2), 303–319 (2003)
16. Hillman, A.R., Efimov, I., Ryder, K.S.: Time-scale- and temperature-dependent mechanical properties of viscoelastic poly (3,4-ethylenedioxythlophene) films. J. Am. Chem. Soc. **127**(47), 16611–16620 (2005)

Chapter 4
Modeling the Resonator as a Parallel Plate

Abstract After an introduction to complex resonance frequencies, the chapter provides a thorough discussion of the acoustic impedance, the acoustic wave impedance, and other types of impedances with relevance to either the QCM itself or to related problems. The load impedance (the complex ratio of the amplitudes of periodic stress and periodic velocity, both evaluated at the resonator surface) is, what the QCM measures on a fundamental level. The description continues with three separate but equivalent ways of modeling the viscoelastic plate and its resonances. All three formulations have their benefits and drawbacks. Building on these models, it is proven that the complex frequency shift is proportional to the complex load impedance, which is the essence of the small load approximation. The load impedance can be averaged over area and time. The last section deals with samples, which themselves are small resonators with their own resonance frequency. In the presence of such "coupled resonances", the frequency shift may be positive or negative, depending on whether the resonance frequency of the coupled resonator is smaller or larger than the resonance frequency of the crystal.

4.1 Description of Oscillations with Complex Numbers

4.1.1 Euler's Relation and Complex Amplitudes

The modeling section heavily relies on complex numbers and we therefore briefly expand on their merit. There is popular misunderstanding that imaginary numbers were less "real" than real numbers [1]. Admittedly, the nomenclature suggests that this would be the case, but it is not. All numbers are computational tools. Complex numbers are particularly efficient tools, when the computations are concerned with oscillations. This is so because cos(ω*t*) and sin(ω*t*) are the real and the imaginary part of the complex function exp(iω*t*) according to Euler's relation. Any time-harmonic signal (any signal of the form cos(ω*t* + φ) with φ a phase) can be expressed as a sum

D. Johannsmann, *The Quartz Crystal Microbalance in Soft Matter Research*, Soft and Biological Matter, DOI: 10.1007/978-3-319-07836-6_4,

of a sine and a cosine. Consider a function $x(t)$, given as $x_0 \cos(\omega t + \varphi)$. Following one of the addition theorems for trigonometry, one has (Fig. 4.1)

$$\begin{aligned} x_0 \cos(\omega t + \varphi) &= x_0 \cos(\varphi)\cos(\omega t) - x_0 \sin(\varphi)\sin(\omega t) \\ &= x' \cos(\omega t) - x'' \sin(\omega t) \\ \text{with} & \\ x' = x_0 \cos(\varphi), &\quad x'' = x_0 \sin(\varphi) \end{aligned} \tag{4.1.1}$$

The first and the second term on the right-hand side are the "in-phase component" and the "out-of-phase component" (also: "quadrature component") of the signal. Using complex numbers, the signal can be expressed as

$$\begin{aligned} x_0 \cos(\omega t + \varphi) &= \mathrm{Re}(\exp(\mathrm{i}(\omega t + \varphi))) \\ &= \mathrm{Re}(x_0 \exp(\mathrm{i}\varphi)\exp(\mathrm{i}\omega t)) = \mathrm{Re}(\hat{x}\exp(\mathrm{i}\omega t)) \\ \text{with} & \\ \hat{x} = x_0 \exp(\mathrm{i}\varphi) &= x_0 \cos(\varphi) + \mathrm{i}\, x_0 \sin(\varphi) \\ \text{and} & \\ x_0 = |\hat{x}|, &\quad \varphi = \arg(\hat{x}) \end{aligned} \tag{4.1.2}$$

$\hat{x}$ is a complex amplitude. In this book, complex amplitudes of time-harmonic signals carry a hat (^). It is *much* easier to turn around $\exp(\mathrm{i}\omega t)$ in calculations of various kinds, than it is to chew around on $\cos(\omega t)$, $\sin(\omega t)$, or $\cos(\omega t + \varphi)$. In principle, oscillatory motion can be fully described with the sine and the cosine only. If one wants to avoid complex numbers, one can do so. In this book, we go the other way. We always replace $\cos(\omega t)$ by $\mathrm{Re}(\exp(\mathrm{i}\omega t))$, we replace $\sin(\omega t)$ by $\mathrm{Im}(\exp(\mathrm{i}\omega t))$, we do the entire calculation complex, and we take the real part at the very end.

There is a limitation though: The calculation is easy as long as one sticks to linear response. This implies to never look at something like $\cos^2(\omega t)$. While one might hope so, it is not generally true that $\cos^2(\omega t)$ would be equal to $\mathrm{Re}(\exp(\mathrm{i}\omega t)^2)$; $\cos^2(\omega t)$ is equal to $(\mathrm{Re}(\exp(\mathrm{i}\omega t)))^2$; the square is outside the brackets, which makes the calculations tedious. Fortunately, the QCM mostly obeys linear acoustics. Squares of $\cos(\omega t)$ are discussed in Sect. 13.2. For the time being, the QCM is viewed as a linear device. The tangential stress at the resonator surface is almost always proportional to the tangential velocity.

4.1.2 Sign Conventions

In complex formulation, any oscillatory quantity has a time-dependence of the form

$$x(t) = \mathrm{Re}(\hat{x}\ \exp(\mathrm{i}\omega t)) = \frac{1}{2}(\hat{x}\ \exp(\mathrm{i}\omega t) + \hat{x}^*\ \exp(-\mathrm{i}\omega t)) \tag{4.1.3}$$

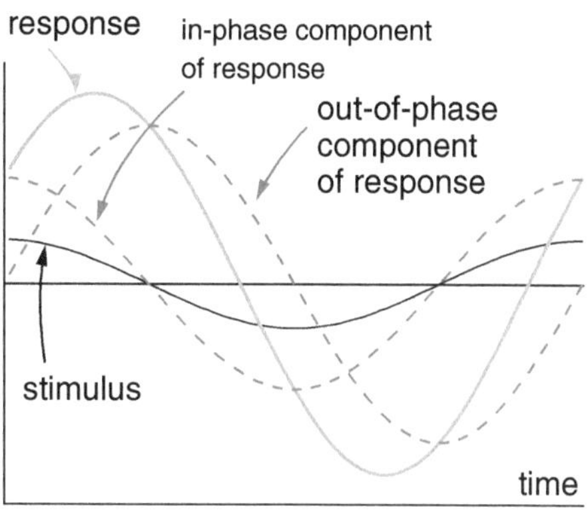

Fig. 4.1 If the response of some system to a time-harmonic excitation of the form $\cos(\omega t)$ is also time-harmonic, it can be written as $|\hat{x}|\cos(\omega t+\varphi)$ with $|\hat{x}|$ a prefactor and φ a phase. One can also write it as $|\hat{x}|\cos(\phi)\cos(\omega t) - |\hat{x}|\sin(\phi)\sin(\omega t)$. The latter two terms are the in-phase and the out-of-phase component of the response

$\hat{x}$ is a complex amplitude of a time-harmonic variable, the star denotes complex conjugation. As long as one is only concerned with quantities that are linear in all trigonometric functions, one may ignore the complex conjugate. One may carry out all calculations only once and eventually take the real part of the result. Alternatively, one may consider $x(t)$ as a complex function, in the first place, and write:

$$\tilde{x}(t) = \hat{x}\exp(i\omega t) \quad (4.1.4)$$

The fact that only the real part of $\tilde{x}(t)$ is observed in experiment is then always tacitly understood.

In the following, complex numbers carry a tilde ($\sim$). The real and the imaginary part of a complex variable carry a prime (′) and a double prime (″). Often, certain parameters are complex, in principle, but the imaginary part is much smaller than the real part. An example is the acoustic wave impedance of AT-cut quartz, $\tilde{Z}_q$. $\tilde{Z}_q$ is complex, but the imaginary part may well be neglected. In such cases the prime for the real part is often omitted. Z_q then is to be understood as $Z_q' = \mathrm{Re}(\tilde{Z}_q)$.

There is a subtlety concerning the argument of the exponential. Rather than writing $\tilde{x}(t) = \hat{x}\exp(i\omega t)$, we might have equally well written $\tilde{x}(t) = \hat{x}\exp(-i\omega t)$. At first glance, the minus sign does not appear to make a difference because the real parts of $\exp(i\omega t)$ and $\exp(-i\omega t)$ are the same. Put differently, it does not make a difference whether ω (if real) is positive or negative.

It is advantageous to stick to the plus sign in the exponent, once the frequencies turn complex. Write $\tilde{\omega} = 2\pi(f + i\Gamma)$ and consider the two functions:

$$\begin{aligned} x(t) &= \hat{x}\exp(2\pi i(f+i\Gamma)t) = \hat{x}\exp(2\pi i f)\exp(-2\pi\Gamma t) \\ x_-(t) &= \hat{x}\exp(-2\pi i(f+i\Gamma)t) = \hat{x}\exp(-2\pi i f)\exp(2\pi\Gamma t) \end{aligned} \quad (4.1.5)$$

The first function decays in time while the second grows. In the context of the QCM, amplification is not an issue. Oscillations rarely grow in time. Of course there are a few exceptions: For instance, the amplitude grows in time when a resonator is turned on. Oscillations growing in time are not of much relevance in this book. With the convention from Eq. 4.1.4, ($\exp(i\tilde{\omega}t)$ rather than $\exp(-i\tilde{\omega}t)$), damped oscillations imply $\Gamma > 0$.

A similar argument holds for a number of other parameters dealing with dissipative processes. The most important ones are G'', J'', η', ε'', and k''. $\tilde{G}$, $\tilde{J}$, $\tilde{\eta}$, $\tilde{\varepsilon}$, and $\tilde{k}$ are the shear modulus, the shear compliance, the viscosity, the dielectric permittivity, and the wavenumber, respectively. If we, firstly, stick to expressing the time dependence as $\exp(i\tilde{\omega}t)$ with $\mathrm{Re}(\tilde{\omega}) > 0$ and $\mathrm{Im}(\tilde{\omega}) > 0$ and, secondly, write:

$$\begin{aligned}\tilde{G} &= G' + iG'' \\ \tilde{J} &= J' - iJ'' \\ \tilde{\eta} &= \eta' - i\eta'' \\ \tilde{\varepsilon} &= \varepsilon' - i\varepsilon'' \\ \tilde{k} &= k' - ik''\end{aligned} \tag{4.1.6}$$

then G'', J'', ε'' η', η'' and k'' are all positive. One wants G'', J'', η', ε'', and k'' to be positive because the second law of thermodynamics states that dissipative processes cannot be reversed. If one sticks to the above sign conventions and arrives at a negative value for G'', J'', η', ε'', or k'' in some calculation, the calculation contains a mistake (or disproves the second law of thermodynamics).

Why does G'' have a plus sign and J'' have a minus sign in Eq. 4.1.6? This the consequence of $\tilde{J}$ being the inverse of $\tilde{G}$. When taking the inverse of a complex number, the imaginary part reverses sign. More generally, response functions like $\tilde{G}$, $\tilde{J}$, and $\tilde{\varepsilon}$ come in the form of either susceptibilities or inverse susceptibilities. Susceptibilities and inverse susceptibilities have a minus and plus sign before the parameter with the double prime, respectively. For k'', the negative sign follows from $\tilde{k} = \omega(\rho\tilde{J})^{1/2}$ (Eq. 4.2.3). The viscosity is a special case. One has $\tilde{\eta} = \tilde{G}/i\omega$. If η' and η'' shall both be positive, one must write $\tilde{\eta} = \eta - i\eta$.

4.1.3 Complex Resonance Frequencies

Figure 4.2 again shows the simple harmonic resonator. This model was belittled in the introduction as being the too-simple harmonic resonator, but for the motivation of complex resonance frequencies, it is good enough. Slightly deviating from the introduction (Sect. 1.5.1). there now is a dashpot in parallel to the spring. The dashpot accounts for damping.

First, consider resonances in the time-domain. The simple harmonic resonator obeys the ordinary differential equation:

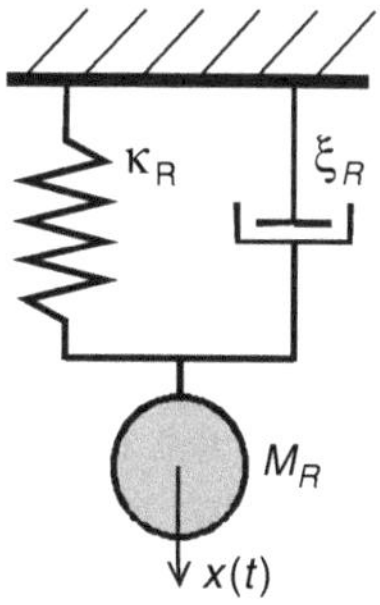

Fig. 4.2 A simple harmonic resonator. The model is one-dimensional; all movement occurs along x. A mass (M_R) is connected to a rigid wall across a Hookean spring with spring constant κ_R and a dashpot with drag coefficient ξ_R. The index R denotes the resonator. The dynamics of this system follows Eq. 4.1.7

$$M_R \frac{\mathrm{d}^2 x(t)}{\mathrm{d}t^2} = -\xi_R \frac{\mathrm{d}x(t)}{\mathrm{d}t} - \kappa_R\, x(t) \tag{4.1.7}$$

$x(t)$ is the displacement, M_R is the mass of the *r*esonator, κ_R is the spring constant, and ξ_R is the drag coefficient of the dashpot. ξ_R is sometimes called "friction coefficient", but the term "drag coefficient" is preferable here because it avoids the confusion with the friction coefficient in tribology, which is a ratio of tangential force and normal force. The drag coefficient is the ratio of a force and a velocity. Equation 4.1.7 expresses Newton's third law. Mass × acceleration is equal to the force (on the right-hand side), where the latter has a viscous and an elastic contribution.

After dividing Eq. 4.1.7 by M_R and slightly rearranging, one arrives at:

$$\begin{aligned} &\frac{\mathrm{d}^2 x(t)}{\mathrm{d}t^2} + \frac{\xi_R}{M_R}\frac{\mathrm{d}x(t)}{\mathrm{d}t} + \frac{\kappa_R}{M_R} x(t) = 0 \qquad \text{(a)} \\ &\frac{\mathrm{d}^2 x(t)}{\mathrm{d}t^2} + 2\gamma_D \frac{\mathrm{d}x(t)}{\mathrm{d}t} + \omega_0^2 x(t) = 0 \qquad \text{(b)} \end{aligned} \tag{4.1.8}$$

In line 2, the parameters ω_0 and γ_D have been introduced. ω_0 is the undamped angular resonance frequency. The undamped resonance frequency will be named f_r, later. $f_r = \omega_0/(2\pi)$ is of central importance. The parameter $\gamma_D = \xi_R/(2M_R)$ is the *D*amping factor [2]. γ_D has dimension of a frequency. It is the rate, by which the oscillation decays.

Dividing the resonator equation by the mass brings it into a form, where all coefficients are frequencies. There is a deeper reason for doing this. Frequencies and shifts thereof are at the core of all sensing processes employing a QCM. Rephrasing an equation such that its essence is contained in one or more frequencies just about always is a progress.

Let the resonator be at some position x_0, initially, and allow it to oscillate freely from thereon. The target of the computation is position as a function of time. In mathematical terms, this is an initial value problem, where the initial conditions are

$$\begin{aligned} x(t=0) &= x_0 \\ \frac{\mathrm{d}x}{\mathrm{d}t}(t=0) &= 0 \end{aligned} \tag{4.1.9}$$

With initial value problems, there are two ways to proceed. The more general one is separation of variables and formal integration. There is a second option: One may guess the solution and prove that the guess is correct. The guess may contain free parameters, the values of which are determined while proving that the guess as such is correct. The guess is also called "ansatz". We try our luck with an ansatz of the form

$$x(t) = \mathrm{Re}(\tilde{x}(t)) = \mathrm{Re}\left(x_0 \exp\left(2\pi\,\mathrm{i}\tilde{f}_d t\right)\right) \tag{4.1.10}$$

$\tilde{f}_d$ is the "*d*amped frequency" or "ringing frequency". It will turn out that f_d is slightly different from the undamped resonance frequency, f_r. However, the difference is so small in QCM experiments, that it is irrelevant in practice.

At this point, the use of a complex resonance frequency is not necessarily an advantage. $\tilde{f}_d$ was made complex, because complex *shifts* of resonance frequencies will be convenient later. Since x_0 is real, Eq. 4.1.10 can be expressed as

$$\begin{aligned} x(t) &= x_0 \cos\left(2\pi f_d' t\right) \exp\left(-2\pi f_d'' t\right) \\ &= x_0 \cos(2\pi f_d t) \exp(-2\pi\,\Gamma t) \end{aligned} \tag{4.1.11}$$

In line 2, f_d'' was renamed as Γ. A side remark on terminology: For the QCM, the resonance frequencies (damped and undamped) have a small imaginary part and the prime behind f_r or f_d is often omitted for the real part of $\tilde{f}_r$ or $\tilde{f}_d$. f_r is synonymous to f_r'; f_d is synonymous to f_d'.

The function dx(t)/dt from Eq. 4.1.12 is, what is observed in a ring-down experiment (Sect. 2.3), hence the name "ringing frequency" (It is the time derivative, rather than $x(t)$ itself, because the instrument measures the current, not the charge). One infers the real part of $\tilde{f}_d$ from the oscillation period and the imaginary part from the rate by which the oscillation decays in time.

In order to prove that Eq. 4.1.10 indeed is the solution to Eq. 4.1.7, one first has to compute the time derivatives. In complex formulation these are

$$\begin{aligned} \frac{\mathrm{d}\tilde{x}(t)}{\mathrm{d}t} &= 2\pi\,\mathrm{i}(f_d + \mathrm{i}\Gamma) x_0 \exp\left(2\pi\,\mathrm{i}\tilde{f}_d t\right) \\ \frac{\mathrm{d}^2\tilde{x}(t)}{\mathrm{d}t^2} &= -4\pi^2\left(f_d^2 - \Gamma^2 + 2\mathrm{i}\Gamma f_d\right) x_0 \exp\left(2\pi\,\mathrm{i}\tilde{f}_d t\right) \end{aligned} \tag{4.1.12}$$

Inserting these results into the complex extension of Eq. 4.1.8b yields

$$\begin{aligned}&-4\pi^2\left(f_d^2-\Gamma^2+2\mathrm{i}\Gamma f_d\right)x_0\exp\left(2\pi\,\mathrm{i}\tilde{f}_d t\right)\\&+2\gamma_D 2\pi\,\mathrm{i}(f_d+\mathrm{i}\Gamma)x_0\exp\left(2\pi\,\mathrm{i}\tilde{f}_d t\right)+\omega_0^2 x_0\exp\left(2\pi\,\mathrm{i}\tilde{f}_d t\right)=0\end{aligned}\tag{4.1.13}$$

Equation 4.1.10 solves Eq. 4.1.7 if the coefficients to the exponential add up to zero. All terms have the same time dependence. One can divide by $\tilde{x}(t)$ and is left with certain relations between the various parameters. Since f_d and Γ are free, they can be chosen to satisfy these relations. Dividing Eq. 4.1.13 by $\tilde{x}(t)$, one finds

$$-4\pi^2\left(f_d^2-\Gamma^2+2\mathrm{i}\Gamma f_d\right)+2\gamma_D 2\pi\mathrm{i}(f_d+\mathrm{i}\Gamma)+\omega_0^2=0\tag{4.1.14}$$

Both the real and the imaginary part of the left-hand side must vanish. Consider the imaginary part first. It is given as

$$-8\pi^2\Gamma f_d+4\pi\gamma_D f_d=0\tag{4.1.15}$$

which implies

$$\Gamma=\frac{\gamma_D}{2\pi}\tag{4.1.16}$$

The real part of Eq. 4.1.15 is

$$-4\pi^2\left(f_d^2-\Gamma^2\right)-4\pi\gamma_D\Gamma+\omega_0^2=-4\pi^2\left(f_d^2-\Gamma^2\right)-8\pi^2\Gamma^2+4\pi^2 f_r^2=0\tag{4.1.17}$$

The relation $\gamma_D = 2\pi\Gamma$ (Eq. 4.1.16) was used in step 1. Also, ω_0 was renamed as $2\pi f_r$ with f_r the undamped resonance frequency. The real part of the ringing frequency turns out to be

$$f_d=\sqrt{f_r^2-\Gamma^2}\tag{4.1.18}$$

f_d is unequal to f_r, in principle, but the difference is small. A typical value of Γ for a 5 MHz resonator immersed in water is 700 Hz. Using this value, one finds

$$\begin{aligned}f_d-f_r&=\sqrt{f_r^2-\Gamma^2}-f_r=f_r\sqrt{1-\frac{\Gamma^2}{f_r^2}}-f_r\\&\approx f_r\left(1-\frac{1}{2}\frac{\Gamma^2}{f_r^2}\right)-f_r=-\frac{\Gamma^2}{2f_r}=-0.049\ \text{Hz}\end{aligned}\tag{4.1.19}$$

The relation $(1-\varepsilon)^{1/2}\approx 1-\varepsilon/2$ was used in line 2. A difference in frequency of 1/20th of a Hertz is at the noise level and one may therefore approximate $(f_r^2-\Gamma^2)^{1/2}$ by f_r. Of course this approximation is not strictly needed. One can always convert between f_r and f_d, using Eq. 4.1.19 and the known value of Γ.

Given that f_d and f_r are so close: Can the entire discussion be based on the ringing frequency, rather than the undamped resonance frequency? Yes, certainly. However, matters become simpler with the undamped resonance frequency, when discussing them in the frequency domain. It will turn out that the maximum of the conductance curve occurs a the undamped resonance frequency. $\tilde{f}_r$ is the parameter determined impedance analysis (see below).

The above remarks had somewhat of a philosophical edge. It is matter of taste, which frequency to use. Users of the QCM-D can base the analysis on the ringing frequency. The following remarks are not at all philosophical; they are immensely practical. A golden rule in QCM-based sensing (and not only there) is: "Measure a frequency, if you can" (Try to find an arrangement of your experiment, where the information you seek is contained in a frequency). The strength of the QCM is in the ease and precision by which the experimentalist determines frequencies and shifts thereof. Γ and γ_D here also count as a frequencies. A more precise term would be "rates". Γ quantifies the rate, by which the oscillation decays in time. Γ is determined with about the same precision as the frequency (which, depending on the details, is about 0.1 Hz).

The mass of the resonator, M_R, can also be determined from experiment (Sects. 4.5.3 and 7.4), but this measurement is much less precise than a frequency measurement (Note: M_R is the total mass of the resonator plate, not the mass per unit area). The determination of M_R relies on the measurement of an electrical impedance. The measurement of an electrical impedance, in principle, is not a problem at all. However, the uncertainty in these measurement is around 1 %, which exceeds the uncertainty in frequency by about 5 decades (Of course one can improve on the 1 % with an effort, but the extraordinary precision achieved in the frequency measurements requires no effort). Being able to determine M_R is a rather irrelevant side aspect of QCM-based sensing. It is the frequency (the set of frequencies), which matters.

This concludes the discussion of resonators in the time-domain. In the frequency-domain, everything becomes *much* easier with complex resonance frequencies. In frequency-domain experiments, there is a time-periodic force exciting the resonator. One adds a periodic driving force of the form $\hat{F}_{ext}\exp(\mathrm{i}\omega t)$ to the right-hand side of equation Eq. 4.1.7. The displacement then is of the form $\hat{x}\exp(\mathrm{i}\omega t)$. After dividing by M_R and making the substitutions $\mathrm{d}/\mathrm{d}t \to \mathrm{i}\omega$ and $\mathrm{d}^2/\mathrm{d}t^2 \to -\omega^2$, one finds:

$$-\omega^2\hat{x}\exp(\mathrm{i}\omega t) + 2\mathrm{i}\omega\gamma_D\,\hat{x}\exp(\mathrm{i}\omega t) + \omega_0^2\,\hat{x}\exp(\mathrm{i}\omega t) = \frac{\hat{F}_{ext}}{M_R}\exp(\mathrm{i}\omega t) \quad (4.1.20)$$

Note: ω is real here. Equation 4.1.20 describes a steady oscillation; there is no decay. Since ω is among the parameters on the left-hand side, $\hat{x}(\omega)$ depends on the frequency of excitation. $\hat{x}(\omega)$ is given by:

$$\hat{x}(\omega) = \frac{1}{\omega_0^2 - \omega^2 + 2\mathrm{i}\gamma_D\omega}\frac{\hat{F}_{ext}(\omega)}{M_R} \tag{4.1.21}$$

In the following, $\hat{F}_{ext}$ is assumed to be real and independent of ω. Also, ω is assumed to be close to ω_0, which implies $\omega_0^2 - \omega^2 \approx 2\omega_0(\omega_0 - \omega)$. Equation 4.1.21 simplifies to

$$\hat{x}(\omega) = \frac{1}{\omega_0^2 - \omega^2 + 2\mathrm{i}\gamma_D\omega}\frac{\hat{F}_{ext}}{M_R} \approx \frac{1}{\omega_0 - \omega + \mathrm{i}\gamma_D}\frac{\hat{F}_{ext}}{2\omega_0 M_R} \tag{4.1.22}$$

In the following, we switch from angular frequencies to frequencies ($\omega \to f$) and define the undamped complex resonance frequency, $\tilde{f}_r$, as

$$\tilde{f}_r = \frac{\omega_0 + \mathrm{i}\gamma_D}{2\pi} = f_r + \mathrm{i}\Gamma \tag{4.1.23}$$

Using $\tilde{f}_r$, Eq. 4.1.22 turns into

$$\hat{x}(f) \approx \frac{1}{\tilde{f}_r - f}\frac{\hat{F}_{ext}(\omega)}{8\pi^2 f_r M_R} \tag{4.1.24}$$

$\hat{x}(f)$ can be separated into real and imaginary parts as

$$\begin{aligned}\hat{x}(f) &= \frac{F_{ext}}{8\pi^2 f_r M_R}\frac{f_r - f - \mathrm{i}\Gamma}{|f_r - f + \mathrm{i}\Gamma|^2} \\ &= \frac{F_{ext}}{8\pi^2 f_r M_R}\frac{f_r - f}{(f_r - f)^2 + \Gamma^2} - \mathrm{i}\frac{F_{ext}}{8\pi^2 f_r M_R}\frac{\Gamma}{(f_r - f)^2 + \Gamma^2}\end{aligned} \tag{4.1.25}$$

The time derivative of the displacement is equal to $\mathrm{i}\omega\hat{x}$, which yields

$$\mathrm{i}\omega\hat{x}(f) = \frac{F_{ext}}{8\pi^2 f_r M_R}\frac{2\pi f\Gamma}{(f_r - f)^2 + \Gamma^2} + \mathrm{i}\frac{F_{ext}}{4\pi f_r M_R}\frac{2\pi f(f_r - f)}{(f_r - f)^2 + \Gamma^2} \tag{4.1.26}$$

A resonance curve as given by Eq. 4.1.26 is, what is measured in impedance analysis. The time derivative is needed in the discussion of impedance analysis (Sect. 4.5.4) because the impedance analyzer measures the current into the electrode. Current differs from charge, which would be the equivalent of $\hat{x}(\omega)$ (The text jumps ahead here in making use of the electromechanical analogy, introduced in detail in Sects. 4.5.1 and 4.7.3. The real part of Eq. 4.1.26 turns into the conductance trace after application of the electromechanical analogy).

The real and the imaginary parts of $\mathrm{i}\omega\hat{x}(\tilde{f})$ are shown in Fig. 4.3. For clarity, the vertical scale was cut-off around the complex resonance frequency. At $\tilde{f} \approx \tilde{f}_d$ the absolute value of $\mathrm{i}\tilde{\omega}\hat{x}(\tilde{f})$ goes to infinity. The fact that the pole is at $\tilde{f}_d$ (rather

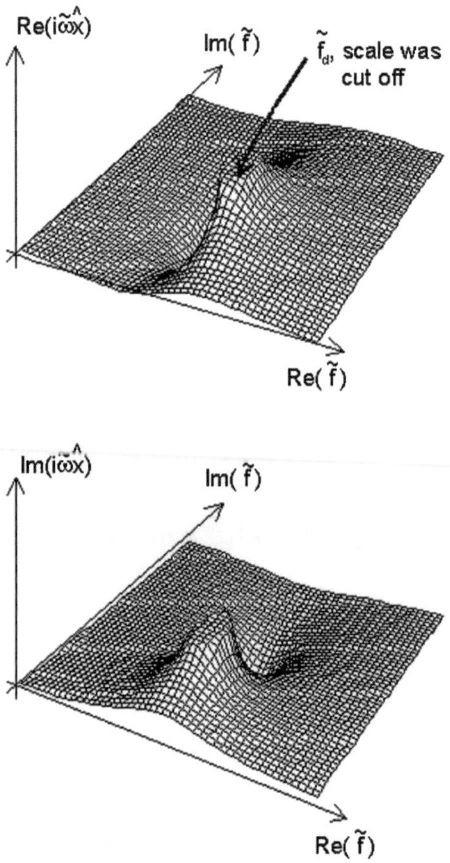

Fig. 4.3 Plots of $\mathrm{Re}(i\omega\hat{x})$ and $\mathrm{Im}(i\omega\hat{x})$ as given by Eq. 4.1.26. $\hat{x}$ becomes infinite at the complex frequency $\tilde{f}_d$ (The vertical scale was cut off around $\tilde{f}_d$). $\tilde{f}_d$ is the damped resonance frequency. With impedance analysis, only real frequencies are accessible. Both $F_{ext}(t)$ and $x(t)$ are oscillatory; they do not decay. Still, the real and the imaginary part of $\tilde{f}_d$ can be determined by fitting $\mathrm{Re}(\hat{x}(f))$ and $\mathrm{Im}(\hat{x}(f))$ with Eq. 4.1.26 at frequencies close to the resonance frequency

than $\tilde{f}_r$) follows from Eq. 4.1.21. Impedance analysis cannot directly access complex frequencies, but the complex resonance frequency can still be inferred from $\mathrm{Re}(i\omega\hat{x}(f))$ and $\mathrm{Im}(i\omega\hat{x}(f))$ with real-valued f by fitting with either Eq. 4.1.21 or 4.1.26. The difference between the two equations is negligible in practice.

The most important fit parameters in impedance analysis (among others as explained in Sect. 4.5.4) are f_r and Γ. These are the real and the imaginary part of $\tilde{f}_r$. There are two more relations to be proven, which are that the peak of conductance curve is at f_r and that Γ is the half-width at half-height of the conductance curve. To see that the peak in the conductance trace is at f_r, start from Eq. 4.1.21 and take the real part of its time derivative (proportional to the conductance) (Note: In going back to Eq. 4.1.21, we do *not* approximate $\omega_0^2 - \omega^2$ by $2\omega_0(\omega_0 - \omega)$.) Substituting angular frequencies by frequencies, one finds

$$\begin{aligned}\mathrm{Re}(\mathrm{i}\omega\hat{x}(f)) &= \frac{\hat{F}_{ext}}{4\pi^2 M_R}\mathrm{Re}\left(2\pi\,\mathrm{i}f\frac{1}{f_r^2-f^2+2\mathrm{i}\Gamma f}\right)\\ &= \frac{\hat{F}_{ext}}{4\pi^2 M_R}\mathrm{Re}\left(2\pi\,\mathrm{i}f\frac{f_r^2-f^2-2\mathrm{i}\Gamma f}{\left(f_r^2-f^2\right)^2+(2\Gamma f)^2}\right)\\ &= \frac{\hat{F}_{ext}}{4\pi^2 M_R}2\pi\frac{2\Gamma f^2}{\left(f_r^2-f^2\right)^2+(2\Gamma f)^2}\end{aligned} \tag{4.1.27}$$

At the peak of the resonance, the frequency derivative of Eq. 4.1.27 turns to zero. Ignoring prefactors, this amounts to

$$0=\frac{\mathrm{d}}{\mathrm{d}f}\left(\frac{f^2}{\left(f_r^2-f^2\right)^2+(2\Gamma f)^2}\right)=\frac{2f\left(f_r^4-f^4\right)}{\left(f_r^4+f^4-2f^2\left(f_r^2-2\Gamma^2\right)\right)^2} \tag{4.1.28}$$

Clearly, Eq. 4.1.28 is fulfilled at $f=f_r$.

To see that Γ is the half-width at half-height of the conductance curve, start from the real part of Eq. 4.1.26 and find the difference in frequency, $\delta f=f-f_r$, where the conductance takes half of the value at the peak:

$$\frac{F_{ext}}{8\pi^2 f_r M_R}\frac{\Gamma f}{\left(f_r-\left(f_r+\Delta f\right)\right)^2+\Gamma^2}=\frac{1}{2}\frac{F_{ext}}{8\pi^2 f_r M_R}\frac{\Gamma f}{\Gamma^2} \tag{4.1.29}$$

Approximating the numerator as constant in the narrow range of the resonance, this relation is fulfilled if $\delta f=\Gamma$. Note: The last statement *does* build on $\omega_0^2-\omega^2\approx 2\omega_0(\omega_0-\omega)$. Γ is the half-bandwidth for *narrow* resonances.

Upon contact with a sample, the resonance frequency and the damping factor change. It will be convenient to view the shifts in f_r and Γ and as measures of the in-phase component and the out-of-phase component of the force exerted by the resonator onto the sample (Sect. 4.6). Consequently, one defines a shift of the complex resonance frequency, $\Delta\tilde{f}$, as

$$\Delta\tilde{f}=\Delta f+\mathrm{i}\Delta\Gamma=f_r-f_{r,ref}+\mathrm{i}\left(\Gamma-\Gamma_{ref}\right) \tag{4.1.30}$$

The index *ref* denotes the reference state. For brevity, the index r was dropped behind Γ, $\Delta\tilde{f}$, and Δf. In principle, one might have called the half-bandwidth Γ_r because it pertains to a *resonance*. Likewise, $\Delta\tilde{f}$ is a difference between two resonance frequencies and one might have labeled it as $\Delta\tilde{f}_r$.

Concluding this section, we bring Eq. 4.1.20 into a form, which emphasizes its relation to the equivalent circuit representation introduced in Sect. 4.5. Again, the text jumps ahead with regard to the electromechanical analogy (Sect. 4.7.3). Consider the electrical circuit shown in Fig. 4.4. Equation 4.1.20 describes this circuit. To see this, multiply Eq. 4.1.20 by M_R, replace $\hat{x}$ by $\hat{v}/\mathrm{i}\omega$ (where $\hat{v}$ is a velocity), and divide by $\exp(\mathrm{i}\omega t)$, arriving at

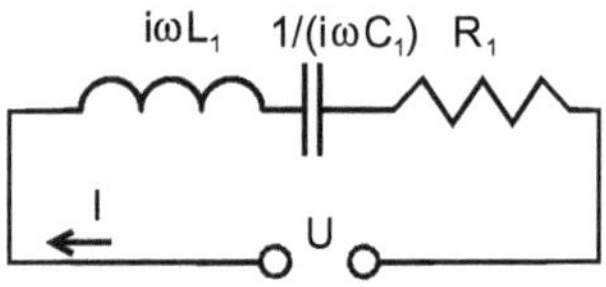

Fig. 4.4 An electrical resonator

$$i\omega M_R \hat{v} + \xi_R \hat{v} + \frac{\kappa_R}{i\omega} \hat{v} = \hat{F}_{ext} \tag{4.1.31}$$

In a next step, apply the electromechanical analogy, that is, replace the mass by a motional inductance ($M_R \to L_1$), replace the dashpot by a motional resistance ($\xi_R \to R_1$), replace the spring constant by an inverse motional capacitance ($\kappa_R \to 1/C_1$), replace the velocity by a current ($\hat{v} \to \hat{I}$), and replace the force by a voltage ($\hat{F}_{ext} \to \hat{U}$). One arrives at

$$\left(i\omega L_1 + R_1 + \frac{1}{i\omega C_1} \right) \hat{I} = \hat{U} \tag{4.1.32}$$

Equation 4.1.32 spells out the current-voltage relation of the electrical circuit shown in Fig. 4.4. At the (complex) resonance frequency, the total electrical impedance of the circuit vanishes, which is the resonance condition. Zero voltage at finite current can only be fulfilled for a complex frequency. This complex frequency is the damped resonance frequency (cf. Fig. 4.3). The reason for showing the equivalent circuit is that some readers know electrical circuits well, but are less familiar with differential equations. Those readers can find the resonance frequency based on their understanding of electric circuits.

4.1.4 Parameters Quantifying Damping

With regard to frequency, consent has been reached in the community not only on how it is defined, but also on what symbol should be used. The frequency is called f, the angular frequency is called ω. No such consent has been reached with regard to damping; the situation is truly multicultural.

A first parameter quantifying damping (the one used in this book) is the half-width at half-height (also: half-bandwidth). Half-height here refers to the conductance curve. In electronic terms, one might also say "at half-amplitude" (to be distinguished from half-power). The half-width at half-height is called Γ here. Further, there is the *full* width a half-height, sometimes called w. Unfortunately, the full width at half-height is sometimes also called Γ (Ref. [3]). Both the half width and the full width have units of frequency, Hz.

Another widely used measure of damping is the quality factor, $Q = f/w = f/(2\Gamma)$. Because the inverse Q-factor is more directly connected to damping than the Q-factor, it at some point acquired its own letter and noun, which are "D" and "dissipation factor" [4]. Note: The abbreviated form "dissipation" differs from the definition of dissipation in Wikipedia. The "dissipation factor" is related to dissipative processes, but it still does denote the process of dissipation itself. One has $D = w/f = 2\Gamma/f = 1/Q$.

The Q-Sense software by default displays the frequency shift as $\Delta f/n$ (labeled ΔF, where ΔF is the "normalized frequency shift") and the shift in the dissipation factor in units of 10^{-6}. n (an odd integer, see text below Eq. 4.3.11) is the overtone order. With 5 MHz-crystals (commonly employed in the QCM-D) one has $\Delta\Gamma/n \approx 2.5 \times \Delta D \times 10^6$ Hz. The complex frequency shift then relates to ΔF and $\Delta D \times 10^6$ as

$$\begin{aligned}\Delta\tilde{f} &= \Delta f + \mathrm{i}\Delta\Gamma \\ &= n\Delta F + \mathrm{i}\frac{f_r}{2}\Delta D \\ &\approx n\Delta F\left(1 + \mathrm{i}\, 2.5 \times 10^6 \times \Delta D\right)\end{aligned} \tag{4.1.33}$$

Again, $f_r \approx n$ 5 MHz was assumed in line 3.

The term mostly used for Γ in this book is "*half-bandwidth*". The term "bandwidth" is meant to encompass Γ, w, D, Q^{-1}. These parameters are fully equivalent to each other and it is entirely a matter of taste which variable to use. There is still one more parameter. The RQCM (distributed by Inficon, formerly Maxtek Inc.), the QCM200 (distributed by SRS), and the Kutner group [5] quantify damping by the motional resistance R_1 (also: "dynamic resistance"). Combining Eqs. 4.5.20, 4.5.8d, 4.3.14, and 4.3.16, one finds

$$R_1 = \frac{d_q^2}{8Ae_{26}^2}\frac{n\pi Z_q}{Q} = \frac{n\pi Z_q}{8Ae_{26}^2}\frac{Z_q^2}{4\rho_q^2 f_0^2}\frac{2\Gamma}{nf_0} = \frac{\pi}{16Ae_{26}^2}\frac{Z_q^3}{\rho_q^2 f_0^3}\Gamma \tag{4.1.34}$$

For the definitions of the variables see the glossary. Unfortunately, Eq. 4.1.34 only establishes a perfect equivalence as long as the effective area, A, is known and—more importantly—remains constant over the duration of the experiment. Experience shows that this is not strictly true. Also, the effective area differs between overtones. In the author's opinion, it is better to stick to frequencies and ratios thereof (All parameters with units of Hz here count as frequencies).

4.2 Acoustic Impedance, Acoustic Wave Impedance, and Acoustic Reflectivity

The acoustic impedance (called $\tilde{Z}_{ac}$ here) is at the core of the physics of the QCM and it is worthwhile to familiarize oneself with the concept in some depth. The word impedance has different meanings, collected in Table 4.1. The impedance is used in acoustics, mechanics, and electrical engineering.

Table 4.1 Different usages of the term impedance

Type of impedance	Definition	Comment
$\tilde{Z}$: acoustic wave impedance	$\tilde{Z} = \rho\tilde{c} = (\rho\tilde{G})^{1/2}$	A materials property, governing the reflectivity at interfaces. In this book always a shear-wave impedance, not a compressional wave impedance
$\tilde{Z}_{ac}$: acoustic impedance	$\tilde{Z}_{ac} = \hat{\sigma}/\hat{v}$	The ratio of tangential stress and tangential velocity, equal to the acoustic wave impedance for a propagating plane wave in a homogeneous medium
		An analogous parameter can be defined for compressional waves
$\tilde{Z}_L$: load impedance	$\tilde{Z}_L = \hat{\sigma}_S/\hat{v}_S$	The acoustic impedance evaluated at the resonator surface
		Sign convention: $\hat{\sigma}_S$ is the stress exerted by the resonator onto the ambient medium
$\tilde{Z}_{L,ijk}$: tensorial load impedance	$\hat{\sigma}_{ij} = \sum_k (\tilde{Z}_{L,ijk}\hat{v}_k)$	For complicated resonator geometries, $\Delta\tilde{f}$ is proportional to a surface integral of $\tilde{Z}_{L,ijk}$ following Eq. 6.1.18
$\tilde{Z}_m$: mechanical impedance	$\tilde{Z}_m = \hat{F}/\hat{v}$	For plane waves, $\tilde{Z}_m = A\tilde{Z}_{ac}$ with A the active area. For discrete objects like springs or dashpots, the force is an integral of the stress
$Z_{W,el}$: electric wave impedance	$Z_{W,el} = (\mathrm{d}L/\mathrm{d}C)^{1/2}$	The electric wave impedance is a property of a cable. Many cables have $Z_{W,el} = 50$ ohm
		For the meaning of dL and dC see Fig. 4.8a
$\tilde{Z}_{el}$: electric impedance	$\tilde{Z}_{el} = \hat{U}/\hat{I}$	$\hat{U}$ and $\hat{I}$ are voltage and current, respectively
$\tilde{Z}_{EM}$: electro-magnetic wave impedance (plane waves)	$\tilde{Z}_{EM} = E/H = (\tilde{\mu}_r\mu_0/\tilde{\varepsilon}_r\varepsilon_0)^{1/2}$	For radio waves in free space one has $\tilde{\mu}_r = \tilde{\varepsilon}_r = 1$. The "wave impedance of free space" is $(\mu_0/\varepsilon_0)^{1/2} = 376.7$ ohm. In principle, $\tilde{Z}_{EM}$ also applies to optical waves, but it is rarely used under that name in that context. At optical frequencies, $\tilde{Z}_{EM}$ is proportional to $1/\tilde{n} = \tilde{\varepsilon}_r^{-1/2}$ with $\tilde{n}$ the refractive index and $\tilde{\varepsilon}_r$ the relative
		E and H are the electric and the magnetic field

The *acoustic impedance* here is defined as the ratio of tangential stress and tangential velocity, where the velocity is the time derivative of the displacement (the "particle velocity", named "response velocity" in Ref. [2]). Since stress and velocity are complex amplitudes, the acoustic impedance is complex, as well.

Although not widely known, the acoustic impedance has its own unit, which is the Rayl (after John Strutt, 3rd Baron Rayleigh). A problem with the Rayl (arguably a serious one) is, that there are *two* definitions, which are kg m^{-2} s^{-1} (the "MKS Rayl") and g cm^{-2} s^{-1} (the "CGS Rayl"). As of 2013, the German edition of Wikipedia considers the Rayl to be outdated.

For a propagating plane wave in homogeneous medium, the acoustic impedance is independent of position and equal to a materials parameter, namely the acoustic wave impedance, $\tilde{Z}$. Unfortunately, the acoustic wave impedance is often called acoustic impedance in the literature. In this book, the acoustic wave impedance and the acoustic impedance have different meanings. Further complicating the matter, the acoustic impedance and the acoustic wave impedance typically are applied to normal stress and compressional waves in the literature, while these terms are used for *tangential* stress and *shear* waves here. The compressional wave impedance will be needed later in the book (Sect. 6.1.3).

To see that $\tilde{Z}$ is a materials property, start from the wave equation in the frequency domain, which is:

$$-\rho\omega^2\hat{u} = \tilde{G}\frac{d^2\hat{u}}{dz^2} \tag{4.2.1}$$

ρ is the density and $\tilde{G}$ is the shear modulus. Equation 4.2.1 is formulated for the one-dimensional case. z is the direction of propagation. The displacement is a vector, in principle, but its direction is everywhere the same and perpendicular to z. Equation 4.2.1 balances inertia against the elastic force density. The force density is the divergence of the stress. For plane shear waves in homogeneous isotropic media, the force density is $\tilde{G}d^2\hat{u}/dz^2$. The solutions of the wave equation are of the form

$$\hat{u}(z) = \hat{u}^{\pm}\exp\left(\pm i\tilde{k}z\right) \tag{4.2.2}$$

$\tilde{k}$ is the wavenumber. $\hat{u}^+$ and $\hat{u}^-$ are amplitudes in time *and space*. Inserting Eq. 4.2.2 into Eq. 4.2.1, one finds

$$\tilde{k} = \frac{\omega}{\tilde{c}} = \omega\sqrt{\frac{\rho}{\tilde{G}}} = \omega\sqrt{\rho\tilde{J}} \tag{4.2.3}$$

$\tilde{c}$ is the speed of sound. One has

$$\tilde{c} = \sqrt{\frac{\tilde{G}}{\rho}} \tag{4.2.4}$$

According to Eq. 4.2.2, the general solution to the wave equation is a superposition of two propagating waves with amplitudes $\hat{u}^-$ (traveling to the right) and $\hat{u}^+$ (traveling to the left).

Now consider a *single* propagating wave. Let the wave travel toward the left ($\hat{u}^+ \neq 0$, $\hat{u}^- = 0$) and evaluate the shear stress, $\hat{\sigma}$, and velocity, $\hat{v}$, at some point inside a homogeneous medium:

$$\begin{aligned}\hat{\sigma}(z) &= \tilde{G}\frac{\mathrm{d}}{\mathrm{d}z}\left(\hat{u}^+ \exp\left(i\tilde{k}z\right)\right) = \tilde{G}i\tilde{k}\,\hat{u}^+\left(i\tilde{k}z\right) \\ &= \tilde{G}i\omega\sqrt{\frac{\rho}{\tilde{G}}}\hat{u}^+ \exp\left(i\tilde{k}z\right) = i\omega\sqrt{\rho\tilde{G}}\,\hat{u}_+ \exp\left(i\tilde{k}z\right) \\ \hat{v}(z) &= \frac{\mathrm{d}}{\mathrm{d}t}\left(\hat{u}^+ \exp\left(i\tilde{k}z\right)\right) = i\omega\hat{u}^+ \exp\left(i\tilde{k}z\right)\end{aligned} \tag{4.2.5}$$

The ratio of stress and velocity is the acoustic impedance:

$$\tilde{Z}_{ac} = \frac{\hat{\sigma}(z)}{\hat{v}(z)} = \sqrt{\rho\tilde{G}} = \rho\tilde{c} = \tilde{Z} \tag{4.2.6}$$

Clearly, the z-dependence has cancelled. The acoustic impedance (the stress-velocity ratio) does not depend on position. It is equal to the acoustic *wave* impedance, $\tilde{Z}$. In Eq. 4.2.5, the wave travels to the left. The acoustic wave impedance is defined to be independent of direction. Should the wave be of the form $\exp(-i\tilde{k}z)$, the minus sign (by definition) does not carry over to the acoustic wave impedance of the medium. The acoustic wave impedance is still given as $(\rho\tilde{G})^{1/2}$; there is no sign change.

Again: The stress-velocity ratio is only equal to $(\rho\tilde{G})^{1/2}$ for a propagating wave. If there are two counter-propagating waves with amplitudes $\hat{u}^-$ and $\hat{u}^+$, the stress-velocity ratio is

$$\tilde{Z}_{ac} = \frac{\hat{\sigma}(z)}{\hat{v}(z)} = \sqrt{\rho\tilde{G}}\frac{\hat{u}^+ \exp\left(i\tilde{k}z\right) - \hat{u}^- \exp\left(-i\tilde{k}z\right)}{\hat{u}^+ \exp\left(i\tilde{k}z\right) + \hat{u}^- \exp\left(-i\tilde{k}z\right)} \tag{4.2.7}$$

For this configuration, the dependence of $\hat{\sigma}/\hat{v}$ on position does *not* cancel. The stress-velocity ratio in Eq. 4.2.7 is an acoustic impedance, but not an acoustic wave impedance. The distinction between the acoustic impedance and the acoustic wave impedance will be of importance in the definition of the load impedance. The load impedance ($\tilde{Z}_L$, Sect. 4.6.1) is an acoustic impedance evaluated at the resonator surface. If the sample is a semi-infinite medium, the load impedance is the same as the acoustic wave impedance of this medium. Otherwise, the stress-velocity ratio must be calculated in whatever way is most efficient.

Now consider an interface between two media labeled i and j (Fig. 4.5). Let the amplitude of the wave coming in from the left be unity ($\hat{u}_i^- = 1$) and let there be no wave coming in from the right ($\hat{u}_j^+ = 0$). From the continuity of velocity and

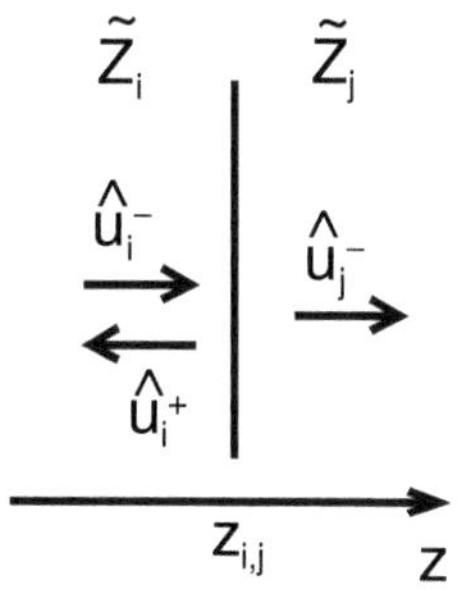

Fig. 4.5 An interface between two acoustic media. The acoustic properties of the two media are specified by their speeds of sound ($\tilde{c}_i$ and $\tilde{c}_j$) and their wave impedances ($\tilde{Z}_i$ and $\tilde{Z}_j$). There is a wave coming in from the left with amplitude $\hat{u}_i^-$. The interface gives rise to a reflected wave with amplitude $\hat{u}_i^+$. The amplitude of the transmitted wave is $\hat{u}_j^-$. The wave vector is always normal to the interface; oblique incidence is outside the scope here

stress at the interface one calculates the reflectivity as $\tilde{r}_{i,j} = \hat{u}_i^+/\hat{u}_i^-$. Continuity of displacement amounts to

$$\hat{u}_i^+ + \hat{u}_i^- = \hat{u}_j^- \tag{4.2.8}$$

Given that the stress is $\tilde{G}\mathrm{d}\hat{u}/\mathrm{d}z = \tilde{G}\mathrm{d}/\mathrm{d}z(\hat{u}^{\pm}\exp(\pm \mathrm{i}\tilde{k}z))$, the continuity of stress reads as

$$\mathrm{i}\tilde{G}_i\tilde{k}_i\left(\hat{u}_i^+ - \hat{u}_i^-\right) = -\mathrm{i}\tilde{G}_j\tilde{k}_j\,\hat{u}_j^- \tag{4.2.9}$$

Using $\tilde{Z} = \left(\rho\tilde{G}\right)^{1/2}$ and $\tilde{k} = \omega/\tilde{c} = \omega\left(\rho/\tilde{G}\right)^{1/2}$, Eq. 4.2.9 can be written as

$$\tilde{Z}_i\left(\hat{u}_i^+ - \hat{u}_i^-\right) = -\tilde{Z}_j\,\hat{u}_j^- \tag{4.2.10}$$

Eliminating $\hat{u}_j^-$ from Eqs. 4.2.8 and 4.2.10 one finds the reflectivity as

$$\tilde{r}_{i,j} = \frac{\hat{u}_i^+}{\hat{u}_i^-} = \frac{\tilde{Z}_i - \tilde{Z}_j}{\tilde{Z}_i + \tilde{Z}_j} \tag{4.2.11}$$

The amplitude transmission coefficient $\tilde{t}_{i,j} = \hat{u}_j^-/\hat{u}_i^-$ comes out as

$$\tilde{t}_{i,j} = \frac{\hat{u}_j^-}{\hat{u}_i^-} = \frac{2\tilde{Z}_j}{\tilde{Z}_i + \tilde{Z}_j} \tag{4.2.12}$$

A note on terminology: Reflectivity here is defined with respect to displacement. $\left(\tilde{r}_{i,j} = \hat{u}_i^+/\hat{u}_i^-\right)$. If the definition had been based on stress, $\left(\hat{\sigma}_i^+/\hat{\sigma}_i^-\right)$, the sign would have been opposite. The reflectivity of compressional waves is sometimes defined with respect to pressure (which is the analog of stress).

Clearly, the wave impedance governs the reflectivity at interfaces. If the wave impedances of two media are much different, acoustic waves are to a large extent reflected at interfaces between the two. "Impedance matching" can be a problem if the goal is to couple acoustic radiation into some medium. Consider medical ultrasound (compressional waves) as an example. The wave impedances (of compressional waves) of solids and liquids, on the one hand, and air, on the other, are vastly different and it is therefore difficult to couple ultrasound into the body across an air gap. One therefore fills the gap between the transducer and the body with a gel. As a second example, consider shear waves at the QCM surface. The *shear-wave* impedances of solids and liquids are much different (contrasting to compressional waves) and therefore the transverse sound produced by the piezo-effect only weakly couples to the liquid. Most of the acoustic energy is reflected back into the resonator. If this was not the case, the resonator would not have a high Q-factor. The resonances would not be nearly as sharp as they are. In this case, the weak coupling is a benefit. The resonator *weakly* couples to the sample.

The discussion above concerned plane waves in an infinite medium. A wave impedance can also be defined for propagating waves supported by some kind of linear structure such as a string or a pipe. For such waves, the wave impedance depends on geometrical parameters (such as the diameter of the pipe). Impedance matching can therefore be achieved by suitable design of the supporting structure. The shape of the trumpet is the outcome of impedance matching optimization. On the one hand, some of the acoustic energy must be fed back into the resonator (the trumpet) in order to establish a resonance. On the other hand, one wants the sound to reach the audience, preferably at impressive amplitude. This is achieved by gradually widening the pipe at the end. The conical end of the trumpet (the "bell") smoothens the interface between the resonator and free space, thereby reducing the reflectivity (The bell has functions beyond impedance matching).

So far, the conditions of continuity were formulated in terms of stress and displacement. They might also be formulated in terms of momentum and energy. The density of momentum flow by definition is the stress. The density of energy flow is the work per unit area and unit time, $1/A\ \mathrm{d}W/\mathrm{d}t$, done on some sheet of material by the neighboring sheet. One has

$$\frac{1}{A}\frac{\mathrm{d}W}{\mathrm{d}t} = \sigma(z,t)\mathrm{v}(z,t) = (\sigma_+(z,t) - \sigma_-(z,t))\mathrm{v}(z,t) \tag{4.2.13}$$

In the last step, the stress was decomposed into contributions from the two waves propagating to the right and to the left. For simplicity, real-valued parameters were used in Eq. 4.2.13. This includes the ratio of stress and velocity, which is the shear-wave impedance, Z. With real Z, (that is, with real shear modulus, G) there is no dissipation in the bulk, which simplifies the discussion. We write $u(z,t) = (\hat{u}^+\cos(\omega t + kz) + \hat{u}^-\cos(\omega t - kz))$ with real $\hat{u}^\pm$. Further, we have $\mathrm{v}(z) = -\omega(\hat{u}^+\sin(\omega t + kz) + \hat{u}^-\sin(\omega t - kz))$ and $\sigma(z,t) = Z(z)(-\omega(\hat{u}^+\sin(\omega t + kz) - \hat{u}^-\sin(\omega t - kz))$. Using $z = 0$ and $\hat{u}_j^+ = 0$, the continuity of energy flow at the interface between layer i and layer j reads as

$$Z_i\omega^2\left(\hat{u}_{i+}^2 - \hat{u}_{i-}^2\right)\sin^2(\omega t) = Z_j\omega^2\hat{u}_{j-}^2\sin^2(\omega t) \tag{4.2.14}$$

or, equivalently

$$\tilde{Z}_i(\hat{u}_{i+} + \hat{u}_{i-})(\hat{u}_{i+} - \hat{u}_{i-}) = \tilde{Z}_j\hat{u}_{j-}^2 \tag{4.2.15}$$

Dividing Eq. 4.2.15 by Eq. 4.2.10 yields Eq. 4.2.8. In other words, if both momentum and energy are to be conserved at an interface, stress and displacement must also be continuous at this interface. Actually, the flow of energy is not necessarily conserved. Energy might be dissipated at the interface if the interface slips. This is just what the combination of Eqs. 4.2.15, 4.2.10, and 4.2.8 says. If energy is not conserved because of slip, the displacement is not continuous either. We come back to slip in Sects. 10.7 and 13.3.

The impedance deserves a few more remarks:

- There also is a "mechanical impedance", which is the ratio of force to velocity. For plane waves the conversion from acoustic to mechanical impedance amounts to a multiplication by the area A. The Mason-circuit (Fig. 4.9) contains mechanical impedances in this sense.
  Force-displacement relations are convenient for contacts between discrete mechanical elements, where stress and strain are confined to some region with the consequence that the stress distribution inside the contact can be left aside in the analysis. An example is the sphere-plate contact (Chap. 11). The force is the stress integrated over the contact area. The displacement is evaluated at a location far away from the deformed region (Fig. 11.1a). Of course the internal details do matter, but since they are complicated, modeling the contacts in terms of force and displacement is a good start.
- The term impedance is well-known from electricity, where it denotes the ratio of voltage and current. The unit of electrical impedance is ohm (volt/ampere). Its significance is the same as that of the force-velocity ratio in mechanics. The electromechanical analogy is discussed in detail in Sect. 4.7.3. Most electrical waves are guided waves in the sense that they travel along a cable. Electrical cables are characterized by a wave impedance, $Z_{W,el}$. Many cables have $Z_{W,el} = 50\ \Omega$, but cables with $Z_{W,el} = 70\ \Omega$ also are around. Non-guided electromagnetic waves are the radio waves traveling in free space. For radio waves, the impedance is the ratio of the electric field, E, to the magnetic field, H. Its value is $(\mu_0/\varepsilon_0)^{1/2} = 376.7\ \Omega$ with $\mu_0 = 4\pi \times 10^{-7}$ Vs/(Am) the magnetic permeability of vacuum and $\varepsilon_0 = 8.854 \times 10^{-12}$ As/(Vm) the electric permittivity of vacuum.
- In principle, an impedance can also be formulated for optical waves, although the term is rarely used. In matter, the wave impedance of electromagnetic radiation is $(\mu/\varepsilon)^{1/2} = (\mu_r\mu_0/\varepsilon_r\varepsilon_0)^{1/2}$ with μ and ε the magnetic permeability and dielectric permittivity. μ_r and ε_r are the respective relative quantities. Since $\mu_r \approx 1$ at optical frequencies, the "optical impedance" scales as $\tilde{\varepsilon}_r^{-1/2} = \tilde{n}^{-1}$

Fig. 4.6 Layer system and definition of variables

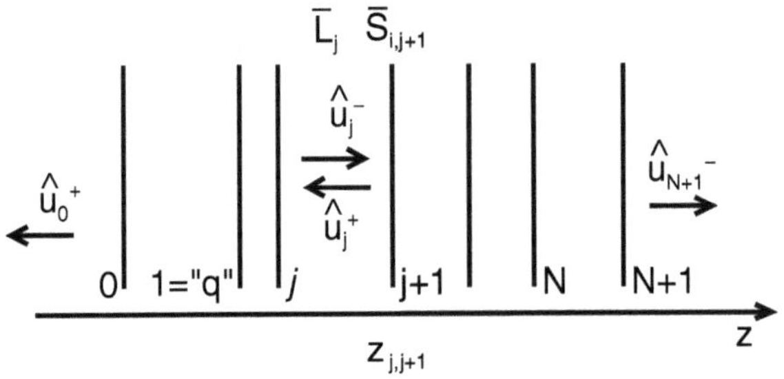

with $\tilde{n}$ the refractive index. The "optical impedance" therefore is proportional to $\tilde{n}^{-1}$ (not to $\tilde{n}$, as one might guess).

- Energy and momentum are also conserved for elastic collisions between two particles of mass M_1 and M_2. Consider the 1D-case and let particle 2 be at rest, initially. The total energy and the total momentum are conserved during the collision. From $p_{tot,fin} = p_{tot,ini}$ and $E_{tot,fin} = E_{tot,ini}$ it follows that the velocity of the particle 1 after the collision is

$$v_{1,fin} = \frac{M_1 - M_2}{M_1 + M_2} v_{1,ini} \tag{4.2.16}$$

The indices *ini* and *fin* stand for initial and final. Note the similarity with Eq. 4.2.11. It is not at all a coincidence.

- Since stress is a tensor and velocity is a vector, the acoustic impedance in full generality is a third-rank tensor (cf. Sect. 6.1.1). Most of the book is concerned with a shear stress at the resonator surface and the corresponding tangential velocity. If the surface normal is the z-direction and if displacement is along x, the relevant component of the stress tensor is $\hat{\sigma}_{xz}$; the relevant component of the velocity is $\hat{v}_x$. Unless stated otherwise, an acoustic impedance without tensor indices is meant to denote $\hat{\sigma}_{xz}/\hat{v}_x$.
- Above, conservation of energy and momentum was invoked to prove continuity; angular momentum was left aside. For conventional shear waves, orientation is strictly coupled to the gradient of translation. If displacement and stress are continuous, torque is continuous as well. This is different for nematic liquid crystals. For nematics, orientation is a separate dynamic variable and the continuity of torque is not trivial. Section 9.2 touches on how nematics behave when brought into contact with a QCM.

The following three sections go through three different ways of modeling the parallel plate. Section 4.3 uses as little physical interpretation as possible. We just solve the equations given to us. These equations are the wave equation and the conditions of continuity at the interfaces. In Sect. 4.4, we take ratios of the amplitudes derived in Sect. 4.3 and interpret these as reflectivities. It will turn out that the QCM can be viewed as an acoustic reflectometer. Section 4.5 introduces equivalent electrical circuits and formulates the theory of the parallel plate in this frame. All three descriptions are equivalent. What description to use is largely a matter of taste, but there are specific benefits and drawbacks to each formulation. The mathematical description is elegant because it needs no physical insight. Although it is minimalistic, it predicts the resonance frequencies of the loaded and the unloaded resonator. The mathematical formulation was chosen in the classical paper by Reed et al. [6] and in the formulation of the "Voigt model" [7]. The description in terms of reflectivities (Sect. 4.4) will appeal to researchers familiar with optics. The concept of the wave being an acoustic beam guides intuition into the right direction. It was, for instance, used in Ref. [8]. Naturally, the formulation with equivalent electrical circuits appeals to electrical engineers. Equivalent circuits were used by the Sandia group (work summarized in Ref. [9]) and in the early papers by Moriizumi et al. [10]. The old work on acoustic resonators (summarized in Mason's book [11]) also heavily relies on equivalent circuits. Employing equivalent circuits, anyone who masters the Kirchhoff rules can write down the resonance condition from looking at the circuit. A further benefit is that the Mason equivalent circuit (Fig. 4.9) covers piezoelectricity. Of course one can also include piezoelectricity into the mathematically inspired and the optically inspired formulation, but this is not done here. Importantly, the equivalent circuit representation moves the stress at the resonator surface into the focus of the modeling. The stress can be averaged over area and time (Sect. 4.6.2). Averaging might also be introduced into the other two models, but, again, this is not done here. In order to emphasize the equivalence of the three formulations, the Gordon–Kanazawa–Mason result (Ref. [12], Eq. 9.1.1) is derived at the end of each section. These derivations will always require the assumption $|\tilde{Z}_{liq}| \ll |\tilde{Z}_q|$, that is, require a small load.

4.3 Acoustic Plane Wave Model: Mathematical Form

The following section derives of the resonance condition for the non-piezoselectric slab in a form, which is mostly based on certain equations and their solution with little physical interpretation.

Figure 4.7 shows the geometry. There are a total of N layers, where the semi-infinite media to the left and to the right of the layer stack have the indices 0 and $N + 1$. Later on, the crystal will usually be layer 1 and the index 1 will be replaced by q. Each layer j is characterized by its thickness, d_j, its shear-wave impedance, $\tilde{Z}_j$, and its speed of sound, $\tilde{c}_j$. Each layer supports two waves of the form $\hat{u}_{j,\pm}(z) = \hat{u}_j^{\pm}\exp(\pm i\tilde{k}_j x)$. Notation: If $\hat{u}$ carries the + or − as a superscript, it is an amplitude in time *and* space. If it carries the sign as a subscript, it is an amplitude in time, only. + and − denote the direction of propagation, where "−" is propagation from left to right (that is, towards positive z). $\hat{u}_j^{+}$ is the amplitude evaluated the *left* border of the layer:

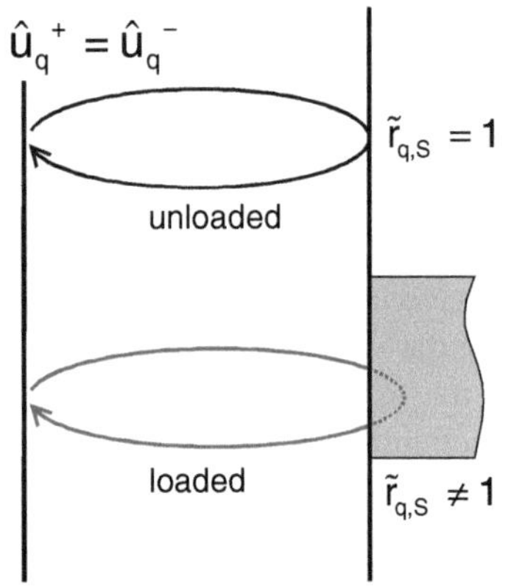

Fig. 4.7 The frequency shift is linked to the reflectivity at the interface between the resonator and the sample. A reflectivity different from unity may be thought of as a configuration, where the wave briefly enters the sample and then returns with a phase shift and, possibly, some degree of attenuation. The composite resonator thereby appears to have become thicker, which lowers the resonance frequency. Attenuation inside the sample increases the bandwidth

$$\hat{u}_{j,\pm}(z) = \hat{u}_j^{\pm} \exp\left(\pm i\tilde{k}_j\left(z - z_{j-1,j}\right)\right) \tag{4.3.1}$$

Exception: The amplitude in layer 0 must be evaluated on the right-hand side, since there is no left border.

The semi-infinite media to the left and to the right of the layer stack each support one wave only, which propagates outwards. The amplitudes $\hat{u}_0^-$ and $\hat{u}_{N+1}^+$ therefore vanish:

$$\begin{aligned} \hat{u}_0^- &= 0 \\ \hat{u}_{N+1}^+ &= 0 \end{aligned} \tag{4.3.2}$$

This leaves a set of $2N + 2$ nontrivial amplitudes. The N layers are bounded by $N + 1$ interfaces. The continuity of displacement is expressed as:

$$\hat{u}_{j,+}^{+}\left(z_{j,j+1}\right) + \hat{u}_{j,-}\left(z_{j,j+1}\right) = \hat{u}_{j+1,+}\left(z_{j,j+1}\right) + \hat{u}_{j+1,-}^{-}\left(z_{j,j+1}\right) \tag{4.3.3}$$

$z_{j,j+1}$ is the location of the interface between layers j and $j + 1$. Since $\hat{u}_j^{\pm}$ is equal to $\hat{u}_{j,\pm}(z)$ at the left border of layer j, one has

$$\hat{u}_j^+ \exp\left(i\tilde{k}_j d_j\right) + \hat{u}_j^- \exp\left(-i\tilde{k}_j d_j\right) = \hat{u}_{j+1}^+ + \hat{u}_{j+1}^- \tag{4.3.4}$$

The continuity of stress is expressed as (cf. Eq. 4.2.10)

$$\tilde{Z}_j\left(\hat{u}_j^+\exp\left(i\tilde{k}_jd_j\right)-\hat{u}_j^-\exp\left(-i\tilde{k}_jd_j\right)\right)=\tilde{Z}_{j+1}\left(\hat{u}_{j+1}^+-\hat{u}_{j+1}^-\right) \tag{4.3.5}$$

Equations 4.3.4 and 4.3.5 form a homogenous system of $2N+2$ linear equations. A nontrivial solution (a solution with nonzero amplitudes) only exists, if the determinant of the coefficient matrix is zero. This is the resonance condition.

For illustration, consider the case of $N = 1$ ($j = 0, 1, 2$), corresponding to the bare crystal. There are four amplitudes, which are $\hat{u}_0^+$, $\hat{u}_1^-$, $\hat{u}_1^+$, and $\hat{u}_2^-$. Since layer 1 is the resonator, the index 1 is replaced by the index q in the following. Equations 4.3.4 and 4.3.5 take the form

$$\begin{aligned}\hat{u}_0^+-\hat{u}_q^--\hat{u}_q^+&=0\\ \tilde{Z}_0\hat{u}_0^++\tilde{Z}_q\hat{u}_q^--\tilde{Z}_q\hat{u}_q^+&=0\\ \exp\left(-i\tilde{k}_qd_q\right)\hat{u}_q^-+\exp\left(i\tilde{k}_qd_q\right)\hat{u}_q^+-\hat{u}_2^-&=0\\ -\tilde{Z}_q\exp\left(-i\tilde{k}_qd_q\right)\hat{u}_q^-+\tilde{Z}_q\exp\left(i\tilde{k}_qd_q\right)\hat{u}_q^++\tilde{Z}_2\hat{u}_2^-&=0\end{aligned} \tag{4.3.6}$$

One can write these relations in matrix form as

$$\begin{bmatrix}1 & -1 & -1 & 0\\ \tilde{Z}_0 & -\tilde{Z}_q & \tilde{Z}_q & 0\\ 0 & \exp\left(-i\tilde{k}_qd_q\right) & \exp\left(i\tilde{k}_qd_q\right) & -1\\ 0 & -\tilde{Z}_q\exp\left(-i\tilde{k}_qd_q\right) & \tilde{Z}_q\exp\left(i\tilde{k}_qd_q\right) & \tilde{Z}_2\end{bmatrix}\begin{bmatrix}\hat{u}_0^+\\ \hat{u}_q^-\\ \hat{u}_q^+\\ \hat{u}_2^-\end{bmatrix}=\begin{bmatrix}0\\0\\0\\0\end{bmatrix} \tag{4.3.7}$$

Equation systems like Eq. 4.3.7 always have the "trivial solution", which is the solution with all amplitudes equal to zero. A nontrivial solution does not always exist. For an arbitrary set of matrix elements, one cannot not usually find a set of amplitudes, $\{\hat{u}_j^\pm\}$, such that the matrix applied to the vector of amplitudes yields zeros on the right-hand side. A nontrivial solution only exists, if the determinant of the matrix vanishes [13]. Remember: The matrix elements contain the wave-number and therefore *depend on frequency* ($\tilde{k}=\tilde{\omega}/\tilde{c}$). The frequencies with vanishing determinant are the resonance frequencies. Since the shear modulus contains a dissipative component ($G''>0$), the resonance frequencies are complex.

For a bare crystal the condition that the determinant be zero amounts to:

$$\begin{vmatrix}-1 & 1 & 1 & 0\\ -\tilde{Z}_0 & \tilde{Z}_q & -\tilde{Z}_q & 0\\ 0 & \exp\left(i\tilde{k}_qd_q\right) & \exp\left(-i\tilde{k}_qd_q\right) & -1\\ 0 & \tilde{Z}_q\exp\left(i\tilde{k}_qd_q\right) & -\tilde{Z}_q\exp\left(-i\tilde{k}_qd_q\right) & \tilde{Z}_2\end{vmatrix}=0 \tag{4.3.8}$$

To further simplify the matter, assume that the adjacent media have a small acoustic wave impedance (as can be assumed if the crystal is in air). One can then take the limits of $\tilde{Z}_0\to 0$ and $\tilde{Z}_2\to 0$. With $\tilde{Z}_0\approx 0$ and $\tilde{Z}_2\approx 0$, Eq. 4.3.8 simplifies to

$$-\tilde{Z}_q^2 \exp\left(\mathrm{i}\tilde{k}_q d_q\right) + \tilde{Z}_q^2 \exp\left(-\mathrm{i}\tilde{k}_q d_q\right) = 0 \tag{4.3.9}$$

Using $\sin(x) = 1/(2\mathrm{i})(\exp(\mathrm{i}x) - \exp(-\mathrm{i}x))$, this reads as

$$\sin\left(\tilde{k}_q d_q\right) = 0 \tag{4.3.10}$$

Equation 4.3.10 is fulfilled if the wavenumber obeys the relation

$$\tilde{k}_{q,OC} = \frac{\tilde{\omega}_{OC}}{\tilde{c}_q} = \tilde{\omega}_{OC}\sqrt{\frac{\rho_q}{\tilde{G}_q}} = \frac{n\pi}{d_q} \tag{4.3.11}$$

The index OC denotes the resonance condition of the unloaded plate under *O*pen-*C*ircuit conditions. If there is no current into the electrodes, the effects of piezoelectric stiffening can simply be accounted for by using the appropriate parameter for $\tilde{G}_q$, which is the "piezoelectrically stiffened shear modulus" (Sect. 5.2). In all other respects, piezoelectric stiffening can be ignored. The overtone order, n, is an integer. For real crystals, it will later turn out that the resonance frequencies at the overtones are not integer multiples of the fundamental frequency (Sect. 7.7). For that reason, these modes are called "overtones", rather than "harmonics", here. "Harmonics" in this book denote signals with a frequency at exact integer multiples of the fundamental (Sect. 13.4). In other fields of science, the terms "Harmonic" and "Overtone" are used synonymously and the "overtones" are called the "higher eigenmodes". In Eq. 4.3.11, the overtone order, n, may be even or odd. However, only odd overtones can be excited piezoelectrically (Sect. 4.5.2). Clearly, $\tilde{k}_{q,OC}$ is real. This happens because $\tilde{\omega}/\tilde{c}_q$ have the same phase. The tilde above $k_{q,OC}$ is omitted in the following.

The displacement pattern on resonance is given by a standing wave with antinodes at the surfaces:

$$\hat{u}(z) = \hat{u}_S \cos\left(n k_{q,OC} z\right) \tag{4.3.12}$$

The origin of the z-scale here is at the back of the crystal. $\hat{u}_S$ is the amplitude at the *S*urface. For the fundamental, the wavelength is twice the crystal thickness. The surfaces are stress-free with vanishing strain ($\mathrm{d}\hat{u}(z)/\mathrm{d}z = 0$).

Separating real and imaginary parts of $\tilde{\omega}_{r,OC} = 2\pi\left(f_{r,OC} + \mathrm{i}\Gamma_{OC}\right)$ one finds

$$f_{OC} = \frac{n c'_q}{2 d_q} \tag{4.3.13}$$

and

$$\Gamma_{OC} = \frac{nc''_q}{2d_q} = f_{OC}\frac{c''_q}{c'_q} = f_{OC}\frac{\mathrm{Im}\left(\sqrt{\tilde{G}_q}\right)}{\mathrm{Re}\left(\sqrt{\tilde{G}_q}\right)} \approx f_{OC}\frac{\mathrm{Im}\left(\sqrt{G'_q}\left(1+\mathrm{i}\frac{1}{2}\frac{G''_q}{G'_q}\right)\right)}{\mathrm{Re}\left(\sqrt{G'_q}\left(1+\mathrm{i}\frac{1}{2}\frac{G''_q}{G'_q}\right)\right)}$$
$$= f_{OC}\frac{1}{2}\frac{G''_q}{G'_q} = f_{OC}\frac{1}{2}\tan\delta_q \approx f_{OC}\frac{\delta_q}{2} \quad (4.3.14)$$

c_q' and c_q'' are the real and the imaginary part of the speed of shear sound. Taylor expansions were applied in step 4 ($(1 + \varepsilon)^{1/2} \approx 1 + \varepsilon/2$) and in step 7 ($\tan(\varepsilon) \approx \varepsilon$). $\tan(\delta_q) = G_q''/G_q'$ is the crystal's loss tangent (Sect. 3.7), δ_q is the loss angle. Γ_{OC} as given by Eq. 4.3.14 only accounts for intrinsic losses inside the crystal. For most QCRs, dissipative processes in the electrodes also contribute to the bandwidth. One might view $\tan(\delta_q)$ as an "effective loss tangent".

From Eq. 4.3.14, it follows that the Q-factor obeys

$$Q^{-1} = D = \frac{2\Gamma}{f} = \delta_q \quad (4.3.15)$$

D is the dissipation factor. For later use, Eq. 4.3.11 is rewritten in two different ways. Calling the frequency of the fundamental f_0, one finds:

$$d_q = \frac{1}{2}\frac{\tilde{c}_q}{\tilde{f}_0} = \frac{1}{2\sqrt{\rho_q}}\frac{\sqrt{\tilde{G}_q}}{\tilde{f}_0} = \frac{1}{2\rho_q}\frac{\tilde{Z}_q}{\tilde{f}_0} = \frac{1}{2\rho_q}\frac{Z_q}{f_0} \quad (4.3.16)$$

The fact that d_q is real implies that $\tilde{c}_q$, $\tilde{Z}_q$, and $\tilde{f}_0$ have the same phase. It is half of the phase angle of the plate's shear modulus (of the loss angle). Since the phases of $\tilde{Z}_q$ and $\tilde{f}_0$ are the same, one can write $\tilde{Z}_q/\tilde{f}_0 = Z_q/f_0$. Equation 4.3.11 can be further rewritten as

$$m_q = \rho_q d_q = \frac{1}{2}\frac{\rho_q \tilde{c}_q}{\tilde{f}_0} = \frac{1}{2}\frac{\tilde{Z}_q}{\tilde{f}_0} = \frac{1}{2}\frac{Z_q}{f_0} \quad (4.3.17)$$

m_q is the mass per unit area of the crystal, and ρ_q is the density. Equations 4.3.16 and 4.3.17 will be frequently needed in calculations of various kinds.

Before closing this section, we derive the Gordon–Kanazawa–Mason result (Eq. 9.1.1) from the mathematically-inspired formulation. Consider a configuration, where the back of the resonator is exposed to air ($\tilde{Z}_0 \approx 0$) and the front is exposed to a liquid ($\tilde{Z}_2 = \tilde{Z}_{liq}$). The resonance condition of this system is Eq. 4.3.8 with $\tilde{Z}_0 = 0$. Spelling out the determinant yields the relation

$$\tilde{Z}_{liq}\tilde{Z}_q\left(\exp\left(\mathrm{i}\tilde{k}_q d_q\right) + \exp\left(-\mathrm{i}\tilde{k}_q d_q\right)\right) + \tilde{Z}_q^2\left(\exp\left(\mathrm{i}\tilde{k}_q d_q\right) - \exp\left(-\mathrm{i}\tilde{k}_q d_q\right)\right) = 0 \quad (4.3.18)$$

Using $\cos(x) = 1/2(\exp(\mathrm{i}x) + \exp(-\mathrm{i}x))$ and $\sin(x) = 1/(2\mathrm{i})(\exp(\mathrm{i}x) - \exp(-\mathrm{i}x))$ yields

$$\mathrm{i}\tan\left(\tilde{k}_q d_q\right) = -\frac{\tilde{Z}_{liq}}{\tilde{Z}_q} \tag{4.3.19}$$

Writing $\tilde{k}_q = k_{q,OC} + \Delta\tilde{k}_q = n\pi/d_q + \Delta\tilde{k}_q$ yields

$$\mathrm{i}\tan\left(\left(\frac{n\pi}{d_q} + \Delta\tilde{k}_q\right)d_q\right) = \mathrm{i}\tan\left(\Delta\tilde{k}_q d_q\right) = -\frac{\tilde{Z}_{liq}}{\tilde{Z}_q} \tag{4.3.20}$$

The relation $\tan(n\pi + x) = \tan(x)$ for n an odd integer was used. Taylor expansion of the tangent ($\tan(\varepsilon) \approx \varepsilon$) leads to

$$\Delta\tilde{k}_q = \frac{\mathrm{i}}{d_q}\frac{\tilde{Z}_{liq}}{\tilde{Z}_q} \tag{4.3.21}$$

Using $\Delta\tilde{k}_q = \Delta\tilde{\omega}/\tilde{c}_q, d_q = \tilde{c}_q/\left(2f_0\right)$, and $\tilde{Z}_{liq} = \left(\mathrm{i}\omega\rho_{liq}\eta_{liq}\right)^{1/2}$ yields

$$\Delta\tilde{f} = \frac{\mathrm{i}\tilde{c}_q}{2\pi\, d_q}\frac{\tilde{Z}_{liq}}{\tilde{Z}_q} \tag{4.3.22}$$

and

$$\frac{\Delta\tilde{f}}{f_0} = \frac{\mathrm{i}}{\pi\tilde{Z}_q}\left(\mathrm{i}\omega\,\rho_{liq}\eta_{liq}\right) \tag{4.3.23}$$

The last relation is the Gordon–Kanazawa–Mason result (Eq. 9.1.1). Cleary, the derivation is a bit technical, but it does lead to the desired result (assuming $|\tilde{Z}_{liq}| \ll |\tilde{Z}_q|$).

4.4 Acoustic Plane Wave Model: Description in Terms of Acoustic Reflectivity

A good motivation for the use of reflectivities is the mathematically inspired formulation for the crystal loaded with a film. There are a total of six amplitudes. In matrix form, the equation system is

$$\begin{bmatrix} -1 & 1 & 1 & 0 & 0 & 0 \\ \tilde{Z}_0 & \tilde{Z}_q & -\tilde{Z}_q & 0 & 0 & 0 \\ 0 & \exp(i\tilde{k}_q d_q) & \exp(-i\tilde{k}_q d_q) & -1 & -1 & 0 \\ 0 & \tilde{Z}_q \exp(i\tilde{k}_q d_q) & -\tilde{Z}_q \exp(-i\tilde{k}_q d_q) & -\tilde{Z}_2 & +\tilde{Z}_2 & 0 \\ 0 & 0 & 0 & \exp(i\tilde{k}_2 d_2) & \exp(-i\tilde{k}_2 d_2) & -1 \\ 0 & 0 & 0 & \tilde{Z}_2 \exp(i\tilde{k}_2 d_2) & -\tilde{Z}_2 \exp(-i\tilde{k}_2 d_2) & -\tilde{Z}_3 \end{bmatrix} \begin{bmatrix} \hat{u}_0^- \\ \hat{u}_q^+ \\ \hat{u}_q^- \\ \hat{u}_2^+ \\ \hat{u}_2^- \\ \hat{u}_3^+ \end{bmatrix} = 0 \tag{4.4.1}$$

That does not look encouraging. One side remark before we continue with an algebraic strategy to simplify the matter: The numeric search of the frequencies, at which the determinant turns to zero, amounts to one line of code in Mathematica. This one line is an answer to the challenge of modeling the QCM (although not the only one and arguably not the best one).

Sticking to algebra: Finding the nontrivial solution of Eq. 4.4.1 is not quite as complicated as one might think because the matrix contains zeros. There is a formalized method of solving Eq. 4.4.1 (and its extension to arbitrary numbers of layers) called "matrix transfer method" [14–16]. The matrix transfer method assigns a physical interpretation to the ratio $\hat{u}_j^+/\hat{u}_j^-$: It is the reflectivity, evaluated at the respective location (It is not a property of a single interface, as in Eq. 4.2.11). In optics, a similar formalism carries the name Abéles matrix method. The Abéles method is mostly employed to calculate the reflectivity of multilayers.

The matrix method is an iterative scheme to compute the amplitudes $\hat{u}_j^\pm$, starting from the last layer. Eventually, one arrives at the 0th layer. There is no incoming wave in medium 0 ($\hat{u}_0^- = 0$). Only for certain values of the matrix elements does $\hat{u}_0^-$ turn to zero. Using reflectivities, the resonance condition states that the reflectivity of the entire resonator evaluated at its left end is infinite. Of course this is impossible for real frequencies. However, it can happen if one allows the vibration to decay in time, that is, for complex resonance frequencies.

When employing the matrix method, one writes the amplitudes as a two-dimensional vector ($\hat{u}_j^-$, $\hat{u}_j^+$). There is no reflected wave in the last, semi-infinite medium on the right-hand side (labeled $N + 1$). The transmitted wave in this medium, $\hat{u}_{N+1}^-$, is normalized to one. The vector of amplitudes in this medium therefore is (1, 0). The vectors ($\hat{u}_j^-$, $\hat{u}_j^+$) at the other interface are related to the amplitudes in layer $N + 1$ by transfer matrices. There are transfer matrices for the layers ($\bar{L}_j$) and for the interfaces ($\bar{S}_j$). The amplitudes are calculated as

$$\begin{pmatrix} \hat{u}_j^- \\ \hat{u}_j^+ \end{pmatrix} = \bar{L}_j \cdot \bar{S}_{j,j+1} \cdots \bar{L}_N \cdot \bar{S}_{N,N+1} \cdot \begin{pmatrix} 1 \\ 0 \end{pmatrix} \tag{4.4.2}$$

The matrix accounting for wave propagation inside layer j is

$$\bar{L}_j = \begin{pmatrix} \exp\left(\mathrm{i}\tilde{k}_j d_j\right) & 0 \\ 0 & \exp\left(-\mathrm{i}\tilde{k}_j d_j\right) \end{pmatrix} \tag{4.4.3}$$

The matrix $\bar{S}_{j,j+1}$ takes care of reflection at interfaces. It is given as

$$\bar{S}_{j,j+1} = \frac{1}{2}\begin{pmatrix} \tilde{Z}_j/\tilde{Z}_{j+1} + 1 & \tilde{Z}_j/\tilde{Z}_{j+1} - 1 \\ \tilde{Z}_j/\tilde{Z}_{j+1} - 1 & \tilde{Z}_j/\tilde{Z}_{j+1} + 1 \end{pmatrix} \tag{4.4.4}$$

One can quickly convince oneself that this form of $\bar{S}_{j,j+1}$ reproduces Eq. 4.2.11. Appling $\bar{S}_{j,j+1}$ to the vector (1, 0) and taking the ratio $\hat{u}_j^+/\hat{u}_j^-$, one finds the reflectivity at $z_{j,j+1}$ as predicted by Eq. 4.2.11.

Eventually, the vector $(\hat{u}_0^-, \hat{u}_0^+)$ is computed as

$$\begin{pmatrix} \hat{u}_0^- \\ \hat{u}_0^+ \end{pmatrix} = \bar{S}_{0,q} \cdot \bar{L}_q \cdot \bar{S}_{q,2} \cdot \bar{L}_2 \cdot \bar{S}_{2,3} \cdot \ldots \bar{L}_N \cdot \bar{S}_{N,N+1} \cdot \begin{pmatrix} 1 \\ 0 \end{pmatrix} \tag{4.4.5}$$

Only for the resonance frequencies is the condition $\hat{u}_0^- = 0$ (Eq. 4.3.2) fulfilled. Inserting $\hat{u}_0^- = 0$ into Eq. 4.4.5 leads to an equation, which is equivalent to a vanishing determinant of the coefficient matrix in Sect. 4.3.

In addition to being more physical, the formulation in terms of reflectivities also has the benefit that a central parameter is easily identified. It is the reflectivity at the interface between the resonator and the sample, called $\tilde{r}_{q,S}$ (Note: $\tilde{r}_{q,S}$ is different from $\tilde{r}_{q,2}$ as evaluated with Eq. 4.2.11. $\tilde{r}_{q,2}$ is only a function of Z_q and $\tilde{Z}_2$. $\tilde{r}_{q,S}$ is the ratio $\hat{u}^+/\hat{u}^-$ evaluated at the resonator surface and calculated with the matrix formalism. The entire layer stack enters into this calculation). $\tilde{r}_{q,S}$ embodies the acoustic properties of the sample as seen by the resonator. For the amplitudes evaluated at the interface between the resonator and the sample one has

$$\hat{u}_q^+ \exp\left(\mathrm{i}\tilde{k}_q d_q\right) = \tilde{r}_{q,S}\hat{u}_q^- \exp\left(-\mathrm{i}\tilde{k}_q d_q\right) \tag{4.4.6}$$

The condition $\hat{u}_0^- = 0$ amounts to

$$\begin{aligned} 0 &= \frac{1}{2}\left(\frac{\tilde{Z}_0}{\tilde{Z}_q} + 1\right)\hat{u}_q^- + \left(\frac{\tilde{Z}_0}{\tilde{Z}_q} - 1\right)\hat{u}_q^+ \\ &= \frac{\hat{u}_q^-}{2}\left(\left(\frac{\tilde{Z}_0}{\tilde{Z}_q} + 1\right) + \left(\frac{\tilde{Z}_0}{\tilde{Z}_q} - 1\right)\tilde{r}_{q,S}\exp\left(-2\mathrm{i}\tilde{k}_q d_q\right)\right) \end{aligned} \tag{4.4.7}$$

Assuming $\tilde{Z}_0 \approx 0$ leads to

$$0 = \frac{\hat{u}_q^-}{2}\left(1 - \tilde{r}_{q,S}\exp\left(-2\mathrm{i}\tilde{k}_q d_q\right)\right) \tag{4.4.8}$$

and

$$\exp\left(2\mathrm{i}\tilde{k}_q d_q\right) = \tilde{r}_{q,S} \tag{4.4.9}$$

Using $\tilde{k}_q = \omega/\tilde{c}_q = 2\pi\left(f_{OC} + \Delta\tilde{f}\right)/\tilde{c}_q$ one finds

$$\begin{aligned}\exp\left(\frac{4\pi\,\mathrm{i}\left(f_{OC}+\Delta\tilde{f}\right)}{\tilde{c}_q}d_q\right) &= \exp\left(\frac{4\pi\,\mathrm{i}f_{OC}}{\tilde{c}_q}d_q\right)\exp\left(\frac{4\pi\,\mathrm{i}\Delta\tilde{f}}{\tilde{c}_q}d_q\right)\\ &= \exp\left(\frac{4\pi\,\mathrm{i}\Delta\tilde{f}}{\tilde{c}_q}d_q\right) = \exp\left(2\pi\,\mathrm{i}\frac{\Delta\tilde{f}}{\tilde{f}_0}\right)\tilde{r}_{q,S}\end{aligned} \tag{4.4.10}$$

The relations $\exp\left(4\pi\,\mathrm{i}f_{OC}d_q/\tilde{c}_q\right) = 1$ and $\tilde{f}_0 = \tilde{c}_q/\left(2d_q\right)$ were used in line 2. Inverting for $\Delta\tilde{f}/\tilde{f}_0$ one finds

$$\frac{\Delta\tilde{f}}{\tilde{f}_0} \approx \frac{\Delta\tilde{f}}{f_0} \approx \frac{1}{2\pi\mathrm{i}}\ln\left(\tilde{r}_{q,S}\right) \tag{4.4.11}$$

The parameter $\tilde{r}_{q,S}$ is often close to one. This happens if the sample is either soft or thin (or both). In the former case, most of the wave is reflected at the interface between the resonator and the sample because of the impedance mismatch. In the latter case, a significant portion of the acoustic wave enters the sample and returns to the resonator after having been reflected at the back of the sample. Since the sample is thin, the phase shift and the attenuation acquired while traversing the sample are small. If $\tilde{r}_{q,S} \approx 1$, the logarithm in Eq. 4.4.11 can be Taylor-expanded (following $\ln(1 + \varepsilon) \approx \varepsilon$), which leads to

$$\frac{\Delta\tilde{f}}{f_0} \approx \frac{1}{2\pi\,\mathrm{i}}\left(\tilde{r}_{q,S} - 1\right) = \frac{\mathrm{i}}{\pi}\frac{1}{2}\left(1 - \tilde{r}_{q,S}\right) \tag{4.4.12}$$

Making use of Eq. 4.4.12, a frequency shift found in a QCM experiment can be compared to reflectivities as obtained with other forms of ultrasonic shear-wave reflectometry (Refs. [17, 18], Sect. 15.1). Put differently, the QCM is a *shear-wave reflectometer*. The fact that a film thickness can be inferred from a measurement of the reflectivity is known from optical techniques such as ellipsometry [19, 20] or surface plasmon resonance (SPR) spectroscopy [21]. The same principles apply in acoustics.

For later use, we compute the stress-velocity ratio at the interface between the resonator and the sample. This ratio, $\tilde{Z}_L$, will be of importance in the context of the small load approximation. One must pay attention to a sign problem here: The stress inserted into the calculation of $\tilde{Z}_L$ is exerted *by* the resonator *onto* the sample. It acts into the direction of positive z and therefore is given as $\tilde{\sigma}_S = -\tilde{G}\,\mathrm{d}\hat{u}/\mathrm{d}z$. The load impedance is

$$\begin{aligned}\tilde{Z}_L = \frac{\hat{\sigma}_S}{\hat{v}_S} &= \frac{-i\omega\tilde{Z}_q\left(-\hat{u}^- \exp\left(-i\tilde{k}_q d_q\right) + \hat{u}^+ \exp\left(i\tilde{k}_q d_q\right)\right)}{i\omega\left(\hat{u}^- \exp\left(-i\tilde{k}_q d_q\right) + \hat{u}^+ \exp\left(i\tilde{k}_q d_q\right)\right)} \\ &= \frac{-i\omega\tilde{Z}_q\left(-\hat{u}^- \exp\left(-2i\tilde{k}_q d_q\right)\left(1 - \tilde{r}_{q,S}\right)\right)}{i\omega\left(\hat{u}^- \exp\left(-2i\tilde{k}_q d_q\right)\left(1 + \tilde{r}_{q,S}\right)\right)} \\ &\approx \tilde{Z}_q \frac{1}{2}\left(1 - \tilde{r}_{q,S}\right)\end{aligned} \tag{4.4.13}$$

The relation $1 + \tilde{r}_{q,S} \approx 2$ was used in line 3.

As an example, assume that the crystal is coated with a thin film of thickness d_f. Let the film be loss-less and let it have the same acoustic wave impedance as the crystal. In this case, the entire acoustic wave enters the sample and returns. Evaluating $\tilde{r}_{q,S}$ as $\hat{u}_q^+/\hat{u}_q^-$, the absolute value of $\tilde{r}_{q,S}$ is found to be unity $|(|\tilde{r}_{q,S}| = 1)$. There is a phase shift $\Delta\varphi = -2k_f d_f$, which the wave acquires while traversing the film. Following Eq. 4.4.12, the frequency shift obeys (Fig. 4.8)

$$\frac{\Delta\tilde{f}}{f_0} \approx \frac{i}{2\pi}\left(1 - \exp\left(-2i\tilde{k}_f d_f\right)\right) \approx \frac{-2\tilde{k}_f d_f}{2\pi} = \frac{-\tilde{\omega}_r d_f}{\pi\,\tilde{c}_q} \approx \frac{-2nf_0 d_f}{c_q} = \frac{-2nf_0}{Z_q} m_f \tag{4.4.14}$$

Taylor-expansion of the exponential ($\exp(\varepsilon) \approx 1 + \varepsilon$) as well as the relations $\tilde{c}_q = \tilde{Z}_q/\rho_q$ and $m_f = \rho_f d_f$ were used. m_f is the mass per unit area of the film. Equation 4.4.14 is the Sauerbrey relation (Eq. 8.1.2).

As a second example, assume that the crystal is in contact with a homogeneous semi-infinite medium with a wave impedance, $\tilde{Z}_{liq}$. Let $\tilde{Z}_{liq}$ be much smaller than Z_q (meaning $|\tilde{Z}_{liq} / Z_q| \ll 1$). According to Eq. 4.2.11, the reflectivity at the crystal surface can be approximated as

$$\begin{aligned}\tilde{r}_{q,S} = \tilde{r}_{q,2} &= \frac{\tilde{Z}_q - \tilde{Z}_{liq}}{\tilde{Z}_q + \tilde{Z}_{liq}} = \frac{1 - \tilde{Z}_{liq}/Z_q}{1 + \tilde{Z}_{liq}/Z_q} \\ &\approx \left(1 - \frac{\tilde{Z}_{liq}}{Z_q}\right)^2 \\ &\approx 1 - \frac{2\tilde{Z}_{liq}}{Z_q}\end{aligned} \tag{4.4.15}$$

The relations $1/(1 + \varepsilon) \approx 1 - \varepsilon$ and $(1 - \varepsilon)^2 \approx 1 - 2\varepsilon$ were used in lines 2 and 3. Inserting Eq. 4.4.15 into Eq. 4.4.12 one finds

$$\frac{\Delta\tilde{f}}{f_0} \approx \frac{i}{\pi}\frac{\tilde{Z}_{liq}}{Z_q} \tag{4.4.16}$$

This is the Gordon–Kanazawa–Mason result (Sect. 9.1).

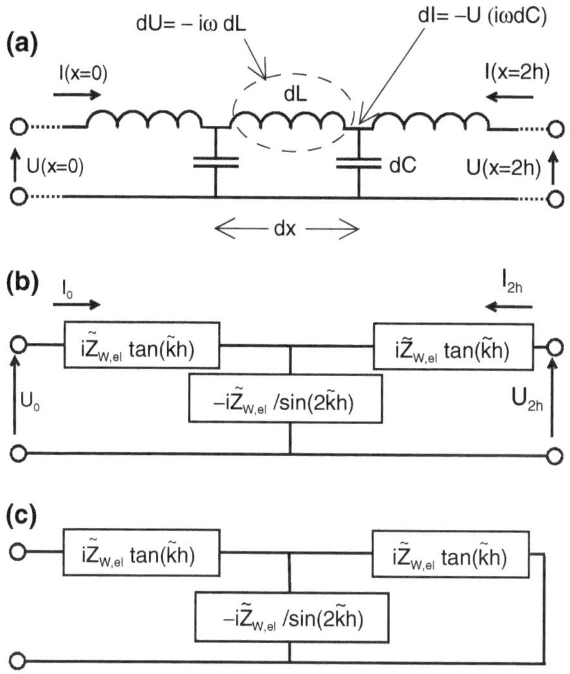

Fig. 4.8 a A distributed-element network. An electrical cable may be represented as an arrangement of infinitely many infinitely small inductances and capacitances. Applying the Kirchhoff rules to the loops and nodes, one finds the wave equation. **b** If one is only interested in the voltage-current relations at the ends of the cable, a faithful representation can be achieved with only three discrete elements. h is half the cable's length, $\tilde{k}$ is the wavenumber, and $\tilde{Z}_{W,el} = (\mathrm{d}L/\mathrm{d}C)^{1/2}$ is the wave impedance. Such networks are also termed "lumped-element networks". Applying the Kirchhoff rules to the network, one reproduces the reflectivity and the transmissivity (more generally: the "S-matrix") of the device. **c** The impedance of a cable short-circuited at the rear end is calculated by applying the Kirchhoff rules to this network (cf. Eq. 4.5.6)

4.5 Equivalent Electrical Circuits

An electrical equivalent circuit (equivalent circuit, for short) is an arrangement of discrete electrical elements (resistors, capacitors, inductances, or possibly elements with a complicated dependence of the impedance on frequency, cf. Fig. 4.8c) representing the resonator. The link between the circuit and the real resonator contains two separate steps. Firstly, a shear wave can be modeled using tools known from electrical engineering and used there to describe an electrical wave in a cable. This use of language and concepts is part of the electromechanical analogy (Sect. 4.7.3). Secondly, an object supporting a continuous distribution of currents and electrical potentials (or a distribution of stress and strain) can be represented by a set of discrete elements, if one limits one's interest to the values of these fields at the ends (the ports) of the respective object. This is the essence of the "lumped-element representation". Of course this lumped-element representation is incomplete. It only captures the stress-velocity ratios at the acoustic port (the surfaces) and the voltage-current ratio at the electrical port (the electrodes). The equivalent circuit hides the internal distribution of current, voltage, stress, and strain from the reader. It is only concerned with the ports. The reader can understand the current-voltage relations and the stress-velocity relations (and also the relations between—let's say—current and stress) *at the ports* from the circuit. All that needs to be known are the Kirchhoff rules (Sect. 4.7.2).

To illustrate the concept of the lumped-element representation, consider an electrical cable. Figure 4.9a shows a representation with infinitely many distributed elements. dL and dC are infinitesimally small inductances and capacitances. To see that the cable transmits electrical waves, apply Kirchhoff's voltage law ("voltages sum to zero around a loop") to the central loop in Fig. 4.8a. Further, apply Kirchhoff's current law ("currents sum to zero at nodes") to the nodes. The following equations result:

$$\begin{aligned} \frac{\mathrm{d}U}{\mathrm{d}x} &= -I\,\mathrm{i}\omega\frac{\mathrm{d}L}{\mathrm{d}x} \qquad \text{(a)} \\ \frac{\mathrm{d}I}{\mathrm{d}x} &= -U\,\mathrm{i}\omega\frac{\mathrm{d}C}{\mathrm{d}x} \qquad \text{(b)} \end{aligned} \tag{4.5.1}$$

U is the voltage across a capacitor; I is the current through an inductor, dU is the difference between the voltages across two neighboring capacitors, and dI is the difference in current between two neighboring inductors. Combining the two equations, one finds

$$\frac{\mathrm{d}^2U}{\mathrm{d}x^2} = -\omega^2\frac{\mathrm{d}L}{\mathrm{d}x}\frac{\mathrm{d}C}{\mathrm{d}x} \tag{4.5.2}$$

For a wave of the form $\exp(\mathrm{i}(\omega t \pm kx))$ one has $\mathrm{d}^2U/\mathrm{d}x^2 = -k^2U$, which implies

$$k = \frac{\omega}{c} = \sqrt{\frac{1}{(\mathrm{d}L/\mathrm{d}x)(\mathrm{d}C/\mathrm{d}x)}} \tag{4.5.3}$$

The denominator defines the speed of propagation. The circuit also predicts the electrical wave impedance of this line. Dividing Eq. 4.5.1b by Eq. 4.5.1a, one finds

$$\frac{U}{I}\frac{\mathrm{d}U}{\mathrm{d}I} = \frac{\mathrm{d}L}{\mathrm{d}C} \tag{4.5.4}$$

Using that the differences in voltage and current (dU and dI) are proportional to voltage and current (U and I), one concludes that

$$Z_{W,el} = \frac{U}{I} = \sqrt{\frac{\mathrm{d}L}{\mathrm{d}C}} \tag{4.5.5}$$

There is an analogy to the acoustic wave impedance, $\tilde{Z} = \left(\rho\tilde{G}\right)^{1/2}$ (Eq. 4.2.6). The role of density is taken by dL/dx and the role of the shear modules is taken by $(\mathrm{d}C/\mathrm{d}x)^{-1}$.

If one is only interested in the current-voltage relations at the two ends, there is no need to deal with infinitely many elements of infinitely small magnitude. The simpler representation shown in Fig. 4.8b suffices. This lumped-element network contains only three elements. An abbreviated proof that Fig. 4.8b indeed describes a cable is given in Sect. 4.7.4. A more in-depth treatment can be found in Refs. [22, 23]. Contrasting to Fig. 4.8a, the representation in Fig. 4.8b only contains a finite number of elements. There are two loops and two nodes. The Kirchhoff rules therefore yield a total of four equations, linking the four external variables (two voltages and two currents). Figure 4.8b is called a "transmission line" in electrical engineering [24].

Figure 4.9c shows an application example. The cable was short-circuited at the rear end, thereby turning a four-terminal network into a two-terminal network. A natural question would be: "What is the electrical impedance at the remaining port?". The Kirchhoff rules provide the answer. Adding impedances according to the Kirchhoff rules, one finds

$$\begin{aligned}\tilde{Z}_{tot} &= \mathrm{i}\tilde{Z}_w\left(\tan\left(\tilde{k}h\right) + \left(\left(-\frac{1}{\sin\left(2\tilde{k}h\right)}\right)^{-1} + \tan\left(\tilde{k}h\right)^{-1}\right)^{-1}\right)\\ &= \mathrm{i}\tilde{Z}_w\tan\left(2\tilde{k}h\right)\end{aligned} \tag{4.5.6}$$

The second step made use of sin(2x) = 2sin(x)cos(x) and tan(2x) = 2sin(x) cos(x)/($\cos^2(x)$ – $\sin^2(x)$). h is $h$alf of the cable's length. By analogy, the acoustic wave impedance of a viscoelastic film of thickness $d_f = 2h_f$ is given as $\mathrm{i}\tilde{Z}_f \tan(\tilde{k}_f d_f)$, where $\tilde{Z}_f$ is the film's shear-wave impedance. The same result is obtained within the wave picture in Sect. 10.1.

4.5.1 The Piezoelectric Transducer

By virtue of the electromechanical analogy, the network shown in Fig. 4.8b not only describes the current-voltage relations at the ends of a cable, but also the stress-velocity relations at the two surfaces of a viscoelastic plate. It can be expected that the piezoelectric plate resembles this network as far as viscoelasticity is concerned. Deviating from Fig. 4.8b, piezoelectric coupling leads to a third port. The third port is electrical in nature; it represents the electrodes. This network is shown in Fig. 4.9. It is called the "Mason-circuit" here.

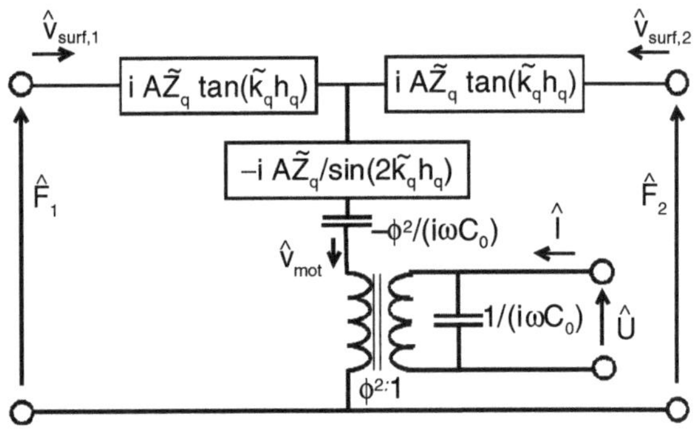

Fig. 4.9 Equivalent circuit of the piezoelectric transducer. The circuit is called "Mason circuit" in this book. A transducer here denotes an element with open acoustic ports. It can detect sound waves from the outside

The most important consequence of piezoelectricity is that stress couples to voltage and vice versa. Electromechanical coupling is depicted as a transformer in Fig. 4.9. Recall that a classical transformer is specified by the numbers of turns on both sides, n_{left} and n_{right}. The turns ratio is n_{right}/n_{left}. The following relations hold:

$$\begin{aligned} \hat{I}_{right} &= \frac{n_{left}}{n_{right}} \hat{I}_{left} \\ \hat{U}_{right} &= \frac{n_{right}}{n_{left}} \hat{U}_{left} \\ \tilde{Z}_{right} &= \left(\frac{n_{right}}{n_{left}}\right)^2 \tilde{Z}_{left} \end{aligned} \tag{4.5.7}$$

Z_{right} here is the impedance of an element connected to the left-hand side of the transformer as measured across the transformer from the right. The transformer acts as an impedance converter.

What is depicted as a transformer in Fig. 4.9, also is an "impedance converter", albeit in a different sense. It converts between electrical and mechanical quantities. The conversion reads as

$$\begin{aligned} \hat{I} &= \phi \hat{v} && \text{(a)} \\ \hat{U} &= \frac{1}{\phi}\hat{F} = \frac{1}{\phi} A\hat{\sigma} && \text{(b)} \\ \tilde{Z}_{el} &= \frac{\hat{U}}{\hat{I}} = \frac{1}{\phi^2}\frac{A\hat{\sigma}}{\hat{v}_{mot}} = \frac{1}{\phi^2} A\tilde{Z}_{mot} && \text{(c)} \\ \phi &= \frac{Ae_{26}}{d_q} && \text{(d)} \end{aligned} \tag{4.5.8}$$

The quantities to the right of the transformer in Fig. 4.9 are electrical in nature, while the ones to the left are mechanical. The first three lines in Eq. 4.5.8 spell out the electromechanical analogy. Line 4 states how the conversion factor, ϕ, depends on the geometry and the material. A is the area of the plate and d_q is its thickness. The piezoelectric stress coefficient, e_{26}, relates the shear stress to the electric field. e_{26} is a component of the e-tensor, discussed in more detail in Sects. 5.2 and 5.4. Its value is 9.65×10^{-2} C/m^2 for AT-cut quartz.

At this point, the focus is on the transformer; the element $-\phi^2/(i\omega C_0)$ located above the transformer is of minor importance. This element is a consequence of piezoelectric stiffening as discussed in Sect. 5.2. Piezoelectric stiffening is viewed as a technical detail here.

The derivation of the network shown Fig. 4.9 is far from trivial and even the correct interpretation of all its elements involves some subtleties or—as Thurston puts it in Ref. [22]—some "traps for the unwary". When discussing piezoelectricity, one must be careful to always remember what the boundary conditions are. For instance, the polarization induced by an external voltage depends on whether the surfaces are mechanically free or clamped (or in some intermediate condition). Similarly, the strain induced by some external stress depends on whether the electrodes are electrically open or short-circuited. This phenomenon is referred to as piezoelectric stiffening (also: "electromechanical coupling", see Sect. 5.2).

There are two further subtleties with Fig. 4.9, mentioned here without going into the details. Firstly, in interpreting Fig. 4.9, one applies *electrical* Kirchhoff rules everywhere, rather than mechanical Kirchhoff rules (Sect. 4.7.3). Impedances arranged in series are additive. Secondly, Fig. 4.9 assumes an ideal voltage source. The driving electronics supplies a fixed voltage, regardless of how low the impedance of the network is. This is unrealistic. A typical output resistance of a network analyzer is 50 Ω. This value is comparable to the motional resistance, R_1, (Eq. 4.5.20), which is a few ohms in air and a few hundred ohms in liquids. Because of the finite output resistance, the voltage at the electrodes is lower than the nominal voltage at the output of the driving electronics. This will lead to the topic of electric and dielectric loads, discussed in more depth in Sect. 5.2.

4.5.2 The Piezoelectric Resonator

The circuit shown in Fig. 4.9 represents a transducer. A transducer detects acoustic waves from the outside; it might be operated as a shear-wave microphone, in principle (only in principle because shear waves do not propagate in air and liquids). In QCM-based sensing, the sample itself does not emit a signal of any kind. It only responds to the stress exerted by the crystal. There is no need for acoustic ports in the modeling process. After closing the acoustic ports to the left and right in Fig. 4.9, one arrives at Fig. 4.10a. This device (a resonator) no longer listens to the outside world. The sample's properties enter the formalism as an acoustic impedance, not as an acoustic source. Its properties are fully contained in the stress-velocity ratio at the interface between the crystal and the sample. This is an essential (if not the most essential) step in the modeling process.

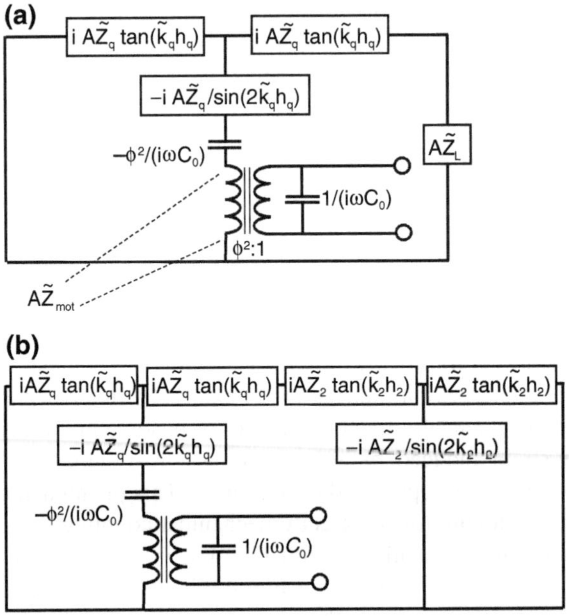

Fig. 4.10 Equivalent circuit of the piezoelectric resonator. In panel **a**, the sample is represented by the load impedance, $A\tilde{Z}_L$. $\tilde{Z}_L$ is multiplied by the area in order to turn the element into a mechanical (rather than an acoustic) impedance. $\tilde{Z}_L$ represents any type of sample and may be computed in a number of different ways. Panel **b** shows the piezoelectric resonator loaded with a viscoelastic film. Here, $\tilde{Z}_L$ is calculated based on the film's equivalent circuit. A multilayer might be represented by adding more T*s* to the right-hand side of the network. The classical work by Granstaff and Martin [26] made use of a similar circuit. Electrical Kirchhoff rules apply everywhere

Again, the circuit predicts the behavior at the ports, only. For this reason, there is no intuitive interpretation of the motional impedance, $A\tilde{Z}_{mot}$, as defined in Eq. 4.5.8c and Fig. 4.10. $A\tilde{Z}_{mot}$ is the mechanical impedance acting onto the left-hand side of the transformer. One can experimentally determine the quantity $A\tilde{Z}_{mot}/\phi^2$ from the electrical port. On a formal level, $A\tilde{Z}_{mot}$ is a force-velocity ratio, but there is no location, where the respective force and the correspondingly velocity, $\hat{v}_{mot}$, could be measured. This slightly awkward situation is the consequence of the fact that the circuit hides the stress distribution inside the plate.

Since the resonator is stress-free at the back, the port on the left-hand side is just short-circuited. The sample is located at the front (on the right-hand side in the diagram). It is represented by the element $A\tilde{Z}_L$, inserted between the two terminals. The load impedance, $\tilde{Z}_L$, is multiplied by the area, A, because the Mason circuit is written down in terms of mechanical impedances, not acoustic impedances. With the acoustic ports closed, the resonator looks like an electrical device from the outside. A resonance occurs if the total electrical impedance across the electrodes,

$\tilde{Z}_{el}$, vanishes. This can only happen for complex frequencies. Since the mechanical impedance to the left of the transformer, $A\tilde{Z}_{mot}$, acts in parallel to the electrical capacitance, the condition $\tilde{Z}_{el} = 0$ is equivalent to $\tilde{Z}_{mot} = 0$. In the following, it is left open what the load is; the sample is represented as $A\tilde{Z}_L$, regardless of its nature. From application of the Kirchhoff laws to Fig. 4.10a one finds the $\tilde{Z}_{mot}$ as

$$\tilde{Z}_{mot} = -\frac{\phi^2}{\mathrm{i}\omega C_0} - \frac{\mathrm{i}\tilde{Z}_q}{\sin\left(2\tilde{k}_q h_q\right)} + \left(\left(\mathrm{i}\tilde{Z}_q \tan\left(\tilde{k}_q h_q\right)\right)^{-1} + \left(\mathrm{i}\tilde{Z}_q \tan\left(\tilde{k}_q h_q\right) + \tilde{Z}_L\right)^{-1}\right)^{-1} \tag{4.5.9}$$

Again: Once the equivalent network is written down correctly, the resonance condition simply follows from repeated application of the Kirchhoff laws and requiring $A\tilde{Z}_{mot} = 0$. Putting equivalent networks to work is simple. Equivalent networks sketch a physical situation and at the same time quantitatively predict the resonance frequency. Admittedly, the search for the zeros of $\tilde{Z}_{mot}(\tilde{\omega})$ will usually require a numeric algorithm. But this is a technicality.

The calculation of $\tilde{Z}_L$ can occur in a number of different ways. For layered systems, $\tilde{Z}_L$ may be computed within the framework of equivalent circuits. A circuit representing a resonator coated with a viscoelastic layer is shown in Fig. 4.10b. The film is represented by a second "T" on the right-hand side. The film is stress-free at the film-air interface and the circuit is therefore closed to the right of the second T. Applying the Kirchhoff rules, the load induced by the second T is found as $A\tilde{Z}_L = \mathrm{i}A\tilde{Z}_2 \tan\left(2\tilde{k}_2 h_2\right)$ (cf. Eq. 4.5.6). If the resonator is immersed in a liquid, one does not short-circuit the second T on the right-hand side, but closes it across an element of the form $A\tilde{Z}_{liq} = A(\mathrm{i}\omega\rho_{liq}\tilde{\eta}_{liq})^{1/2}$. $\tilde{Z}_{liq}$ is the shear-wave impedance of the liquid. The resulting overall load impedance then is given by Eq. 10.2.3. Multilayers are represented as sequences of Ts. These can be short-circuited at the rear end or be closed across an impedance $A\tilde{Z}_{liq}$, depending on whether the measurement occurs in air or liquid.

The load, $\tilde{Z}_L$, can (for layered samples) be computed from equivalent networks, but it can also be computed in others ways. It might be computed with the matrix method (Sect. 4.3). For point contacts, $\tilde{Z}_L$ will be derived from contact mechanics (Chap. 11). For samples with a complicated geometry, one can use numerical techniques like the finite element method (Chap. 12). The Mason circuit creates a platform for modeling. The user is free to calculate $\tilde{Z}_L$ in whatever way is most convenient. Once $\tilde{Z}_L$ is known, the user inserts the value into the network (Fig. 4.10a), calculates $\tilde{Z}_{mot}$, finds the zeros of $\tilde{Z}_{mot}(\tilde{\omega})$, and thereby obtains the resonance frequencies (Note: The load impedance may be an area-average, see Sect. 4.6.2. It must be an area-average for all samples which are not laterally homogeneous).

A remark on mathematics: The equivalent circuit amounts to a treatment of a boundary value problem (BVP), where the boundary conditions are of the "Robin type" [25]. These boundary conditions are also called "impedance boundary condition" in electromagnetism. Clearly, they deserve the same name in acoustics. Neither the displacement, nor the gradient of the displacement are specified at the surface (which would be Dirichlet boundary conditions and the Neumann boundary conditions, respectively). Rather, the boundary condition is of the form

$$\tilde{G}_q \frac{\mathrm{d}\hat{u}}{\mathrm{d}z} - \tilde{Z}_L \mathrm{i}\omega\hat{u} = 0 \tag{4.5.10}$$

The boundary condition amounts to a relation between the displacement and the shear strain, specified by the parameter $\tilde{Z}_L$. Above, the back of the crystal was assumed to be stress-free. If it is not, then there must be a second parameter $\tilde{Z}_{L,back} \neq 0$, taking care of the load at the back. If the small load approximation holds, the total load impedance is the sum of the loads at the front and the back.

In the following, approximations are invoked, which lead to an explicit analytical expression for the resonance frequency. Firstly, the resonator is assumed to be unloaded $\left(\tilde{Z}_L = 0\right)$. Also, the element $-\phi^2/(\mathrm{i}\omega C_0)$ is assumed to be small, which amounts to neglecting piezoelectric stiffening. The latter simplification of course is unrealistic. The assumption is only made to show that the equivalent-network formalism is equivalent to the matrix method as discussed in Sect. 4.4 (if piezoelectric stiffening is ignored). Ignoring the element $-\phi^2/(\mathrm{i}\omega C_0)$ and using $\tilde{Z}_L \approx 0$, one finds

$$\begin{aligned}\tilde{Z}_{mot,OC} &= \mathrm{i}\tilde{Z}_q\left(-\frac{1}{\sin\left(2\tilde{k}_{q,OC}h_q\right)} + \left(\tan\left(\tilde{k}_{q,OC}h_q\right)^{-1} + \left(\tan\left(\tilde{k}_{q,OC}h_q\right)\right)^{-1}\right)^{-1}\right)\\ &= \mathrm{i}\tilde{Z}_q\left(-\frac{1}{\sin\left(2\tilde{k}_{q,OC}h_q\right)} + \frac{1}{2}\tan\left(\tilde{k}_{q,OC}h_q\right)\right)\\ &= -\mathrm{i}\tilde{Z}_q\cot\left(\tilde{k}_{q,OC}h_q\right)\end{aligned} \tag{4.5.11}$$

The relation $\tan(x) + \cot(x) = 2/(\sin(2x))$ was used in line 3. The index *OC* stands for the *O*pen-*C*ircuit condition. If there is a large resistance immersed between the electrodes (that is, if there is almost no electrical current into the electrodes), the impedance produced by the electrical branch of the network exactly compensates the element $-\phi^2/(\mathrm{i}\omega C_0)$ and one may ignore both. For the proof, see Sect. 5.2. Equation 4.5.11 is solved by

$$\begin{aligned}\tilde{k}_{q,OC}h_q &= \frac{2\pi\tilde{f}_{OC}}{\tilde{c}_q}h_q = \frac{\pi\tilde{f}_{OC}}{\tilde{c}_q}d_q = \frac{n\pi}{2}, \quad n = 1,3,5,\ldots\\ \tilde{f}_{OC} &= \frac{\tilde{\omega}_{r,OC}}{2\pi} = n\tilde{f}_0\end{aligned} \tag{4.5.12}$$

Note the difference between Eqs. 4.3.11 and 4.5.12. The equivalent circuit representation shows explicitly that the overtone order must be odd in order to ensure an antisymmetric pattern of motion. For the symmetric deformation pattern, the piezoelectrically-induced surface polarization is the same at the two surfaces and there is no current between the electrodes.

4.5.3 The Four-Element Circuit

When $-\phi^2/(i\omega C_0)$ and $\tilde{Z}_L$ are nonzero, there is no explicit analytical expression for the resonance frequency as a function of the $\tilde{Z}_L$. Actually, a closed expression for the resonance frequency is not what is needed. What is needed, is a closed expressions for the frequency shift, $\Delta\tilde{f}$, as a function of $\tilde{Z}_L$. In order to derive such an approximate relation, one linearizes $\tilde{Z}_{mot}$ in $\tilde{\omega} - \tilde{\omega}_{r,ref}$. This linearization will contain $\tilde{Z}_L$ as a linear term. Once there is a linear relation between $\tilde{Z}_L$ and $\tilde{\omega} - \tilde{\omega}_{r,ref}$, this relation can be inverted, which solves the problem (Sect. 4.6.1).

In Ref. [22] the reader is guided through a series of circuit transformations, which eventually lead to the famous four-element circuit (also: "Butterworth-van Dyke circuit" [27, 28]). A different route is taken here, based on the Taylor expansion of Eq. 4.5.9 in $\tilde{k}_q$ around $\tilde{k}_{q,OC} = n\pi/(2h_q)$. (Remember: $\tilde{k}_q = \omega/\tilde{c}_q$). The Taylor expansion leads to

$$\begin{aligned}\tilde{Z}_{mot} &\approx -\frac{\phi^2}{i\omega C_0} + \frac{i\tilde{Z}_q}{2}\left(\tilde{k}_q h_q - \frac{n\pi}{2}\right) + \frac{\tilde{Z}_L}{4} \\ &= -\frac{\phi^2}{i\omega C_0} + i\frac{n\pi}{2}\frac{\tilde{Z}_q}{2}\left(\frac{2\tilde{k}_q h_q}{n\pi} - 1\right) + \frac{\tilde{Z}_L}{4} \\ &= \frac{1}{4}\left(-\frac{4\phi^2}{i\omega C_0} + in\pi\tilde{Z}_q\frac{\tilde{\omega} - \tilde{\omega}_{OC}}{\tilde{\omega}_{OC}} + \tilde{Z}_L\right)\end{aligned} \tag{4.5.13}$$

This is a good place to derive the Gordon–Kanazawa–Mason result from the equivalent circuit. Use Eq. 4.5.13 and set $\tilde{Z}_{mot}$ to zero (the resonance condition), assume $-\phi^2/(i\omega C_0) \approx 0$ (neglect piezoelectric stiffening), and assume $\tilde{Z}_L \approx \tilde{Z}_{liq}$. The Gordon–Kanazawa–Mason result follows from solving Eq. 4.5.13 for $\tilde{\omega} - \tilde{\omega}_{OC}$.

Equation 4.5.13 will be of central importance in the derivation of the small load approximation (Sect. 4.6.1). The small load approximation relates the frequency shift to the load. For now, we stick to the resonator itself. The following discussion shows that the motional branch of the resonator can be depicted as an assembly of a capacitor, an inductor, and a resistor (Fig. 4.11). This will make the QCM look like an entirely electrical device. For the moment, let again be $-\phi^2/(i\omega C_0) \approx 0$, $\tilde{Z}_L \approx 0$, and $\tilde{\omega} \approx \tilde{\omega}_{OC}$. Viewed across the transformer, the electrical equivalent of $A\tilde{Z}_{mot}$ is (cf. Eq. 4.5.13)

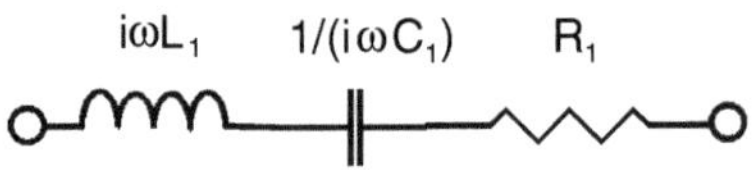

Fig. 4.11 An electrical resonator, characterized by an inductance L_1, a capacitance, C_1, and a resistance, R_1

$$AZ_{mot,el,OC} = \frac{1}{\phi^2} AZ_{mot,OC} \approx \frac{1}{4\phi^2}\, in\pi A\tilde{Z}_q \frac{\tilde{\omega} - \tilde{\omega}_{OC}}{\tilde{\omega}_{OC}} \tag{4.5.14}$$

The right-hand side has three free parameters, which are $n\pi A\tilde{Z}_q/(4\phi^2)$, the real part of $\tilde{\omega}_{OC}$, and the imaginary part of $\tilde{\omega}_{OC}$. Now consider the electrical LCR-circuit shown in Fig. 4.11. Bringing its impedance into the same algebraic form as Eq. 4.5.14 allows to formulate the correspondences between its three parameters L_1, C_1, and R_1 and the three free parameters of Eq. 4.5.14.

The electrical impedance of the LCR-circuit is

$$\begin{aligned}
\tilde{Z}_{LCR} &= \mathrm{i}\omega L_1 + \frac{1}{\mathrm{i}\omega C_1} + R_1 = \mathrm{i}\sqrt{\frac{L_1}{C_1}}\left(\omega\sqrt{L_1 C_1} - \frac{1}{\omega\sqrt{L_1 C_1}}\right) + R_1 \\
&= \mathrm{i}\sqrt{\frac{L_1}{C_1}}\left(\frac{\omega}{\omega_{LC}} - \frac{\omega_{LC}}{\omega}\right) + R_1 = \mathrm{i}\sqrt{\frac{L_1}{C_1}}\frac{(\omega + \omega_{LC})(\omega - \omega_{LC})}{\omega_{LC}\omega} + R_1 \\
&\approx 2\mathrm{i}\sqrt{\frac{L_1}{C_1}}\frac{\omega - \omega_{LC}}{\omega_{LC}} + R_1
\end{aligned} \tag{4.5.15}$$

$\omega_{LC} = (L_1 C_1)^{-1/2}$ is a frequency. In the last line, the relations $\omega \approx \omega_{LC}$ and $\omega + \omega_{LC} \approx 2\omega_{LC}$ were applied in the denominator and in the numerator, respectively. This is permissible because the frequency was assumed to be close to the resonance frequency.

The complex resonance frequency of the LCR-circuit, $\tilde{\omega}_r = \omega_r' + \mathrm{i}\,\omega_r''$, follows from demanding $\tilde{Z}_{LCR}(\tilde{\omega}_r) = 0$:

$$0 = 2\mathrm{i}\sqrt{\frac{L_1}{C_1}}\frac{\tilde{\omega}_r - \omega_{LC}}{\omega_{LC}} + R_1 \tag{4.5.16}$$

Note that $\tilde{\omega}_r$ is now complex, while ω was real as in Eq. 4.5.15. Separating the real and the imaginary part yields

$$\begin{aligned}
\omega_r' &= \omega_{LC} = \frac{1}{\sqrt{L_1 C_1}} && \text{(a)} \\
\omega_r'' &= \omega_{LC}\frac{R_1}{2}\sqrt{\frac{C_1}{L_1}} = \frac{1}{2}\frac{R_1}{L_1} && \text{(b)}
\end{aligned} \tag{4.5.17}$$

$\omega_r'/2\pi$ as given in Eq. 4.5.17 is also termed "series resonant frequency" (ω_r' is the *un*damped resonance frequency of the LCR circuit). Solving Eq. 4.5.17b for R_1 and inserting the result into Eq. 4.5.15 yields

$$\begin{aligned}\tilde{Z}_{LCR} &\approx 2\mathrm{i}\sqrt{\frac{L_1}{C_1}}\frac{\omega-\omega_r'}{\omega_{LC}}+2\sqrt{\frac{L_1}{C_1}}\frac{\omega_r''}{\omega_{LC}}\\ &=2\mathrm{i}\sqrt{\frac{L_1}{C_1}}\frac{\omega-\tilde{\omega}_r}{\omega_{LC}}\\ &\approx 2\mathrm{i}\sqrt{\frac{L_1}{C_1}}\frac{\omega-\tilde{\omega}_r}{\tilde{\omega}_r}\end{aligned} \tag{4.5.18}$$

In line 3, ω_{LC} in the denominator was replaced by $\tilde{\omega}_r$, which is permissible if $\omega_r'' \ll \omega_r'$. Equations 4.5.18 and 4.5.14 have the same algebraic form and one can identify the coefficients as:

$$\begin{aligned}\frac{1}{\sqrt{L_1C_1}} &= 2\pi nf_0\\ \sqrt{\frac{L_1}{C_1}} &= \frac{n\pi AZ_q}{8\phi^2}\\ R_1 &= \sqrt{\frac{L_1}{C_1}}\frac{2\omega_r''}{\omega_r'} = \frac{n\pi AZ_q}{4\phi^2}\frac{\omega_r''}{\omega_r'} = \frac{n\pi Z_q d_q^2}{4Ae_{26}^2}\frac{1}{2Q} = \frac{n\pi AZ_q}{8\phi^2}\Delta_q\end{aligned} \tag{4.5.19}$$

Equation 4.3.14 was used in the last step in line 3. Using $f_0 = Z_q/(2m_q)$, the elements L_1, C_1, and R_1 are found as

$$\begin{aligned}L_1 &= \sqrt{\frac{L_1}{C_1}}\frac{1}{\omega_{LC}} = \frac{AZ_q}{16\phi^2 f_0} = \frac{1}{4\phi^2}\frac{1}{2}\rho_q A d_q = \frac{1}{4\phi^2}A\frac{m_q}{2}\\ C_1 &= \sqrt{\frac{C_1}{L_1}}\frac{1}{\omega_{LC}} = \frac{8\phi^2}{n\pi AZ_q}\frac{1}{2\pi nf_0} = 4\phi^2\frac{2}{(n\pi)^2}\frac{d_q}{AG_q}\\ R_1 &= \frac{1}{8\phi^2}n\pi AZ_q\Delta_q\end{aligned} \tag{4.5.20}$$

Typical values for a 5 MHz crystal on the fundamental are a few tens of milliHenry for L_1, a few tens of femtoFarad for C_1, and a few ohm for R_1. Note: All these values depend on the area.

The relation for R_1 is sometimes expressed in terms of a "viscosity" of the crystal, η_q. Of course η_q is not a Newtonian viscosity. It is called "elastic viscosity" in Ref. [30]. The terminology makes sense in so far, as the crystal's loss modulus, G_q'', scales about linearly with ω in the MHz range. If one writes $G_q'' = \omega\eta_q$, η_q is experimentally found to be roughly independent of frequency, while G_q'' is not. The relation connecting R_1 and η_q is

$$R_1 = \frac{1}{8\phi^2} n\pi A Z_q \frac{\omega\eta_q}{G'_q} = \frac{n\pi\, d_q}{8Ae_{26}^2} \frac{d_q}{c_q} \omega\eta_q = \frac{(n\pi)^2 d_q}{8Ae_{26}^2} \eta_q \tag{4.5.21}$$

The relations $\phi = Ae_{26}/d_q$, $\omega = n\pi c_q/d_q$, $Z_q = (G_q\rho_q)^{1/2}$, and $c_q = (G_q/\rho_q)^{1/2}$ were used.

In the derivation of Eq. 4.5.20 the element $-\phi^2/(i\omega C_0)$ had been set to zero. Repeating the derivation of Eq. 4.5.20 with the element $-\phi^2/(i\omega C_0)$ included, one finds the motional capacitance as (Sect. 5.2):

$$\bar{C}_1 = 4\phi^2 \frac{2}{(n\pi)^2} \frac{d_q}{AG_q} \left(1 - \frac{8}{(n\pi)^2} \frac{e_{26}^2}{\varepsilon_q \varepsilon_0 G_q}\right)^{-1} = 4\phi^2 \frac{2}{(n\pi)^2} \frac{d_q}{AG_q} \left(1 - \frac{8k_t^2}{(n\pi)^2}\right)^{-1} \tag{4.5.22}$$

k_t is the piezoelectric coupling coefficient. Note the term n^2 in the denominator. The effects of the element $-\phi^2/(i\omega C_0)$ decrease with overtone order.

Along similar lines, one can (and should) lump the electrical equivalent of the load impedance, $A\tilde{Z}_L/4\phi^2$, into the elements of the LCR-circuit. How, exactly, this changes the values of $\overline{L}_1$, $\overline{C}_1$, and $\overline{R}_1$, evidently depends on the load. If the load is a Sauerbrey film, it will increase the motional inductance. This happens because the Sauerbrey film increases the resonator's mass, but leaves the resonator's stiffness and dissipation unchanged. Since the Sauerbrey film is characterized by $\tilde{Z}_L = i\omega m_f$ (m_f the mass of the film, Chap. 8), incorporating a thin rigid film into the parameter $\overline{L}_1$ leads to

$$\bar{L}_1 = \frac{A}{4\phi^2} \left(\frac{m_q}{2} + m_f\right) \tag{4.5.23}$$

Without lumping $-\phi^2/(i\omega C_0)$ and $A\tilde{Z}_L$ into the elements $\overline{L}_1$, $\overline{C}_1$, and $\overline{R}_1$, the equivalent electrical circuit contains six elements (Fig. 4.12a). These six elements all have their own physical significance, but they cannot be determined individually based on an electrically acquired admittance curve (Sect. 4.5.4). The admittance curve is specified by the four elements of the four-element circuit shown in Fig. 4.12b. The circuit also carries the name "Butterworth-van Dyke circuit", abbreviated as BvD circuit.

Again: The four-element circuit contains four parameters, but only two of them are exploited in most experiments. These are $\omega_r' = (L_1C_1)^{-1/2}$ and $\omega_r'' = R_1/(2L_1)$. This is so because frequencies are determined with a better precision than the impedance itself.

In some of the older literature, the results of QCM measurements are expressed in terms of shifts of the motional resistance ("ΔR", see also Sect. 4.1.4) and the corresponding reactance ("ΔX"). ΔX here replaces Δf. The conversion is [29]

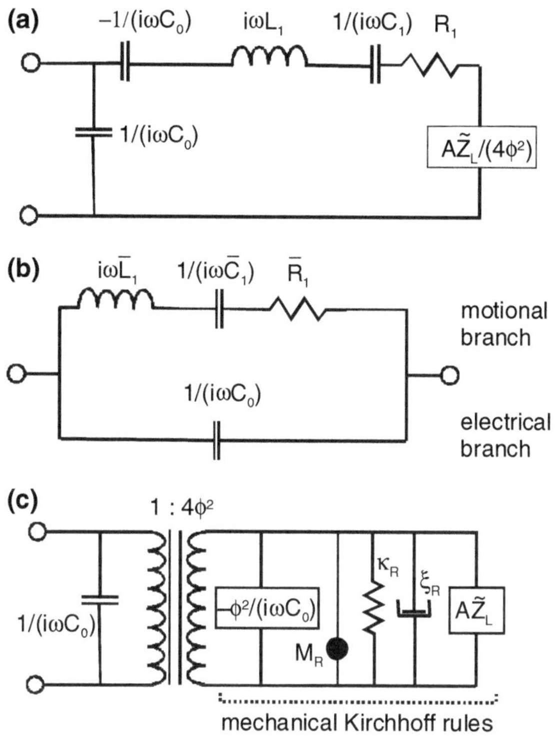

Fig. 4.12 Approximations of the Mason circuit, valid close to the resonances. **a** Electrical representation showing piezoelectric stiffening and the sample as separate elements. **b** Representation, where the element $-\phi^2/(i\omega C_0)$ has been subsumed under the capacitor (now named $\bar{C}_1$) and where the load, $A\tilde{Z}_L$, was subsumed under the inductance (now named $\bar{L}_1$). The latter step only makes sense if the sample is a Sauerbrey film. R_1 was named $\bar{R}_1$ because it may contain contributions from the sample. This happens if the samples dissipates energy. **c** Representation, where the motional branch is approximated by mechanical elements. In mechanics, elements must be arranged in parallel, if the impedances are supposed to be additive. This is the consequence of the mechanical Kirchhoff rules (Sect. 4.7.3)

$$\Delta R + i\Delta X = \frac{A}{4\phi^2}\tilde{Z}_L = -i\frac{\pi}{16}\frac{Z_q^3}{Ae_{26}^2\rho_q^2 f_0^3}(\Delta f + i\Delta\Gamma) \tag{4.5.24}$$

ΔR and ΔX are obtained from directly fitting the complex admittance with the circuit elements from Fig. 4.13 as fit parameters. Note: The conversion contains the active area of the resonator. Converting ΔX to (for instance) a film thickness needs knowledge of A. Interpreting Δf is simpler in this regard.

The four-element circuit is widely used. Apart from $\bar{L}_1, \bar{C}_1$, and $\bar{R}_1$, which constitute the "motional branch", there is the "parallel capacitance", C_0. The ratio of C_0 and C_1 (the "capacitance ratio") has a special significance. For the parallel plate, one has

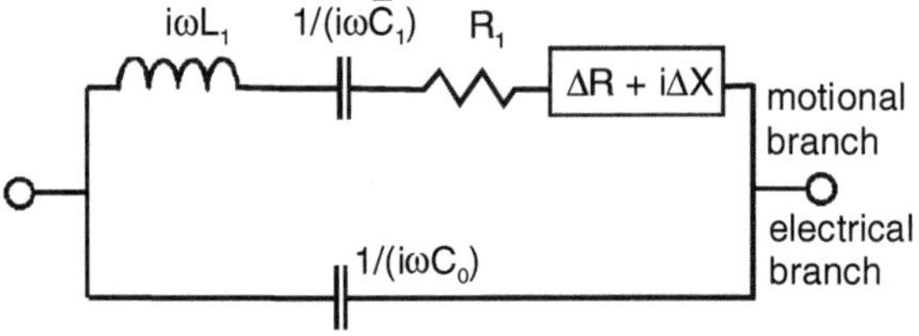

Fig. 4.13 Equivalent circuit, where the sample is represented as an electrical element (ΔR + iΔX)

$$\frac{C_0}{C_1} = \frac{(n\pi)^2}{8}\frac{1}{k_t^2} = \frac{(n\pi)^2}{8}\frac{G'_q\varepsilon_0\varepsilon_q}{e_{26}^2} \tag{4.5.25}$$

with $k_t = (e_{26}^2/(\varepsilon_q\varepsilon_0 G_q'))^{1/2}$ the piezoelectric coupling coefficient (Chap. 5). Equation 4.5.25 is only approximate for real crystals. The geometry (the electrode shape in particular) enters the capacitance ratio. Equation 4.5.25 can be used to define an effective coupling coefficient as $k_{t,eff}^2 = C_1/C_0\ (n\pi)^2/8$. $k_{t,eff}$ is a function of geometry. The larger the coupling coefficient, the stronger the effects of piezoelectric stiffening. If effects of piezoelectric stiffening are problematic (they often are), small coupling is better than large coupling. Of course small coupling at the same time implies a correspondingly small current into the electrodes. Materials or resonator geometries with poor coupling lead to intrinsically more stable resonators, but the read-out is more difficult. Conversely: If one actually wants to shift a resonance frequency making use of piezoelectric stiffening, a large k_t is beneficial. The effective coupling coefficient determines the frequency range, over which the resonance frequency can be "pulled" with an external capacitor.

The use of the four-element circuit is not limited to quartz plates, it describes a large class of mechanical resonators. This includes the MEMS resonators (Sect. 15.5). The transformer may represent types of electromechanical coupling other than piezoelectricity. k_t then is the "electromechanical coupling coefficient" in a wider sense. Its definition is based on the capacitance ratio; one writes $k_t^2 = (n\pi)^2/8\ C_1/C_0$. C_1 and C_0 then are experimental parameters, determined with impedance analysis.

Given that $\tilde{Z}_{mot}$ is a genuinely mechanical entity, one may also represent it by a mass, a spring, and a dashpot. More technically, one may write

$$\tilde{Z}_{mot,el,0} = \frac{1}{4\phi^2}\left(\mathrm{i}\omega M_R + \frac{\mathrm{i}\kappa_R}{\omega} + \xi_R\right) \tag{4.5.26}$$

The index R stands for "resonator". Comparison with Eq. 4.5.20 leads to

$$\begin{aligned} M_R &= \frac{A\rho_q d_q}{2} = \frac{1}{2} A m_q && \text{(a)} \\ \kappa_R &= \frac{A G_q}{d_q} \frac{(n\pi)^2}{2} = \frac{(n\pi)^2}{2} \kappa_{q,stat} && \text{(b)} \\ \xi_R &= \frac{n\pi}{2} \delta_q A Z_q && \text{(c)} \end{aligned} \qquad (4.5.27)$$

$\kappa_{q,stat}$ is the static shear stiffness of the crystal. As expected, κ_R, is related to the static stiffness, but the relation is not trivial. Similarly, M_R is related to the mass of the crystal, Am_q, but it differs by a factor of 1/2. The factor of 1/2 occurs because the mass close to the nodal plane rests in place and therefore does not contribute to the plate's inertia.

M_R, κ_R, and ξ_R might also be called "modal mass", "modal stiffness", and "modal damping". In the introduction, the resonance frequency was written as $f_r = 1/(2\pi)\ (\kappa_R/M_R)^{1/2}$ (Eq. 1.5.1). This model was found to be inapplicable because a 1 % increase in mass resulted in an 0.5 % decrease in resonance frequency, contradicting the Sauerbrey result. When making this argument, the modal mass had been naively identified with the true mass of the plate. This was the mistake, leading to a wrong frequency shift. Had the modal mass been identified with one half of the true mass (Eq. 4.5.27a), the frequency shift would have come out correctly. In a way, Eq. 4.5.27a fixes the problem, but one still should not put too much trust in it because it relies on the parallel plate model. Extensions are needed as discussed in Sect. 6.1.4.

Clearly, the three modal parameters pertain to a certain resonator geometry *and* to a certain mode of vibration. The relations from Eq. 4.5.27 pertain to the parallel plate in the thickness-shear mode. This raises the question of whether all three motional parameters can be determined from experiment, if the mode of vibration is poorly known (It is poorly known for the QCM because of energy trapping). It would be desirable to know M_R well because the modal mass enters the Sauerbrey equation.

Experiment easily provides two relations between the three modal parameters, which are the resonance frequency ($\omega_r = (\kappa_R/M_R)^{1/2}$) and the Q-factor ($Q = (\kappa_R M_R)^{1/2}/\xi_R$). One might think that M_R itself was also known. M_R follows from the mass per unit area (m_q, accessible from $f_0 = Z_q/(2m_q)$ and the area, A. The (effective) area can be obtained experimentally from the conductance curves, using Eq. 7.4.7. However, both the relation $f_0 = Z_q/(2m_q)$ and Eq. 7.4.7 are based on the parallel plate model and the target is to go beyond this model in order to improve on the accuracy of the Sauerbrey equation (Sect. 6.1.4).

There *is* a rigorous solution to the problem, at least in principle. It is based on thermal noise. The integrated noise power spectral density (integrated over the frequencies around the resonance) of a resonator in the absence of external excitation is given as $\int |x^2(\omega)| d\omega = 1/2\ k_B T/\kappa_R$ with x the dynamical variable and $k_B T$ the thermal energy [30]. The above statement is general, it follows from the equipartition theorem. The dynamical variable, x, of course must be identified.

It enters the definition. u_S is the dynamical variable here. With energy trapping applied, a suitable choice for x would be the displacement at the surface in the center of the plate (termed u_c in Sect. 7.4). By measuring the thermal noise, one gets access to κ_R from the equipartition theorem and from there, one obtains M_R and ξ_R, using ω_r and Q. Unfortunately, this argument is a bit of an academic exercise for the QCM because the integrated noise is around $(10^{-15}\text{ m})^2$ and the instrument capable of measuring such small displacements still waits to be invented. In other words: Thermal noise is of no help for macroscopic resonators. Still, the argument as such holds, and it is applied to MEMS resonators [31]. Again: All three parameters depend on the definition of the dynamical variable (chosen as u_S in Eq. 4.5.27). Once the dynamical variable is defined, M_R, κ_R, and ξ_R can be determined from experiment based on thermal noise, ω_r, and Q, even if the mode shape is poorly known.

Figure 4.12c shows a circuit with mechanical elements to the right of the transformer. There is a subtlety with regard to the graphical representation of mechanical circuits (Sect. 4.7.3). When two mechanical elements are placed in series, the force onto these two elements is the same and the displacements (as well as the velocities) are additive. This contrasts to electrical impedances arranged in series, in which case the current through both elements is the same, while the voltage is additive. Since the electromechanical analogy maps force onto voltage and velocity onto current, there is an inequivalence. For two mechanical impedances placed in series, the *inverse* impedances are additive ($1/Z_{tot} = 1/Z_1 + 1/Z_2$). If the impedances themselves are supposed to be additive ($Z_{tot} = Z_1 + Z_2$), one must arrange the respective elements in parallel as shown in Fig. 4.12c. Again: When evaluating this circuit, one must apply the mechanical Kirchhoff rules. In mechanics, the rule of nodes and the rule of loops are interchanged compared to the electrical case. Forces sum to zero at nodes, velocities sum to zero around loops. The circuits in Fig. 4.12a and c are fully equivalent.

Some authors draw mechanical circuits like electrical circuits. John Vig takes this approach in his widely distributed tutorial [32]. Vig places mechanical elements in series if the total impedance is supposed to be the sum of the individual impedances. He applies the electrical Kirchhoff rules to mechanical elements. Figure 4.12c illustrates that he has a point. Because equivalent circuits hide the internal deformation pattern, Fig. 4.12c cannot be understood intuitively, anyway, and there is no benefit in connecting the mechanical elements to each other "properly". In the context of the equivalent circuits for thickness shear resonators, a representation using mechanical Kirchhoff rules mostly creates confusion. Still, it is preferable to use mechanical Kirchhoff rules for mechanical elements in this book because mechanical circuits will be drawn later, which *are* amenable to intuition. These circuits will show a sphere making contact to the resonator across a spring and a dashpot and the circuits will aid the interpretation of contact mechanics experiments (Chap. 11). The equivalent circuits will be both predictive and easy to understand—if drawn according to the mechanical Kirchhoff rules. In the author's opinion, the best option is to accept the fact that there are two sets of rules on how to add circuit elements. The rules are different in mechanics and electrical engineering.

A side remark: Mechanical Kirchhoff rules are also used in rheology and in polymer science. Examples are the Voigt element (Sect. 11.4) and the Maxwell element. The term "Kirchhoff rules" is not used in that context, but the rules, by which elements are added, are, what is called the mechanical Kirchhoff rules here.

4.5.4 The Admittance Circle and the Series Resonance Frequency

In impedance analysis, one fits the impedance spectra with the prediction following from the four-element circuit, which is

$$\tilde{Z}_{el} = \left(\mathrm{i}\omega\, C_0 + \left(\mathrm{i}\omega\, \bar{L}_1 - \frac{\mathrm{i}}{\omega \bar{C}_1} + \bar{R}_1 \right)^{-1} \right)^{-1} \tag{4.5.28}$$

The fit yields the four resonance parameters $C_0, \overline{L}_1, \overline{C}_1$, and $\overline{R}_1$. These are easily converted to the parameters f_r, Γ, $\overline{R}_1$, and C_0. Of course one might as well directly fit an equation formulated in terms of f_r, Γ, $\overline{R}_1$, and C_0 to the data.

It turns out that the discussion flows more smoothly, if it is based on the electrical admittance, $\overline{R}_1$, rather than the impedance, $\tilde{Z}_{el}(\omega)$. The admittance is the inverse of the impedance. Plots of $\tilde{Y}_{el}(\omega)$ are more transparent than plots of $\tilde{Z}_{el}(\omega)$ because the electrical capacitance, C_0, acts in parallel to the motional branch. One has

$$\tilde{Y}_{el} = \mathrm{i}\omega\, C_0 + \left(\mathrm{i}\omega \bar{L}_1 - \frac{\mathrm{i}}{\omega \bar{C}_1} + \bar{R}_1 \right)^{-1} \approx \mathrm{i}\omega_r C_0 + \left(\mathrm{i}\omega \bar{L}_1 - \frac{\mathrm{i}}{\omega \bar{C}_1} + \bar{R}_1 \right)^{-1} \tag{4.5.29}$$

Since only frequencies close to the resonance are of interest, one may approximate the term $\mathrm{i}\omega C_0$ by $\mathrm{i}\omega_r C_0$. The contribution of the parallel capacitance then turns into a constant imaginary offset. This is the benefit of using $\tilde{Y}_{el}(\omega)$.

For reasons, which are not easily explained without diving into theory of complex functions, the admittance of the motional branch is a circle when displayed in polar form. A polar diagram displays the imaginary part of a complex parameter versus the real part. The impedance of the motional branch is $\tilde{Z}_{mot,el} = \mathrm{i}\omega\overline{L}_1 - \mathrm{i}/(\omega\overline{C}_1) + \overline{R}_1$. When displayed as $Z_{mot,el}''$ versus $Z_{mot,el}'$, it forms a straight vertical line. Straight lines in the complex plane turn into circles upon taking the inverse. More generally, circles remain circles upon inversion and straight lines are equivalent to circles with infinite radius. A line turns into a circle

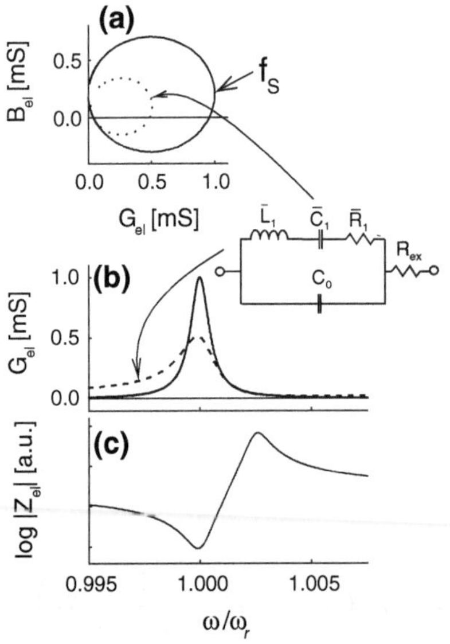

Fig. 4.14 **a** Polar diagram of the admittance $\tilde{Y}_{el}(\omega)$ as predicted by the four-element circuit. The conductance, G_{el} (ω), and the susceptance, B_{el} (ω), are the real and the imaginary part of the complex admittance as given in Eq. 4.5.29. f_s is the series resonant frequency. The *dashed line* corresponds to a network of the form shown in the *inset*. It contains an external ohmic resistance. **b** Plot of conductance, G_{el}, versus normalized frequency ω/ω_r. The *solid line* is the same as in Fig. 2.4. The dashed line is predicted by the circuit in the *inset*. An external resistance induces an asymmetry. **c** Plot of $\log|\tilde{\tilde{Z}}_{el}| = \log |1/\tilde{Y}_{el}|$ versus normalized frequency. The absolute value of the electrical impedance shows a minimum and a maximum. The minimum occurs close to (but not exactly at) the series resonance frequency. The peak in $|\tilde{\tilde{Z}}_{el}|$ is sometimes called "anti-resonance". The corresponding frequency is the "parallel resonance frequency". mS stands for milliSiemens

upon inversion because the complex inversion is among the "Möbius transformations" (see below).

One quickly convinces oneself that the thick solid lines in Fig. 4.14a, b are the same as what is plotted as $G_{el}(\omega)$ and $B_{el}(\omega)$ in Fig. 2.4. The real part of the resonance curve, $G_{el}(\omega)$, is symmetric. The imaginary, $B_{el}(\omega)$ becomes antisymmetric after subtracting the offset originating from the parallel capacitance, $i\omega_r C_0$.

In experiment, $G_{el}(\omega)$ often is slightly asymmetric. Most of the time, the reason for asymmetry is imperfect calibration. Perfect calibration is impossible. Calibration of the impedance analyzer cannot occur right at the electrodes, but rather must occur at the electrical connectors. If the electrodes and the wires connecting to them have a finite ohmic resistance (they do), this resistance will affect $G_{el}(\omega)$ and $B_{el}(\omega)$ as measured from the outside. A network accounting for an ohmic resistance of the electrodes is shown in the inset in Fig. 4.14. The element R_{ex} tilts the circle and also moves it in the admittance plane.

Interestingly, the addition of more circuit elements (where the addition of R_{ex} in Fig. 4.14 is just one example) usually leaves the shape of the circle intact. The reason is that a wide class of transformations in the complex plane transform circles into circles. These are the Möbius transformations. Möbius transformations are of the form $\tilde{y} \rightarrow (\tilde{a}\tilde{y} + \tilde{b})/(\tilde{c}\tilde{y} + \tilde{d})$, where $\tilde{y}$ is the complex function to be transformed and $\tilde{a}$, $\tilde{b}$, $\tilde{c}$, and $\tilde{d}$ are constant coefficients. The polar admittance diagram obtained after adding the external resistance, R_{ex}, into the circuit diagram still is a circle. However, the conductance curve it asymmetric (thin dashed line in Fig. 4.14a, b).

From a practical point of view, it is helpful to fit the admittance curves with the following expanded function:

$$\begin{aligned} G_{fit} &= G_{max}\frac{f}{f_r}\left(\frac{\Gamma}{(f_r-f)^2+\Gamma^2}\cos\varphi + \frac{f_r-f}{(f_r-f)^2+\Gamma^2}\sin\varphi\right) + G_{off} \\ B_{fit} &= G_{max}\frac{f}{f_r}\left(-\frac{f_r-f}{(f_r-f)^2+\Gamma^2}\sin\varphi + \frac{\Gamma}{(f_r-f)^2+\Gamma^2}\cos\varphi\right) + B_{off} \end{aligned} \tag{4.5.30}$$

For the special case of $\varphi = 0$, $G_{off} = 0$, and $B_{off} = 0$, Eq. 4.5.30 essentially is the resonance curve as in Eq. 4.1.26. The term B_{off} accounts for the parallel capacitance. There are two new parameters, which are G_{off} (an offset of the circle along the real axis) and φ (a tilt angle in the complex plane). Both should be zero, ideally, but are nonzero in practice. When fitting experimental admittance curves with Eq. 4.5.30, one usually achieves a fit, where the errors are predominantly statistical. The scatter in the derived fit parameters for f_r and Γ is much reduced compared to fits with the four-element circuit. Of course this comes at a price. The two new fit parameters have unclear physical meaning. The possibility of artifacts in the determination of f_r and Γ exists, but this possibility must be weighed against the reduction in scatter.

The emphasis in QCM measurements is on the resonance frequency, not on the electrical impedance itself. For this reason, calibration of the impedance analyzer is not a major problem. Impedance analyzers (also called network analyzers) infer the impedance of the device-under-test from the ratio of two complex amplitudes, which are the voltage at the detector and the output voltage of the analyzer. There are different ways of wiring the resonator. Regardless of what the configuration is, there are always cables (and possibly more elements) between the device-under-test and the detector. These are accounted for by a calibration procedure. A popular procedure applied to devices wired in the reflection mode is the so-called "3-term calibration". The user places a "short", an "open" and a "load" (a 50 ohm resistor) to where the device-under-test will later be. The instrument measures the amplitude at the detector produced by these elements and uses this information to correct for the influence of the cables. Calibration is simple, but it is never perfect because the position of the cables may have shifted between calibration and experiment or because the calibration standards cannot be placed at the exact same location as the resonator. If calibration occurs at a position away from the resonator (for instance outside a cell), the resulting inaccuracy will mostly affect the values of the impedances. f_r and Γ are affected less strongly. Note: When cables are inserted or removed, this will influence the effects of piezoelectric stiffening and can thereby change the series resonance frequency. These effects are unrelated to calibration.

In this book, the term "resonance frequency" (labeled f_r) is synonymous to what is called *series resonance frequency*, (labeled f_s) in electrical engineering. This usage slightly deviates from the conventions in electronics. For details see Fig. 9 in Ref. [33]. If the four-element circuit applies, f_s is strictly equal to the frequency, where $G_{el}(\omega)$ attains its maximum. Γ then is the half-bandwidth at half-height of the conductance curve. If there is an external resistance, f_s differs from the frequency where $G_{el}(\omega)$ attains its maximum. f_s and Γ then have to be determined by fitting with Eq. 4.5.30, they cannot be read from the graph. f_s is the frequency where the *amplitude of oscillation* goes through a maximum. Should the admittance circle be tilted for some reason, this does not affect f_s. If fitting is not an option, one can still guess f_s based on the rate-of-change of the admittance as a function of frequency. The rate of change is defined as $\left|\mathrm{d}\tilde{Y}_{el}/\mathrm{d}f\right|$. The frequency, where $\left|\mathrm{d}\tilde{Y}_{el}/\mathrm{d}f\right|$ reaches its maximum, can be determined from the raw data without resorting to a fit. The thus-determined value for f_s can serve as a starting guess for a nonlinear least-square fit.

In the context of oscillator circuits, the impedance (rather than the admittance) often is the more relevant parameter. Figure 4.14c plots the absolute value of the impedance $\left|\tilde{Z}_{el}(\omega)\right|$ versus ω. Of course there also is a phase (not displayed). $\left|\tilde{Z}_{el}(\omega)\right|$ has a minimum close to the series resonance frequency. There also is a maximum, sometimes called "antiresonance". The corresponding frequency is close to the "parallel resonance frequency" [33]. The matter is discussed in most textbooks on electronics [34, 35].

4.6 The Small Load Approximation

4.6.1 Derivation of the Small Load Approximation from the Four-Element Circuit

For the derivation of the SLA, go back to Eq. 4.5.13. Let $\tilde{\omega}_{ref}$ be the resonance frequency with $\tilde{Z}_L = 0$ (but not $-\phi^2/(\mathrm{i}\omega C_0) \approx 0$) and let $\tilde{\omega}_r$ be the frequency of the loaded resonator. The frequency shift obeys the relation [8]

$$\frac{\Delta\tilde{f}}{f_0} = \frac{1}{2\pi}\frac{\tilde{\omega}_r - \tilde{\omega}_{ref}}{f_0} = \frac{\mathrm{i}}{\pi Z_q}\tilde{Z}_L \tag{4.6.1}$$

Equation 4.6.1 is the small load approximation. Since Eq. 4.5.13 only holds for frequencies close to the resonance frequency, the SLA also requires $\left|\Delta\tilde{f} \ll f_0\right|$.

Following Eq. 4.6.1, $|\tilde{Z}_L|$ must be smaller than Z_q, hence the name "small load approximation". The frequency shift just about always is much smaller than the frequency itself and the SLA therefore is widely applicable.

To be precise, the SLA entails two separate approximations. The first one is the linearization of $\tilde{Z}_{mot}$ in $\omega - \omega_{ref}$ (Eq. 4.5.13). Secondly the load impedance is evaluated at the reference frequency, rather than the resonance frequency of loaded resonator.

There are two caveats to the use of the SLA: Firstly, the condition $|\tilde{Z}_L| \ll Z_q$ applies to the parallel plate. For arbitrary geometries, the relation to be satisfied is Eq. 6.1.23. Secondly, the SLA has some short-comings even if $|\Delta\tilde{f}| \ll f_0$. Section 6.2 discusses these and also shows, how the resulting problems can be solved.

Equation 4.6.1 is the single one most important equation in the physics of the QCM. As long as the frequency shift is small compared to the frequency itself, the frequency shift is proportional to the load impedance at the crystal surface. At this point, no statement is made on the nature of the sample. The SLA connects the frequency shift to the stress–velocity ratio. Importantly, the SLA also holds in an average sense (Sect. 4.6.2). Assume that the sample does not consist of planar layers, but instead of a sand pile, a froth, an assembly of vesicles, a cell culture, a droplet, or any other complicated, heterogeneous material. The frequency shift induced by all such objects can be computed from the area-averaged ratio of stress and velocity at the resonator–sample interface. The concept of the load impedance tremendously broadens the range of applicability of the QCM. It is the conceptual link between the QCM and complex samples. If the stress-velocity ratio can be predicted in one way or another from the properties of the sample, a quantitative analysis of the QCM experiment is in reach. Otherwise, the discussion has to remain qualitative.

The SLA has a long history. It is implicitly contained in Mason's book [11], but only implicitly. The tensorial version (Sect. 6.1) was published in 1959 by Pechhold [36]. Moriizumi analyzed the problem based on the equivalent circuit in 1990 [10]. The author has elaborated on the SLA in Ref. [8]. The Sandia group in collaboration with Hillman has a number of publications on the same topic, summarized in Ref. [9]. A similar concept was also put forward by Lucklum. Lucklum calls the SLA the "acoustic load concept" in Chap. 9 of Ref. [37].

The SLA implies that the frequency shift—to first order—measures the oscillatory stress at the resonator surface. One has

$$\frac{\Delta\tilde{f}}{f_0} = \frac{i}{\pi Z_q}\tilde{Z}_L = \frac{i}{\pi Z_q}\frac{\hat{\sigma}_S}{\hat{v}_S} = \frac{i}{\pi Z_q}\frac{\hat{\sigma}_S}{\hat{v}_{S,ref} + \Delta\hat{v}_S} \approx \frac{i}{\pi Z_q}\frac{\hat{\sigma}_S}{\hat{v}_{S,ref}}\left(1 - \frac{\Delta\hat{v}_S}{\hat{v}_{S,ref}}\right) \approx \frac{i}{\pi Z_q}\frac{\hat{\sigma}_S}{\hat{v}_{S,ref}} \tag{4.6.2}$$

$\hat{v}_{S,ref}$ is the velocity in the reference state. Taylor expansion $(1/(1+\varepsilon) \approx 1 - \varepsilon)$ was applied in step 4. To first order, the presence of sample does not change denominator in the above equation. The QCM therefore essentially determines an *area-averaged periodic stress* [38]. Note: The stress in Eq. 4.6.2 by convention is exerted *by* the resonator *onto* the sample. This convention will necessitate a few sign reversals later on.

Using $Z_q = 2m_q/f_0$, Eq. 4.6.1 can be rewritten as

$$\Delta\tilde{f} = \frac{\mathrm{i}}{2\pi\, m_q}\tilde{Z}_L \tag{4.6.3}$$

Equation 4.6.3 shows that the mass per unit area of the crystal is the only parameter entering the conversion from the acoustic load to frequency shift as long as the latter is small. The stiffness of the crystal (and piezoelectric stiffening, in particular) is of no influence at this level of approximation.

As emphasized before: The parameter m_q is well defined within the parallel plate model. Because of energy trapping (and other complications), Eq. 4.6.3 is only approximate. Even the Sauerbrey equation (which follows from Eq. 4.6.3 with $\tilde{Z}_L = \mathrm{i}\omega\, m_f$, Sect. 8.1) is approximate. The QCM is tremendously precise, but unfortunately, it is not equally accurate. There is no easy way to solve this problem.

Equation 4.6.1 is of such fundamental importance that we briefly re-derive it in the frame of the plane-wave formalism. As expressed in Eqs. 4.2.8 and 4.2.10, the displacement and the stress must both be continuous at the crystal surface. Continuity of displacement implies the relation

$$\hat{u}_{q,+}(z_{q,2}) + \hat{u}_{q,-}(z_{q,2}) = \exp(\mathrm{i}\tilde{k}_q d_q)\hat{u}_q^+ + \exp(-\mathrm{i}\tilde{k}_q d_q)\hat{u}_q^- = \hat{u}_S \tag{4.6.4}$$

$z_{q,2}$ is the crystal surface and $\hat{u}_S$ is the displacement at the crystal surface. Remember that $\hat{u}_{j,\pm}(z)$ was defined to be equal to $\hat{u}_j^{\pm}$ at the *left border* of layer j (Eq. 4.2.2, Fig. 4.6). The condition that the stress at the back of the crystal vanishes, amounts to

$$\mathrm{i}\tilde{G}_q\tilde{k}_q\left(\hat{u}_{q,+}(z_{q,0}) - \hat{u}_{q,-}(z_{q,0})\right) = \mathrm{i}\omega\tilde{Z}_q\left(\hat{u}_q^+ - \hat{u}_q^-\right) = 0 \tag{4.6.5}$$

$z_{q,0}$ is the back of the crystal. The continuity of stress at the front surface is expressed as

$$\mathrm{i}\omega\tilde{Z}_q\left(\exp(\mathrm{i}\tilde{k}_q d_q)\hat{u}_q^+ - \exp(-\mathrm{i}\tilde{k}_q d_q)\hat{u}_q^-\right) = \mathrm{i}\omega\tilde{Z}_L\left(\exp(\mathrm{i}\tilde{k}_q d_q)\hat{u}_q^+ + \exp(-\mathrm{i}\tilde{k}_q d_q)\hat{u}_q^-\right) \tag{4.6.6}$$

Remember: $\tilde{Z}_q$ is an acoustic wave impedance, while $\tilde{Z}_L$ is an acoustic impedance. For that reason, the left-hand side in Eq. 4.6.6 contains a difference, while right-hand side contains a sum. From Eq. 4.6.5 it follows that $\hat{u}_q^+ = \hat{u}_q^-$. Using this result, Eq. 4.6.6 can be simplified to

$$\tilde{Z}_q\left(\exp\left(\mathrm{i}\tilde{k}_q d_q\right) - \exp\left(-\mathrm{i}\tilde{k}_q d_q\right)\right) = \tilde{Z}_L\left(\exp\left(\mathrm{i}\tilde{k}_q d_q\right) + \exp\left(-\mathrm{i}\tilde{k}_q d_q\right)\right) \tag{4.6.7}$$

Using tan(x) = −i (exp(ix) − exp(−ix))/(exp(ix) + exp(−ix)), this reads as

$$\begin{aligned}\frac{\tilde{Z}_L}{\tilde{Z}_q} &= -\mathrm{i}\tan\left(\tilde{k}_q d_q\right) = -\mathrm{i}\tan\left(2\pi\left(\tilde{f}_{ref} + \Delta\tilde{f}\right)\frac{d_q}{\tilde{c}_q}\right) = -\mathrm{i}\tan\left(2\pi\Delta\tilde{f}\frac{d_q}{\tilde{c}_q}\right)\\ &= -\mathrm{i}\tan\left(\pi\frac{\Delta\tilde{f}}{\tilde{f}_0}\right)\end{aligned} \tag{4.6.8}$$

The relations $d_q = \tilde{c}_q/\left(2\tilde{f}_0\right)$ and tan($n\pi$ + ε) = tan(ε) with n an odd integer were used. Taylor-expanding tan(x) as tan(x) ≈ x, one finds Eq. 4.6.1 confirmed. The third-order perturbation analysis in Sect. 6.2 will start out from Eq. 4.6.8.

4.6.2 Area-Averaging, Time-Averaging, and Averaging over Displacement

In Sect. 4.6.1 the SLA was derived for the parallel plate geometry. $\tilde{Z}_L$ was assumed to be constant over the entire resonator surface. Typical samples obeying lateral homogeneity would be films and other stratified layer systems. One might also be interested in structured samples. For example, the sample might be a layer of adsorbed nanoparticles. Since the SLA is linear in the load, one can average the load over the surface of the resonator and write

$$\frac{\Delta\tilde{f}}{f_0} = \frac{\mathrm{i}}{\pi Z_q}\left\langle\tilde{Z}_L\right\rangle_{area} \tag{4.6.9}$$

For the proof, see Sect. 6.1. Applied to a layer of adsorbed nanoparticles, area-averaging implies that the frequency shift is proportional to the average mass per unit area. A complication arises if the lateral heterogeneity occurs on a scale comparable to the width of the resonator. This is, for instance, the case, if a viscoelastic sample touches the center of the crystal, only (Sect. 9.4). For such configurations, a statistical weight proportional to the square of the local amplitude of oscillation must be applied. The statistical weight is discussed in Sect. 6.1.5. Also, one must have an eye on possible changes in the degree of energy trapping caused by localized loads (Sect. 9.4).

Averaging can also occur in time. This is of interest if the stress is not a sinusoidal function of time, but still time-periodic with the frequency of the oscillation. Such stresses lead to a frequency shift given by

$$\frac{\Delta\tilde{f}}{f_0} = \frac{i}{\pi Z_q}\frac{2}{\hat{v}_S}\langle\sigma(t)\exp(i\omega t)\rangle_{area,time}$$
$$= \frac{2\langle\sigma(t)\cos(\omega t)\rangle_{area,time}}{\pi Z_q \omega \hat{u}_S} + i\frac{2\langle\sigma(t)\sin(\omega t)\rangle_{area,time}}{\pi Z_q \omega \hat{u}_S} \tag{4.6.10}$$

Equation 4.6.10 is proven in Sect. 13.3.

Since the stress is time-periodic with the frequency of oscillation, there is not only a relation between stress and time, but also a relation between stress and displacement. If the stress is small in the sense of the SLA, the displacement is approximately time-harmonic, that is $u_S(t) \approx u_0 \cos(\omega t)$. u_0 is the absolute value of $\hat{u}_S$. The average over time can then be transformed into an average over displacement. The relation proven in Sect. 13.3 is

$$\Delta f(u_0) = \frac{1}{n\pi^2 Z_q}\frac{1}{2\pi u_0^2}\int_{-u_0}^{u_0}\left(\langle\sigma_+(u_S)\rangle_{area}+\langle\sigma_-(u_S)\rangle_{area}\right)\frac{u_S/u_0}{\sqrt{1-(u_S/u_0)^2}}du_S$$
$$\Delta\Gamma(u_0) = \frac{1}{n\pi^2 Z_q}\frac{1}{2\pi u_0^2}\int_{-u_0}^{u_0}\left(\langle\sigma_+(u_S)\rangle_{area}-\langle\sigma_-(u_S)\rangle_{area}\right)du_S \tag{4.6.11}$$

The indices + and − denote a direction of motion towards positive and negative u_S. The notation $\Delta f(u_0)$ and $\Delta\Gamma(u_0)$ on the left-hand side is meant to emphasize that Δf and $\Delta\Gamma$ depend on amplitude in the general case.

4.6.3 Coupled Resonances

Coupled resonances occur at a few different places in the book; some common features are collected in the following.

The simplest example of a coupled resonance is a sphere on the resonator surface discussed in Sect. 11.4. Figure 4.15a shows a mechanical representation of a large resonator, contacted by a small sphere. The indices R and P denote the large and the small resonator. The large resonator usually is the crystal and the small resonator often (but not always) is an adsorbed particle. $\tilde{\kappa}_P = \kappa_P + i\omega\xi_P$ and M_P are the complex stiffness of the contact and the mass of the *P*article, respectively. The allowing assumptions and approximations are needed:

(a) The mass of the particle is much smaller than the mass of the resonator ($M_P \ll M_R$). As always, the mass entering the calculation of the resonance frequency of a system containing two masses and a spring is the reduced mass.

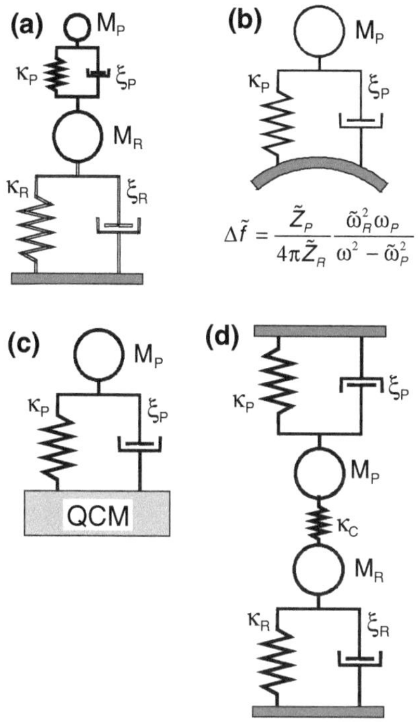

Fig. 4.15 **a** A main resonator (the crystal, index *R*) is coupled to a second, smaller resonator (index *P* for *P*article) across a spring and a dashpot. This system displays two modes (Fig. 4.16). **b**, **c** Focusing on the movement of the large resonator, only, and, also, assuming that the force exerted by the second resonator onto the first one is small, one can simplify the dynamic equations. One arrives at relations equivalent to the SLA. **d** A similar behavior is observed if the second resonator itself is not small, but is coupled to the main resonator across a soft spring, κ_C . This type of coupled resonance is found when anharmonic side bands or standing compressional waves in a liquid cell couple to the quartz crystal (Sects. 7.5 and 7.6)

One has $1/M_{red} = 1/M_P + 1/M_R$. Since $M_P \ll M_R$, one can approximate the reduced mass by the mass of the particle (see Eq. 4.6.12).

(b) The characteristic impedance of the coupled resonator, $\tilde{Z}_P$, is much smaller than the characteristic impedance of the main resonator, $\tilde{Z}_R$. The characteristic impedance is defined as $(\kappa M)^{1/2}$ (The characteristic impedance is a mechanical impedance; it has units of Ns/m. One might have included it in Table 4.1, but it is only needed here and of little relevance elsewhere in the book.)

(c) The possibility of particle rotation is ignored (Sect. 11.5). The model is one-dimensional. There is only one dynamical variable, called x.

The particle linked to the main resonator has its own resonance frequency, given as

$$\tilde{\omega}_P = \sqrt{\frac{\tilde{\kappa}_P}{M_{red}}} \approx \sqrt{\frac{\tilde{\kappa}_P}{M_P}} \tag{4.6.12}$$

$\tilde{\omega}_P$ is also called "particle resonance frequency" in the following. A word of caution: $\tilde{\omega}_P$ is complex like the resonance frequency of the crystal. Differing from resonance of the crystal, the coupled resonance may be strongly damped. Some of the relations proven in Sect. 4.1.3 carry over to the coupled resonance, but others do not because of the large damping.

The frequency of the coupled resonance cannot be measured directly because the bond itself is not piezoelectric. However, one can infer some properties of the small resonator from the shifts, which the small "coupled" resonator induces to the frequency of the crystal. Expressed differently, the system is a composite resonator with two modes, only one of which is visible most of the time. The frequency of the visible mode is affected by the invisible mode and this effect can be put to use.

Let the position of the large sphere and the small sphere be x_R and x_P. The equations of motion are

$$\begin{aligned} M_R \frac{d^2 x_R}{dt^2} &= -\kappa_R x_R - \xi_R \frac{dx_R}{dt} + \kappa_P (x_P - x_R) + \xi_P \left(\frac{dx_P}{dt} - \frac{dx_R}{dt} \right) \\ M_P \frac{d^2 x_P}{dt^2} &= -\kappa_P (x_P - x_R) - \xi_P \left(\frac{dx_P}{dt} - \frac{dx_R}{dt} \right) \end{aligned} \tag{4.6.13}$$

In the frequency domain, this reads as

$$\begin{aligned} -\tilde{\omega}^2 M_R \hat{x}_R &= -\kappa_R \hat{x}_R - i\omega\xi_R \hat{x}_R + \kappa_P (\hat{x}_P - \hat{x}_R) + i\omega\xi_P (\hat{x}_P - \hat{x}_R) \\ -\tilde{\omega}^2 M_P \hat{x}_P &= -\kappa_P (\hat{x}_P - \hat{x}_R) - i\omega\xi_P (\hat{x}_P - \hat{x}_R) \end{aligned} \tag{4.6.14}$$

Using complex spring constants ($\tilde{\kappa}_M = \kappa_M + i\omega\xi_M$ and $\tilde{\kappa}_P = \kappa_P + i\omega\xi_P$) as well as matrix notation, this yields

$$\begin{bmatrix} \tilde{\omega}^2 - \frac{\tilde{\kappa}_R}{M_R} - \frac{\tilde{\kappa}_P}{M_R} & \frac{\tilde{\kappa}_P}{M_R} \\ \frac{\tilde{\kappa}_P}{M_P} & \tilde{\omega}^2 - \frac{\tilde{\kappa}_P}{M_P} \end{bmatrix} \begin{pmatrix} \hat{x}_R \\ \hat{x}_P \end{pmatrix} = 0 \tag{4.6.15}$$

This linear homogeneous equation system has a nontrivial solution if the determinant vanishes:

$$\begin{vmatrix} \tilde{\omega}^2 - \tilde{\omega}_R^2 - \frac{\tilde{Z}_P}{\tilde{Z}_R} \tilde{\omega}_R \tilde{\omega}_P & \frac{\tilde{Z}_P}{\tilde{Z}_R} \tilde{\omega}_R \tilde{\omega}_P \\ \tilde{\omega}_P^2 & \tilde{\omega}^2 - \tilde{\omega}_P^2 \end{vmatrix} = 0 \tag{4.6.16}$$

The relations $\tilde{\omega}_R^2 = \tilde{\kappa}_R/M_R$ and $\tilde{\omega}_P^2 = \tilde{\kappa}_P/M_P$ were used. The variables $\tilde{Z}_R = (\tilde{\kappa}_R M_R)^{1/2}$ and $\tilde{Z}_P = (\tilde{\kappa}_P M_P)^{1/2}$ were introduced. $\tilde{Z} = (\tilde{\kappa} M)^{1/2}$ is the characteristic impedance. Spelling out the definition of the determinant, one finds:

$$\begin{aligned}&\left(\tilde{\omega}^2 - \tilde{\omega}_R^2 - \frac{\tilde{Z}_P}{\tilde{Z}_R}\tilde{\omega}_R\tilde{\omega}_P\right)\left(\tilde{\omega}^2 - \tilde{\omega}_P^2\right) - \frac{\tilde{Z}_P}{\tilde{Z}_R}\tilde{\omega}_R\tilde{\omega}_P^3 \\ &= \left(\tilde{\omega}^2 - \tilde{\omega}_R^2\right)\left(\tilde{\omega}^2 - \tilde{\omega}_P^2\right) - \frac{\tilde{Z}_P}{\tilde{Z}_R}\tilde{\omega}_R\tilde{\omega}_P\tilde{\omega}^2 \\ &= \tilde{\omega}^4 + \tilde{\omega}^2\left(-\tilde{\omega}_P^2 - \tilde{\omega}_R^2 - \frac{\tilde{Z}_P}{\tilde{Z}_R}\tilde{\omega}_R\tilde{\omega}_P\right) + \tilde{\omega}_P^2\tilde{\omega}_R^2 = 0\end{aligned} \quad (4.6.17)$$

This equation has two solutions for $\tilde{\omega}^2$, which are:

$$\begin{aligned}&\tilde{\omega}_r^2 = \frac{C}{2} \pm \sqrt{\frac{C^2}{4} - \tilde{\omega}_P^2\tilde{\omega}_R^2} \\ &\text{with} \\ &C = \tilde{\omega}_P^2 + \tilde{\omega}_R^2 + \frac{\tilde{Z}_P}{\tilde{Z}_R}\tilde{\omega}_R\tilde{\omega}_P\end{aligned} \quad (4.6.18)$$

For the sake of display, consider real spring constants and real frequencies ω_r, ω_R, and ω_P. Figure 4.16 shows the two positive solution of ω versus ω_P/ω_R for $Z_P/Z_R = 0.01$. When ω_R and ω_P are much different, the corresponding vibrations are readily interpreted. In the horizontal part of the two branches (where $\omega_r \approx \omega_R$), the larger sphere moves with respect to the wall (that is, the crystal undergoes a thickness-shear oscillation). The smaller sphere either moves with the crystal surface (which is the case if $\omega_P \gg \omega_R$), or it rests in space (if $\omega_P \ll \omega_R$). On the inclined parts of the two branches (where $\omega_r \approx \omega_P$), the particle vibrates relative to the crystal. However, these latter modes are not detected in experiment. Only the crystal's motion produces an electrical current. The link between the crystal and the particle is non-piezoelectric. The presence of the particle is only noticed in so far, as it perturbs the motion of the crystal, hence the name "coupled resonance". As Fig. 4.16 shows, the shift in crystal's resonance frequency is strongest when $\omega_P \approx \omega_R$. Figure 4.16b shows an enlargement of the range where $\omega_P \approx \omega_R$. The dashed lines are the prediction from Eq. 4.6.26.

Only those vibrations, which can be detected in experiment, are of interested in sensing. These are the horizontal portions of the two branches. One can write $\tilde{\omega}_r = \tilde{\omega}_R + \Delta\tilde{\omega}$ with $|\Delta\tilde{\omega}| \ll |\tilde{\omega}_R|$. If $\tilde{\omega}_r \approx \tilde{\omega}_R$, one can approximate $\tilde{\omega}_r^2 - \tilde{\omega}_R^2$ as $2\tilde{\omega}_R(\tilde{\omega}_r - \tilde{\omega}_R)$. Equation 4.6.17 approximates as

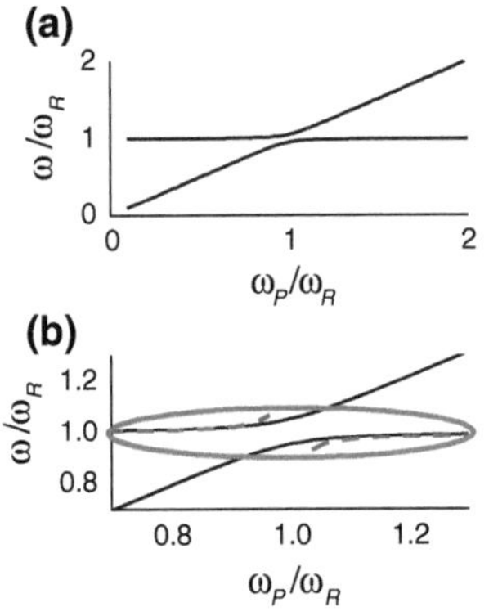

Fig. 4.16 Solution to Eq. 4.6.17 for real spring constants. There are two modes. Panel *b* shows an enlargement of the range where $\omega_P{}' \approx \omega_R{}'$. The horizontal parts of both branches correspond to patterns of motion, where the main resonator moves relative to the wall supporting it (Fig. 4.15a). This movement represents the thickness-shear vibration of the crystal. The second, small resonator perturbs this resonance. *Dashed line* Prediction from Eq. 4.6.21

$$\begin{aligned}&2\tilde{\omega}_R\Delta\tilde{\omega}\left(\tilde{\omega}_R^2+2\Delta\tilde{\omega}\tilde{\omega}_R+\Delta\tilde{\omega}^2-\tilde{\omega}_P^2\right)-\frac{\tilde{Z}_P}{\tilde{Z}_R}\tilde{\omega}_R\tilde{\omega}_P\left(\tilde{\omega}_R^2+2\Delta\tilde{\omega}\tilde{\omega}_R+\Delta\tilde{\omega}^2\right)\\&\approx 2\tilde{\omega}_R\Delta\tilde{\omega}\left(\tilde{\omega}_R^2-\tilde{\omega}_P^2\right)-\frac{\tilde{Z}_P}{\tilde{Z}_R}\tilde{\omega}_R\tilde{\omega}_P\left(\tilde{\omega}_R^2+2\Delta\tilde{\omega}\tilde{\omega}_R\right)\approx 0\end{aligned} \tag{4.6.19}$$

Quadratic terms in $\Delta\tilde{\omega}$ were neglected in line 2. The next step makes use of assumption b) above ($|\tilde{Z}_P/\tilde{Z}_R| \ll 1$). Omitting the term containing $\tilde{Z}_P/\tilde{Z}_R\Delta\tilde{\omega}$, Eq. 4.6.19 approximates as

$$2\tilde{\omega}_R\Delta\tilde{\omega}\left(\tilde{\omega}_R^2-\tilde{\omega}_P^2\right)-\frac{\tilde{Z}_P}{\tilde{Z}_R}\tilde{\omega}_R^3\tilde{\omega}_P\approx 0 \tag{4.6.20}$$

This relation can be solved for $\Delta\tilde{f}=\Delta\tilde{\omega}/(2/\pi)$ as:

$$\Delta\tilde{f}=\frac{1}{4\pi}\frac{\tilde{Z}_P}{\tilde{Z}_R}\frac{\tilde{\omega}_R^2\tilde{\omega}_P}{\tilde{\omega}_R^2-\tilde{\omega}_P^2}=\frac{1}{4\pi M_R}\tilde{\kappa}_P\frac{\tilde{\omega}_R}{\tilde{\omega}_R^2-\tilde{\omega}_P^2} \tag{4.6.21}$$

In step 2, the relations $\tilde{Z}_R/\tilde{\omega}_R=M_R$ and $\tilde{Z}_P\tilde{\omega}_P=\tilde{\kappa}_P$ were used. Equation 4.6.21, is similar to Eq. 4.1.21. The complex frequency shift as a function of $\tilde{\omega}_P$ itself forms a resonance curve.

Equation 4.6.21 could have been derived in a simpler way. One could have directly started from Fig. 4.15c. Applying the mechanical Kirchhoff rules from Sect. 4.7.3 to the diagram in Fig. 4.15c and inserting the resulting interfacial traction into the SLA directly leads to $\Delta\tilde{f}$ as given in Eq. 4.6.21. This will be done in Sect. 11.4. Equation 11.4.5 is equivalent to Eq. 4.6.21, realizing that $M_R = Am_q/2 = AZ_q/(4f_0)$ for the parallel plate. Equation 4.6.21 is more general. In addition to adsorbed particles, it also describes the film resonance in air (Sect. 10.1) and liquids (Sect. 10.2), resonances produced by standing compressional waves (Sect. 7.6), resonances produced by coupling to anharmonic side bands (Sect. 7.5), and laterally coupled resonators [39].

A formalism similar to the derivation above was used by Dybwad in his paper on the determination of contact stiffness [40]. Dybwad did not use the SLA, though, and his final result therefore is more complicated than Eq. 4.6.21. In addition to being simpler, the small load approximation has the benefit that it can be adapted to the case of many particles. If one goes though Dybwad's treatment and adapts it to two particles touching the resonator (rather than one), one has three dynamical variables and three solutions to the condition that the determinant of the dynamic systems vanishes. This situation basically is known from molecular physics. The more atoms a molecule contains, the larger is the number of vibrational modes. When using the SLA, the focus is on the one mode, which is visible in experiment. In order to predict the changes to the this mode induced by the sample, one just adds the forces exerted by the different adsorbed spheres. This is much easier than the solution to the full dynamical problem.

In the following, we consider three limits of Eq. 4.6.21, which are

(a) $\omega'_P \gg \omega'_R$ (inertial loading)
(b) $\omega'_P \ll \omega'_R$ (elastic coupling)
(c) $\omega_P' \approx \omega_R'$ combined with $\omega_P'' \gg |\omega_P' - \omega_R'|$ (zero crossing).

(a) *Inertial Loading* ($\omega'_P \gg \omega'_R$)

With $\omega'_P \gg \omega'_R$, Eq. 4.6.21 is approximated as:

$$\Delta\tilde{f} \approx -\frac{1}{4\pi}\frac{\tilde{\omega}_R^2}{\tilde{Z}_R}\frac{\tilde{Z}_P}{\tilde{\omega}_P} = -\frac{1}{4\pi M_R}\tilde{\omega}_R M_P \tag{4.6.22}$$

Substituting M_R by $Am_q/2$ (Eq. 4.5.27a), substituting M_P/A by m_f (m_f the mass of the "film", where the "film" consists of all matter rigidly attached to the resonator), and using $m_q = Z_q/(2f_0)$ (Eq. 4.3.17), Eq. 4.6.22 reduces to the Sauerbrey result (Eq. 4.4.14). If $\omega'_P \gg \omega'_R$, the adsorbate is coupled so tightly to the resonator that it follows the resonator's movement and increases its mass.

(b) *Elastic Coupling* ($\omega'_P \gg \omega'_R$)

With $\omega'_P \gg \omega'_R$, Eq. 4.6.21 is approximated as:

$$\Delta\tilde{f} \approx \frac{1}{4\pi}\frac{\tilde{Z}_P\tilde{\omega}_P}{\tilde{Z}_R} = \frac{1}{4\pi M_R}\frac{\tilde{\kappa}_P}{\tilde{\omega}_R} \tag{4.6.23}$$

Substituting M_R by $Am_q/2$ and using $m_q = Z_q/(2f_0)$, Eq. 4.6.23 reduces to the case discussed in Sect. 11.3. The sphere is heavy enough to be clamped in space by inertia. It exerts a restoring force onto the resonator, thereby increasing the stiffness of the composite system.

(c) *Particle Resonance Frequency Comparable to the Resonance Frequency of the Main Resonator*

When the two resonance frequencies are about equal ($\tilde{\omega}_P \approx \tilde{\omega}_R$ and $\tilde{\omega}_P + \tilde{\omega}_R \approx 2\omega_P$), Eq. 4.6.21 turns into:

$$\Delta\tilde{f} \approx \frac{1}{8\pi}\frac{\tilde{\omega}_R}{\tilde{Z}_R}\tilde{Z}_P\tilde{\omega}_P\frac{1}{\tilde{\omega}_R - \tilde{\omega}_P} = \frac{1}{8\pi}\frac{\tilde{\kappa}_P}{M_R}\frac{1}{\tilde{\omega}_R - \tilde{\omega}_P} \tag{4.6.24}$$

Since Δf and $\Delta\Gamma$ follow a Lorentzian, they yield a circle when displayed in a polar diagram as shown in Fig. 11.8 (They form a circle when approximating $\tilde{\omega}_P \ll \tilde{\omega}_R$ as $2\tilde{\omega}_P$. Otherwise, they form a spiral). Such polar plots are discussed in Sect. 11.4.

Importantly, the SLA may not be applicable around the coupled resonance, because the denominator may come close to zero. Using $M_R = AZ_q/(4f_0)$, Eq. 4.6.24 reads as

$$\frac{\Delta\tilde{f}}{f_0} \approx \frac{1}{2\pi}\frac{\tilde{\kappa}_P}{AZ_q}\frac{1}{\tilde{\omega}_R - \tilde{\omega}_P} \tag{4.6.25}$$

The absolute value of the left-hand side, $|\Delta\tilde{f}/f_0|$, must be small in order for the SLA to hold. This may be true even if the real parts of $\tilde{\omega}_P$ and $\tilde{\omega}_R$ are equal. Since the main resonator is the crystal, its damping is small ($\tilde{\omega}_R \approx \omega_R$). The damping of the coupled resonator, on the other hand, may be large. The SLA holds as long as one has

$$\frac{1}{2\pi}\frac{\tilde{\kappa}_P}{AZ_q}\frac{1}{\tilde{\omega}''_P} \ll 1 \tag{4.6.26}$$

If this is not true one has to go back to Fig. 4.16 and live with the fact that there are two solutions, that is, two resonant modes. If the damping of the coupled resonator is large, then these two modes are so broad that they merge into one peak. This condition is equivalent to the applicability of the SLA.

4.7 Appendices

4.7.1 Broadening of a Resonance Curve by Damping

Resonance curves become broader when the damping is increased. On a formal level, this is the consequence of the Fourier transform. The resonance curve is the Fourier transform of the decaying cosine. The decay rate of the oscillation in the time domain turns into the width of the resonance curve upon Fourier transformation.

There is less formal explanation for this correspondence, which is based on the efficiency, by which a resonance can be excited if the driving frequency is slightly different from the resonance frequency. The larger the damping, the less critical is such a detuning. If the resonator is driven at a frequency different from f_r, the driving frequency beats against the frequency of the resonator. At $t = 0$ the external force feeds energy into the system and the resonator starts to vibrate. At some later time, the driving force and resonator fall out of synchronization. The force now works against the vibration, it *withdraws* energy from the resonance. Still later, the relative phase between the source and the resonator is correct again and energy is again supplied to the system. On average, the external force does not energize the resonance efficiently, because the flow of energy periodically reverts sign.

If the resonator is highly damped, more energy is needed to drive the vibration, in the first place. However, if the driving frequency slightly differs from the resonance frequency, this hardly reduces the amplitude because the resonator quickly forgets about the energy supplied earlier. At the time, when the source and the resonator have fallen out of synchronization, the energy supplied earlier has been dissipated, anyway. There is not much of a vibration, against which the source would beat. As long as source and the resonator differ in frequency by less than the decay rate, the excitation is about as efficient as it would be for optimum frequency match.

The above argument shows that broadening of a resonance by damping is not caused by the resonator internally having a number of slightly different resonance frequencies. That situation would be called "heterogeneous broadening". Heterogeneous broadening of a resonance is different from damping (Sect. 11.5).

4.7.2 Kirchhoff Rules

Most readers know the Kirchhoff rules, but some may not know them under that name. Consider the two networks shown Fig. 4.17. The first one has two resistors placed in series, while the second one has the same two resistors placed in parallel. The total resistance of the former circuit is given as

$$R_{tot} = R_A + R_B \tag{4.7.1}$$

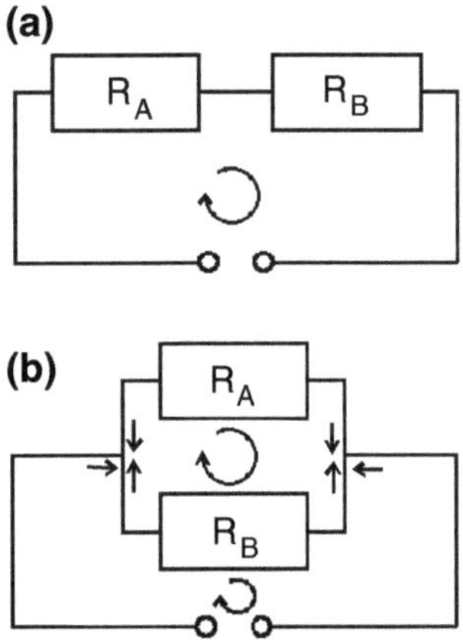

Fig. 4.17 Two resistors arranged in series (**a**) and in parallel (**b**)

while in the latter case, one has

$$\frac{1}{R_{tot}} = \frac{1}{R_A} + \frac{1}{R_B} \tag{4.7.2}$$

These, basically, are the Kirchhoff rules.

For more complicated circuits, one applies a more general version of Eqs. 4.7.1 and 4.7.2, called the "loop rule" and the "node rule". The loop rule states that the voltages across circuit elements arranged in a loop add to zero. The node rule states that the currents flowing into a vertex likewise add to zero. Equations 4.7.1 and 4.7.2 can be derived from the loop rule and the node rule. For example, the current is constant everywhere in Fig. 4.17a because there are no nodes. The voltage drop across the source must equal the sum of the voltages at the two resistors. This amounts to $R_{tot} = R_A + R_B$.

Understanding equivalent circuits requires an additional bit of information. The capacitor and the inductor fit into this scheme, but differing from the ohmic resistors shown in Fig. 4.17, the capacitor and the inductor have a complex impedance. One has $\tilde{Z}_{cap} = 1/(\mathrm{i}\omega C)$ and $\tilde{Z}_{ind} = \mathrm{i}\omega L$ with C the capacitance and L the inductance. Accepting complex impedances (possibly with a dependence on ω more complicated than for the capacitor and the inductor, cf. Fig. 4.8c) and remembering how to add up circuit elements (arranged in series or in parallel), one is all set to predict resonance frequencies from a circuit.

4.7.3 The Electromechanical Analogy

Underlying the representation of acoustic devices by electrical circuits is the electromechanical analogy [41]. The rules, by which electrical circuits are mapped onto mechanical circuits, are summarized in Table 4.2. The electromechanical

Table 4.2 The electromechanical analogy

Electrical		Mechanical	
Voltage		Force	
Current		Velocity	
Resistor	$\tilde{Z}_{el} = R$	Dashpot	$\tilde{Z}_m = \xi$
Capacitor	$\tilde{Z}_{el} = \frac{1}{i\omega C}$	Spring[a)]	$\tilde{Z}_m = \frac{\kappa}{i\omega}$
Inductor	$\tilde{Z} = i\omega L$	Mass	$\tilde{Z}_m = i\omega M$
Two elements in parallel[b)]	$\frac{1}{\tilde{Z}_{el,tot}} = \frac{1}{\tilde{Z}_{el,A}} + \frac{1}{\tilde{Z}_{el,B}}$		$\tilde{Z}_{m,tot} = \tilde{Z}_{m,A} + \tilde{Z}_{m,B}$
Two elements in series[b)]	$\tilde{Z}_{el,tot} = \tilde{Z}_{el,A} + \tilde{Z}_{el,B}$		$\frac{1}{\tilde{Z}_{m,tot}} = \frac{1}{\tilde{Z}_{m,A}} + \frac{1}{\tilde{Z}_{m,B}}$
Ground	$U = 0$	Open end	$F = 0$
Open end	$I = 0$	Wall	$v = 0$

[a] Coincidentally, the spring is depicted with the same symbol as the resistor. The context usually clarifies the element's nature

[b] The diagrams were drawn with resistors and dashpots. The Kirchhoff rules (electrical and mechanical) apply to all other elements in the same way

analogy looks intriguing, at first glance, but it can easily create confusion because the rules by which different elements are *connected* to each other, differ between the two types of circuit representations. The problem was already mentioned in Sect. 4.5.3. When two mechanical elements are arranged in parallel, their elongations (and the time derivatives thereof, which are the velocities) are the same, while the total force is the sum of the forces acting onto the two elements individually. In consequence, the impedance is additive. Along the same line, the inverse impedance is additive for elements placed in series. These rules are called "mechanical Kirchhoff rules" in this book. The role of parallel and serial arrangement is reversed between mechanical and electrical networks. A similar problem occurs with open ends and connections to ground. When an electrical element is connected to ground, the voltage is strictly zero and the current takes whatever value is needed to achieve $U = 0$. The corresponding mechanical situation is an open end, where the force is zero and the velocity is unconstrained. An open end in electricity is characterized by vanishing current and unconstrained voltage. It is often emphasized that an "open end" most of the time amounts to a poorly controlled stray capacitance. The corresponding mechanical element is a rigid wall, where the displacement (the velocity) is strictly zero and the force is unconstrained. In analogy to the electrical case, one must be careful with rigid walls because there are no perfectly rigid materials. Finally, there are problems with sign changes, resurfacing at various points (see for example the comment above Eq. 4.7.8). Given these complications, it is clear why many researchers prefer to stick to electrical Kirchhoff rules applied to the entire circuit (including the motional branch).

A few further comments:

- In both the mechanical and the electrical representation, the circuit elements are discrete objects without internal structure. This situation is also expressed by the term "lumped-element circuit". For instance, a "mass" is a rigid object, the behavior of which is fully specified by Newton's second law ($F = M\ddot{x}$ with $\ddot{x}$ the acceleration). The model is one-dimensional in the sense that there is only translation along x. Translation along y or z (or rotation, Sect. 11.5) are not part of the picture. A similar limitation applies to electrical circuits. A resistor is fully described by the ratio of voltage and current as determined at the wires connecting to it. The internal current distribution is ignored.
- For the parallel plate, (more generally, for plane waves) there is an electro-*acoustic* analogy. In the electro-acoustic analogy, the analog of voltage is stress, not force.
- What is called mechanical Kirchhoff rules here, is also applied in rheology. This concerns, for example, the Voigt element and the Maxwell element [42]. (The rules are not called Kirchhoff rules in that context).
- The opto-acoustic analogy is touched upon in Sect. 16.2. It can be formulated, but is not widely used.
- There is a version of the electromechanical analogy, which directly leads to the electrical Kirchhoff rules. It is called the "mobility analog" in Ref. [43]. In this version, a spring is mapped onto an inductor and a mass is mapped onto a capacitor. In the author's opinion, this way of formulating the (imperfect) analogy only shifts the problem to somewhere else.
- Given that QCM experiments often occur in Newtonian liquids, one may wonder if there is a circuit element representing a liquid. The liquid's impedance is given by $\tilde{Z}_L = (i\omega\rho_{liq}\eta_{liq})^{1/2}$ (Sect. 9.1). None of the familiar electrical elements has a square-root dependence on frequency. Interestingly, there is a similar element in electrochemical impedance spectroscopy (EIS) [44]. Its name is "Warburg impedance" (often depicted as –W–). The Warburg impedance represents current-voltage relations governed by the diffusion of ions close to an electrode surface. Its frequency dependence is given by $\tilde{Z}_W = A_W/(i\omega)^{1/2}$. The Warburg parameter, A_W, is real-valued and independent of ω. It depends on the concentration and the diffusivity of the ions. For details see the textbooks on electrochemistry [44]. Evidently, the analogy between the Warburg impedance and the term $(i\omega\rho_{liq}\eta_{liq})^{1/2}$ describing the Newtonian liquid is imperfect because the Warburg impedance and the Newtonian liquid have the term $(i\omega)^{1/2}$ in the denominator and in the numerator. respectively. Still, both elements describe a diffusional wave. The wave transports ions in the case of the Warburg impedance, while it transports momentum in the Newtonian liquid.

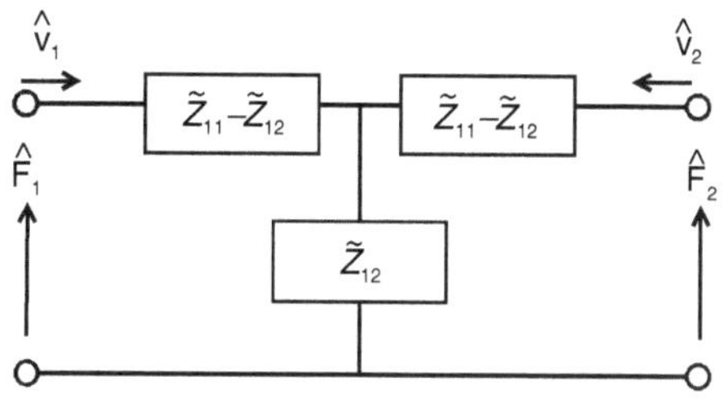

Fig. 4.18 Representation of a transmission line as a "T"

4.7.4 Derivation of the Mason Circuit

In the following, we derive the Mason circuit, but still cut some technicalities short. For more extended treatments see Refs. [22, 23].

The first step in the derivation is the proof that the viscoelastic plate can be represented by a "T" as in Fig. 4.18. A side remark: Representing the plate as a T is a choice. One might also have chosen a Π instead of a T. More generally, equivalent circuit representations are not unique. Multiple circuits can represent the same set of force-velocity (or voltage-current) relations. Two circuits can look different when displayed as a diagram, but still be equivalent with regard to their behavior at the ports.

Application of the Kirchhoff rules to Fig. 4.18 shows that forces and velocities are related by

$$\begin{aligned}\hat{F}_1 &= \tilde{Z}_{11}\hat{v}_1 + \tilde{Z}_{12}\hat{v}_2\\ \hat{F}_2 &= \tilde{Z}_{12}\hat{v}_1 + \tilde{Z}_{11}\hat{v}_2\end{aligned} \tag{4.7.3}$$

For later use, we express Eq. 4.7.3 in matrix form as:

$$\begin{pmatrix}\hat{F}_1\\ \hat{F}_2\end{pmatrix} = \begin{bmatrix}\tilde{Z}_{11} & \tilde{Z}_{12}\\ \tilde{Z}_{21} & \tilde{Z}_{22}\end{bmatrix}\begin{pmatrix}\hat{v}_1\\ \hat{v}_2\end{pmatrix} = \begin{bmatrix}\tilde{Z}_{11} & \tilde{Z}_{12}\\ \tilde{Z}_{12} & \tilde{Z}_{11}\end{bmatrix}\begin{pmatrix}\hat{v}_1\\ \hat{v}_2\end{pmatrix} \tag{4.7.4}$$

The second step exploits symmetry. Since the indices 1 and 2 can be interchanged, one must have $\tilde{Z}_{11} = \tilde{Z}_{22}$ and $\tilde{Z}_{12} = \tilde{Z}_{21}$. As such, neither Fig. 4.18 nor Eq. 4.7.3 contain a physical statement. The goal is to assign values to $\tilde{Z}_{11}$ and $\tilde{Z}_{12}$, which depend on the properties of the plate (see Eq. 4.7.12). Below, the relations between forces and velocities at the surfaces of a plate will be brought into matrix form as in Eq. 4.7.3. This will allow to relate the elements $\tilde{Z}_{11}$ and $\tilde{Z}_{12}$ to the parameters of the plate.

The displacement, $\hat{u}(z)$, is a superposition of a sine and a cosine

$$\hat{u}(z) = \hat{u}_\alpha \sin\left(\tilde{k}z\right) + \hat{u}_\beta \cos\left(\tilde{k}z\right) \tag{4.7.5}$$

$\hat{u}_\alpha$ and $\hat{u}_\beta$ are amplitudes to be determined by the boundary conditions. $\tilde{k}$ is the wavenumber. At this point, the resonance condition is not invoked. $\tilde{k}$ can take any value. The wavenumber, $\tilde{k}$, is complex for the sake of generality. For the resonant mode, $\tilde{k}$ will be equal to $\tilde{\omega}_r/\tilde{c}$. $\tilde{\omega}_r$ and $\tilde{c}$ will have the same phase angle and $\tilde{k}$ will be real ($\tilde{k} = k$).

The stress is given as $\hat{\sigma}(z) = \tilde{G}\mathrm{d}\hat{u}(z)/\mathrm{d}z$. For the stress field, one finds

$$\begin{aligned}\hat{\sigma}(z) &= \tilde{G}\tilde{k}\left(\hat{u}_\alpha \cos\left(\tilde{k}z\right) - \hat{u}_\beta \sin\left(\tilde{k}z\right)\right)\\ &= \omega\tilde{Z}\left(\hat{u}_\alpha \cos\left(\tilde{k}z\right) - \hat{u}_\beta \sin\left(\tilde{k}z\right)\right)\end{aligned} \tag{4.7.6}$$

In line 2 the relations $\tilde{k} = \omega/\tilde{c}$, $\tilde{c} = \left(\tilde{G}/\rho\right)^{1/2}$, and $\tilde{Z} = \left(\rho\tilde{G}\right)^{1/2}$ were applied. With regard to the forces, there is a sign problem. See Ref. [23] for more details. The forces in the equivalent circuit representation are exerted by the crystal onto the environment, which leads to the relation $F_1 = -A\sigma(z = -h)$ and $F_2 = -A\sigma(z = h)$. Also using $\sin(-\tilde{k}h) = -\sin(\tilde{k}h)$, the forces at the ports are found to be

$$\begin{aligned}\hat{F}_1 &= A\omega\tilde{Z}\left(-\hat{u}_\alpha \cos\left(\tilde{k}h\right) - \hat{u}_\beta \sin\left(\tilde{k}h\right)\right)\\ \hat{F}_2 &= A\omega\tilde{Z}\left(-\hat{u}_\alpha \cos\left(\tilde{k}h\right) + \hat{u}_\beta \sin\left(\tilde{k}h\right)\right)\end{aligned} \tag{4.7.7}$$

The velocity is given as $\hat{v}(z) = \mathrm{i}\omega\hat{u}(z)$. There again is a complication with regard to the signs [23]. It has to do with the fact that positive currents flow *into* the circuit. For that reason, we have $\hat{v}_2 = -\hat{v}(z = h)$. The velocities at the two ports obey

$$\begin{aligned}\hat{v}_1 + \hat{v}_2 &= -2\mathrm{i}\omega\hat{u}_\alpha \sin\left(\tilde{k}h\right)\\ \hat{v}_1 - \hat{v}_2 &= 2\mathrm{i}\omega\hat{u}_\beta \cos\left(\tilde{k}h\right)\end{aligned} \tag{4.7.8}$$

The amplitudes $\hat{u}_\alpha$ and $\hat{u}_\beta$ are given as

$$\begin{aligned}\hat{u}_\alpha &= \mathrm{i}\frac{\hat{v}_1 + \hat{v}_2}{2\omega \sin\left(\tilde{k}h\right)}\\ \hat{u}_\beta &= -\mathrm{i}\frac{\hat{v}_1 - \hat{v}_2}{2\omega \cos\left(\tilde{k}h\right)}\end{aligned} \tag{4.7.9}$$

Inserting this result into Eq. 4.7.7 yields

$$\begin{aligned}\hat{F}_1 &= \frac{\mathrm{i}A\tilde{Z}}{2}\left(-\cot\left(\tilde{k}h\right)(\hat{v}_1 + \hat{v}_2) + \tan\left(\tilde{k}h\right)(\hat{v}_1 - \hat{v}_2)\right)\\ \hat{F}_2 &= \frac{\mathrm{i}A\tilde{Z}}{2}\left(-\cot\left(\tilde{k}h\right)(\hat{v}_1 + \hat{v}_2) - \tan\left(\tilde{k}h\right)(\hat{v}_1 - \hat{v}_2)\right)\end{aligned} \tag{4.7.10}$$

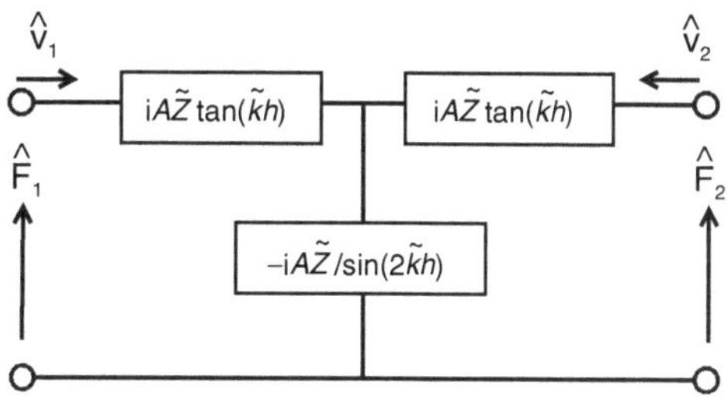

Fig. 4.19 Equivalent circuit of the viscoelastic plate. Same as Fig. 4.18 with Eq. 4.7.12 applied

In matrix form, this reads as

$$\begin{pmatrix} \hat{F}_1 \\ \hat{F}_2 \end{pmatrix} = \frac{\mathrm{i}A\tilde{Z}}{2}\begin{bmatrix} +\tan(\tilde{k}h) - \cot(\tilde{k}h) & -\tan(\tilde{k}h) - \cot(\tilde{k}h) \\ -\tan(\tilde{k}h) - \cot(\tilde{k}h) & +\tan(\tilde{k}h) - \cot(\tilde{k}h) \end{bmatrix}\begin{pmatrix} \hat{v}_1 \\ \hat{v}_2 \end{pmatrix} \quad (4.7.11)$$

Comparison with Eq. 4.7.4 shows that

$$\begin{aligned} \tilde{Z}_{11} &= \frac{\mathrm{i}A\tilde{Z}}{2}\left(\tan(\tilde{k}h) - \cot(\tilde{k}h)\right) \\ \tilde{Z}_{12} &= -\frac{\mathrm{i}A\tilde{Z}}{2}\left(\tan(\tilde{k}h) + \cot(\tilde{k}h)\right) = \frac{-\mathrm{i}A\tilde{Z}}{\sin(2\tilde{k}h)} \end{aligned} \quad (4.7.12)$$

The trigonometric identity $\tan(x) + \cot(x) = 2/(\sin(2x))$ was used in line 2. The element $\tilde{Z}_{11} - \tilde{Z}_{12}$ in Fig. 4.18 turns into

$$\tilde{Z}_{11} - \tilde{Z}_{12} = \mathrm{i}A\tilde{Z}\tan(\tilde{k}h) \quad (4.7.13)$$

This brings the equivalent circuit into the form shown in Fig. 4.19.

We now add piezoelectricity into the picture. Piezoelectricity turns a two-port network into a three-port network as shown in Fig. 4.20. The electrical port is connected to the mechanical branch across a *transformer*. In a first step, let there be a large resistance, $\tilde{Z}_{ex}$, between the electrical source and the electrodes. Because $\tilde{Z}_{ex} = \infty$, the current into the electrodes is small. This is the open-circuit configuration (also called open-loop configuration). As discussed in Sect. 5.3, the piezoelectric plate under open-circuit conditions is represented by the same T as the non-piezoelectric plate. The only difference is that the shear modulus must be replaced by the piezoelectrically stiffened modulus. The parameter $\tilde{G}_q$ in this book is to be understood as this stiffened modulus.

Of course real resonators can only be near to the open-circuit condition. With electrodes not connected (with $\tilde{Z}_{ex}$ in Fig. 4.20 infinite), the resonator cannot be interrogated electrically. The element $-\phi^2/(\mathrm{i}\omega C_0)$ accounts for the finite current as shown below. For now, ignore the element $-\phi^2/(\mathrm{i}\omega C_0)$.

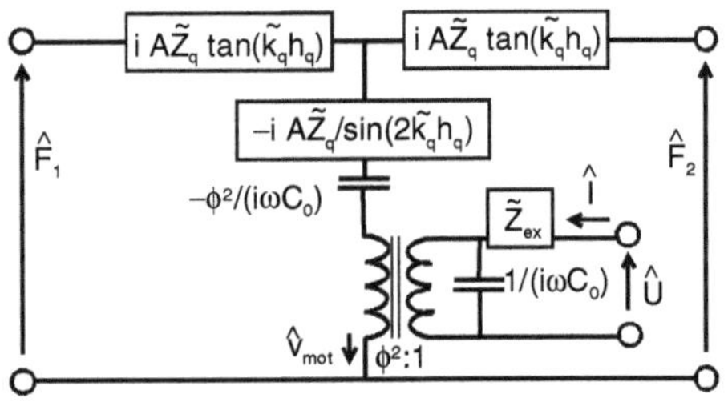

Fig. 4.20 A Mason circuit with a large electrical impedance inserted between the electrodes and the external electrical power supply. Because $\tilde{Z}_{ex}$ is large, there is almost no current into the electrodes. The crystal operates close to the open-circuit condition

The plate must satisfy the constitutive relations of piezoelectricity (Table 5.2.1). Most convenient is the stress-voltage form. Adapted to the parallel plate, the piezoelectric equations in stress-voltage form amount to (Eq. 5.3.1):

$$\begin{aligned} \hat{\sigma} &= \tilde{G}_q \frac{\mathrm{d}\hat{u}}{\mathrm{d}z} - \frac{\tilde{e}_{26}}{\tilde{\varepsilon}_q \varepsilon_0} \hat{D} && \text{(a)} \\ \hat{E} &= -\frac{\tilde{e}_{26}}{\tilde{\varepsilon}_q \varepsilon_0} \frac{\mathrm{d}\hat{u}}{\mathrm{d}z} + \frac{1}{\tilde{\varepsilon}_q \varepsilon_0} \hat{D} && \text{(b)} \end{aligned} \tag{4.7.14}$$

In the following, we show that the electrical Kirchhoff rules applied to the Mason circuit reduce to Eq. 4.7.14 in the limit of $\omega \to 0$. At $\omega \approx 0$, the elements in the horizontal bar at the top of the circuit turn to zero ($\tilde{k}_q$ is proportional to ω and $\tan(0) = 0$). The "current" through the vertical bar of the T, $\hat{v}_{mot}$, is twice is the current at the surface:

$$\hat{v}_{mot} = 2\,\mathrm{i}\omega\,\hat{u}_S \tag{4.7.15}$$

The "voltage drop" at the element $-\mathrm{i}A\tilde{Z}_q/\sin(2\tilde{k}_q h_q)$ is

$$\frac{-\mathrm{i}A\tilde{Z}_q}{\sin\left(2\tilde{k}_q h_q\right)} 2\,\mathrm{i}\omega\hat{u}_S = \frac{2A\omega\tilde{Z}_q\hat{u}_S}{\sin\left(\left(2\omega/\tilde{c}_q\right)h_q\right)} \approx A\tilde{c}_q Z_q \frac{\hat{u}_S}{h_q} = A\tilde{G}_q \frac{\mathrm{d}\hat{u}}{\mathrm{d}z} \tag{4.7.16}$$

Taylor-expansion in ω and the relations $\tilde{c}_q = \left(\tilde{G}_q/\rho_q\right)^{1/2}$ and $\tilde{Z}_q = \left(\tilde{G}_q\rho_q\right)^{1/2}$ were used in step 2. Under static conditions, the strain in the plate is the same everywhere and is given as $\mathrm{d}\hat{u}/\mathrm{d}z = \hat{u}_S/h_q$. This relation was applied in step 3. The loop rule applied to the right-hand side of the Mason circuit (containing the load impedance, the element $-\mathrm{i}A\tilde{Z}_q/\sin(2\tilde{k}_q h_q)$, and the left-hand side of the transformer) leads to

$$A\hat{\sigma}_S - A\tilde{G}_q \frac{\mathrm{d}\hat{u}}{\mathrm{d}z} + \frac{A\tilde{e}_{26}}{d_q}\hat{U} = 0 \tag{4.7.17}$$

The relation $\phi = A\tilde{e}_{26}/d_q$ was used in the third term. Remembering that $\hat{D} = \tilde{\varepsilon}_q\varepsilon_0\hat{E} = \tilde{\varepsilon}_q\varepsilon_0\hat{U}/d_q$, Eq. 4.7.17 reduces to Eq. 4.7.14a.

The node rule applied to the junction above the electrical capacitance in Fig. 4.20 yields

$$\begin{aligned} &\frac{2A\tilde{e}_{26}}{d_q}\mathrm{i}\omega\,\hat{u}_S + \mathrm{i}\omega\,C_0\hat{U} - \hat{I} = 0 \\ &\mathrm{i}\omega A\tilde{e}_{26}\frac{\mathrm{d}\hat{u}}{\mathrm{d}z} + \mathrm{i}\omega A\tilde{\varepsilon}_q\varepsilon_0\frac{\hat{U}}{d_q} - \mathrm{i}\omega A\hat{D} = 0 \end{aligned} \tag{4.7.18}$$

$\mathrm{d}\hat{u}/\mathrm{d}z = \hat{u}_S/h_q$ was applied as before. The electrical capacitance is given as $C_0 = A\tilde{\varepsilon}_q\varepsilon_0/d_q$. Remembering that $\hat{D} = \tilde{\varepsilon}_q\varepsilon_0\hat{E}$, Eq. 4.7.18 reduces to Eq. 4.7.14b.

We now turn to the element $-\phi^2/(\mathrm{i}\omega C_0)$. The plate under open-circuit conditions is represented by a T, if the piezoelectrically stiffened modulus is used in the calculation of $\tilde{Z}_q$ and $\tilde{k}_q$. However, the transformer not only couples stress to voltage and strain to charge, it also introduces a new mechanical impedance into the circuit. The left-hand side of the transformer exerts a mechanical stress. The corresponding impedance is $\phi^2/\mathrm{i}\omega C_0$ according to Fig. 4.12c. An impedance of $\phi^2/\mathrm{i}\omega C_0$ (which is what the transformer amounts to) inserted into the vertical bar of the circuit introduces a mistake. The mistake is corrected by the element $-\phi^2/(\mathrm{i}\omega C_0)$ above the transformer.

When short-circuiting the crystal ($\tilde{Z}_{ex} \approx 0$), the impedance across the transformer turns to zero. Short-circuiting does not affect the element $-\phi^2/(\mathrm{i}\omega C_0)$. Since the element $-\phi^2/(\mathrm{i}\omega C_0)$ has negative sign, the short-circuited resonator is *softer* than the resonator with the electrodes open.

The case where $\tilde{Z}_{ex}$ takes some finite value different from both 0 and ∞ is discussed in Sect. 5.3.

Glossary

Variable	Definition (Comments)
$\langle . \rangle$	Average (over area or time, as indicated by an index)
A	(Effective) area of the resonator plate (see Sect. 7.4)
B_{el}	Electric susceptance ($B_{el} = \mathrm{Im}(\tilde{Y}_{el})$)
B_{off}	Offset of the susceptance in a data trace of $B_{el}(f)$ (fit parameter in Eq. 4.5.30)
$\tilde{c}$	Speed of (shear) sound $\left(\tilde{c} = \left(\tilde{G}/\rho\right)^{1/2}\right)$

C_1	Motional capacitance
$\overline{C}_1$	Motional capacitance of the four-element circuit (Piezoelectric stiffening and the load have been taken into account)
d	Thickness of a layer
d_q	Thickness of the resonator ($d_q = m_q/\rho_q = Z_q/(2\rho_q f_0)$)
$\mathrm{d}C$	Infinitesimal capacitance (Fig. 4.8a)
$\mathrm{d}L$	Infinitesimal inductance (Fig. 4.8a)
D	Dissipation factor ($D = 1/Q = 2\Gamma/f_r$)
E	Electric field
E_{tot}	Total kinetic energy involved in a collision (see text above Eq. 4.2.16)
e_{26}	Piezoelectric stress coefficient ($e_{26} = 9.65 \times 10^{-2}$ C/m^2 for AT-cut quartz)
f	As an index: *f*ilm
f	Frequency
$\tilde{f}$	Complex resonance frequency ($\tilde{f} = f_r + i\Gamma$)
$\tilde{f}_d$	Damped resonance frequency ($\tilde{f}_d = \left(f_r^2 - \Gamma^2\right)^{1/2} + i\Gamma$, also: "ringing frequency")
$\tilde{f}_r$	Undamped resonance frequency
f_s	Series resonance frequency ($f_r = f_s$ by definition in this book)
f_0	Resonance frequency at the fundamental $f_0 = Z_q/(2m_q) = Z_q/(2\rho_q d_q)$ Might also have been called f_1; we follow the literature in calling it f_0
$\hat{F}$	Force
$\hat{F}_{ex}$	An external force driving a resonator
$\tilde{G}$	Shear modulus
G_e	Electric conductance ($G_{el} = \mathrm{Re}(\tilde{Y}_{el})$)
G_{max}	Conductance on the peak of the resonance (fit parameter in Eq. 4.5.30)
G_{off}	Offset of the conductance in a data trace of $G_{el}(f)$ (fit parameter in Eq. 4.5.30)
G_q	Shear modulus of AT-cut quartz ($G_q \approx 29 \times 10^9$ Pa, often called μ_q in the literature. G_q is the piezoelectrically stiffened modulus (Sect. 5.3))
$\tilde{J}$	Shear compliance

h	Half the thickness of a plate or layer, half the length of a cable
H	Magnetic field
$\hat{I}$	Electric current
$\tilde{k}$	Wavenumber ($\tilde{k} = \tilde{\omega}/\tilde{c}$)
k_t	Piezoelectric coupling coefficient ($k_t^2 = e_{26}/(\varepsilon_q \varepsilon_0 G_q)^2$)
liq	As an index: *liq*uid
$\bar{L}_j$	Transfer matrix for layer j (See Eq. 4.4.2)
L_1	Motional inductance
$\bar{L}_1$	Motional inductance of the four-element circuit (Piezoelectric stiffening and the load have been taken into account)
mot	As a index: *mot*ional (Applies to either the motional branch of the four-element circuit (Fig. 4.12b) or to the left-hand side of the transformer in the Mason-circuit (Fig. 4.10))
M	Mass
m_q	Mass per unit area of the resonator ($m_q = \rho_q d_q = Z_q/(2f_0)$)
M_R	Mass of a *R*esonator
M_{red}	Reduced mass of a coupled two-body system ($1/M_{red} = 1/M_A + 1/M_B$)
n	Overtone order
n_{right}, n_{left}	Number of turns of a transformer (Eq. 4.5.7)
n, $\tilde{n}$	Refractive index ($\tilde{n} = \tilde{\varepsilon}_r^{1/2}$)
OC	As an index: resonance condition of the unloaded plate under *O*pen-*C*ircuit conditions. With no current into the electrodes (more precisely, with vanishing electric displacement everywhere), piezoelectric stiffening is fully accounted for by using the piezoelectrically stiffened shear modulus
P	As an index: *P*article (Sect. 4.6.3)
p_{tot}	Total momentum involved in a collision (see text above Eq. 4.2.16)
q	As an index: *q*uartz resonator
Q	Q-factor ($Q = 1/D = f_r/(2\Gamma)$)
ref	As an index: *ref*erence state of a crystal in the absence of a load
$\tilde{r}$	Amplitude reflection coefficient (reflectivity, for short) Not to be confused with the power reflectance coefficient Defined with regard to displacement (not stress)

$\tilde{r}_{q,S}$	Reflectivity evaluated at the resonator surface (see text above Eq. 4.4.6)
R	As an index: *R*esonator
R_1	Motional resistance
$\overline{R}_1$	Motional resistance of the four-element circuit (piezoelectric stiffening and the load have been taken into account)
$\overline{S}_{i,j}$	Transfer matrix the interface between layer i and j (see Eq. 4.4.2)
S	As an index: *S*urface
t	Time
$\tilde{t}$	Amplitude transmission coefficient
$\hat{u}$	(Tangential) displacement
$\hat{U}$	Electric voltage
$\hat{v}$	Velocity ($\hat{v} = i\omega\hat{u}$)
w	Half-band full-width of a resonance ($w = 2\Gamma$)
$x(t)$, $\tilde{x}(t)$	Some time-harmonic function (real or complex) $(xt) = x_0\cos(\omega t + \varphi)$, $\tilde{x}(t) = \hat{x}\exp(i\omega t)$
$\hat{x}$	The function's amplitude ($\hat{x} = x_0\cos(\varphi) - ix_0\sin(\varphi)$)
z	Spatial coordinate perpendicular to the surface
$\tilde{Y}_{el}$	Electric admittance ($\tilde{Y}_{el} = 1/\tilde{Z}_{el}$)
$\tilde{Z}$	Acoustic wave impedance (Table 4.1, mostly a shear-wave impedance)
$\tilde{Z}_{ac}$	Acoustic impedance (Table 4.1)
$\tilde{Z}_{el}$	Electric impedance (Table 4.1)
$\tilde{Z}_{EM}$	Electromagnetic wave impedance of plane waves (Table 4.1)
$\tilde{Z}_L$	Load impedance (Table 4.1)
$\tilde{Z}_{L,ijk}$	Load impedance in tensor form (Table 4.1)
Z_{LCR}	Electric impedance of an LCR-circuit (Eq. 4.5.15)
$\tilde{\tilde{Z}}_{liq}$	Acoustic wave impedance of a liquid ($\tilde{Z}_{liq} = \left(i\omega\rho_{liq}\eta_{liq}\right)^{1/2}$)
$\tilde{Z}_m$	Mechanical impedance (Table 4.1)
Z_P	Characteristic impedance of a particle coupled to the main resonator ($Z_P = (\kappa_P M_P)^{1/2}$, Sect. 4.6.3)
Z_q	Acoustic wave impedance of AT-cut quartz ($Z_q = 8.8 \times 10^6$ kg m^{-2} s^{-1})

Z_R	Characteristic impedance of a resonator ($Z_R = (\kappa_R M_R)^{1/2}$, Sect. 4.6.3)
$\tilde{Z}_{W,et}$	Electric wave impedance (Table 4.1)
Z_{11}, Z_{12}	Elements of the transmission line in Fig. 4.18
γ_D	Damping factor (has units of Hz, $\gamma_D = 2\pi\Gamma$)
Γ	Imaginary part of a resonance frequency
δf	A difference in frequency from the resonance frequency (Eq. 4.1.29)
δ_L	Loss angle ($\tan(\delta_L) = G''/G' = J''/J'$ often called $\tan(\delta)$ in rheology)
δ_q	Loss angle of the quartz plate (can be an effective parameter capturing losses other than the intrinsic viscous dissipation)
Δ	As a prefix: A shift induced by the presence of the sample
ε	A small quantity (In Taylor expansions)
$\tilde{\varepsilon}$	Dielectric permittivity ($\tilde{\varepsilon} = \tilde{\varepsilon}_r \varepsilon_0$)
ε_q	Dielectric constant of AT-cut quartz ($\varepsilon_q = 4.54$)
$\tilde{\varepsilon}_r$	Relative dielectric permittivity (Also: "dielectric constant")
ε_0	Dielectric permittivity of vacuum ($\varepsilon_0 = 8.854 \times 10^{-12}$ C/(Vm))
φ	Phase
ϕ	Factor converting between mechanical and electric quantities in the Mason circuit ($\phi = Ae_{26}/d_q$)
$\tilde{\eta}$	Viscosity ($\tilde{\eta} = \tilde{G}/(i\omega)$)
η_q	The "elastic viscosity" of AT-cut quartz, defined as $\eta_q = G_q''/\omega$. η_q is roughly independent of frequency (G_q'' is not). η_q depends on the defect density
κ_R	Spring constant of a *R*esonator
$\tilde{\mu}$	Magnetic permeability
μ_0	Magnetic permeability of vacuum ($\mu_0 = 4\pi \times 10^{-7}$ Vs/Am)
μ_r, $\tilde{\mu}_r$	Relative magnetic permeability
ρ	Density
ρ_q	Density of crystalline quartz ($\rho_q = 2.65$ g/cm^3)
σ	(Tangential) stress
σ_S	Tangential stress at the resonator surface (Also: "traction")
ξ_R	Drag coefficient of a *R*esonator (sometimes called "friction coefficient", not to be confused with the friction coefficient in tribology)
ω	Angular frequency

ω_0 Undamped angular resonance frequency ($\omega_0 = (\kappa_R/M_R)^{1/2}$)

ω_{LC} Undamped resonance frequency of an LCR-circuit ($\omega_{LC} = (LC)^{-1/2}$)

ω_P Particle resonance frequency of a coupled resonance ($\omega_P \approx (\kappa_P/M_P)^{1/2}$)

$\tilde{\omega}_r$ Angular resonance frequency ($\tilde{\omega}_r = 2\pi\tilde{f}_r = 2\pi(f_r + i\Gamma)$)

References

1. Musil, R.: The Confusions of Young Törless. Penguin Classics (2001) (first published 1906)
2. Woan, G.: The Cambridge Handbook of Physics Formulas. Cambridge University Press, Cambridge (2000)
3. Daikhin, L., Gileadi, E., Katz, G., Tsionsky, V., Urbakh, M., Zagidulin, D.: Influence of roughness on the admittance of the quartz crystal microbalance immersed in liquids. Anal. Chem. **74**(3), 554–561 (2002)
4. Rodahl, M., Hook, F., Krozer, A., Brzezinski, P., Kasemo, B.: Quartz-crystal microbalance setup for frequency and Q-factor measurements in gaseous and liquid environments. Rev. Sci. Instrum. **66**(7), 3924–3930 (1995)
5. Kochman, A., Krupka, A., Grissbach, J., Kutner, W., Gniewinska, B., Nafalski, L.: Design and performance of a new thin-layer radial-flow holder for a quartz crystal resonator of an electrochemical quartz crystal microbalance. Electroanalysis **18**(22), 2168–2173 (2006)
6. Reed, C.E., Kanazawa, K.K., Kaufman, J.H.: Physical Description of a Viscoelastically Loaded at-Cut Quartz Resonator. J. Appl. Phys. **68**(5), 1993–2001 (1990)
7. Voinova, M.V., Rodahl, M., Jonson, M., Kasemo, B.: Viscoelastic acoustic response of layered polymer films at fluid-solid interfaces: continuum mechanics approach. Phys. Scr. **59**(5), 391–396 (1999)
8. Johannsmann, D., Mathauer, K., Wegner, G., Knoll, W.: Viscoelastic properties of thin-films probed with a quartz-crystal resonator. Phys. Rev. B **46**(12), 7808–7815 (1992)
9. Bandey, H.L., Martin, S.J., Cernosek, R.W., Hillman, A.R.: Modeling the responses of thickness-shear mode resonators under various loading conditions. Anal. Chem. **71**(11), 2205–2214 (1999)
10. Nakamoto, T., Moriizumi, T.: A theory of a quartz crystal microbalance based upon a mason equivalent-circuit. Jpn. J. Appl. Phys. Part 1 Regul. Pap. Short Notes Rev. Pap. **29**(5), 963–969 (1990)
11. Mason, W.P.: Piezoelectric crystals and their applications to ultrasonics. Princeton, Van Nostrand (1948)
12. Kanazawa, K.K., Gordon, J.G.: Frequency of a quartz microbalance in contact with liquid. Anal. Chem. **57**(8), 1770–1771 (1985)
13. http://en.wikibooks.org/wiki/Linear_Algebra. Accessed 14 Feb 2013
14. Yeh, P.: Optical Waves in Layered Media. Wiley, New York (2005)
15. Katsidis, C.C., Siapkas, D.I.: General transfer-matrix method for optical multilayer systems with coherent, partially coherent, and incoherent interference. Appl. Opt. **41**(19), 3978–3987 (2002)
16. Salomaki, M., Loikas, K., Kankare, J.: Effect of polyelectrolyte multilayers on the response of a quartz crystal microbalance. Anal. Chem. **75**(21), 5895–5904 (2003)
17. Mason, W.P., Baker, W.O., McSkimin, H.J., Heiss, J.H.: Measurement of shear elasticity and viscosity of liquids at ultrasonic frequencies. Phys. Rev. **75**(6), 936–946 (1949)

18. Alig, I., Lellinger, D., Sulimma, J., Tadjbakhsch, S.: Ultrasonic shear wave reflection method for measurements of the viscoelastic properties of polymer films. Rev. Sci. Instrum. **68**(3), 1536–1542 (1997)
19. Azzam, R.M.A., Bashara, N.M.: Ellipsometry and Polarized Light. Springer, New York (1987)
20. Lekner, J.: Theory of Reflection of Electromagnetic and Particle Waves. Springer, Dordrecht (1987)
21. Homola, J.: Present and future of surface plasmon resonance biosensors. Anal. Bioanal. Chem. **377**(3), 528–539 (2003)
22. Thurston, R.N.: Piezoelectrically excited vibrations. In: Truesdell, C. (ed.) Mechanics of Solids, vol. 4, p. 257. Springer, Heidelberg (1984)
23. Royer, D., Dieulesaint, E.: Elastic Waves in Solids II: Generation, Acousto-optic Interaction, Applications. Springer, New York (1999)
24. Ulaby, F.T.: Fundamentals of Applied Electromagnetics. Prentice Hall, Upper Saddle River (2010)
25. http://en.wikipedia.org/wiki/Robin_boundary_condition. Accessed 28 Dec 2013
26. Granstaff, V.E., Martin, S.J.: Characterization of a thickness-shear mode quartz resonator with multiple nonpiezoelectric layers. J. Appl. Phys. **75**(3), 1319–1329 (1994)
27. Butterworth, S.: On electrically-maintained vibrations. Proc. Phys. Soc. London **27**, 410 (1914)
28. Dyke, K.V.: The piezo-electric resonator and its equivalent network. Proc. Inst. Radio Engin. **16**(6), 742 (1928)
29. Calvo, E.J., Etchenique, R., Bartlett, P.N., Singhal, K., Santamaria, C.: Quartz crystal impedance studies at 10 MHz of viscoelastic liquids and films. Faraday Discuss. **107**, 141–157 (1997)
30. Rast, S., Wattinger, C., Gysin, U., Meyer, E.: The noise of cantilevers. Nanotechnology **11**(3), 169–172 (2000)
31. Lee, I., Lee, J.: Measurement uncertainties in resonant characteristics of MEMS resonators. J. Mech. Sci. Technol. **27**(2), 491–500 (2013)
32. http://www.am1.us/Local_Papers/U11625%20VIG-TUTORIAL.pdf. Accessed 18 June 2014
33. http://txccrystal.com/term.html. Accessed 10 Aug 2013
34. Horowitz, P., Hill, W.: The Art of Electronics. Cambridge University Press, Cambridge (1989)
35. Gottlieb, I.M.: Practical Oscillator Handbook. Newnes, Oxford (1997)
36. Pechhold, W.: Zur Behandlung von Anregungs- und Störungsproblemen bei akustischen Resonatoren. Acustica **9**, 48–56 (1959)
37. Arnau, A.: Piezoelectric Transducers and Applications. Springer, Heidelberg (2004)
38. Johannsmann, D.: Viscoelastic, mechanical, and dielectric measurements on complex samples with the quartz crystal microbalance. Phys. Chem. Chem. Phys. **10**(31), 4516–4534 (2008)
39. Berg, S., Johannsmann, D.: Laterally coupled quartz resonators. Anal. Chem. **73**(6), 1140–1145 (2001)
40. Dybwad, G.L.: A sensitive new method for the determination of adhesive bonding between a particle and a substrate. J. Appl. Phys. **58**(7), 2789–2790 (1985)
41. Mason, W.P.: Electrical and mechanical analogies. Bell Syst. Tech. J. **20**, 405–414 (1941)
42. Meyers, M.A., Chawla, K.K.: Mechanical Behavior of Materials. Cambridge University Press, Cambridge (2008)
43. http://en.wikibooks.org/wiki/Engineering_Acoustics/Electro-Mechanical_Analogies. Accessed 26 July 2012
44. Gileadi, E.: Physical Electrochemistry: Fundamentals. A Textbook for Students of Science and Engineering, Techniques and Applications. Wiley, New York (2011)

Chapter 5
Piezoelectric Stiffening

Abstract The electrical impedance between the front and the back electrode of the resonator affects the frequency shift by a mechanism called piezoelectric stiffening. If the front electrode is not grounded well (that is, if electric fringe fields permeate the sample), the sample's electric and dielectric properties enter this impedance. They can be probed by switching between a grounded front electrode and a grounded back electrode. For the parallel plate, the formalism can be cast into a form, where the effects of a nonzero electric displacement are equivalent to a stress exerted onto the two resonator surface. Its influence on the frequency shift can be treated within the frame of the small load approximation (SLA).

5.1 General

To the normal QCM user, piezoelectricity is mostly a convenience, allowing for the detection of acoustic resonances by electrical means. If one owns an instrument with a well-grounded front electrode, there is no need to immerse oneself into the intricacies of piezoelectricity. Certain precautions need to be taken (Sect. 14.2), but once these safeguards are in place, one may happily ignore piezoelectric stiffening. The following section matters for readers, who are setting up their own instrument or who want to study the electrical properties of the sample by acoustical means.

Two notes:

- Piezoelectric stiffening is sometimes also discussed under the term "electro-mechanical coupling". The latter term also encompasses the action of the transformer in Fig. 4.9, which converts a voltage to a stress and vice versa. Piezoelectric stiffening denotes the dependence of a body's mechanical stiffness on the electrical boundary conditions.
- If piezoelectricity shall be avoided altogether, one can resort to magnetic actuation. This principle is implemented in the MARS resonator, [1, 2] where

D. Johannsmann, *The Quartz Crystal Microbalance in Soft Matter Research*, Soft and Biological Matter, DOI: 10.1007/978-3-319-07836-6_5,

"MARS" stands for "magnetic-acoustic resonator sensor". MARS resonators are more flexible than the piezoelectric resonators in that the plate can be made of arbitrary materials. Generally speaking, magnetic coupling is weaker than piezoelectric coupling. Also, MARS resonators require a static magnetic field. Coupling occurs by means of a coil evaporated onto the resonator surface.

Piezoelectricity does more than coupling stress to voltage. When a piezoelectric object is deformed, the electric field makes a contribution to the overall energy of deformation. In consequence, it also contributes to the crystal's shear stiffness. Under open-circuit conditions (more precisely, under conditions where the electric displacement, $\hat{D}$, vanishes everywhere), the fractional contribution of the electric field to the crystal's stiffness is equal to the square of the piezoelectric coupling coefficient, k_t^2 (Eq. 5.3.2). k_t^2 is about 0.8 % for AT-cut quartz. Roughly speaking, the influence of the resonator's electrical environment onto the resonance frequency is of the same order of magnitude. These effects are not at all small. They can amount to a frequency shift of a kHz and more.

From an application point of view, piezoelectric stiffening has the following consequences:

- The electrical boundary conditions must be carefully controlled. In frequency control applications, the resonator is usually encased in a metal housing. In sensing, one grounds the front electrode and shields the back electrode by a similar casing. A so-called pi-network may also be employed in order to diminish electrical influences (Sect. 14.2).
- If one wishes to tune the resonance frequency by electrical means, one can do so with an external capacitor. One can "pull" the resonance frequency. [3, 4] This phenomenon is the basis of the voltage-controlled crystal oscillator (VCXO).
- If one is interested in the electric impedance of a sample, one can access this impedance from the change of the resonance frequency upon switching between a grounded front electrode and a grounded back electrode (Sect. 14.1).

A side remark concerns the series resonance frequency and the oscillation frequency of an oscillator. A capacitance close to the resonator has two effects. It firstly shifts the series resonance frequency, f_s, because of piezoelectric stiffening. It secondly translates the admittance circle in the complex plane. In impedance analysis, all interpretation is based on f_s. The value of f_s is not affected by a translation of the admittance circle in the complex plane. This is different for oscillator circuits. The oscillation frequency is governed by Barkhausen condition; it is different from f_s (Sect. 2.1). A translation of the admittance cycle shifts the Barkhausen condition and thereby would change the oscillation frequency even if the series resonance frequency were unaffected. The two effects are to be distinguished.

5.2 Formal Structure of the Piezoelectric Equations

Piezoelectricity in its full generality fills entire books just by itself. [5, 6] It is a complicated matter because it couples two response functions (stiffness and polarizibility), which are both tensors. [7] Before explaining piezoelectric stiffening (in limited detail), the reader is reminded of a more widely known system with a similar type of coupling, which is the gas subjected to heat and pressure. The state of a gas depends on *two* external variables, which are temperature, *T*, and pressure, *p*. More precisely, these two variables can be *p* and *T*, but other variables can be chosen, as well. The state of the gas corresponds to a point in the *p*-*T* plane. A change of state amounts to a path in this plane. There are two conjugate variables, which are volume, *V*, and entropy, *S* (Do not confuse temperature with the stress tensor, named *T* in Table 5.1, and do not confuse entropy with the infinitesimal strain tensor, named *S* in Table 5.1). For reversible processes, entropy and volume are unique functions of temperature and pressure. For that reason, the state of a gas can also be specified by prescribing values for—let's say—*p* and *V*, rather than *p* and *T*. Any two of the four state variables *p*, *V*, *S*, and *T* can serve to specify the state of the system.

One can increase the temperature of a gas by supplying heat from the outside. Complicating the matter, the amount of heat needed to induce a certain change in temperature differs between conditions of constant volume and conditions of constant pressure. When under constant pressure, the gas expands upon heating, thereby doing mechanical work on its environment. The corresponding energy must be supplied as additional heat. There are two heat capacities, called "c_p" (at constant pressure) and "c_V" (at constant volume). Likewise, the compressibility depends on whether the compression occurs at constant temperature or at constant entropy.

Piezoelectricity has this same formal structure. The state of a piezoelectric material is specified by two external variables, which can be the stress, *T*, and electric field, **E**. The conjugate variables are the infinitesimal strain, *S*, and the electric displacement, **D** ("Infinitesimal" here means small. All strains are assumed to be small in this sense). As with the gas, there is a choice with regard to the state variables. Also, the amount of stress needed to achieve a certain deformation depends on whether the deformation occurs at constant electric field or at constant electric displacement (or under some other electrical boundary condition). There a two different stiffness tensors (labeled *c* in Table 5.1). The tensor c_E applies under conditions of constant electric field, while the tensor c_D applies under conditions of constant electric displacement. Likewise, the dielectric permittivity, ε, depends on whether the electric field is applied under conditions of constant strain or of constant stress (or under some other mechanical boundary condition).

Again: This is only a formal analogy. c_E and c_D in piezoelectricity are different because the stiffness depends on whether the electric field or the electric displacement remain constant during the deformation. Likewise, the heat capacity in thermodynamics (c_P or c_V) depends on whether pressure or volume are maintained constant. Apart from the analogy, there is no relation between the heat capacity in gases and mechanical stiffness in piezoelectric materials.

Table 5.1 Different forms of the constitutive equations of a piezoelectric material and conversions as given in Ref. [8]

Constitutive equations	Conversions
Strain-Charge Form $S = s_E \cdot T + d^t \cdot \mathbf{E}$ $\mathbf{D} = d + \varepsilon_T \cdot \mathbf{E}$	Strain-Charge to Stress-Charge $c_E = s_E^{-1}, \qquad e = d \cdot s_E^{-1}$ $\varepsilon_S = \varepsilon_T - d \cdot s_E^{-1} \cdot d^t$
Strain-Voltage Form $S = s_D \cdot T + g^t \cdot \mathbf{D}$ $\mathbf{E} = -g \cdot T + \varepsilon_T^{-1} \cdot \mathbf{D}$	Stress-Charge to Stress-Voltage $h = \varepsilon_S^{-1} \cdot e$ $c_D = c_E + e^t \cdot \varepsilon_S^{-1} \cdot e$
Stress-Charge Form $T = c_E \cdot S - e^t \cdot \mathbf{E}$ $\mathbf{D} = e \cdot S + \varepsilon_S \cdot \mathbf{E}$	Strain-Charge to Strain-Voltage $g = \varepsilon_T^{-1} \cdot d$ $s_D = s_E - d^t \cdot \varepsilon_T^{-1} \cdot d$
Stress-Voltage Form $T = c_D \cdot S - h^t \cdot \mathbf{D}$ $\mathbf{E} = -h \cdot S + \varepsilon_S^{-1} \cdot \mathbf{D}$	Strain-Volage to Stress-Voltage $c_D = s_D^{-1}, \qquad h = g \cdot s_D^{-1}$ $\varepsilon_S^{-1} = \varepsilon_T^{-1} + g \cdot s_D^{-1} \cdot g^t$

S: Infinitesimal strain tensor (in 1D: du/dz) T: Stress tensor (in 1D: σ)
E: Electric field **D**: Electric displacement
c: Stiffness tensor (in 1D: G) s: Compliance tensor (in 1D: J)
ε: Dielectric permittivity (a tensor)
d, e, h, g: Piezoelectric coefficients (tensors)
(in 1D: d_{26}: Piezoelectric strain coefficient, e_{26}: Piezoelectric stress coefficient)
Subscripts E, D, S, or T: Quantities remaining constant
Center-dot ($\cdot$): Matrix multiplication in the Voigt sense (Sect. 5.4)
Superscript t: Matrix transpose in the Voigt sense

The naming of the variables and the signs are as in the source. Hats (for amplitudes) and tildes (for complex numbers) have been omitted

Table 5.1 collects the constitutive equations of piezoelectricity in four different forms. Figure 5.1 sketches the various interrelations in graphical form. The reader may wonder why the coefficients of the cross-terms (d, g, e, and h) all appear twice. For instance, the piezoelectric strain coefficient, d, (part of strain-charge form, first entry in Table 5.1) appears both on the upper right and the lower left of the equation system. The fact that the two cross coefficients are the same, is an analogy to the Maxwell relations in thermodynamics.

In the context of piezoelectric stiffening, the most noteworthy feature in Table 5.1 is the difference between the piezoelectrically stiffened stiffness tensor, c_D, (the modulus at constant electric displacement, D) and the stiffness tensor at constant field, c_E. The two are related as

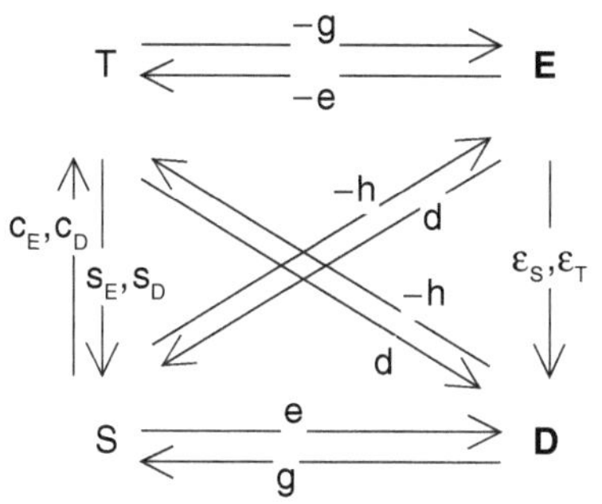

Fig. 5.1 Diagrammatic representation of the constitutive equations of piezoelectricity. Any two parameters out of *T*, *S*, **E**, and **D** (stress, strain, electric field, and electric displacement) can specify the state of the system. The two other state variables are fixed by the constitutive equations. As an exercise, the reader may draw the corresponding diagram for gases, where the state variables are pressure, temperature, volume, and entropy

$$c_D = c_E + e^t \cdot \varepsilon_S^{-1} \cdot e \tag{5.2.1}$$

e is the piezoelectric stress coefficient (a tensor). ε_S is the dielectric permittivity at constant strain. ε_S^{-1} also is a tensor. The difference between c_D and c_E is the essence of piezoelectric stiffening.

Having emphasized the formal analogy between the piezoelectric solid and the ideal gas, it is worthwhile to outline conceptual differences:

- For solids, there is the possibility of vanishing coupling between electrical and mechanical quantities. Actually, this is more often the case than not. Piezoelectric coupling is symmetry-forbidden for all materials with inversion symmetry. A gas, which would not heat up when adiabatically compressed, does not exist.
- The distinction between reversible and irreversible processes (a time-honored complication in thermodynamics) is of less relevance for piezoelectric solids. For piezoelectric solids, linear response usually holds and irreversible processes are covered by making the coupling coefficients (*d, e, h,* and *g* in Table 5.1) complex. The coupling coefficients are complex, in principle, but the imaginary parts are small and we ignore them, whenever ignoring them makes the discussion simpler. For example, the piezoelectric coupling coefficient (see Eq. 5.3.2) is viewed as a real parameter.
- For the piezoelectric solid, the response functions (all lower-case parameters in Table 5.1) are tensor, while they are scalars for the gas. Tensor algebra is avoided here as much as possible by basing most of the discussion on the parallel plate.
- For the piezoelectric solid, the state variables may depend on position. That is the case if the object of interest vibrates. Stress not only is a tensor, it is a tensor *field* ($T_{ij}(x,y,z)$). Thi sounds complicated (and it is), but the matter is well understood.

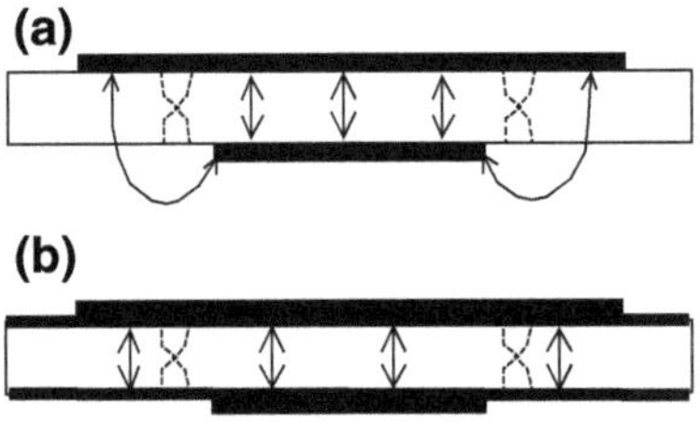

Fig. 5.2 With regard to acoustics, approximating the crystal as a parallel plate is successful because most of the strain occurs along z. The electric field, on the other hand, has strong non-normal components (Panel **a**). This might be changed, in principle. One can metalize the entire resonator and apply a second, thick metal layer to the center of the back electrode, which will provide for energy trapping. In this way, the electric field is made to point along the vertical and the stray capacitances are reduced

5.3 Piezoelectric Stiffening of the Parallel-Plate

A caveat at the beginning: The parallel-plate model is less appropriate for electric fields than for deformation, because the electric field usually has strong non-normal components. The electrodes usually do not cover the entire crystal; there are "fringe fields" (Fig. 5.2). Fortunately, the formalism covering fringe fields (Sect. 6.1.7) is not as difficult as one might think. It does require a finite element calculation, but the effort is moderate.

The following section derives the shift of the resonance frequency of a parallel plate caused by piezoelectric stiffening. For the parallel-plate geometry, the constitutive relations of piezoelectricity in the stress-voltage form are

$$\begin{aligned} \hat{\sigma} &= \tilde{G}_q \frac{\mathrm{d}\hat{u}}{\mathrm{d}z} - \frac{\tilde{e}_{26}}{\tilde{\varepsilon}_q \varepsilon_0}\hat{D} && \text{(a)} \\ \hat{E} &= -\frac{\tilde{e}_{26}}{\tilde{\varepsilon}_q \varepsilon_0}\frac{\mathrm{d}\hat{u}}{\mathrm{d}z} + \frac{1}{\tilde{\varepsilon}_q \varepsilon_0}\hat{D} && \text{(b)} \end{aligned} \qquad (5.3.1)$$

Variables have been renamed compared to Table 5.1 in order to conform with the notation in the rest of the book. The tangential stress T_{xz} has turned into $\hat{\sigma}$; the strain is $\mathrm{d}\hat{u}/\mathrm{d}z$ (Sect. 5.4); the relevant component of the stiffness tensor is $\tilde{G}_q$, and the relevant component of the permittivity tensor is $\tilde{\varepsilon}_q\ \varepsilon_0.\tilde{\varepsilon}_q$ is dielectric constant at constant strain ($A\tilde{\varepsilon}_q\ \varepsilon_0/d_q$ is also called "clamped capacitance" in order to express the constant-strain condition). $\tilde{h}_{26}$ (the relevant component of the h-tensor) was replaced by $\tilde{e}_{26}/(\tilde{\varepsilon}_q\ \varepsilon_0)$. Hats and tildes were re-inserted for amplitudes and complex response functions. $\hat{D}$ and $\hat{E}$ are scalars, equal to the z-components of the corresponding vectors.

Equation 5.2.1 simplifies to

$$\tilde{G}_q = \tilde{G}_E + \frac{\tilde{e}_{26}^2}{\tilde{\varepsilon}_q \varepsilon_0} = \tilde{G}_E \left(1 + \frac{\tilde{e}_{26}^2}{\tilde{G}_E \tilde{\varepsilon}_q \varepsilon_0}\right) \approx G_E \left(1 + \frac{e_{26}^2}{G_E \varepsilon_q \varepsilon_0}\right) = G_E\left(1 + k_t^2\right) \tag{5.3.2}$$

The index E denotes constant electric field. The corresponding tensor is c_E (G_q might also be called G_D, where the index D would denote constant electric displacement). The combination of parameters $e_{26}/(G_E\varepsilon_q\varepsilon_0)^{1/2}$ is the piezoelectric coupling coefficient, k_t (More precisely, k_t is defined as $e_{26}/(G_q\varepsilon_q\varepsilon_0)^{1/2} = e_{26}/(G_D\varepsilon_q\varepsilon_0)^{1/2}$, but the difference is of little relevance here. See Ref. [9] for a discussion). k_t^2 is about 0.8 % for AT-cut quartz. Since k_t usually is viewed as real, the tilde was omitted for complex parameters to the right in Eq. 5.3.2.

The coupling coefficient can also be expressed in terms of energies. If a piezoelectric device is excited mechanically, k_t^2 is the fraction of the mechanical energy converted to electrical energy. Conversely, if a piezoelectric device is excited electrically, k_t^2 is the fraction of the electric energy converted to mechanical energy. The two effects are sometimes distinguished as the "piezoelectric" and the "inverse piezoelectric" effect. Since the coupling coefficients are the same, one might as well not talk of an "inverse" effect and remember that the coupling works both ways.

The stress-voltage representation was chosen above because the model of the parallel plate becomes particularly simple with $\hat{u}$ and $\hat{D}$ being the independent variables. Most resonators are insulators and the charge density therefore vanishes everywhere in the bulk. By virtue of Maxwell's first law, this implies vanishing divergence of the electric displacement ($\nabla \cdot \hat{D} = 0$), which in the parallel-plate geometry amounts to $\mathrm{d}\hat{D}/\mathrm{d}z = 0$ and $\hat{D} = const.$ (Actually, Langasite *does* have a finite ionic conductance at high temperature. [10] This case is left aside). The force density is the divergence of the stress, which in 1D is given as

$$\hat{f} = \frac{\partial \hat{\sigma}}{\partial z} = \frac{\mathrm{d}}{\mathrm{d}z}\left(\tilde{G}_q \frac{\mathrm{d}\hat{u}}{\mathrm{d}z} - \frac{\tilde{e}_{26}}{\tilde{\varepsilon}_q \varepsilon_0}\hat{D}\right) \tag{5.3.3}$$

Do not confuse force density with frequency, both named f. In Eq. 5.3.3, the response functions ($\tilde{G}_q$ and $\tilde{e}_{26}/(\tilde{\varepsilon}_q\varepsilon_0)$) are under the outer derivative. If the response functions vary with position, their derivative must be included in the calculation of the force density. This will be essential later, when evaluating Eq. 5.3.3 at the surface. At the surface $\tilde{h}_{26} = \tilde{e}_{26}/(\varepsilon_q\ \varepsilon_0)$ varies rapidly.

In the bulk, $\tilde{G}_q$ and $\tilde{e}_{26}/(\tilde{\varepsilon}_q\ \tilde{\varepsilon}_0)$ are constant and one can write

$$\hat{f} = \tilde{G}_q \frac{\mathrm{d}^2\hat{u}}{\mathrm{d}z^2} - \frac{\tilde{e}_{26}}{\tilde{\varepsilon}_q \varepsilon_0}\frac{\mathrm{d}\hat{D}}{\mathrm{d}z} = \tilde{G}_q \frac{\mathrm{d}^2\hat{u}}{\mathrm{d}z^2} \tag{5.3.4}$$

Vanishing divergence of the electric displacement was used in step 2. The wave equation is

$$-\omega^2 \hat{u} = \frac{\hat{f}}{\rho_q} = \frac{\tilde{G}_q}{\rho_q} \frac{d^2 \hat{u}}{dz^2} \quad (5.3.5)$$

Clearly, the wave equation is unaffected by $\hat{D}$. This simplification motivates the use of the stress-voltage form. Again, the modulus to be used in the wave equation is the *piezoelectrically stiffened* modulus. The piezoelectrically stiffened modulus must also be used in the calculation of the speed of sound and the shear-wave impedance.

The following discussion distinguishes three cases, which are the open circuit condition, short-circuited electrodes, and the intermediate case with a finite electric impedance between the electrodes.

5.3.1 Open-Circuit Condition

In a first step, assume $\hat{D} = 0$ everywhere. For the sake of this argument, let electrodes be absent. One might also consider them to be very thin and not connected to the outside. This corresponds to Fig. 5.3a with infinite $\tilde{Z}_{ex}$. When requiring $\hat{D} = 0$ everywhere, this includes the space outside the crystal. Outside the crystal one has $\hat{D} = \tilde{\varepsilon}_q \, \varepsilon_0 \hat{E}$ and the condition $\hat{D} = 0$ therefore implies $\hat{E} = 0$, as well. The condition $\hat{E} = 0$ everywhere outside the crystal is somewhat artificial. In realistic configurations, there will always be an electric field between the resonator surface and its housing. If there are neither electrodes nor an electric field outside the crystal, the resonator does not communicate with the outside. It still has a resonance frequency. $\hat{D} = 0$ is a good starting point for the discussion, it is not a practical reference state. With $\hat{D} = 0$, the plate behaves as if it were non-piezoelectric. The index *OC* (open circuit) in Sect. 4 denotes a state with all piezoelectric effects absent.

For later use we calculate the electric potential at the surface of the plate. Note: $\hat{E}$ is nonzero inside the resonator even if it is zero outside. Inside the resonator, one has $\hat{D} = \tilde{\varepsilon}_q \, \varepsilon_0 \, \hat{E} + \hat{P}$ with $\hat{P}$ the piezoelectrically-induced polarization. Inside the crystal, $\hat{D} = 0$ implies that the electric field compensates the polarization $(\hat{E} = -\hat{P}/(\tilde{\varepsilon}_q \, \varepsilon_0))$. For the following discussion, assume the acoustic load to be small. The displacement pattern then is almost antisymmetric in z. It is given as $\hat{u}(z) \approx \hat{u}_S \sin(k_q z)$. From Eq. 5.3.1 it follows that the electric field is

$$\hat{E} = -\frac{\tilde{e}_{26}}{\tilde{\varepsilon}_q \varepsilon_0} \frac{d}{dz} \left(\hat{u}_S \sin(k_q z) \right) \quad (5.3.6)$$

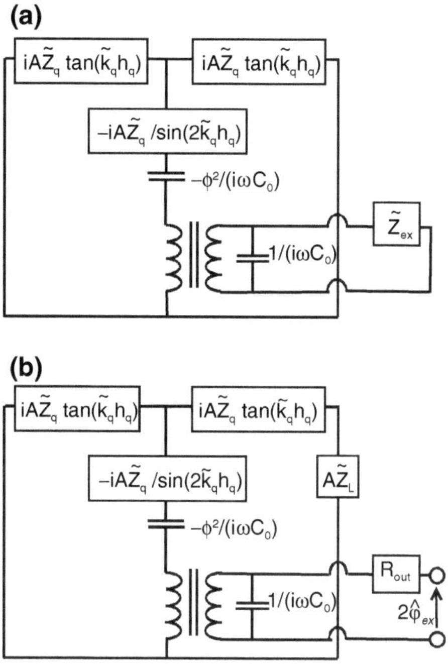

Fig. 5.3 **a** The load-free Mason circuit in ring-down configuration. The electrical port is closed, but it is closed across an external impedance $\tilde{Z}_{ex}$. **b** The Mason circuit with an acoustic load and, also, an output resistance inserted between the voltage source and the transformer. *Note* The external electrical impedance might comprise elements other than R_{out}

Integration and the relation $\hat{E} = -\mathrm{d}\hat{\varphi}/\mathrm{d}z$ lead to

$$\hat{\varphi}_S = \frac{\tilde{e}_{26}}{\tilde{\varepsilon}_q \varepsilon_0} \hat{u}_S \tag{5.3.7}$$

The relation $\sin(k_q h_q) = 1$ was used. The integration constant is zero for reasons of symmetry.

5.3.2 *Short-Circuited Electrodes*

With the electrodes short-circuited, charge flows into the electrodes until the voltage between the electrodes has turned to zero. For now, consider the ring-down configuration, where there is no external voltage (Fig. 5.3a with $\tilde{Z}_{ex} = 0$). A voltage source will be introduced later (Fig. 5.3b). The total charge, $\tilde{Q}_S$, follows from $\hat{\varphi}_S$, Eq. 5.3.7, and the electrical capacitance, C_0, as

$$\hat{Q}_S = 2C_0 \hat{\varphi}_S = \frac{2\tilde{e}_{26} A}{d_q} \hat{u}_S = \frac{\tilde{e}_{26} A}{h_q} \hat{u}_S \tag{5.3.8}$$

Step 1 used that the voltage across the electrodes under open-circuit conditions is given by $2\hat{\varphi}_S$. The charge terminates the electric displacement and the charge per unit area is therefore equal to $\hat{D}$:

$$\hat{D} = \frac{\hat{Q}_S}{A} = \tilde{e}_{26}\frac{\hat{u}_S}{h_q} \tag{5.3.9}$$

At this point, one has to remember that the parameter $\tilde{h}_{26} = \tilde{e}_{26}/(\tilde{\varepsilon}_q\, \varepsilon_0)$ in Eq. 5.3.1 is discontinuous at the surface. The nonzero displacement gives rise to a piezoelectrically generated surface traction of

$$\hat{\sigma}_{PESC} = -\int_{h_q-\varepsilon}^{h_q+\varepsilon} -\frac{d}{dz}\left(\frac{\tilde{e}_{26}}{\tilde{\varepsilon}_q\varepsilon_0}\right)\hat{D}dz = \left[\frac{\tilde{e}_{26}}{\tilde{\varepsilon}_q\varepsilon_0}\right]_{h_q-\varepsilon}^{h_q+\varepsilon}\hat{D} = -\frac{\tilde{e}_{26}^2}{\tilde{\varepsilon}_q\varepsilon_0 h_q}\hat{u}_S \tag{5.3.10}$$

The index *PESC* stands for *P*iezo*E*lectric stiffening with the electrodes *S*hort-*C*ircuited. The first minus sign occurs because the stress to be inserted into the SLA is exerted by the resonator onto the sample (that is, into the direction of positive z). The integration is carried out from a location just below the interface to a location just above the interface, where ε is a small distance.

The stress divided by the velocity, $\hat{v}_S$, yields an acoustic impedance, termed $\tilde{Z}_{PESC}$. A factor of 2 is incorporated into its definition in order to account for the fact that the stress $\hat{\sigma}_{PESC}$ acts at *both* surfaces:

$$\tilde{Z}_{PESC} = \frac{2\hat{\sigma}_{PESC}}{\hat{v}_S} \tag{5.3.11}$$

With this definition, the piezoelectrically-induced frequency shift takes the form familiar from the SLA:

$$\frac{\Delta\tilde{f}_{PESC}}{f_0} = \frac{\tilde{f}_{rPESC} - f_{OC}}{f_0} = \frac{i}{\pi Z_q}\tilde{Z}_{PESC} = \frac{i}{\pi Z_q}\frac{2\hat{\sigma}_{PESC}}{\hat{v}_S} \tag{5.3.12}$$

Inserting Eq. 5.3.10 yields

$$\frac{\Delta\tilde{f}_{PESC}}{f_0} = \frac{i}{\pi Z_q}\tilde{Z}_{PESC} = \frac{i}{\pi Z_q}\frac{-4\tilde{e}_{26}^2}{i\omega\tilde{\varepsilon}_q\varepsilon_0 d_q} = \frac{i}{\pi Z_q}\frac{1}{A}\frac{-4\phi^2}{i\omega C_0} \tag{5.3.13}$$

The relations $\phi = Ae_{26}/d_q$ and $C_0 = A\hat{\varepsilon}_q\, \varepsilon_0/d_q$ were used in step 3. Equation 5.3.13 shows that the element to be inserted into the vertical bar of the Mason circuit is $-\phi^2/(i\omega C_0)$. Short-circuiting the electrodes *lowers* the resonance frequency. The electric field induced by the charge in the electrodes counteracts the piezo-electrically generated electric field inside the crystal and thereby lowers the electric contribution to the energy of deformation.

The effects of piezoelectric stiffening (where "piezoelectric softening" would arguably be the more appropriate term) can also be expressed in terms of the piezoelectric coupling coefficient as

$$\begin{aligned}\frac{\Delta\tilde{f}_{PESC}}{f_0} &= \frac{\mathrm{i}}{\pi Z_q}\frac{-4\tilde{e}_{26}^2}{\mathrm{i}\omega\tilde{\varepsilon}_q\varepsilon_0 d_q} = \frac{-1}{\pi Z_q}\frac{4\tilde{e}_{26}^2}{2\pi n f_0\tilde{\varepsilon}_q\varepsilon_0}\frac{2\rho_q f_0}{Z_q} = \frac{-1}{n\pi^2\tilde{G}_q\rho_q}\frac{4\tilde{e}_{26}^2}{\tilde{\varepsilon}_q\varepsilon_0}\rho_q \\ &= \frac{-4k_t^2}{n\pi^2}\end{aligned} \tag{5.3.14}$$

On the level of the four-element circuit, the effects of short-circuiting can be lumped into the motional capacitance. The thus-corrected motional capacitance is called $\bar{C}_1$ in Sect. 4.5.3. One finds

$$\begin{aligned}\frac{1}{\mathrm{i}\omega\bar{C}_1} &= \frac{1}{4\phi^2}\left(\frac{\kappa_R}{\mathrm{i}\omega} - \frac{4\phi^2}{\mathrm{i}\omega C_0}\right) = \frac{1}{4\phi^2}\frac{\kappa_R}{\mathrm{i}\omega}\left(1 - 4\phi^2\frac{1}{C_0}\frac{1}{\kappa_R}\right) \\ &= \frac{1}{4\phi^2}\frac{\kappa_R}{\mathrm{i}\omega}\left(1 - 4\frac{A^2e_{26}^2}{d_q^2}\frac{d_q}{A\varepsilon_q\varepsilon_0}\frac{2d_q}{(n\pi)^2 A G_q}\right) \\ &= \frac{1}{4\phi^2}\frac{\kappa_R}{\mathrm{i}\omega}\left(1 - \frac{8k_t^2}{(n\pi)^2}\right)\end{aligned} \tag{5.3.15}$$

This result leads to Eq. 4.5.22.

5.3.3 Finite Electrical Impedance Between Electrodes

For real resonators, there always is an external electrical impedance. Figure 4.10 is misleading in this regard because it draws the voltage source as being ideal. An ideal voltage source would have vanishing output resistance. It would supply whatever current is needed in order to let the voltage across the device-under-test be equal to the prescribed value. Impedance analyzers are never ideal. Most of the time, the output resistance is in the range of 50 Ω. R_1 of a crystal in air can be less than this (typically a few ohms to a few tens of ohms, depending on size). On resonance, a large part of the voltage nominally supplied by the analyzer drops off at the output resistance. Accounting for the output resistance is important.

With a finite impedance between the electrodes, $\tilde{Z}_{ex}$, the relation between surface charge and surface potential is given as

$$2\hat{\varphi}_S = \tilde{Z}_{ex}\hat{I} = \tilde{Z}_{ex}\mathrm{i}\omega\hat{Q}_S = \tilde{Z}_{ex}\mathrm{i}\omega A\hat{D} \tag{5.3.16}$$

$\tilde{Z}_{ex}$ might be dominated by the output resistance of the impedance analyzer (Fig. 5.3b), but stray capacitances and the inductance of the cables also play a role. From Eq. 5.3.1, it follows that the electric field is

$$\hat{E} = -\frac{\tilde{e}_{26}}{\tilde{\varepsilon}_q \varepsilon_0} \frac{\mathrm{d}}{\mathrm{d}z} \left(\hat{u}_S \sin\left(k_q z\right)\right) + \frac{1}{\tilde{\varepsilon}_q \varepsilon_0} \hat{D} \tag{5.3.17}$$

Integration and the relation $\hat{E} = -\mathrm{d}\hat{\varphi}/\mathrm{d}z$ lead to

$$\hat{\varphi}_S = \frac{\tilde{e}_{26}}{\tilde{\varepsilon}_q \varepsilon_0} \hat{u}_S - \frac{h_q}{\tilde{\varepsilon}_q \varepsilon_0} \hat{D} \tag{5.3.18}$$

The constants of integration are zero for reasons of symmetry. Equating $\hat{\varphi}_S$ from Eq. 5.3.16 to $\hat{\varphi}_S$ from Eq. 5.3.18, one finds

$$\frac{1}{2} \tilde{Z}_{ex} \mathrm{i}\omega A \hat{D} = \frac{\tilde{e}_{26}}{\tilde{\varepsilon}_q \varepsilon_0} \hat{u}_S - \frac{h_q}{\tilde{\varepsilon}_q \varepsilon_0} \hat{D} \tag{5.3.19}$$

and

$$\hat{D} = \frac{\tilde{e}_{26}}{h_q + \frac{\tilde{Z}_{ex} \mathrm{i}\omega A \tilde{\varepsilon}_q \varepsilon_0}{2}} \hat{u}_S = \frac{1}{1 + \tilde{Z}_{ex} \mathrm{i}\omega C_0} \frac{\tilde{e}_{26}}{h_q} \hat{u}_S \tag{5.3.20}$$

The relation $C_0 = A\varepsilon_q \varepsilon_0/(2h_q)$ was used in step 2. As in the short-circuited configuration, the electric displacement generates a surface traction equal to $\hat{\sigma}_{PE} = -\tilde{h}_{26}\hat{D} = -\tilde{e}_{26}/(\tilde{\varepsilon}_q\, \varepsilon_0)\hat{D}$. With $\hat{D}$ from Eq. 5.3.20, the surface traction becomes

$$\begin{aligned} \hat{\sigma}_{PE} &= \frac{\tilde{e}_{26}^2}{\mathrm{i}\omega h_q \tilde{\varepsilon}_q \varepsilon_0} \frac{1}{\tilde{Z}_{ex} \mathrm{i}\omega C_0 + 1} \mathrm{i}\omega \hat{u}_S \\ &= \frac{2}{A} \phi^2 \left(\frac{1}{\mathrm{i}\omega C_0} \frac{1}{\tilde{Z}_{ex} \mathrm{i}\omega C_0 + 1} \right) \hat{v}_S \\ &= \frac{2}{A} \phi^2 \left(-\frac{1}{\mathrm{i}\omega C_0} + \left(\mathrm{i}\omega C_0 + \tilde{Z}_{ex}^{-1} \right)^{-1} \right) \hat{v}_S \end{aligned} \tag{5.3.21}$$

The relations $\phi = Ae_{26}/d_q$ and $C_0 = A\varepsilon_q \varepsilon_0/d_q$ were used in line 2. In line 3, the term in brackets was rearranged, such that one can recognize the corresponding elements in the equivalent circuit (Fig. 5.3a).

As before, the stress divided by the velocity yields an acoustic impedance and a factor of 2 is included into its definition, so that the $\Delta \tilde{f}$ later takes the familiar form:

$$\tilde{Z}_{PE} = \frac{2\hat{\sigma}_{PE}}{\hat{v}_S} \tag{5.3.22}$$

The piezoelectrically-induced frequency shift is

$$\frac{\tilde{f}_{rPE} - f_{OC}}{f_0} = \frac{\mathrm{i}}{\pi Z_q} \tilde{Z}_{PE} = \frac{\mathrm{i}}{\pi Z_q} \frac{2\hat{\sigma}_{PE}}{\hat{v}_S} \tag{5.3.23}$$

Note that the term $\tilde{f}_{r,PE} - f_{OC}$ was not called $\Delta\tilde{f}_{PE}$ because the open circuit condition is not a practical reference state. Accounting for this complication, $\Delta\tilde{f}_{PE}$ is defined as

$$\begin{aligned} \frac{\Delta\tilde{f}_{PE}}{f_0} &= \frac{\Delta\tilde{f}_{rPE} - \Delta\tilde{f}_{rPE,ref}}{f_0} = \frac{\mathrm{i}}{\pi Z_q} \frac{2\left(\hat{\sigma}_{PE} - \hat{\sigma}_{PE,ref}\right)}{\hat{v}_S} \\ &= \frac{\mathrm{i}}{A\pi Z_q} 4\phi^2 \left(\left(\mathrm{i}\omega C_0 + \tilde{Z}_{ex}^{-1}\right)^{-1} - \left(\mathrm{i}\omega C_0 + \tilde{Z}_{ex,ref}^{-1}\right)^{-1} \right) \end{aligned} \tag{5.3.24}$$

Equation 5.3.24 is the basis for the frequency-based determination of a sample's electrical impedance (Sect. 14.1).

With nonzero $\tilde{Z}_{ex}$ the motional capacitance becomes

$$\bar{C}_1 = C_1 \left(1 - \frac{8k_t^2}{(n\pi)^2} \frac{1}{\tilde{Z}_{ex} \mathrm{i}\omega C_0 + 1} \right)^{-1} \tag{5.3.25}$$

Note: $\bar{C}_1$ depends on elements *outside* of the resonator.

Of special relevance is the situation, where the analyzer's output resistance is high and where there is a second capacitor close to the resonator. The second capacitor might be a load capacitor inserted with the aim of pulling the resonance frequency. It also might be a double layer capacitance, occurring when the crystal is loaded with a conductive liquid. With $\tilde{Z}_{ex} = 1/(\mathrm{i}\omega C_L)$ and $\tilde{Z}_{ex,ref} = \mathbb{N}$, Eq. 5.3.24 reads as

$$\begin{aligned} \frac{\Delta\tilde{f}_{PE}}{f_0} &= \frac{\mathrm{i}}{A\pi Z_q} 4\phi^2 \left(\left(\mathrm{i}\omega(C_0 + C_L)\right)^{-1} - \left(\mathrm{i}\omega C_0\right)^{-1} \right) \\ &= \frac{1}{A\pi Z_q} \frac{4A^2 e_{26}^2}{d_q^2} \frac{1}{\omega} \frac{-C_L}{(C_0 + C_L) C_0} \\ &= -\frac{1}{\pi Z_q} \frac{4A}{d_q^2} \tilde{G}_q \tilde{\varepsilon}_q \varepsilon_0 \frac{e_{26}^2}{\tilde{\varepsilon}_q \varepsilon_0 \tilde{G}_q} \frac{1}{\omega} \frac{C_L}{(C_0 + C_L) C_0} \\ &= -\frac{4}{n\pi^2} \frac{\tilde{G}_q}{2 d_q Z_q} k_t^2 \frac{1}{\omega} \frac{C_L}{C_0 + C_L} = -\frac{4k_t^2}{n\pi^2} \frac{C_L}{C_0 + C_L} \end{aligned} \tag{5.3.26}$$

Equation 5.3.26 is often shown with the frequency shift normalized to the reference frequency at the respective overtone order, rather than the fundamental. Further, one often converts k_t^2 to the capacitance ratio, using Eq. 4.5.25. One finds

$$\frac{\Delta \tilde{f}_{PE}}{f} = \frac{4k_t^2}{(n\pi)^2} \frac{-C_L}{C_0 + C_L} = -\frac{1}{2} \frac{C_1}{C_0} \frac{C_L}{C_0 + C_L} \tag{5.3.27}$$

Because n^2 appears in the denominator, overtone crystals are not suitable for pulling with a load capacitance. The relation is further simplified if the reference state is the state with infinite load capacitance. The widely quoted result is

$$\frac{\Delta \tilde{f}_{PE}}{f} - \frac{\Delta \tilde{f}_{PE,inf}}{f} = \frac{1}{2} \frac{C_1}{C_0 + C_L} \tag{5.3.28}$$

5.4 Appendix: Voigt Notation

Stress and infinitesimal strain are $2nd$-rank tensors, in principle; and the stiffness tensor therefore is a $4th$-rank tensor. The piezoelectric tensors (d, g, e, and h, see Table 5.1) are $3rd$-rank tensors. However, stress and strain are symmetric tensors. They only have six independent components, not nine. Voigt proposed to collect these six values in a one-dimensional array, following the scheme

$$\begin{aligned} T &= (T_1, T_2, T_3, T_4, T_4, T_6) = \left(T_{xx}, T_{yy}, T_{zz}, T_{yz}, T_{xz}, T_{xy}\right) \\ S &= (S_1, S_2, S_3, S_4, S_4, S_6) = \left(S_{xx}, S_{yy}, S_{zz}, 2S_{yz}, 2S_{xz}, 2S_{xy}\right) \end{aligned} \tag{5.4.1}$$

Note the factor of 2 in line 2. The factor of 2 is the consequence of the fact that the *symmetrized* infinitesimal strain enters the calculation of the stress (cf. Sects. 6. 1.1 and 6.1.8). The infinitesimal strain tensor is defined as

$$S_{ij} = \frac{1}{2} \left(\frac{\partial u_i}{\partial x_j} + \frac{\partial u_j}{\partial x_i} \right) \tag{5.4.2}$$

In Voigt notation the stiffness tensor turns into a 6 × 6 matrix. This matrix should not be called a tensor because is does not transform like a tensor when the coordinate system is changed. For the same reason, T and S as written in Eq. 5.4.1 are not vectors in the narrow sense of the word. The piezoelectric tensors turn into 3 × 6 matrices. For example, e_{26} connects the y-component of the electric field to the xy-component of the stress tensor (which is a linear array in Voigt notation). Voigt notation is used in Table 5.1. Tensor notation is used in Sects. 6.1.1 and 6.1.8.

Glossary

Variable	Definition (Comments)
26	As an index: component of piezoelectric coupling tensor applicable to AT-cut quartz (Table 5.1)
A	(Effective) area of the resonator plate
c	Stiffness tensor (Table 5.1)
c_p, c_V	Specific heat capacity at constant pressure and constant volume
C_L	Load capacitance (inserted with the aim of pulling the resonance)
C_0	Parallel capacitance
C_1	Motional capacitance
$\bar{C}_1$	Motional capacitance taking piezoelectric stiffening into account
d	One of the tensors quantifying piezoelectric coupling (Table 5.1, piezoelectric strain coefficient)
d_{26}	Relevant component of the d-tensor (piezoelectric strain coefficient, $d_{26} = 3.1 \times 10^{-12}$ m/V for AT-cut quartz)
d_q	Thickness of the resonator ($d_q = m_q/\rho_q = Z_q/(2\rho_q f_0)$)
$\hat{D}$, D	Electric *D*isplacement (if bold: a vector; if not bold: z-component of $\hat{\mathbf{D}}$ or $\mathbf{D}$)
D	As an index: at constant electric *D*isplacement
e	One of the tensors quantifying piezoelectric coupling (Table 5.1 (piezoelectric stress coefficient))
e_{26}	Relevant component of the e-tensor (piezoelectric stress coefficient, $e_{26} = 9.65 \times 10^{-2}$ C/m^2 for AT-cut quartz)
$\hat{E}$, E	Electric field (If bold: a vector. if not bold: z-component of $\hat{\mathbf{E}}$ or $\mathbf{E}$)
E	As an index: at constant *E*lectric field
$\hat{f}$	Force density (Do not confuse with frequency)
f	Frequency
f_r	Resonance frequency
f_s	Series resonance frequency
f_0	Resonance frequency at the fundamental ($f_0 = Z_q/(2m_q) = Z_q/(2\rho_q d_q)$)
g	One of the tensors quantifying piezoelectric coupling (Table 5.1)
G	Shear modulus

G_q	Shear modulus of AT-cut quartz ($G_q \approx 29 \times 10^9$ Pa G_q is the piezoelectrically stiffened modulus)
h	One of the tensors quantifying piezoelectric coupling (Table 5.1)
h_q	Half the thickness of a the resonator plate
k	Wavenumber ($k = \omega/c$)
k_t	Piezoelectric coupling coefficient ($k_t = (e_{26}/(\varepsilon_q\varepsilon_0 G_q))^{1/2}$)
$\hat{I}$	Electric current
m_q	Mass per unit area of the resonator ($m_q = \rho_q d_q = Z_q/(2f_0)$)
n	Overtone order
OC	As an index: the resonance condition of the unloaded plate under *O*pen-*C*ircuit conditions. With no current into the electrodes (more precisely, with vanishing electric displacement everywhere), piezoelectric stiffening is fully accounted for by using the piezoelectrically stiffened shear modulus.
p	Pressure
$\hat{P}$	Electric polarization
PE	As an index: *P*iezo*E*lectric stiffening
PESC	As an index: *P*iezo*E*lectric stiffening with *S*hort-*C*ircuited electrodes
q	As an index: quartz resonator
$\hat{Q}_S$	Electrical surface charge density (Unit: C/m^2)
ref	As an index: reference
R	Resistance
R_{ex}	External resistance
R_{out}	Output resistance of the driving electronics
s	Compliance tensor (Table 5.1)
S	Entropy (Do not confuse with infinitesimal strain tensor)
S	As an index: *S*urface or at constant *S*train (At constant strain: in Table 5.1)
S_{ij}	Infinitesimal strain tensor (Table 5.1, Eq. 5.4.2)
T	Temperature
T	As an index: at constant stress (Table 5.1)
T_{ij}	Stress tensor (Table 5.1)
u	Tangential displacement

u_i	Displacement (component of vector)
$\hat{u}_S$	Tangential displacement at the resonator surface
V	Volume
x_i	Spatial coordinate (component of vector)
z	Spatial coordinate perpendicular to the surface of a layer or to the resonator plate
$\tilde{Z}_{ex}$, Z_{ex}	External electrical impedance
$\tilde{Z}_{PE}$	Stress-velocity ratio at the two resonator surfaces resulting from of *P*iezo*E*lectric stiffening
$\tilde{Z}_{PESC}$	Same as $\tilde{Z}_{PE}$ with *S*hort-*C*ircuited electrodes
$\tilde{Z}_q$, Z_q	Acoustic wave impedance of AT-cut quartz ($Z_q = 8.8 \times 10^6$ kg m^{-2}s^{-1})
Δ	As a prefix: A shift induced by the presence of the sample
ε	A small distance (Eq. 5.3.10)
$\tilde{\varepsilon}$, ε	Dielectric permittivity (A tensor, Table 5.1)
$\tilde{\varepsilon}_q$, ε_q	Dielectric constant of AT-cut quartz (ε_q is the clamped dielectric constant. $\varepsilon_q = 4.54$ for AT-cut quartz)
ε_0	Dielectric permittivity of vacuum ($\varepsilon_0 = 8.854 \times 10^{-12}$ C/(Vm))
φ	Electric potential
ϕ	Factor converting between mechanical and electric quantities in the Mason circuit ($\phi = Ae_{26}/d_q$)
κ_R	Resonator's motional stiffness
ρ_q	Density of crystalline quartz ($\rho_q = 2.65$ g/cm^3)
$\tilde{\sigma}, \sigma$	Tangential stress
$\tilde{\sigma}_{PE}$, σ_{PE}	Piezoelectrically-induced apparent tangential stress
ω	Angular frequency

References

1. Stevenson, A.C., Lowe, C.R.: Magnetic-acoustic-resonator sensors (MARS): a new sensing methodology. Sens. Actuators a-Phys. **72**(1), 32–37 (1999)
2. Thompson, M., Nisman, R., Hayward, G.L., Sindi, H., Stevenson, A.C., Lowe, C.R.: Surface energy and the response of transverse acoustic wave devices in liquids. Analyst **125**(9), 1525–1528 (2000)
3. Driscoll, M.M., Healey, D.J.: Voltage-controlled crystal oscillators. IEEE Trans. Electron Dev. **ED18**(8), 528 (1971)

4. Neubig, B., Briese, W.: Das Grosse Quarzkochbuch. Franzis-Verlag, Feldkirchen (1997)
5. Cady, W.G.: Piezoelectricity. McGraw-Hill Book Company Inc, New York (1946)
6. Ikeda, T.: Fundamental of Piezoelectricity. Oxford University Press, Oxford (1990)
7. Tichý, J., Erhart, J., Kittinger, E., Prívratská, J.: Fundamentals of Piezoelectric Sensorics: Mechanical, Dielectric, and Thermodynamical Properties of Piezoelectric Materials. Springer (2010)
8. http://www.efunda.com/materials/piezo/piezo_math/transforms.cfm. Accessed on 28 Mar 2013
9. Thurston, R.N.: Piezoelectrically excited vibrations. In: Truesdell, C. (ed.) Mechanics of Solids, vol. 4, p. 257. Springer, Heidelberg (1984)
10. Seh, H., Tuller, H.L., Fritze, H.: Defect properties of langasite and effects on BAW gas sensor performance at high temperatures. J. Eur. Ceram. Soc. **24**(6), 1425–1429 (2004)

Chapter 6
The Small Load Approximation Revisited

Abstract The chapter contains two extensions of the SLA. The first (Sect. 6.1) generalizes the SLA to arbitrary resonator shapes and modes of vibration. The load impedance in this formulation is a 3rd-rank tensor. The formalism shows that the statistical weight in area-averaging is the square of the local amplitude. The second extension (Sect. 6.2) is a perturbation analysis, applied to the model of the parallel plate. The perturbation is carried to 3rd order. The 3rd-order result fixes an inconsistency obtained when treating viscoelastic thin films in air with the conventional SLA.

6.1 Small Load Approximation in Tensor Form

6.1.1 Derivation

In most of the book the load impedance is viewed as a scalar quantity. $\tilde{Z}_L$ is the ratio of the shear stress (in the *xz*-plane) to the tangential velocity at the crystal surface (along *x*). One can also view $\tilde{Z}_L$ as the *xzx*–component of a third-rank tensor, defined by the relation

$$\hat{\sigma}_{S,ij} = \sum_k \tilde{Z}_{L,ijk} \hat{v}_{S,k} \tag{6.1.1}$$

The local velocity may have components into different directions. Also, there may be stress components other than the *xz*-component. Note: Eq. 6.1.1 assumes that the stress *locally* depends on the strain. Nonlocal stress-strain relations will be needed for piezoelectrically generated stresses (Sect. 6.1.7).

In the following, the SLA is extended to arbitrary geometries, displacement fields, and stress distributions. A perturbation analysis is applied, similar to what is done in quantum mechanics [1]. In addition to being more general than Eq. 4.6.1,

D. Johannsmann, *The Quartz Crystal Microbalance in Soft Matter Research*, Soft and Biological Matter, DOI: 10.1007/978-3-319-07836-6_6,

the calculation also leads to a more precise definition of what a "small load" in the sense of the SLA should be. Further, it proves that the statistical weight to be applied in area-averaging is $|\hat{\mathbf{u}}_S^2(\mathbf{r}_S)|$ (Sect. 6.1.5).

The problem is formulated as an eigenvalue problem. In the bulk, the material obeys the wave equation

$$\frac{\hat{f}_i}{\rho_q} = \sum_j \frac{1}{\rho_q}\frac{\partial \hat{\sigma}_{ij}}{\partial r_j} = \sum_{jkl} \frac{1}{\rho_q}\frac{\partial}{\partial r_j}\left(\tilde{c}_{ijkl}\frac{1}{2}\left(\frac{\partial \hat{u}_l}{\partial r_k} + \frac{\partial \hat{u}_k}{\partial r_l}\right)\right) = -\tilde{\omega}_r^2 \hat{u}_i \tag{6.1.2}$$

$\hat{\mathbf{f}}$ is the force density. c_{ijkl} is the stiffness tensor. For the stiffness tensor of α-quartz see Refs. [2, 3].

The force density exerted at the resonator surface is the stress tensor contracted with the surface normal. It is linked to the displacement as

$$\sum_j n_j \hat{\sigma}_{S,ij}(\mathbf{r}_S) = \sum_{jk} \mathrm{i}\omega n_j \tilde{Z}_{L,ijk}(\mathbf{r}_S)\, \hat{u}_{S,k}(\mathbf{r}_S) \tag{6.1.3}$$

$\mathbf{n}$ is the surface normal. The two above equations can be condensed into one by use of the Dirac δ-function:

$$\frac{1}{\rho_q}\sum_{jkl}\frac{\partial}{\partial r_j}\left(\tilde{c}_{ijkl}\frac{1}{2}\left(\frac{\partial \hat{u}_l}{\partial r_k} + \frac{\partial \hat{u}_k}{\partial r_l}\right) + \delta(\mathbf{r}-\mathbf{S})\mathrm{i}\omega n_j \tilde{Z}_{L,ijk}\, \hat{u}_k\right) = -\tilde{\omega}_r^2 \hat{u}_i \tag{6.1.4}$$

$\mathbf{S}$ is a position on the resonator surface. $\delta(\mathbf{r} - \mathbf{S})$ is zero for locations away from the surface and infinite at the surface. $\delta(\mathbf{r} - \mathbf{S})$ integrates to unity if the integration is carried out along a path perpendicular to the surface.

In mathematical language, Eq. 6.1.4 constitutes an eigensystem. The vector field $\hat{\mathbf{u}}(\mathbf{r})$ is the eigenfunction and $-\tilde{\omega}_r^2$ is the eigenvalue. The left-hand side defines the "ω^2-operator", termed $\underline{A}$. Deviating from quantum mechanics, the eigenvalue is complex. In the spirit of perturbation theory, the operator $\underline{A}$ is separated into a larger, "unperturbed" part, $\underline{A}_0$, and a smaller part, $\underline{A}_1$ (the "perturbation"):

$$\underline{A} = \underline{A}_0 + \underline{A}_1 \tag{6.1.5}$$

Later, the operator $\underline{A}_0$ will have to be self-adjoint. This entails a complication because operators with complex eigenvectors are not self-adjoint. In order to make $\underline{A}_0$ self-adjoint, the complex-valued stiffness tensor must be replaced by its real part c'_{ijkl}. The imaginary part of the stiffness tensor, c''_{ijkl}, needs to be part of the perturbation. Since the unperturbed stiffness tensor is defined as real, the displacement field solving the unperturbed eigensystem, $\hat{\mathbf{u}}_0(\mathbf{r})$, can also be chosen as real. For a real operator $\underline{A}_0$, self-adjointness implies that that any two test functions $\hat{\mathbf{u}}^\alpha(\mathbf{r})$ and $\hat{\mathbf{u}}^\beta(\mathbf{r})$ obey the relation

$$\int \hat{\mathbf{u}}^{\alpha}\underline{A}_0\hat{\mathbf{u}}^{\beta}\mathrm{d}^3\mathbf{r} = \int \hat{\mathbf{u}}^{\beta}\underline{A}_0\hat{\mathbf{u}}^{\alpha}\mathrm{d}^3\mathbf{r} \tag{6.1.6}$$

In mathematical language: Upon contraction of the operator with two test-functions, the result is symmetric under exchange of the test functions.

The unperturbed operator and the perturbation are given by

$$\begin{aligned}(\underline{A}_0\hat{\mathbf{u}})_i &= \sum_{jkl}\frac{1}{\rho_q}\frac{\partial}{\partial r_j}\left(c'_{ijkl}\frac{1}{2}\left(\frac{\partial \hat{u}_l}{\partial r_k}+\frac{\partial \hat{u}_k}{\partial r_l}\right)\right)\\ (\underline{A}_1\hat{\mathbf{u}})_i &= \sum_{jkl}\frac{1}{\rho_q}\left(\frac{\partial}{\partial r_j}c''_{ijkl}\frac{1}{2}\left(\frac{\partial \hat{u}_l}{\partial r_k}+\frac{\partial \hat{u}_k}{\partial r_l}\right)+\delta(\mathbf{r}-\mathbf{S})\mathrm{i}\omega n_j\tilde{Z}_{L,ijk}\,\hat{u}_k\right)\end{aligned} \tag{6.1.7}$$

The proof that $\underline{A}_0$ defined this way is indeed self-adjoint is given in Sect. 6.1.8.

With real eigenvalues of the unperturbed operator, one can apply perturbation theory in the same way as in quantum mechanics. One can do that, even though the perturbation operator is *not* self-adjoint. The frequency *shift* may be complex. The formalism only requires self-adjointness of the unperturbed operator. This statement holds for the 1st-order perturbation analysis. If a 2nd-order calculation is to be carried out as in quantum mechanics, the 1st-order perturbation operator must be self-adjoint. For a simplified version of the 2nd-order calculation see Ref. [4].

In perturbation theory, the strength of the perturbation is proportional to a small perturbation parameter. One expands all terms in the perturbation parameter and requires that the equations hold for all powers of the perturbation parameter, separately. Following this line of reasoning, the solution, $\hat{\mathbf{u}}(\mathbf{r})$, and the eigenvalue, $-\omega_r^2$, are expanded as

$$\hat{\mathbf{u}} = \hat{\mathbf{u}}_0 + \hat{\mathbf{u}}_1 + \cdots \tag{6.1.8}$$

and

$$-\tilde{\omega}_r^2 = -\omega_{r,0}^2 - \tilde{\omega}_{r,1}^2 - \cdots \tag{6.1.9}$$

Again: $\omega_{r,0}^2$ is real. The eigensystem now reads as

$$(\underline{A}_0 + \underline{A}_1)(\hat{\mathbf{u}}_0 + \hat{\mathbf{u}}_1 + \cdots) = \left(-\omega_{r,0}^2 - \tilde{\omega}_{r,1}^2 - \cdots\right)(\hat{\mathbf{u}}_0 + \hat{\mathbf{u}}_1 + \cdots) \tag{6.1.10}$$

We collect the terms at the different orders and write down separate equations for the 0th and the 1st order. The 0th-order equation is

$$\underline{A}_0\hat{\mathbf{u}}_0 = -\omega_{r,0}^2\hat{\mathbf{u}}_0 \tag{6.1.11}$$

Finding the solution to the 0th-order equation will in general be far from trivial. $\hat{\mathbf{u}}_0(\mathbf{r})$ is the mode of vibration. $\hat{\mathbf{u}}_0(\mathbf{r})$ is a standing plane wave for the parallel plate.

For more realistic resonators (this includes the plate with energy trapping), the mode shapes and the corresponding frequencies will usually be calculated numerically [5, 6].[1]

Assume that $\hat{\mathbf{u}}_0(\mathbf{r})$ has been found. In the next step, one collects all terms which are linear in the small perturbation. The 1st-order equation is

$$\underline{A}_0\hat{\mathbf{u}}_1 + \underline{A}_1\hat{\mathbf{u}}_0 = -\tilde{\omega}_{r,1}^2\hat{\mathbf{u}}_0 - \omega_{r,0}^2\hat{\mathbf{u}}_1 \tag{6.1.12}$$

$-\tilde{\omega}_{r,1}^2$ is the shift of the eigenvalue induced by the perturbation. The corresponding frequency shift is calculated from $\tilde{\omega}_{r,1}^2$ and $\omega_{r,0}^2$ as

$$2\pi\Delta\tilde{f}_1 = \sqrt{\omega_{r,0}^2 + \tilde{\omega}_{r,1}^2} - \omega_{r,0} \approx \omega_{r,0}\left(1 + \frac{1}{2}\frac{\tilde{\omega}_{r,1}^2}{\omega_{r,0}^2}\right) - \omega_{r,0} = \frac{1}{2}\frac{\tilde{\omega}_{r,1}^2}{\omega_{r,0}} \tag{6.1.13}$$

Taylor expansion ($(1 + \varepsilon)^{1/2} \approx 1 + \varepsilon/2$) was applied in step 2. Note: $\Delta\tilde{f}_1$ encompasses the intrinsic dissipation (originating from c''_{ijkl}) as well as the contribution from the sample.

As in quantum mechanics, we multiply Eq. 6.1.12 by $\hat{\mathbf{u}}_0(\mathbf{r})$ from the left and integrate over the volume

$$\int \hat{\mathbf{u}}_0 \cdot \underline{A}_0\hat{\mathbf{u}}_1 d^3\mathbf{r} + \int \hat{\mathbf{u}}_0 \cdot \underline{A}_1\hat{\mathbf{u}}_0 d^3\mathbf{r} = -\tilde{\omega}_{r,1}^2 \int \hat{\mathbf{u}}_0 \cdot \hat{\mathbf{u}}_0 d^3\mathbf{r} - \omega_{r,0}^2 \int \hat{\mathbf{u}}_0 \cdot \hat{\mathbf{u}}_1 d^3\mathbf{r} \tag{6.1.14}$$

The next step makes use of the fact the operator $\underline{A}_0$ is self-adjoint. The self-adjointness of $\underline{A}_0$ implies that

$$\int \hat{\mathbf{u}}_0\underline{A}_0\hat{\mathbf{u}}_1 d^3\mathbf{r} = \int \hat{\mathbf{u}}_1\underline{A}_0\hat{\mathbf{u}}_0 d^3\mathbf{r} = -\omega_{r,0}^2 \int \hat{\mathbf{u}}_1\hat{\mathbf{u}}_0 d^3\mathbf{r} \tag{6.1.15}$$

The first and the last term in Eq. 6.1.14 cancel, which leads to

$$\tilde{\omega}_{r,1}^2 = -\frac{\int \hat{\mathbf{u}}_0 \cdot \underline{A}_1\hat{\mathbf{u}}_0 d^3\mathbf{r}}{\int \hat{\mathbf{u}}_0 \cdot \hat{\mathbf{u}}_0 d^3\mathbf{r}} \tag{6.1.16}$$

---

[1] Among the resonators having been subjected to modal analysis is the sun. It displays a group of acoustic resonances at frequencies between 2 and 4 mHz [5]. The amplitudes are a few hundred kilometers. Readers may also consult a certain J. W. Goethe on that matter. In his prolog to "Faust" he lets Raphael say:

The Sun sings out, in ancient mode [!],
His note among his brother-spheres,
And ends his pre-determined road
With peals of thunder for our ears.

In this case, J. W. got it right: There indeed is a lot of noise on the sun [6].

Inserting Eq. 6.1.16 into 6.1.13 leads to

$$\begin{aligned}
\Delta\tilde{f}_1 &\approx \frac{1}{4\pi\omega_{r,0}} \frac{\int \hat{\mathbf{u}}_0 \cdot \underline{A}_1 \hat{\mathbf{u}}_0 \mathrm{d}^3\mathbf{r}}{\int \hat{\mathbf{u}}_0 \cdot \hat{\mathbf{u}}_0 \mathrm{d}^3\mathbf{r}} \\
&= \frac{1}{4\pi\rho_q\omega_{r,0} \sum\limits_k \int\limits_{Volume} \hat{u}_{0,k} \cdot \hat{u}_{0,k} \mathrm{d}^3\mathbf{r}} \\
&\times \left[\sum_{ijkl} \left(\int\limits_{Volume} \hat{u}_{0,i} \frac{\partial}{\partial r_j} c''_{ijkl} \frac{1}{2} \left(\frac{\partial \hat{u}_{0,l}}{\partial r_k} + \frac{\partial \hat{u}_{0,k}}{\partial r_l} \right) \mathrm{d}^3\mathbf{r} + \int\limits_{Surface} \hat{u}_{0,S,i} \mathrm{i}\omega n_j \tilde{Z}_{L,ijk} \hat{u}_{0,S,k} \mathrm{d}^2\mathbf{r}_S \right) \right]
\end{aligned} \tag{6.1.17}$$

$\mathbf{r}_S$ is a vector on the surface. The first term in the square brackets originates from the intrinsic dissipation. We call the respective (purely imaginary) frequency shift $\Delta\tilde{f}_{diss}$ and write $\Delta\tilde{f} = \Delta\tilde{f}_1 - \Delta\tilde{f}_{diss}$. $\Delta\tilde{f}$ is the frequency shift originating from the sample. The next step exploits the second approximation inherent to the SLA, which is $\omega \approx \omega_{r,0}$. The load impedance is evaluated at the reference frequency rather than the shifted frequency, which leads to

$$\Delta\tilde{f} \approx \frac{\mathrm{i}}{4\pi\rho_q} \frac{\sum\limits_{ijk} \int\limits_{Surface} \hat{u}_{0,S,i} n_j \tilde{Z}_{L,ijk} \hat{u}_{0,S,k} \mathrm{d}^2\mathbf{r}_S}{\sum\limits_k \int\limits_{Volume} \hat{u}_{0,k} \hat{u}_{0,k} \mathrm{d}^3\mathbf{r}} \tag{6.1.18}$$

Equation 6.1.18 is the small load approximation in tensor form.

Equation 6.1.18 can be used to calculate the complex frequency shift of resonators with arbitrary shape from the stress-velocity ratio (in its tensor form) at the resonator surface. As before, the SLA entails two separate approximations: The dependence of resonance frequency on the load is linearized in the load and the stress is calculated at the reference frequency. Those two approximations are separate.

Using $\hat{\sigma}_{ij} = \hat{Z}_{L,ijk} \mathrm{i}\omega \hat{u}_k$, one can express the frequency shift in terms of the surface stress:

$$\Delta\tilde{f} \approx \frac{1}{4\pi\rho_q\omega_{r,0}} \frac{\sum\limits_{ij} \int\limits_{Surface} \hat{u}_{0,S,i} n_j \left(\hat{\sigma}_{ij} - \hat{\sigma}_{ref,ij} \right) \mathrm{d}^2\mathbf{r}_S}{\sum\limits_k \int\limits_{Volume} \hat{u}_{0,k} \hat{u}_{0,k} \mathrm{d}^3\mathbf{r}} \tag{6.1.19}$$

This is the SLA in the form written down by Pechhold in Ref. [7]. Note: Eq. 6.1.19 is more general than Eq. 6.1.18. Equation 6.1.18 assumes that the local surface traction is connected to the local velocity by the tensor $\tilde{Z}_{L,ijk}$. In the case of piezoelectric stiffening (Sect. 6.1.7), the electric displacement depends *nonlocally* on the strain of the entire crystal because electromagnetic waves travel much faster than sound.

In the following, Eqs. 6.1.18 and 6.1.19 are applied to various geometries, all related to plates in one way or another.

6.1.2 Parallel Plate

For the parallel plate, the tensor form of the SLA must reduce to the standard form, which is proven in the following. For the parallel plate, the displacement field has only one component, $\hat{u}_x$. $\hat{u}_x(z)$ is a standing wave with antinodes at the surface. One has

$$\hat{u}_{0,x}(z) = \hat{u}_S \sin\left(\frac{n\pi z}{d_q}\right) \tag{6.1.20}$$

$z = 0$ is in the center of the plate. The only nontrivial term on the right-hand side in Eq. 6.1.18 is the xzx-term. Renaming $\tilde{Z}_{L,xzx}$ as $\tilde{Z}_L$ (assumed as constant over the entire resonator surface) and inserting Eq. 6.1.20 into Eq. 6.1.18 yields

$$\begin{aligned}\Delta\tilde{f} &= \frac{\mathrm{i}}{4\pi\rho_q}\frac{\int\limits_{Surface}\hat{u}_S\tilde{Z}_L\hat{u}_S\mathrm{d}^2\mathbf{r}_S}{\int\limits_{Volume}\hat{u}_S^2\sin^2\left(\frac{n\pi z}{d_q}\right)\mathrm{d}^3\mathbf{r}} \\ &= \frac{\mathrm{i}}{4\pi}\frac{1}{\rho_q}\frac{\hat{u}_S^2A\tilde{Z}_L}{\hat{u}_S^2Ad_q\frac{1}{2}} = \frac{\mathrm{i}}{\pi}\frac{f_0}{Z_q}\tilde{Z}_L\end{aligned} \tag{6.1.21}$$

Line 2 used that the integral of the sin^2-term evaluates to 1/2. Line 3 used $f_0 = Z_q/(2\rho_q d_q)$ (Eq. 4.3.16). Equation 6.1.21 is the conventional SLA (Sect. 4.6.1).

6.1.3 Parallel Plate with Energy Trapping

In the following Eq. 6.1.18 is applied to a parallel plate with energy trapping. The plate undergoes flexural deformations, in addition to thickness-shear (Sect. 7.6). First consider the plate *in air*.

Let the displacement have a vertical component, $\hat{u}_z$, caused by bending. Equation 6.1.18 then reads as

$$\Delta\tilde{f} = \frac{\mathrm{i}}{4\pi\rho_q}\frac{\int\limits_{Surface}\hat{u}_{0,x}^2\tilde{Z}_{L,xzx}+\hat{u}_{0,z}^2\tilde{Z}_{L,zzz}\mathrm{d}^2\mathbf{r}_S}{\int\limits_{Volume}\hat{u}_{0,x}^2+\hat{u}_{0,z}^2\mathrm{d}^3\mathbf{r}}$$
$$= \frac{\mathrm{i}}{4\pi\rho_q}\left(\frac{\int\limits_{Surface}\hat{u}_{0,x}^2\tilde{Z}_{L,xzx}\mathrm{d}^2\mathbf{r}_S}{\int\limits_{Volume}\hat{u}_{0,x}^2+\hat{u}_{0,z}^2\mathrm{d}^3\mathbf{r}}+\frac{\int\limits_{Surface}\hat{u}_{0,z}^2\tilde{Z}_{L,zzz}\mathrm{d}^2\mathbf{r}_S}{\int\limits_{Volume}\hat{u}_{0,x}^2+\hat{u}_{0,z}^2\mathrm{d}^3\mathbf{r}}\right) \tag{6.1.22}$$

One hopes that the effects of normal stress are constant over the time of the experiment and that they therefore disappear from the frequency shifts.

Now consider the plate immersed in a liquid. This situation deserves special attention because it will turn out that the SLA cannot be applied to the flexural components in this case. The perturbation scheme requires the perturbation operator, $\underline{A}_1$, to be much smaller than the unperturbed operator, $\underline{A}_0$. This is not the case if the resonator surface radiates compressional waves into a liquid.

Using $\hat{\mathbf{u}}_0$ as a test function, the small-load condition amounts to

$$\int\limits_{Volume}\hat{\mathbf{u}}_0\underline{A}_1\hat{\mathbf{u}}_0\mathrm{d}^3\mathbf{r} \ll \int\limits_{Volume}\hat{\mathbf{u}}_0\underline{A}_0\hat{\mathbf{u}}_0\mathrm{d}^3\mathbf{r} \quad \text{(a)}$$
$$\int\limits_{Surface}\sum_{ijk}\hat{u}_{0,S,i}n_j\frac{\mathrm{i}\omega\tilde{Z}_{L,ijk}}{\rho_q}\hat{u}_{0,S,k}\mathrm{d}^2\mathbf{r}_S \tag{6.1.23}$$
$$\ll \int\limits_{Volume}\sum_{ijkl}\hat{u}_{0,i}\frac{c'_{ijkl}}{2\rho_q}\frac{\partial}{\partial r_j}\left(\frac{\partial\hat{u}_{0,l}}{\partial r_k}+\frac{\partial\hat{u}_{0,k}}{\partial r_l}\right)\mathrm{d}^3\mathbf{r} \quad \text{(b)}$$

The volume integral in (a) is meant to include the surface. The integration boundaries are outside the resonator surface. In the line 2, the intrinsic dissipation was neglected. For the parallel plate in the thickness shear mode Eq. 6.1.23b reads as

$$\left|\int\limits_{Surface}\hat{u}_{0,S,x}^2\mathrm{i}\omega\tilde{Z}_L\,\mathrm{d}^2\mathbf{r}_S\right| \ll \left|\int\limits_{Volume}\hat{u}_{0,x}G_q\frac{\partial^2\hat{u}_{0,x}}{\partial z^2}\mathrm{d}^3\mathbf{r}\right| = \left|\int\limits_{Volume}G_qk_z^2\hat{u}_{0,x}^2\mathrm{d}^3\mathbf{r}\right| \tag{6.1.24}$$

Applying the relations $k_z^2 = (\omega/c_q)^2 = \omega^2\rho_q/G_q$, $Z_q = (G_q\rho_q)^{1/2}$, $d_q = c_q/(2f_0)$ and, further, ignoring numerical factors of order one, this reduces to the familiar condition $|\tilde{Z}_L| \ll Z_q$.

Importantly, the small-load condition differs between thickness-shear deformation and flexural deformation. In the following, let all deformation be independent of y. For a flexural deformation of a plate, the vertical force density is given by the Euler–Bernoulli equation, which is [8, 9]

$$\hat{f}_{0,S,z} \approx E_q I \frac{\partial^4 \hat{u}_{0,z}}{\partial x^4} \tag{6.1.25}$$

$\hat{f}_{0,S,z}$ is normalized to area, not to volume. It has dimensions of N/m^2. The corresponding bulk force density is given as $\hat{f}_{0,bulk,z} = \hat{f}_{0,s,z}/d_q$. E_q is the Young's modulus (of an isotropic material), and I is the second moment of area. For a plate, the second moment of area is $I = d_q^3/12$. Since the force density does not depend on z, the right-hand side in Eq. 6.1.23 is

$$\int_{Volume} \hat{\mathbf{u}}_0 \underline{A}_0 \hat{\mathbf{u}}_0 \mathrm{d}^3\mathbf{r} = \int_{Volume} \hat{\mathbf{u}}_0 \frac{\mathbf{f}_{0,bulk}}{\rho_q} \mathrm{d}^3\mathbf{r} = \frac{E_q I}{\rho_q} \int_{Surface} \hat{u}_{0,z} \frac{\partial^4 \hat{u}_{0,z}}{\partial x^4} \mathrm{d}^2\mathbf{r}_S \tag{6.1.26}$$

Inserting $I = d_q^3/12$ and approximating $\partial^4 \hat{u}_{0,z}/\partial x^4$ by $\hat{u}_{0,z}/L_q^4$ with L_q a length of the order of radius of the active area, one finds

$$\int_{Volume} \hat{\mathbf{u}}_0 \underline{A}_0 \hat{\mathbf{u}}_0 \mathrm{d}^3\mathbf{r} \approx \frac{1}{12} \frac{1}{\rho_q} \frac{d_q^3}{L_q^4} E_q \int_{Surface} \hat{u}_{0,z}^2 \mathrm{d}^2\mathbf{r}_S \tag{6.1.27}$$

The right-hand side in Eq. 6.1.23 is small because of the term d_q^3/L_q^4. The left-hand side in Eq. 6.1.23, on the other hand, is large because the perturbation is caused by compressional waves. The parameter $\tilde{Z}_{L,zzz}$ is the equal to the compressional wave impedance, $\tilde{Z}_{comp} = c_{comp}\rho_{liq}$ (Eq. 4.2.6) with c_{comp} the speed of compressional ultrasound. The left-hand side in Eq. 6.1.23 is

$$\frac{i\omega \tilde{Z}_{comp}}{\rho_q} \int_{Surface} \hat{u}_{0,z}^2 \mathrm{d}^2\mathbf{r}_S \tag{6.1.28}$$

If the right-hand side in Eq. 6.1.23a is supposed to be smaller than the left-hand side, this requires

$$\left| \frac{\omega \tilde{Z}_{comp}}{\rho_q} \right| \ll \frac{1}{12} \frac{1}{\rho_q} \frac{d_q^3}{L_q^4} E_q \tag{6.1.29}$$

However, inserting values ($\omega = 2\pi$ 5 MHz, $c_{liq} \approx 1500$ m/s, $\rho_{liq} = 10^3$ kg/m^3, $d_q = 330$ µm, $E_q \approx 70$ GPa, $L_q \approx 3$ mm) one finds the left-hand side to be *larger* than the right-hand side by a factor of about 10^4. Equation 6.1.29 is not fulfilled; the SLA does not apply to compressional waves in liquids.

What are the consequences of this argument? This is good news, in the first place, because it means that the QCM can actually be used in liquids. This is what Mason and McSkimin had missed, when they discarded thickness-shear resonators for use in liquids in 1949 (Sect. 1.4, Ref. [10]). In air, the amplitude of normal motion of the plate can be many times larger than the tangential component. Bending translates to large amplitudes of motion in the vertical direction. If the amplitude was as large in liquids as it is in air, compressional waves would be detrimental. The compressional load impedance would govern the frequency response. However, the liquid itself much reduces the flexural component (Actually, the liquid may revert the sign of the flexural component as discussed in Sect. 7.6. The sign is irrelevant here).

Of course the fact, that the SLA does not apply to the flexural components, is a drawback in the context of modeling. Note, however, that the SLA still applies to the thickness-shear component. The *transverse* shear wave impedance of the liquid is small and the *shear* stiffness of the crystal is large. Because the liquid reduces the flexural components, one may model the QCM as a thickness-shear device and disregard energy trapping. The flexural components do exist and they do have an influence, but we can live with them. For example, viscosity measurements on Newtonian liquids yield values within about ±10 % of the expected value. If flexural modes were not suppressed by the liquid, the results would be much different.

For reasons similar to the ones discussed above, the SLA does not apply to cantilever resonators in liquids [11]. In the latter case, this is actually evident from experiment. When immersing a cantilever into the liquid, its resonance frequency sharply drops. Even without going through the mathematics, one sees that $|\Delta f/f| \approx 1$, which violates the SLA.

The above discussion does not apply to Lamb-wave sensors [12]. Lamb wave sensors *exploit* flexural motion, while thickness shear has little relevance. That configuration is different from the QCM.

6.1.4 The Modal Mass and Limits of the Sauerbrey Equation

Experiment shows that the Sauerbrey equation is not strictly obeyed even for those films, which are too thin to let viscoelastic effects originating from the film itself be significant. There are different reasons, discussed in Sect. 8.1. Among them is a dependence of the modal mass (Eq. 4.5.27a) on the overtone order. Such a dependence can be explained with flexural modes.

Let the surface normal be along z as before ($n_x = 0$, $n_y = 0$, $n_z = 1$) and let the stress at the surface be of entirely inertial origin ($\tilde{Z}_{L,izk} = \mathrm{i}\omega m_f \delta_{ik}$ with δ_{ik} the Kronecker δ). Equation 6.1.18 becomes

$$\Delta\tilde{f} \approx \frac{\mathrm{i}}{4\pi}\frac{\int\limits_{Surface} \mathrm{i}m_f(\mathbf{r}_S)\omega \sum\limits_k \hat{u}_{0,S,k}\hat{u}_{0,S,k}\mathrm{d}^2\mathbf{r}_S}{\rho_q \sum\limits_k \int\limits_{Volume} \hat{u}_{0,k}\hat{u}_{0,k}\mathrm{d}^3\mathbf{r}} = \frac{-m_f\omega}{4\pi}\left[\frac{\int\limits_{Surface} |u_{0.S}^2|\mathrm{d}^2\mathbf{r}_S}{\rho_q \int\limits_{Volume} |u_0^2|\mathrm{d}^3\mathbf{r}}\right] \quad (6.1.30)$$

For the parallel plate (only for the parallel plate), the term in square brackets evaluates to $2/m_q$, which leads to the Sauerbrey result. The modal mass is $Am_q/2$, which is half of the resonator's true mass.

With flexural contributions present, the situation becomes more complicated. The frequency shift is still negative and proportional to m_f, but the prefactor is only approximately equal to f/m_q. Also, the prefactor differs between overtones. The prefactor is not vastly different from f/m_q, but its dependence on overtone order can be seen in experiment. The gravimetric QCM for that reason is less accurate than it is precise. In order to let it be accurate, the true modal mass (for the given crystal and as a function of overtone order) would have to be known. (As far as the n-dependence is concerned, calibration with thin films of course solves the problem.)

6.1.5 Statistical Weight in Area Averaging

The load impedance might be a function of position. Following the literature, we make two assumptions (which actually are slightly inconsistent with each other). We consider a plate with a nontrivial amplitude distribution, but we ignore flexural motion nevertheless. Equation 6.1.22 simplifies as

$$\Delta\tilde{f} = \frac{\mathrm{i}}{4\pi\rho_q}\frac{\int\limits_{Surface}\hat{u}_{0,x}^2(\mathbf{r}_S)\tilde{Z}_{L,xzx}(\mathbf{r}_S)\mathrm{d}^2\mathbf{r}_S}{\int\limits_{Volume}\hat{u}_{0,x}^2(\mathbf{r})\mathrm{d}^3\mathbf{r}} = \frac{\mathrm{i}}{2\pi d_q\rho_q}\frac{\int\limits_{Surface}\hat{u}_S^2(\mathbf{r}_S)\tilde{Z}_L(\mathbf{r}_S)\mathrm{d}^2\mathbf{r}_S}{\int\limits_{Surface}\hat{u}_S^2(\mathbf{r}_S)\mathrm{d}^2\mathbf{r}_S} \tag{6.1.31}$$

$\tilde{Z}_{L,xzx}(\mathbf{r}_S)$ was renamed as $\tilde{Z}_L(\mathbf{r}_S)$ and $\hat{u}_{0,x}(\mathbf{r}_S)$ was renamed as $\hat{u}_S(\mathbf{r}_S)$ in step 2. Also, we used that $\hat{u}_{0,x}(z)$ is a sine wave and performed the integration (see Sect. 6.1.2). Equation 6.1.31 shows that the statistical weight in area-averaging is the square of the local amplitude.

6.1.6 Parallel Plate with Surface Roughness

For a rough surface, the calculation proceeds along similar lines as for the parallel plate. The difference is that the surface normal does not point along z, locally, but has three components, (n_x, n_x, n_z), where $n_x^2 + n_y^2 + n_z^2 = 1$. Let the displacement be entirely along x. All terms other than the surface integral in Eq. 6.1.18 then turn into $\mathrm{i}f_0/(\pi Z_q)$ as in Sect. 6.1.2. In the surface integral, one needs to take all components of the stress tensor into account. Equation 6.1.18 turns into

$$
\begin{aligned}
\Delta\tilde{f} &= \frac{if_0}{\pi Z_q}\left\langle n_x\tilde{Z}_{L,xxx} + n_y\tilde{Z}_{L,xyx} + n_z\tilde{Z}_{L,xzx}\right\rangle_{area} \\
&= \frac{if_0}{\pi Z_q}\frac{\left\langle n_x\hat{\sigma}_{xx} + n_y\hat{\sigma}_{xy} + n_z\hat{\sigma}_{xz}\right\rangle_{area}}{\hat{v}_{0,x}}
\end{aligned}
\qquad (6.1.32)
$$

At this point, the stress at the surface of a rough plate is not computed. Hydrodynamics would be needed to do that.

6.1.7 Piezoelectric Stiffening

The discussion of piezoelectric stiffening in Sect. 5.3 was limited to the parallel plate. The external electrical impedance in Eq. 5.3.24 was a discrete element, called $\tilde{Z}_{ex}$; the distribution of the electric field inside the sample entered the value of $\tilde{Z}_{ex}$, but how this happened, was left open. Also, the electric field was strictly vertical inside the plate. In the following, we relax these assumptions. In particular, we allow for non-normal components of the electric field inside the crystal.

At this point, the SLA again comes to help. There is a formalism leading from the electric field distribution (more precisely, from the distribution of the electric displacement $D(\mathbf{r})$) to the complex frequency shift. It is not necessary to solve the coupled equations (elastic and electrostatic, coupled by the piezoelectric tensor). This has been done [13, 14], but the procedure is tedious. Making use of the SLA, the problem can be reduced to an entirely electrostatic calculation, where the vibration pattern of the resonator enters as a source term. The SLA separates the elastic problem from the electrostatic problem.

Start from the constitutive equations of piezoelectricity in the stress-voltage form (Table 5.2.1):

$$
\begin{aligned}
T &= c_D \cdot S - h^t \cdot \mathbf{D} && \text{(a)} \\
\mathbf{E} &= -h \cdot S + \varepsilon_S^{-1} \cdot \mathbf{D} && \text{(b)}
\end{aligned}
\qquad (6.1.33)
$$

The superscript t denotes a transposed matrix in the Voigt sense (Sect. 5.4). In a first step, calculate the 0th-order solution (that is, the mode of vibration) from Eq. 6.1.33a assuming open circuit conditions. With $\mathbf{D} = 0$, the problem is entirely elastic in nature. The calculation might be done using the finite element method, but one might also guess the solution. For instance, one might assume a sine wave along z and a Gaussian amplitude distribution in the plane (Eq. 7.1.3). The Gaussian should be elongated along x by 25 % (Sect. 7.3). The width of Gaussian can be adapted to experiment by calculating the effective area as in Eq. 7.4.9 and matching this area to the experimentally determined area, which follows from the motional resistance and Eq. 7.4.7.

In the second step, calculate the electric displacement field from Maxwell's first law, the boundary conditions, and Eq. 6.1.33b. The term $-h\cdot S$ in Eq. 6.1.33b is a source of polarization (a "remanent polarization"). Viscoelasticity does *not* enter this calculation. The partial differential equation to be solved is Maxwell's first law. The resonator's mode of vibration enters as a fixed source term, the strain field itself is not a dynamical variable. There are numerous software packages which solve electrostatics problems of this kind. Among them is COMSOL's electrostatics module.

Once the electric displacement field, $\hat{\mathbf{D}}(\mathbf{r})$, is known, one calculates the associated force density and derives $\Delta\tilde{f}$, again treating the force density as a small perturbation. Following Eq. 6.1.33a, the force density resulting from a nonzero electric displacement in the bulk is

$$\hat{f}_{PE,bulk,i}(\mathbf{r}) = \frac{\partial \hat{T}_{PE,ij}(\mathbf{r})}{\partial r_j} = -\sum_{jk} \tilde{h}_{kij} \frac{\partial \hat{D}_k(\mathbf{r})}{\partial r_j} \tag{6.1.34}$$

Note that the piezoelectrically-induced force density in the bulk does not vanish as in the case of the parallel plate because of the tensorial nature of $\tilde{h}$. The divergence of $\hat{\boldsymbol{D}}$ vanishes ($\Sigma_k \partial \hat{D}_k(\mathbf{r})/\partial r_k = 0$), but unless the h-tensor is isotropic, this does not mean that the divergence of the stress tensor would also vanish. (For the parallel plate, the condition $\mathrm{div}\hat{\boldsymbol{D}} = 0$ implies constant dielectric displacement inside the resonator and the effects of piezoelectric stiffening can therefore all be lumped into a surface term, see Eq. 5.3.10.) As before (Sect. 5.3), there is a surface term (in addition to the bulk term), given as

$$\hat{f}_{PE,S,i}(\mathbf{r}_S) = -\sum_{jk} n_j \tilde{h}_{kij} \hat{D}_k(\mathbf{r}_S). \tag{6.1.35}$$

The surface term goes back to the fact that the h-tensor is discontinuous at the surface. The surface term and the bulk term can be condensed, using the Dirac δ-function as

$$\hat{f}_{PE,i}(\mathbf{r}) = -\sum_{jk} \tilde{h}_{kij} \frac{\partial \hat{D}_k(\mathbf{r})}{\partial r_j} - \sum_{jk} n_j \tilde{h}_{kij} \hat{D}_k(\mathbf{r}) \delta(\mathbf{r} - \mathbf{S}) \tag{6.1.36}$$

This force density gives rise to a perturbation operator (cf. Eq. 6.1.7) of the form

$$\begin{aligned}\left(\underline{A}_{1,PE}\{\hat{\mathbf{u}}\}\right)_i &= \frac{1}{\rho_q} \hat{f}_{PE,i}(\{\hat{\mathbf{u}}\}, \mathbf{r}) \\ &= \frac{1}{\rho_q} \left(\sum_{jk} -\tilde{h}_{kij} \frac{\partial \hat{D}_k(\{\hat{\mathbf{u}}\}, \mathbf{r})}{\partial r_j} - \sum_{jk} n_j \tilde{h}_{kij} \hat{D}_k(\{\hat{\mathbf{u}}\}, \mathbf{r}) \delta(\mathbf{r} - \mathbf{S}) \right)\end{aligned} \tag{6.1.37}$$

The notation in Eq. 6.1.37 is somewhat compact. The displacement in curly brackets ($\{\hat{\mathbf{u}}\}$) is meant to imply that the acceleration depends on the *entire field* $\hat{\mathbf{u}}(\mathbf{r})$. Stress depends *nonlocally* on strain. There is no such thing as a wave impedance (even tensorial), which would connect the local piezoelectrically-induced stress to a local velocity (Remember: The piezoelectrically-induced force density, $\hat{f}_{PE}$, is the divergence of the piezoelectrically-induced stress). When saying that stress depends nonlocally on velocity, the speed of light is approximated as infinite. Put differently, electrostatics (as opposed to electrodynamics) was applied to calculate $\hat{D}(r)$.

With the perturbation operator as given by Eq. 6.1.37, the frequency shift follows as

$$\begin{aligned}\Delta\tilde{f}_{PE,OC} &\approx \frac{1}{4\pi\omega_{r,0}} \frac{\int \hat{\mathbf{u}}_0 \cdot \underline{A}_{1,PE}\{\hat{\mathbf{u}}_0\} d^3\mathbf{r}}{\int \hat{\mathbf{u}}_0 \cdot \hat{\mathbf{u}}_0 d^3\mathbf{r}} \\ &= \frac{1}{4\pi\rho_q\omega_{r,0} \sum_k \int_{Volume} \hat{u}_{0,k} \cdot \hat{u}_{0,k} d^3\mathbf{r}} \\ &\times \left[\sum_{ijk}\left(\int_{Volume} -\hat{u}_{0,i}\tilde{h}_{kij}\frac{\partial \hat{D}_k}{\partial r_j} d^3\mathbf{r} + \int_{Surface} -\hat{u}_{0,S,i} n_j \tilde{h}_{kij} \hat{D}_k d^2\mathbf{r}_S\right)\right]\end{aligned} \tag{6.1.38}$$

The index OC indicates that the frequency shift is calculated with the open circuit condition being the reference state. Since the experimental reference state is not usually given by the open circuit condition, Eq. 6.1.38 must be applied twice, once for the state with the sample present and once for the reference state. The frequency shift originating from the sample is the difference between these two results.

6.1.8 Appendix: Self-Adjointness of the Unperturbed ω^2-Operator

The relation to be proven is Eq. 6.1.6:

$$\int \hat{\mathbf{u}}^{\alpha}\underline{A}_0\hat{\mathbf{u}}^{\beta} d^3\mathbf{r} = \int \hat{\mathbf{u}}^{\beta}\underline{A}_0\hat{\mathbf{u}}^{\alpha} d^3\mathbf{r} \tag{6.1.39}$$

$\hat{\mathbf{u}}^{\alpha}(\mathbf{r})$ and $\hat{\mathbf{u}}^{\beta}(\mathbf{r})$ are two displacement fields.

To be precise, Eq. 6.1.39 expresses symmetry under exchange of $\hat{\mathbf{u}}^{\alpha}$ and $\hat{\mathbf{u}}^{\beta}$. In mathematical terms, symmetry is not strictly equivalent to self-adjointness, but we skip the details.

$\underline{A}_0$ is the unperturbed ω^2-operator, given as

$$(\underline{A}_0\hat{\mathbf{u}})_i = \frac{\hat{\mathbf{f}}'_i}{\rho_q} = \sum_j \frac{1}{\rho_q}\frac{\partial\hat{\sigma}'_{ij}}{\partial r_j} = \sum_{jkl}\frac{1}{\rho_q}\frac{\partial}{\partial r_j}\left(c'_{ijkl}\left[\frac{1}{2}\left(\frac{\partial\hat{u}_l}{\partial r_k}+\frac{\partial\hat{u}_k}{\partial r_l}\right)\right]\right) \quad (6.1.40)$$

c'_{ijkl} is the real part of the stiffness tensor. The quantity in square brackets is the infinitesimal strain tensor.

The proof of self-adjointness rests on that, firstly, the resonator is stress-free at the surfaces, and that, secondly, the stiffness tensor obeys certain symmetries. More specifically, the stiffness tensor is unchanged when exchanging indices as $i\leftrightarrow j$, as $k\leftrightarrow l$, and as $ij\leftrightarrow kl$. It is the last exchange, which requires time-reversal symmetry. Dissipation violates the time-reversal symmetry. If the stiffness tensor has dissipative components, the exchange $ij\leftrightarrow kl$ cannot be made. For this reason, we had to make the dissipative part of the resonator's stiffness part of the perturbation in Sect. 6.1.1. It cannot be part of $\underline{A}_0$.

As in quantum mechanics, the self-adjointness follows from partial integration. At the surface, the unperturbed resonator is stress-free:

$$\hat{\sigma}_{S,zx} = \hat{\sigma}_{S,zy} = \hat{\sigma}_{S,zz} = 0 \quad \text{with } z \text{ the surface normal} \quad (6.1.41)$$

Remember that the stress tensor is symmetric. Vanishing stress at the surface will make the first term in a partial integration go to zero. The integration can be confined to a cube because the total resonator volume can be decomposed into small cubes of this kind. When joining two cubes together, the stresses exerted at the interface are equal and opposite by the principle of reaction. The newly-formed contact between the cubes does not change the integral over the force density (cf. Eq. 6.1.40).

In a first step, integrate $\hat{\mathbf{u}}^\beta\underline{A}_0\hat{\mathbf{u}}^\alpha$ from Eq. 6.1.39 along r_j as

$$\begin{aligned}
\int_0^L \mathrm{d}r_j\left\{\sum_i \hat{u}_i^\alpha\frac{\hat{f}_i^{\beta\prime}}{\rho_q}\right\} &= \int_0^L \mathrm{d}r_j\left\{\frac{1}{\rho_q}\sum_{ij}\hat{u}_i^\alpha\frac{\partial\hat{\sigma}_{ij}^{\beta\prime}}{\partial r_j}\right\}\\
&= \left[\sum_{ij}\hat{u}_i^\alpha\frac{\hat{\sigma}_{ij}^{\beta\prime}}{\rho_q}\right]_0^L - \int_0^L \mathrm{d}r_j\left\{\sum_{ij}\frac{\partial\hat{u}_i^\alpha}{\partial r_j}\frac{\hat{\sigma}_{ij}^{\beta\prime}}{\rho_q}\right\}\\
&= -\int_0^L \mathrm{d}r_j\left\{\sum_{ijkl}\frac{\partial\hat{u}_i^\alpha}{\partial r_j}\frac{c'_{ijkl}}{\rho_q}\frac{1}{2}\left(\frac{\partial\hat{u}_l^\beta}{\partial r_k}+\frac{\partial\hat{u}_k^\beta}{\partial r_l}\right)\right\}
\end{aligned} \quad (6.1.42)$$

In the last line, vanishing stress at the surface (Eq. 6.1.41) was used. The next step exploits the symmetry of the stiffness tensor

$$
\begin{aligned}
-\int_0^L \mathrm{d}r_j \left\{ \sum_{ijkl} \frac{\partial \hat{u}_i^\alpha}{\partial r_j} \frac{c'_{ijkl}}{\rho_q} \frac{1}{2} \left(\frac{\partial \hat{u}_l^\beta}{\partial r_k} + \frac{\partial \hat{u}_k^\beta}{\partial r_l} \right) \right\} &= -\int_0^L \mathrm{d}r_j \left\{ \sum_{ijkl} \frac{c'_{ijkl}}{2\rho_q} \left(\frac{\partial \hat{u}_i^\alpha}{\partial r_j} \frac{\partial \hat{u}_l^\beta}{\partial r_k} + \frac{\partial \hat{u}_i^\alpha}{\partial r_j} \frac{\partial \hat{u}_k^\beta}{\partial r_l} \right) \right\} \\
&= -\int_0^L \mathrm{d}r_j \left\{ \sum_{ijkl} \frac{c'_{ijkl}}{2\rho_q} \left(\frac{\partial \hat{u}_i^\alpha}{\partial r_j} \frac{\partial \hat{u}_l^\beta}{\partial r_k} + \frac{\partial \hat{u}_j^\alpha}{\partial r_i} \frac{\partial \hat{u}_l^\beta}{\partial r_k} \right) \right\} \\
&= -\int_0^L \mathrm{d}r_j \left\{ \sum_{ijkl} \frac{c'_{klij}}{2\rho_q} \left(\frac{\partial \hat{u}_i^\alpha}{\partial r_j} + \frac{\partial \hat{u}_j^\alpha}{\partial r_i} \right) \frac{\partial \hat{u}_l^\beta}{\partial r_k} \right\} \\
&= -\int_0^L \mathrm{d}r_j \left\{ \sum_{kl} \frac{\hat{\sigma}_{kl}^{\alpha\prime}}{\rho_q} \frac{\partial \hat{u}_l^\beta}{\partial r_k} \right\}
\end{aligned}
\tag{6.1.43}
$$

Symmetry of the stiffness tensor under $i \leftrightarrow j$ and $k \leftrightarrow l$ was used in step 3. Symmetry of the stiffness tensor under $ij \leftrightarrow kl$ was used in step 4.

In the following, the cases of $k = j$ and $k \neq j$ must be treated separately First consider the case of $k \neq j$. Integrating over r_k and exchanging the outer and the inner integration, one finds

$$
\begin{aligned}
-\int_0^L \mathrm{d}r_j \int_0^L \mathrm{d}r_k \left\{ \sum_{kl} \frac{\hat{\sigma}_{kl}^{\alpha\prime}}{\rho_q} \frac{\partial \hat{u}_l^\beta}{\partial r_k} \right\} &= -\int_0^L \mathrm{d}r_j \left[\sum_{kl} \frac{\hat{\sigma}_{kl}^{\alpha\prime}}{\rho_q} \hat{u}_l^\beta \right]_0^L + \int_0^L \mathrm{d}r_j \int_0^L \mathrm{d}r_k \left\{ \sum_{kl} \frac{1}{\rho_q} \frac{\partial \hat{\sigma}_{kl}^{\alpha\prime}}{\partial r_k} \hat{u}_l^\beta \right\} \\
&= \int_0^L \mathrm{d}r_j \int_0^L \mathrm{d}r_k \left\{ \sum_{kl} \frac{1}{\rho_q} \frac{\partial \hat{\sigma}_{kl}^{\alpha\prime}}{\partial r_k} \hat{u}_l^\beta \right\} \\
&= \int_0^L \mathrm{d}r_j \int_0^L \mathrm{d}r_k \left\{ \sum_{l} \frac{\hat{f}_l^{\alpha\prime}}{\rho_q} \hat{u}_l^\beta \right\}
\end{aligned}
\tag{6.1.44}
$$

Vanishing stress at the surface was used in step 3. After integration over the third dimension, one finds Eq. 6.1.39 confirmed.

Now go back to Eq. 6.1.43 and consider the case of $k = j$. In this case, one can again perform partial integration as

$$
\begin{aligned}
-\int_0^L \mathrm{d}r_j \left\{ \sum_{kl} \frac{\hat{\sigma}_{kl}^{\alpha\prime}}{\rho_q} \frac{\partial \hat{u}_l^{\beta}}{\partial r_k} \right\} &= -\left[\sum_{kl} \frac{\hat{\sigma}_{kl}^{\alpha\prime}}{\rho_q} \hat{u}_l^{\beta} \right]_0^L + \int_0^L \mathrm{d}r_k \left\{ \sum_{kl} \frac{\partial}{\partial r_k} \frac{\hat{\sigma}_{kl}^{\alpha\prime}}{\rho_q} \hat{u}_l^{\beta} \right\} \\
&= \int_0^L \mathrm{d}r_k \left\{ \sum_{kl} \frac{\partial}{\partial r_k} \frac{\hat{\sigma}_{kl}^{\alpha\prime}}{\rho_q} \hat{u}_l^{\beta} \right\} \\
&= \int_0^L \mathrm{d}r_k \left\{ \sum_{l} \frac{\hat{f}_l^{\alpha\prime}}{\rho_q} \hat{u}_l^{\beta} \right\}
\end{aligned}
\tag{6.1.45}
$$

Vanishing stress at the surface was used in step 3. Again, Eq. 6.1.39 is confirmed after integration over the two remaining dimensions.

6.2 Third-Order Perturbation Analysis

The derivation of the SLA had relied on a linearization of the parameter $\tilde{Z}_{mot}$ in $\tilde{\omega} - \tilde{\omega}_{OC}$ (Eq. 4.5.13). Also, the load impedance was evaluated at the reference frequency rather than the actual frequency. Can one do better? Yes, but one needs to keep track of both approximations, consistently. This is what is done in the perturbation calculation below. Improving on the SLA is not quite as academic as one might think. The limits of the conventional SLA are noticed not only when large frequency shifts are encountered, but also when one looks into the details of the dependence of the $\Delta\tilde{f}$ on n. This, one would do in order to derive J' and J'' following Eq. 10.1.9. The viscoelastic correction Eq. 10.1.9 is a typical second-order effect in the sense that the SLA can produce incorrect results.

For an illustration of the short-comings of the SLA, consider a thin film in air and assume that the film has the exact same acoustic properties as the quartz crystal. In this case the fractional frequency shift, $\Delta\tilde{f}/f_{ref}$, must be strictly the same on all overtones. This follows from the fact that the resonance frequency is inversely proportional to the total thickness of the plate, including the film. One has

$$
\frac{\Delta f}{f_{ref}} = \frac{d_q}{d_q + d_f} - 1 = d_q\left(\frac{1}{d_q + d_f} - \frac{1}{d_q}\right) = \frac{1}{\left(1 + \frac{d_f}{d_q}\right)} - 1
\tag{6.2.1}
$$

There is no dependence on overtone order. One may go on and apply Taylor expansion in d_f ($(1 + \varepsilon)^{-1} \approx 1 - \varepsilon$), arriving at

$$\frac{\Delta f}{f_{ref}} \approx -\frac{d_f}{d_q} \tag{6.2.2}$$

Regardless of whether or not the Taylor expansion in d_f is applied: The fractional frequency shift is independent of n. Contrasting to this result, an n-dependence *is* found when applying the SLA and inserting the load impedance of the thin viscoelastic film according to Eq. 10.1.9. For the fractional frequency shift, Eq. 10.1.9 predicts

$$\frac{\Delta \tilde{f}}{f_{ref}} = -\frac{m_f}{m_q}\left(1 + \frac{(n\pi)^2}{3}\frac{\tilde{J}_f}{\rho_f}Z_q^2\left(\frac{m_f}{m_q}\right)^2\right) \tag{6.2.3}$$

The relations $f_{ref} \approx nf_0$ and $m_q = Z_q/(2f_0)$ were used. Unless the compliance of the film, $\tilde{J}_f$, is zero (which is impossible), there is a nontrivial dependence of $\Delta\tilde{f}$ on overtone order, which is in conflict with the simple and general result from Eq. 6.2.1. The error occurs because of the approximations leading to the SLA are not accurate enough for this particular problem.

Requiring that the mechanical impedance, $\tilde{Z}_{mot}$, be zero on resonance (Fig. 4.10 and Eq. 4.5.9) amounts to an implicit equation in the resonance frequency, f_r. In principle, this implicit equation has to be solved twice (with and without the load) in order to determine $\Delta\tilde{f}$. The SLA avoids the implicit equation, making two separate approximations:

- The term $\cot(\tilde{k}_q h_q)$ is expanded to first order in $\tilde{k}_q - n\pi/(2h_q)$, leading to Eq. 4.5.13. Equation 4.5.13 is linear in both $\tilde{\omega} - \tilde{\omega}_{OC}$ and $\tilde{Z}_L$. It can explicitly solved for $\tilde{\omega} - \tilde{\omega}_{OC}$. The expansion to first order in $\tilde{k}_q - n\pi/(2h_q)$ contrasts to Eq. 10.1.9, where the load impedance was expanded to 3rd order in thickness. There is an inconsistency.
- The load impedance to be inserted on the right-hand side of the SLA (Eq. 4.6.1) was evaluated at the reference frequency, as opposed to the actual frequency, which is $f_{ref} + \Delta f$.

These inconsistencies can be removed with an iterative scheme described in detail in Ref. [15]. In the following, we sketch the argument and cite the central results. We call the calculation a "perturbation analysis". A perturbation analysis in this sense deals with an implicit equation, which cannot be solved analytically. Assume that the parameter to be found is called y. One starts from a guess (call it y_0) and expresses the true result as $y = y_0 + \Delta y$. In a second step, one linearizes the implicit equation in Δy. One solves this approximate relation for Δy and calls the result Δy_0. A second guess is thereby obtained, equal to $y_1 = y_0 + \Delta y_0$. One again expresses the true result as $y = y_1 + \Delta y$, linearizes the resulting implicit equation in Δy, solves for Δy, and arrives at a new, improved guess. One iterates and hopes that the sequence of guesses converges to the true result. The procedure may be worth the effort because the intermediate results are *explicit* equations. Using perturbation analysis, one obtains of sequence of analytical expressions, which become more and more accurate as the perturbation order increases (and become more and more complicated at the same time). Using perturbation analysis, one can improve on the SLA without resorting to numerical solutions.

In the following, piezoelectric stiffening is neglected. The analysis starts out from Eq. 4.6.8. Slightly rearranged, Eq. 4.6.8 reads as

$$\tilde{Z}_L\left(\tilde{f}_{ref}+\Delta\tilde{f}\right)=-\mathrm{i}\tilde{Z}_q\tan\left(\pi\frac{\Delta\tilde{f}}{\tilde{f}_0}\right)=-\mathrm{i}\tilde{Z}_q\tan\left(n\pi\frac{\Delta\tilde{f}}{\tilde{f}_{ref}}\right)\equiv\tilde{Z}_{cr}\left(\Delta\tilde{f}\right)\qquad(6.2.4)$$

Equation 6.2.4 is the implicit equation to be solved for $\Delta\tilde{f}$. $\tilde{Z}_{cr}(\Delta\tilde{f})$ is the stress-velocity ratio at the interface, where the stress is exerted from the side of the *cr*ystal. Equation 6.2.4 expresses the fact that the stress exerted by the crystal onto the sample must be same as reverse stress (exerted by the sample onto the crystal). This is the principle of reaction.

For notational brevity, the normalized frequency shift is defined as $\Delta\tilde{f}_N=\Delta\tilde{f}/f_{ref}$. Both sides of Eq. 6.2.4 can be Taylor-expanded in the small perturbation parameter $\Delta\tilde{f}_N$ as

$$\begin{aligned}&\tilde{Z}_L+\frac{\mathrm{d}\tilde{Z}_L}{\mathrm{d}\Delta\tilde{f}_N}\Delta\tilde{f}_N+\frac{1}{2}\frac{\mathrm{d}^2\tilde{Z}_L}{\mathrm{d}\Delta\tilde{f}_N^2}\Delta\tilde{f}_N^2+\frac{1}{6}\frac{\mathrm{d}^3\tilde{Z}_L}{\mathrm{d}\Delta\tilde{f}_N^3}\Delta\tilde{f}_N^3\\&\approx\frac{\mathrm{d}\tilde{Z}_{cr}}{\mathrm{d}\Delta\tilde{f}_N}\Delta\tilde{f}_N+\frac{1}{2}\frac{\mathrm{d}^2\tilde{Z}_{cr}}{\mathrm{d}\Delta\tilde{f}_N^2}\Delta\tilde{f}_N^2+\frac{1}{6}\frac{\mathrm{d}^3\tilde{Z}_{cr}}{\mathrm{d}\Delta\tilde{f}_N^3}\Delta\tilde{f}_N^3\end{aligned}\qquad(6.2.5)$$

There is no 0th-order term on the right-hand side because $\tilde{Z}_{cr}(f_{ref})=0$. All derivatives are evaluated at the reference frequency. The Taylor expansion is terminated after the 3rd term because the perturbation analysis will also terminate after the 3rd iteration.

In the first step, one evaluates the leading orders in Eq. 6.2.5 and solves for $\Delta\tilde{f}/f_{ref}$, considering only the leading terms. The 0th-order term on the right-hand side vanishes. Solving Eq. 6.2.5 to 1st order in $\Delta\tilde{f}_N$, one finds

$$\begin{aligned}\tilde{Z}_L+\frac{\mathrm{d}\tilde{Z}_L}{\mathrm{d}\Delta\tilde{f}_N}\Delta\tilde{f}_N&\approx\frac{\mathrm{d}\tilde{Z}_{cr}}{\mathrm{d}\Delta\tilde{f}_N}\Delta\tilde{f}_N\\\Delta\tilde{f}_N^{[1]}=\frac{\tilde{Z}_L}{\frac{\mathrm{d}\tilde{Z}_{cr}}{\mathrm{d}\Delta\tilde{f}_N}-\frac{\mathrm{d}\tilde{Z}_L}{\mathrm{d}\Delta\tilde{f}_N}}&=\frac{\tilde{Z}_L}{-\mathrm{i}Z_q n\pi-\frac{\mathrm{d}\tilde{Z}_L}{\mathrm{d}\Delta\tilde{f}_N}}\end{aligned}\qquad(6.2.6)$$

Superscripts in square brackets denote the perturbation order. If the second term in the denominator is neglected, the result is equivalent to the SLA.

In the next step one writes $\Delta\tilde{f}_N$ as $\Delta\tilde{f}_N^{[1]} + \Delta\tilde{f}_N^{[2]}$ and inserts this expression into Eq. 6.2.5:

$$\begin{aligned}\tilde{Z}_L + \frac{d\tilde{Z}_L}{d\Delta\tilde{f}_N}\left(\Delta\tilde{f}_N^{[1]} + \Delta\tilde{f}_N^{[2]}\right) + \frac{1}{2}\frac{d^2\tilde{Z}_L}{d\Delta\tilde{f}_N^2}\left(\Delta\tilde{f}_N^{[1]} + \Delta\tilde{f}_N^{[2]}\right)^2 \\ + \frac{1}{6}\frac{d^3\tilde{Z}_L}{d\Delta\tilde{f}_N^3}\left(\Delta\tilde{f}_N^{[1]} + \Delta\tilde{f}_N^{[2]}\right)^3 \\ \approx \frac{d\tilde{Z}_{cr}}{d\Delta\tilde{f}_N}\left(\Delta\tilde{f}_N^{[1]} + \Delta\tilde{f}_N^{[2]}\right) + \frac{1}{2}\frac{d^2\tilde{Z}_{cr}}{d\Delta\tilde{f}_N^2}\left(\Delta\tilde{f}_N^{[1]} + \Delta\tilde{f}_N^{[2]}\right)^2 \\ + \frac{1}{6}\frac{d^3\tilde{Z}_{cr}}{d\Delta\tilde{f}_N^3}\left(\Delta\tilde{f}_N^{[1]} + \Delta\tilde{f}_N^{[2]}\right)^3\end{aligned} \quad (6.2.7)$$

As before, the equation is linearized in $\Delta\tilde{f}_N^{[2]}$ and the linearized equation is solved for $\Delta\tilde{f}_N^{[2]}$. Necessarily, the resulting expressions become clumsy. One repeats this procedure one more time, that is, one writes $\Delta\tilde{f}_N$ as $\Delta\tilde{f}_N^{[1]} + \Delta\tilde{f}_N^{[2]} + \Delta\tilde{f}_N^{[3]}$ and solves for $\Delta\tilde{f}_N^{[3]}$. We stop after the 3rd order.

Fortunately, the resulting equations are not as complicated as one might think. If one Taylor-expands all results to 3rd order in the mass of the film and to 3rd order in the viscosity of the liquid, many terms cancel. Taylor expansion to 3rd order is consistent with a perturbation analysis carried to 3rd order because the normalized frequency shift is of the same order of magnitude as the normalized mass and the normalized shear wave impedance of the liquid (see below).

For the film in air, the perturbation analysis leads to:

$$\begin{aligned}\frac{\Delta\tilde{f}}{f_{ref}} &= -\frac{m_f}{m_q}\left[1 + \frac{(n\pi)^2}{3}\left(\frac{Z_q^2}{\tilde{Z}_f^2} - 1\right)\left(\frac{m_f}{m_q}\right)^2\right] \\ &= -\frac{m_f}{m_q}\left[1 + \frac{(n\pi)^2}{3}\left(\frac{\tilde{J}_f}{\rho_f}Z_q^2 - 1\right)\left(\frac{m_f}{m_q}\right)^2\right]\end{aligned} \quad (6.2.8)$$

Clearly, there is a difference between Eqs. 6.2.3 and 6.2.8. The difference is much welcomed, because the coefficient to the n^2-term now vanishes if the properties of the film and the crystal are the same (if $\tilde{J}_f/\rho_f = Z_q^{-2}$). There is no n–dependence for this case, as demanded by Eq. 6.2.1. The difference between Eqs. 10.1.9 and 6.2.8 is important for all experiments on films with an acoustic wave impedance comparable to Z_q. Many solid films have a shear-wave impedance comparable to Z_q. Naively applied to experiments on such samples, Eq. 10.1.9 can easily lead to negative shear moduli.

We quote the results from the 3rd-order perturbation analysis for a few different configurations below. The following nondimensional variables are introduced for notational convenience:

$$\mu_e = \frac{m_e}{m_q}, \quad \mu_f = \frac{m_f}{m_q}, \quad \tilde{\xi}_{liq}(\omega) = \frac{\tilde{Z}_{liq}(\omega)}{Z_q}$$
$$\tilde{\zeta}_e(\omega) = \frac{Z_q^2}{\tilde{Z}_e^2(\omega)} - 1 = \frac{\tilde{J}_e(\omega)}{\rho_e} Z_q^2 - 1, \quad \tilde{\zeta}_f(\omega) = \frac{Z_q^2}{\tilde{Z}_f^2(\omega)} - 1 = \frac{\tilde{J}_f(\omega)}{\rho_f} Z_q^2 - 1 \tag{6.2.9}$$

The parameter μ is a dimensionless measure of the film thickness. The indices *e* and *f* denote the first and the second layer (the *e*lectrode and the *f*ilm). $\tilde{\xi}_{liq}$ is a dimensionless measure of the viscosity of the liquid. The parameters μ_e, μ_f, and $\tilde{\xi}_{liq}$ are considered as small. Since μ_e and μ_f are small, the calculation only covers thin films, it does not capture the film resonance. The parameter $\tilde{\xi}$ is a dimensionless measure of the shear compliance. $\tilde{\xi}_e$ and $\tilde{\xi}_f$ are not necessarily small. Note that Eqs. 6.2.10–6.2.16 have the quantity $\Delta\tilde{f}_N = \Delta\tilde{f}/f_{ref}$ on the left-hand side, as opposed to $\Delta\tilde{f}/f_0$, which is the left-hand side of the SLA. Also, the reference state always is the bare crystal. Should the crystal in the reference state be loaded already (for instance with an electrode or with a bulk liquid), the respective subtractions have to be carried out.

The relations derived for the various cases are the following:

(a) *Semi-infinite viscoelastic medium*

$$\frac{\Delta\tilde{f}}{f_{ref}} \approx \frac{i}{n\pi}\left(\tilde{\xi}_{liq} + \frac{1}{3}\tilde{\xi}_{liq}^3\right) \tag{6.2.10}$$

The structure of this equation suggests that the second term might explain the experimental result reported around Eq. 12.4.6, but the numbers do not work out. For practical matters, the second term is negligible.

(b) *Single viscoelastic film in air*

$$\frac{\Delta\tilde{f}}{f_{ref}} \approx -\mu_f + \mu_f^2 - \left(1 + \frac{1}{3}(n\pi)^2\tilde{\zeta}_f\right)\mu_f^3 \tag{6.2.11}$$

The coefficient of the term $(n\pi)^2$ is $1/3\left(\tilde{J}_f\big/\rho_f Z_q^2 - 1\right)$, not $1/3\left(\tilde{J}_f\big/\rho_f Z_q^2\right)$ as in Eq. 10.1.9. Again: This statement concerns a single film *without electrodes*.

(c) *Viscoelastic film in a liquid*

$$\frac{\Delta\tilde{f}}{f_{ref}} \approx \frac{\mathrm{i}}{n\pi}\left(\tilde{\xi}_{liq} + \frac{1}{3}\tilde{\xi}_{liq}^{3}\right) - \left(1 + \frac{\mathrm{i}\tilde{\xi}_{liq}}{n\pi} - \tilde{\zeta}_f\tilde{\xi}_{liq}^{2}\right)\mu_f$$
$$+ \left(1 + \left(\frac{\mathrm{i}}{n\pi} + in\pi\tilde{\zeta}_f\right)\tilde{\xi}_{liq}\right)\mu_f^2 - \left(1 + \frac{(n\pi)^2}{3}\tilde{\zeta}_f\right)\mu_f^3 \qquad (6.2.12)$$

For the film in a liquid, the conventional SLA is good enough in most regards. The (small) modification is discussed for the sake of completeness. A typical situation, where Eq. 6.2.12 would be employed, is adsorption from solution. In this case, the reference state usually is the resonator immersed in buffer solution. Also, the film usually is so thin that only the term of 1st order in μ_f is of importance. Subtracting the contribution from the pure liquid and omitting terms proportional to μ_f^2 and μ_f^3, one arrives at:

$$\frac{\Delta\tilde{f}}{f_{ref}} \approx -\left(1 + \frac{\mathrm{i}\tilde{\xi}_{liq}}{n\pi} - \tilde{\zeta}_f\tilde{\xi}_{liq}^{2}\right)\mu_f = -\frac{m_f}{m_q}\left(1 - \frac{\tilde{Z}_{liq}^2}{\tilde{Z}_f^2} + \frac{\tilde{Z}_{liq}^2}{\tilde{Z}_q^2} + \frac{\mathrm{i}\sqrt{\mathrm{i}\omega\rho_{liq}\eta_{liq}}}{n\pi Z_q}\right) \qquad (6.2.13)$$

In order to make the comparison with Eq. 10.2.6, we normalize to the frequency of the fundamental f_0, rather than to f_{ref}. Also, the liquid is assumed to Newtonian. Separating the real and the imaginary parts, one finds:

$$\frac{\Delta f}{f_0} \approx \frac{-\omega m_f}{\pi Z_q}\left[1 - 2\pi\mathrm{i}n\frac{J_f''}{\rho_f}f_0\rho_{liq}\eta_{liq} + \frac{\sqrt{\omega\rho_{liq}\eta_{liq}}}{\sqrt{2}n\pi Z_q}\right]$$
$$\frac{\Delta\Gamma}{f_0} \approx \frac{\omega m_f}{\pi Z_q}\left[2\pi n\frac{J_f'}{\rho_f}f_0\rho_{liq}\eta_{liq} - \frac{\sqrt{\omega\rho_{liq}\eta_{liq}}}{\sqrt{2}n\pi Z_q} - \frac{\omega\rho_{liq}\eta_{liq}}{Z_q^2}\right] \qquad (6.2.14)$$

Comparing to Eq. 10.2.6, one does find a nontrivial contribution from the liquid. The contribution is small, though. At 5 MHz in water, the magnitude of the term containing square roots is 1.5×10^{-4}. The last term in line 2 is even smaller.

(d) *Two viscoelastic films in air*

$$\frac{\Delta\tilde{f}}{f_{ref}} \approx -\mu_e + \mu_e^2 - \left(1 + \frac{(n\pi)^2}{3}\tilde{\zeta}_e\right)\mu_e^3 - \left(1 - 2\mu_e + 3\left(1 + \frac{(n\pi)^2}{3}\tilde{\zeta}_e\right)\mu_e^2\right)\mu_f$$
$$+ \left(1 - 3\left(1 + \frac{(n\pi)^2}{3}\tilde{\zeta}_e\right)\mu_e\right)\mu_f^2 - \left(1 + \frac{(n\pi)^2}{3}up\tilde{\zeta}_f\right)\mu_f^3 \qquad (6.2.15)$$

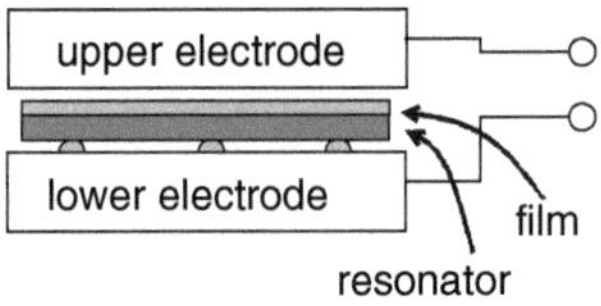

Fig. 6.1 An electrode-less quartz can be excited across an air gap. The blank rests on a metal plate, supported by a few asperities of the rough lower surface. The contact radii of the load-bearing asperities can be so small that the contact to the bottom plate does not affect on the resonance frequency (Sect. 11.3). This configuration avoids an influence of the electrodes on the measurement of the film's shear compliance. However, there are effects of piezoelectric stiffening. The resonance frequency depends on the width of the gap [16]

This case is important because the first "film" may be the electrode. Equation 6.2.15 quantifies the electrode effects. For the viscoelastic film on an electrode, one finds a contribution to the slope in the plot of $\Delta f/f_{ref}$, versus n^2, which is *unrelated to the softness of the film*. It is related to the softness of the *electrode*. The film exerts a shear stress onto the electrode. If the electrode has a finite compliance (it does), this finite compliance induces viscoelastic corrections scaling as $(n\pi)^2$. This electrode effect scales as the mass of the film (see the 4*th* term in Eq. 6.2.15), while the effect of the film's own finite compliance scales as μ_f^3 (last term in Eq. 6.2.15). This electrode effect is among the problems with the Sauerbrey equation outlined below Eq. 8.1.4. The electrode effects also complicate the determination of a film's softness. (A second source of problems is the dependence of the modal mass on overtone order, Sect. 6.1.4.) One would have to know the properties of the electrode well in order to apply Eq. 6.2.15. Rather than trying to gather this information, one will usually employ bare quartz blanks and excite them across an air gap as shown in Fig. 6.1. A second option is to use aluminum electrodes. The shear-wave impedance of aluminum is close to Z_q and the parameter ζ_e in Eq. 6.2.15 then becomes small.

This still leaves the problem that the viscoelastic effects originating from the film scale as the cube of the film thickness. Monomolecular layer are not soft enough to let the finite softness be measurable. Of course one can use Eq. 6.2.15 to one's advantage, by sandwiching the film of interest between the crystal and the electrode [17]. In this geometry, the electrode "clamps" the film from the other side and the viscoelastic effects become linear in film thickness. A similar experiment was reported by Schilling and Pechhold in 1969 [18]. These authors sandwiched the sample between two resonators, rather than placing it between the resonator surface and a second, heavy film above it.

(e) *Two viscoelastic films in liquid*

As for the single film in a liquid, the corrections following from perturbation analysis are small. The result is provided for the sake of completeness:

$$
\begin{aligned}
\frac{\Delta\tilde{f}}{f_{ref}} \approx & \frac{\mathrm{i}}{n\pi}\left(\tilde{\xi}_{liq} + \frac{1}{3}\tilde{\xi}_{liq}^3\right) - \left(1 + \frac{\mathrm{i}}{n\pi}\tilde{\xi}_{liq} - \tilde{\zeta}_e\tilde{\xi}_{liq}^2\right)\mu_e \\
& + \left(1 + \left(\frac{\mathrm{i}}{n\pi} + in\pi\tilde{\zeta}_e\right)\tilde{\xi}_{liq}\right)\mu_e^2 - \left(1 + \frac{1}{3}(n\pi)^2\tilde{\zeta}_e\right)\mu_e^3 \\
& - \left(1 + \frac{\mathrm{i}}{n\pi}\tilde{\xi}_{liq} - \tilde{\zeta}_f\tilde{\xi}_{liq}^2 - 2\left(1 + \left(\frac{\mathrm{i}}{n\pi} + in\pi\tilde{\zeta}_e\right)\tilde{\xi}_{liq}\right)\mu_e + 3\left(1 + \frac{1}{3}(n\pi)^2\tilde{\zeta}_e\right)\mu_e^2\right)\mu_f \\
& + \left(1 + \left(\frac{\mathrm{i}}{n\pi} + in\pi\tilde{\zeta}_f\right)\tilde{\xi}_{liq} - 3\left(1 + \frac{1}{3}(n\pi)^2\tilde{\zeta}_e\right)\mu_e\right)\mu_f^2 - \left(1 + \frac{1}{3}(n\pi)^2\tilde{\zeta}_f\right)\mu_f^3
\end{aligned}
\tag{6.2.16}
$$

Glossary

Variable	**Definition**
$\langle . \rangle$	Average
[.]	As a superscript: perturbation order
A	Effective area of the resonator plate
$\underline{A}$	ω^2-operator
c_{ijkl}	Stiffness tensor
comp	As an index: related to *comp*ressional waves
c_D	Stiffness tensor at constant electric displacement (Eq. 6.1.33)
cr	As an index: exerted by the *cr*ystal
d_f	Film thickness
diss	As an index: caused by *diss*ipative processes
d_q	Thickness of the resonator
$\hat{\mathbf{D}}, \mathbf{D}$	Electric displacement
e	As an index: electrode
$\hat{\mathbf{E}}, \mathbf{E}$	Electric field
f_n	Resonance frequency at overtone order n
f_r	Resonance frequency
f_0	Resonance frequency at the fundamental ($f_0 = Z_q/(2m_q) = Z_q/(2\rho_q d_q)$)
G_q	Shear modulus of AT-cut quartz ($G_q \approx 29 \times 10^9$ Pa)
h	One of the tensors quantifying piezoelectric coupling (Eq. 6.1.33)

k	Wavenumber
h_q	Half the thickness of the resonator plate
I	Second moment of area
$\tilde{J}$	Shear compliance
L	Width of a resonator plate
liq	As an index: *liq*uid
m_f	Mass per unit area of a *film*
m_q	Mass per unit area of the resonator ($m_q = \rho_q d_q = Z_q/(2f_0)$)
n	Overtone order
$\mathbf{n}$	$n_{i,}$ Surface normal (a vector)
OC	As an index: resonance condition of the unloaded plate under open-circuit conditions. With no current into the electrodes (more precisely, with vanishing electric displacement everywhere), piezoelectric stiffening is fully accounted for by using the piezoelectrically stiffened shear modulus
q	As an index: *q*uartz resonator
$\mathbf{r}, r_i$	Position, a vector and its components
ref	As an index: *ref*erence state of a crystal in the absence of a load
$\mathbf{r}_S$, $r_{S,i}$	Position on the resonator surface, a vector and its components
S	As an index: *S*urface
S	Infinitesimal strain tensor (Eq. 6.1.33)
T	Temperature
T	Stress tensor (Eq. 6.1.33)
$\hat{\mathbf{u}}$	Displacement (a vector)
$\hat{\mathbf{v}}$	Velocity (a vector)
$\tilde{Z}_{L,ijk}$	Load impedance in tensor form
$\tilde{Z}_{mot}$	Impedance to the left of the transformer in the Mason circuit (Fig. 4.10)
Z_q	Acoustic wave impedance of AT-cut quartz ($Z_q = 8.8 \times 10^6$ kg m^{-2} s^{-1})
Γ	Imaginary part of a resonance frequency
$\delta_{\alpha\beta}$	Kronecker δ ($\delta_{\alpha\beta} = 1$ if $\alpha = \beta$, $\delta_{\alpha\beta} = 0$ otherwise)
$\delta(.)$	Dirac δ-function
ε_S^{-1}	Inverse dielectric permittivity at constant strain (a tensor) (Eq. 6.1.33)
Δ	As a prefix: A shift induced by the presence of the sample

$\hat{\sigma}_{S,ij}$	Stress tensor
η_{liq}	Viscosity
μ	Non-dimensional mass (Eq. 6.2.9)
ρ	Density
ξ_{liq}	Nondimensional shear-wave impedance of the bulk liquid (Eq. 6.2.9)
ω	Angular frequency
ω_r	Angular resonance frequency ($-\omega_r^2$ is the eigenvalue of the ω^2-operator)
ζ	Non-dimensional measure of the inverse square shear-wave impedance (Eq. 6.2.9)

References

1. Sakurai, J.J.: Modern Quantum Mechanics. Addison Wesley, New York (1985)
2. Ohno, I.: Rectangular parallellepiped resonance method for piezoelectric-crystals and elastic-constants of alpha-quartz. Phys. Chem. Miner. **17**(5), 371–378 (1990)
3. Ogi, H., Ohmori, T., Nakamura, N., Hirao, M.: Elastic, anelastic, and piezoelectric coefficients of alpha-quartz determined by resonance ultrasound spectroscopy. J. Appl. Phys. **100**(5), 053511 (2006)
4. Herrscher, M., Ziegler, C., Johannsmann, D.: Shifts of frequency and bandwidth of quartz crystal resonators coated with samples of finite lateral size. J. Appl. Phys. **101**(11), 114909 (2007)
5. Rabello-Soares, M.C., Korzennik, S.G., Schou, J.: The determination of MDI high-degree mode frequencies. ESA Spec. Publ. **464**, 129–136 (2001)
6. http://www.guardian.co.uk/science/2003/jul/24/research.science. Accessed 11 Mar 2013
7. Pechhold, W.: Zur Behandlung von Anregungs- und Störungsproblemen bei akustischen Resonatoren. Acustica **9**, 48–56 (1959)
8. http:/en.wikipedia.org/wiki/Euler%E2%80%93Bernoulli_beam_theory. Accessed 30 Mar 2013
9. Gere, J.M., Timoshenko, S.P.: Mechanics of Materials. PWS Publishing Company, Boston (1997)
10. Mason, W.P., Baker, W.O., McSkimin, H.J., Heiss, J.H.: Measurement of shear elasticity and viscosity of liquids at ultrasonic frequencies. Phys. Rev. **75**(6), 936–946 (1949)
11. Raiteri, R., Grattarola, M., Butt, H.J., Skladal, P.: Micromechanical cantilever-based biosensors. Sens. Actuators B-Chem. **79**(2–3), 115–126 (2001)
12. Ballantine, D.S., White, R.M., Martin, S.J., Ricco, A.J., Zellers, E.T., Frye, G.C., Wohltjen, H.: Acoustic Wave Sensors: Theory, Design and Physico-Chemical Applications. Academic Press, New York (1997)
13. Hempel, U., Lucklum, R., Hauptmann, P.R., EerNisse, E.P., Puccio, D., Diaz, R.F.: Quartz crystal resonator sensors under lateral field excitation—a theoretical and experimental analysis. Measur. Sci. Technol. **19**(5), 055201 (2008)
14. http://www.comsol.com/model/download/177395/models.mems.thickness_shear_quartz_oscillator.pdf. Accessed 20 Feb 2014

15. Johannsmann, D.: Derivation of the shear compliance of thin films on quartz resonators from comparison of the frequency shifts on different harmonics: a perturbation analysis. J. Appl. Phys. **89**(11), 6356–6364 (2001)
16. Wolff, O., Seydel, E., Johannsmann, D.: Viscoelastic properties of thin films studied with quartz crystal resonators. Faraday Discuss. **107**, 91–104 (1997)
17. Wolff, O., Johannsmann, D.: Shear moduli of polystyrene thin films determined with quartz crystal resonators in the sandwich configuration. J. Appl. Phys. **87**(9), 4182–4188 (2000)
18. Schilling, H., Pechhold, W.: Two quartz resonator methods for investigation of complex shear modulus of polymers. Acustica **22**(5), 244 (1969)

Chapter 7
Energy Trapping and Its Consequences

Abstract Most quartz resonators apply energy trapping. By giving the electrodes the shape of a keyhole, the acoustic thickness of the plate is made larger in the center than at the rim. Such a plate can be viewed as an acoustic cavity, where the surfaces are shaped such that they focus the acoustic energy to the center. Energy trapping has numerous consequences, among them the flexural contributions to the vibration pattern, which lead to the emission of compressional waves.

7.1 General

In Chap. 4, the resonators were modeled as parallel plates and the displacement fields were plane waves, in consequence. Unfortunately, parallel plates are impractical because the resonator must be mounted in one way or another and the holder will damp the oscillation if the amplitude of vibration is large at the point of contact with the holder. Even if that problem could be solved, the edges of the plate would still distort the plane wave. Only a laterally infinite parallel plate is a perfectly parallel plate the sense of this model.

A similar problem exists in optics. Light often is depicted as a plane wave, although plane waves do not exist. An ideal plane wave would have to extend to infinity in the lateral direction. If a plane wave is constrained by some kind of aperture, the aperture diffracts the beam. Real beams have a finite extension and, in consequence, a nonzero internal divergence.

If a beam's lateral extension much exceeds its wavelength (more precisely: if all in-plane gradients occur on a scale larger than λ), the beam is almost a plane wave in the sense that the "paraxial approximation" holds [1]. The paraxial approximation leads do a formalism, which is slightly more complicated than the plane-wave formalism, but has the attractive feature that it covers all three dimensions. One writes the optical field (or the acoustic field) as

D. Johannsmann, *The Quartz Crystal Microbalance in Soft Matter Research*, Soft and Biological Matter, DOI: 10.1007/978-3-319-07836-6_7,

$$\hat{u}(r,\varphi,z) = \hat{u}_E(r,\varphi,z)\exp(-ikz) \quad (7.1.1)$$

r is the distance to the symmetry axis and φ is the azimuthal angle. k is a fixed parameter with units of m^{-1}. In the limit of small internal divergence, k is the wavenumber. The rapidly-varying part of the wave is contained in the term $\exp(-ikz)$. The function $\hat{u}_E$ is a slowly varying complex amplitude. The index E stands for "envelope". $\hat{u}_E$ obeys the relation [1]

$$\nabla_\perp^2 \hat{u}_E + 2ik\frac{\partial \hat{u}_E}{\partial z} = 0 \quad (7.1.2)$$

$\nabla_\perp$ is the gradient in the plane perpendicular to the direction of propagation. In cylinder coordinates one has $\nabla_\perp^2 = 1/r\, \partial/\partial r(r\, \partial/\partial r) + 1/r^2\, \partial^2/\partial\varphi^2$.

The simplest solution to Eq. 7.1.2 is the "Gaussian beam". The Gaussian beam has a waist in the focal plane. As the name says, the field distribution in this plane is a Gaussian:

$$\hat{u}_E(r,\varphi,z=0) = \hat{u}_c \exp\left(-\frac{r^2}{2\sigma_G^2}\right) \quad (7.1.3)$$

$\hat{u}_c$ is the amplitude in the center of the plate and σ_G is the standard deviation of the Gaussian. The beam's internal divergence is inversely proportional to the width of the beam at the focus, which is $2\sigma_G$. In optics, it is not particularly difficult to achieve a Gaussian beam.

The recipe followed in acoustic sensing to confine the wave laterally is similar to the one applied in optics. The mechanism is called "energy trapping" [2]. The role of the mirrors of the optical cavity (Fig. 7.1a) is taken by the resonator surfaces. If strong lateral confinement of the wave is required, the resonator surfaces are indeed given a convex shape (Fig. 7.1c). Alternatively, a flat plate is used and focusing of the acoustic energy is achieved with electrodes covering the center of the plate only. The electrodes make the plate thicker in the center, compared to the edge. A flag and possibly an anchor connect the electrode to the electrical contacts at the edge of the plate. Both measures can be combined: Keyhole-shaped electrodes are also employed with convex crystals, and flat crystals are sometimes given a wedge at the edge (they are "beveled" [3]).

Convex surfaces provide for strong energy trapping. For such crystals, the distortion of the plane wave is so strong that these resonators behave poorly on the overtones. Strong energy trapping has the advantage that the crystals can be small. This saves cost on the side of crystal growth and it saves space on the printed circuit boards. For liquid sensing, these crystals cannot be used. They can be (and are) used as film thickness monitors in vacuum.

If the requirements on energy trapping are less stringent, one can use planar crystals and trap the energy with keyhole-shaped electrodes (Figs. 1.1 and 1.3). This type of crystals is applied in liquid sensing. The back electrode is smaller and thicker than the front electrode. It largely defines the amplitude distribution. The

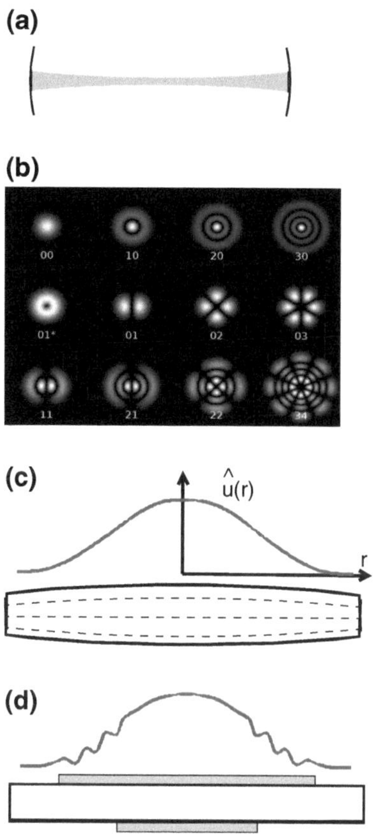

Fig. 7.1 **a** The mirrors at the ends of an optical cavity are curved in order to confine the beam laterally. **b** For spherical mirrors and small internal divergence, the solutions to Eq. 7.1.2 are the different modes of Gauss-Laguerre beam. The simplest mode is the "Gaussian beam". It has no nodal lines in focal plane. There also are modes with a more complicated field distribution. *Notation* The Gaussian beam has the subscript "00" here. Following the literature, the corresponding modes of thickness-shear resonators are called "01-modes". Panel **b** was produced by Bob Mellish and was reprinted from http://commons.wikimedia.org/wiki/File:Laguerre-gaussian.png, accessed 10 June 2014. **c** An amplitude distribution close to the Gaussian is achieved by giving the resonator surfaces a *convex shape*. **d** Applying electrodes only to the center of the crystal is a second method to confine the elastic wave. The resonator still is acoustically thicker in the center than at the rim. Since the front electrode is larger than the back electrode, it covers the entire active area. Because the paraxial approximation does not hold at the edges of the electrodes, the amplitude distribution differs from the Gaussian. It shows ripples on a scale comparable to λ

asymmetry serves two purposes. Firstly, a large, grounded front electrode prevents electric fringe fields from penetrating into the sample. Secondly, the front electrode covers most of the active area and the chemical properties of the sensor surface therefore are the same over the active area.

For reasons which are poorly understood, the fundamental mode of planar resonators with keyhole-shaped electrodes often behaves differently from the overtones. More often than not, all measurements are carried out on the overtones and these crystals therefore are also called overtone crystals. Among the reasons for poor performance on the fundamental presumably is poor energy trapping. When stating that the crystal is thicker in the center than at the edge, one must keep in mind that the yard stick is the wavelength of sound. For energy trapping to be effective, the electrode thickness must be a significant fraction of the wavelength of sound. That this is so, can also be understood from the Gaussian beam. For any given internal divergence of the beam, the diameter of the beam at the waist is proportional to the wavelength. The fundamental mode has the largest wavelength and, in consequence, the broadest amplitude distribution.

Looking at the details, one finds that the in-plane amplitude distribution at the resonator surface differs from a Gaussian [4]. The main reason is the crystal's anisotropy. The mathematical background was discussed by Bechmann [5, 6] and Tiersten [7–10]. However, a Gaussian beam is not actually needed to operate the crystal. A vanishing oscillation amplitude at the edge of the plate suffices, whatever the details of the amplitude distribution are.

All differences between acoustics and optics acknowledged, many features of paraxial optics carry over to the case of thickness-shear resonators with energy trapping:

- There are modes, which have nodal planes perpendicular to the focal plane. In optics, these "transverse modes" are described by the Laguerre polynomials [1]. Figure 7.1b displays a few transverse modes with low indices. The corresponding modes of the QCM are the *anharmonic sidebands* (Sect. 7.2).
- The Gaussian beam has maximum intensity in the center and tails with low intensity at large r. By analogy, the amplitude distribution of the acoustic 01-mode does not have nodal lines in the resonator plane. It has a maximum in the center and decays towards the edge (upper left in Fig. 7.1b, to the left in Figs. 7.2 and 7.3). Amplitude distributions are elaborated on in Sect. 7.3.
- Because of the internal divergence, the electric field vector of the Gaussian beam is no longer strictly perpendicular to the optical axis. There are electric field components along z. By analogy, the surface displacement of resonators with energy trapping applied is not strictly tangential to the resonator surface. There are *flexural deformations* (Sect. 7.6), which emit compressional waves.
- Because the 01-mode is not a plane wave, the resonance frequency at overtone n (n the number of nodal planes parallel to the resonator plane) is not an integer multiple of the fundamental frequency. The in-plane gradients of the displacement pattern increase the resonance frequency (Sect. 7.7).

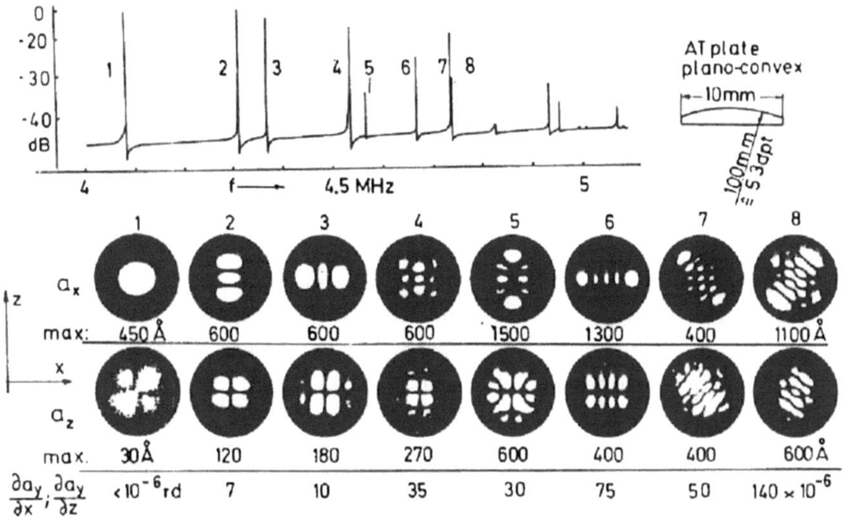

Fig. 7.2 Vibrational patterns of the 01-mode ("1") as well as the anharmonic side bands. The *top* shows the trace of $|\tilde{Y}_{el}|$ versus frequency. Reprinted with permission from Ref. [16]. Copyright 1982 Academic Press

Fig. 7.3 **a** A conductance trace around 15 MHz ($n = 3$), showing the main resonance as well as a few anharmonic sidebands. The sidebands are well separated from the main resonance, as they should be. **b** Same as in **a**, but now at 65 MHz ($n = 13$) for an aged crystal in water. The anharmonic sidebands merge with the 01-mode. Overlap of the 01-mode with a sideband is detrimental for sensing. (Note the imperfect calibration in panel **b**: G_{el} is shown as slightly negative at frequencies below the 01-mode.)

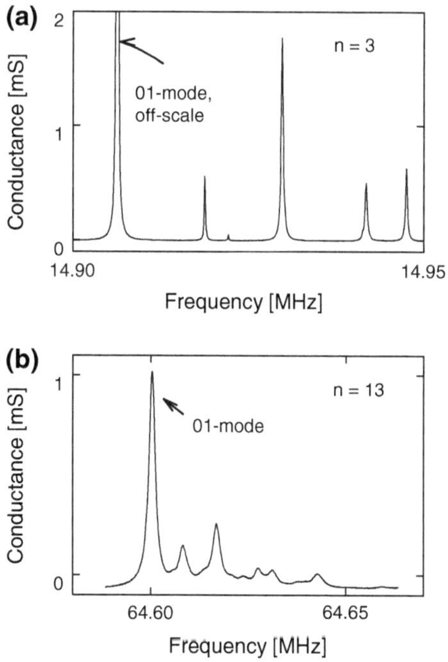

7.2 Anharmonic Sidebands

Similarly to optical cavities, the thickness-shear resonator has vibrational modes with nodal lines in the surface plane. (Strictly speaking, a nodal line should be called a nodal plane intersecting the resonator plane.) These modes are the "anharmonic sidebands" or "spurious modes". The term "anharmonic" is slightly misleading here because there are no overtones, which would be "harmonics" in the narrow sense of the word. All overtones occur at frequencies slightly displaced from the integer multiples of the fundamental. Strictly speaking, all overtones could be called "anharmonic", including the 01-modes. (Some people would not even use the word "overtone", but rather speak of "higher modes".)

Anharmonic sidebands are observed as peaks in the conductance spectrum appearing to the right of the main resonance (Figs. 7.2 and 7.3). The main resonance is called the "01-mode" because the edge (where the amplitude vanishes) is viewed as an azimuthal nodal line. There is no radial nodal line, hence the indices "01". The sidebands have a lower magnitude in the conductance trace than the 01-mode because the piezoelectrically induced polarization changes sign at the nodal lines. A large part of the current flows inside the electrode between the regions of different polarity. These local currents do not contribute to the current measured from the outside.

Different methods have been devised to visualize the deformation patterns of the various modes [11–15]. The book by Bahadur and Parshad gives an overview [16]. Figure 7.2 was taken from the work by Sauerbrey [17]. The line at the top is the electrical admittance, $|\tilde{Y}_{el}|$, versus ω. The bright patterns at the bottom show the various mode shapes. The patterns are similar to the transverse modes in Fig. 7.1b, but they differ in the details.

Sensing experiments are mostly carried out on the 01-modes. For any given overtone order (labeled by number of nodal planes parallel to the resonator surface, n), the 01-mode should display the largest peak in the conductance trace and at the same time have the lowest frequency in a group of peaks at the respective overtone order (Fig. 7.3a). If the peak with the lowest frequency is not the largest peak, this indicates poor crystal quality. If new crystals show conductance traces of this kind, these may go back to a wedge distortion of the plate (an insufficient parallelity of the two surfaces). Old crystals often have suffered to different degrees in different regions of the plate, which also induces asymmetries and enhances the anharmonic sidebands at the expense of the 01-mode.

The anharmonic sidebands must always be well separated in frequency from the 01-mode. If the 01-mode overlaps with a sideband (as in Fig. 7.3b), the results of the experiment cannot be trusted. This requirement poses a limit on the highest usable overtone. At high overtones, the pattern of anharmonic sidebands becomes increasingly dense. Eventually, the sidebands overlap with the 01-mode. Overlap with anharmonic sidebands renders a resonance useless, because the two modes couple to each other and because the degree of coupling is affected by the sample (Sect. 7.5).

There is not much work reported in the literature, where anharmonic sidebands would have been put to use. Goka et al. propose the use of sidebands for multi-channel-gravimetry in Ref. [18]. One might place dots coated with different receptor layers onto a resonator. Since the amplitude distributions of the various sidebands are different, they should respond to the mass adsorbed on the different spots to different extents. There will be a matrix relating the masses on the spots to the frequency shifts on the different sidebands. One should be able to calculate the masses adsorbed to the different spots from the frequency shifts on the different sidebands, using the inverse of this matrix.

A second potential use of the anharmonic sidebands is to monitor the temperature of the crystal. Since the T-f characteristic differs between the 01-mode and the anharmonic sidebands, one can infer temperature from the comparison, assuming that both modes respond in the same way to the deposition of a mass.

7.3 Amplitude Distributions

If the crystal surface has spherical shape with a radius of curvature much larger than the crystal thickness, the paraxial approximation predicts the amplitude distribution of the 01-modes to be a Gaussian. For planar crystals with keyhole-shaped electrodes, the mathematics is somewhat different, but Gaussians are widely assumed anyway. Below are a few further remarks (all concerning the 01-modes).

- Because of anisotropy, the active area is elongated along the x-direction by about 25 % [19, 20].
- The available experimental data suggest that the actual distributions deviate from the Gaussian (even the elliptically distorted Gaussian). For instance, the distributions often do not display mirror symmetry about x and y, although they should. The reasons might have to do with crystal imperfections or with the flags of the electrodes pointing into one particular direction.
- Keyhole shaped electrodes can lead to amplitude distributions showing concentric rings. The rings can be observed when running the resonators at high amplitude and exposing them to particle dispersions [22, 23]. The particles preferentially adsorb in the quiet regions (Fig. 7.4). Similar rings have been seen in the context of electrochemical QCM experiments [21]. The spacing between the rings is about twice the resonator thickness, meaning about equal to the wavelength of sound. Revisiting the assumptions underlying the paraxial approximation, it is clear how these rings come about. The edges of the electrodes are sharp, which violates the condition that the in-plane gradients of all topographical features should be small compared to λ^{-1}. The edges scatter the acoustic beam, such that it has an in-plane component with wavelength λ. Such structures are also reported in old work on convex crystals [17]. However, they were only observed on high-order transverse modes, not on the 01-modes.

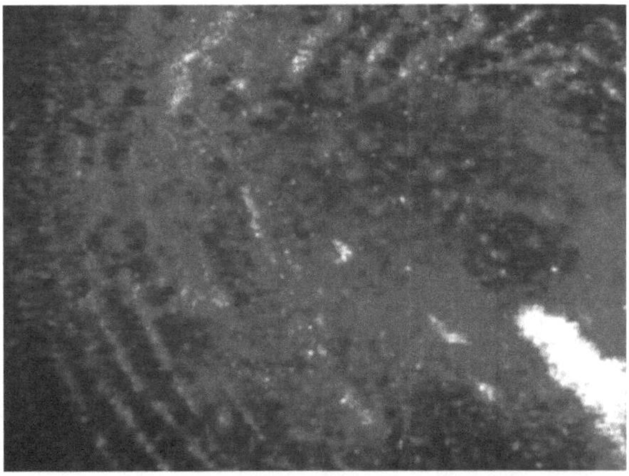

Fig. 7.4 When exposing a resonator driven at high amplitude to dispersion of particles, particles are found to preferentially adsorb in concentric rings. The amplitude of oscillation was about 30 nm. The rings developed after the resonator had been driven at this amplitude for about an hour. Similar ring-shaped structures have also been seen in Refs. [21, 22]

- The width of the Gaussian decreases with overtone order. This happens because the efficiency of energy trapping depends on the ratio of electrode thickness and wavelength. Since the wavelength scales as $1/n$, energy trapping becomes increasingly efficient at high n. When determining the active area of a resonator (for instance by applying droplets of known viscosity and variable radius, cf. Ref. [24]) one needs to repeat the determination for all overtones. The higher the overtone order, the sharper the amplitude distribution.
- The width of the amplitude distribution depends not only on overtone order, but it may also depend on whether or not a sample is present. This may happen when the sample touches the resonator in the center only (Ref. [25], Sect. 9.4).

7.4 Amplitude of Oscillation and Effective Area

In order to estimate the amplitude of oscillation and the effective area of crystals with energy trapping, one first has to go through the same calculation for the parallel plate. The amplitude of oscillation in the following is meant to be the absolute value of the displacement at the surface, $|\hat{u}_S|$. As long as the amplitude of oscillation is low enough, $|\hat{u}_S|$ is not actually relevant for sensing because it does not affect Δf and $\Delta\Gamma$. $|\hat{u}_S|$ is only relevant in the context of nonlinearities.

For the parallel plate, one can estimate the amplitude of oscillation from the Mason circuit. This amplitude is named $\hat{u}_{S,PP}$, where the index PP stands for "*P*arallel *P*late". Start from the equivalent circuit in Fig. 4.10a and set the elements $-\phi^2/(i\omega C_0)$ and $A\tilde{Z}_L$ to zero. Since the circuit is symmetric, the velocity $\hat{v}_{S,PP}$ is the same at both surfaces and given by half the velocity at the left-hand side of the transformer, $\hat{v}_{mot}$. ($\hat{v}_{mot}$ is a virtual quantity, only to be interpreted in the context of this circuit.) Following Eqs. 4.5.8a and d, $\hat{v}_S$ is related to the current into the electrodes by [26]

$$\hat{v}_{S,PP} = \frac{\hat{v}_{mot}}{2} = \frac{\hat{I}}{2\phi} = \frac{d_q}{2Ae_{26}}\hat{I} \tag{7.4.1}$$

Using the relations $\hat{v}_{S,PP} = i\omega\hat{u}_{S,PP}$, one finds [27]

$$\left|\hat{u}_{S,PP}\right| = \frac{d_q}{4\pi\, nf_0 Ae_{26}}\left|\hat{I}\right| \tag{7.4.2}$$

Inserting values ($d_q = 330$ µm, $f_0 = 5$ MHz, $A = 30$ mm^2, and $e_{26} = 9.65 \times 10^{-2}$ C/m^2), the prefactors turns into 1.8/n µm/A. Equation 7.4.2 is to be applied under conditions of current control. Under conditions of voltage control, one makes use of $\hat{I} = \hat{U}/R_1$. Inserting R_1 from Eq. 4.5.20, δ_q from Eq. 4.3.15, and ϕ from Eq. 4.5.8d, one finds [28]

$$\begin{aligned}\hat{v}_{S,PP} &= \frac{\hat{U}}{2\phi}\frac{1}{R_1} = \frac{\hat{U}}{2\phi}\frac{8\phi^2}{n\pi\, AZ_q\delta_q}\\ &= \hat{U}Q\frac{4}{n\pi\, AZ_q}\frac{Ae_{26}}{d_q}\\ &= \hat{U}Q\frac{4}{n\pi}\frac{G_q d_{26}}{Z_q d_q}\end{aligned} \tag{7.4.3}$$

In line 3, the piezoelectric stress coefficient, e_{26}, was replaced by $d_{26}G_q$, where d_{26} is the piezoelectric strain coefficient (more precisely: the relevant tensor component of the d-tensor, Table 5.1). For AT-cut quartz one has $d_{26} = 3.1 \times 10^{-12}$ m/V.

Further using the relations $\hat{v}_{S,PP} = i\omega\hat{u}_{S,PP}$, $Z_q = (G_q\rho_q)^{1/2}$, and $f_0 = Z_q/(2m_q)$, one arrives at

$$\begin{aligned}\hat{u}_{S,PP} &= \frac{\hat{v}_{S,PP}}{i\omega}\\ &= \frac{1}{i}\frac{4}{n\pi}\frac{G_q}{2\pi\, nf_0 Z_q d_q}d_{26}Q\hat{U}\\ &= \frac{1}{i}\frac{4}{(n\pi)^2}\frac{Z_q^2/\rho_q}{2f_0 Z_q(m_q/\rho_q)}d_{26}Q\hat{U}\\ &= \frac{1}{i}\frac{4}{(n\pi)^2}d_{26}Q\hat{U}\end{aligned} \tag{7.4.4}$$

For the absolute value, $\left|\hat{u}_{S,PP}\right|$, one finds

$$\left|\hat{u}_{S,PP}\right| = \frac{4}{(n\pi)^2}Q\, d_{26}\left|\hat{U}\right| = 1.25\ \text{pm}\frac{Q}{n^2}\left|\hat{U}\right| \tag{7.4.5}$$

Note the scaling with overtone order: The amplitude scales as n^{-2} under voltage control, while it scales as n^{-1} under current control.

Before it was stated that the current is equal to $\hat{U}/R_1$ "under conditions of voltage control". Voltage control implies that the driving electronics supplies whatever current is needed to let the voltage at the device be equal to the prescribed value. Voltage control usually breaks down if the resistance of the device to be tested is less than the output resistance of the driving electronics, R_{out}. The voltage $\hat{U}$ is the voltage across the resonator's electrodes, to be distinguished from the nominal output of the impedance analyzer $\hat{U}_{nom}$. The two are related by

$$\hat{U} \approx \hat{U}_{nom} \frac{R_1}{R_1 + R_{out}} \tag{7.4.6}$$

Typical impedance analyzers have R_{out} around 50 Ω; the exact values are always worth a check. Also, there may be elements of influence other than R_{out}. The matter can be complicated; the reader is advised to check carefully if accurate values of the oscillation amplitude are needed.

The area of the parallel plate, A, can be obtained from a calculation along similar lines. The current into the electrodes is proportional to the area. Solving Eqs. 4.5.20 and 4.3.14 for A one finds

$$A = \frac{n\pi}{32 Z_q d_{26}^2 f_0^2} \frac{1}{Q R_1} \tag{7.4.7}$$

Since the mass of the resonator is given by $A\rho_q d_q$, the parameter, M_R (Eq. 4.5.27) can be determined from experiment, using R_1.

In the following, these calculations are adapted to plates with energy trapping. For the sake of this estimate, assume the dependence of amplitude on r to be a Gaussian [29]:

$$|\hat{u}_S(r)| \approx |\hat{u}_c| \exp\left(-\frac{r^2}{2\sigma_G^2}\right) \tag{7.4.8}$$

$\hat{u}_c$ is the amplitude in the center of the plate and σ_G is the standard deviation of the Gaussian. For the plate with energy trapping, the active area can be defined as

$$A = \frac{\int_0^\infty |\hat{u}_S^2(r)| 2\pi\, r\mathrm{d}r}{\left|\hat{u}_{S,PP}^2\right|} \tag{7.4.9}$$

The weight function is the square of the amplitude, because this is the statistical weight entering the area-averaged load impedance (Sect. 6.1.5). The parameters $\hat{u}_c$

and σ_G can be estimated by requiring that the total dissipated power and the total electric current should to be the same for the crystal with energy trapping and the equivalent parallel plate. Since power and current are proportional to $|\hat{u}_S^2|$ and $|\hat{u}_S|$, respectively, one arrives at the relations

$$\begin{aligned}\int_0^\infty \left|\hat{u}_S^2(r)\right| 2\pi\, r\mathrm{d}r &= \left|\hat{u}_{S,PP}^2\right| A \\ \int_0^\infty \left|\hat{u}_S(r)\right| 2\pi\, r\mathrm{d}r &= \left|\hat{u}_{S,PP}\right| A\end{aligned} \tag{7.4.10}$$

Inserting Eq. 7.4.8 for $|\hat{u}_S(r)|$, doing the integrations, and solving for $|\hat{u}_c|$ and σ_G, one finds

$$\begin{aligned}|\hat{u}_c| &\approx 2\left|\hat{u}_{S,PP}\right| \\ \sigma_G &\approx \frac{1}{2}\sqrt{\frac{A}{\pi}}\end{aligned} \tag{7.4.11}$$

Again, these are estimates. Firstly, the amplitude distribution is different from a Gaussian. Secondly, the back electrode does not fully cover the active area.

Two further comments:

- The amplitude of oscillation has been measured by Borovsky and Krim using a scanning tunneling microscope combined with a QCM [30]. The electrode surface showed point-like features, which appeared elongated, once the oscillation had been turned on. The long axis of these elongated objects was identified with twice the oscillation amplitude. Borovsky and Krim report a value on the fundamental of $\hat{u}_{PP}/\hat{U} = Q \times 1.4\,\mathrm{pm/V}$, which is in reasonable agreement with Eq. 7.4.5.
- In water at 5 MHz, a drive level of 15 dBm leads to an amplitude of oscillation at the resonator surface of $\left|\hat{u}_{S,PP}\right| = 1.6$ nm. This length should not be compared to the molecular scale because the liquid does not slip (does not usually slip). One should rather convert $\hat{u}_{S,PP}$ to a shear angle, using the decay length of the shear wave of about $\delta = 250$ nm. Dividing $\hat{u}_{S,PP}$ by δ, one finds a shear angle of 0.6 %. Normal liquids obey linear stress-strain relations at these levels of strain. Using the small load approximation (Sect. 4.6.1), the amplitude of oscillation can be converted to an amplitude of stress. Following Eq. 4.6.2, one finds a stress of $|\hat{\sigma}| \approx 200$ Pa (again for a drive level of 15 dBm at 5 MHz in water, where $\Delta\Gamma \approx 700$ Hz).

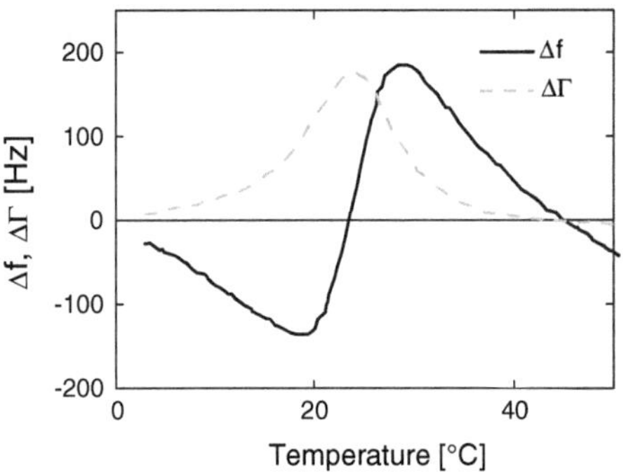

Fig. 7.5 An activity dip. A plane-convex resonator driven at the third overtone was slowly ramped in temperature from 5 to 55 °C. There is weak coupling to another mode, which happens to have the same resonance frequency as the 01-mode at about $T = 22$ °C. Coupling to this other mode withdraws energy from the main resonance; hence the peak in $\Delta\Gamma$ at that temperature. $\Delta f(T)$ and $\Delta\Gamma(T)$ are resonance curves of their own (Sect. 4.6.3). Temperature tunes the difference between the frequencies of the two coupled modes

7.5 Activity Dips

When two modes of vibration are similar in frequency, they often couple to each other. Think of the two modes as two resonators, linked to each other across a soft spring in the sense of Fig. 4.15d. Weak coupling most of the time can be ignored as long as the two modes have well-separated frequencies. However, if the modes are close to being degenerate, coupling shifts the resonance frequencies noticeably and also increases the bandwidths.

These problems are particularly impressive if both frequencies depend on temperature, but with different temperature-frequency coefficients. One can then continuously change the difference between the frequencies with a temperature ramp as shown in Fig. 7.5. Adding to the confusion, the second mode may or may not be visible in the conductance trace. It was not visible in the experiment shown in Fig. 7.5. The second mode behaves like a coupled resonance. The one visible mode went through a maximum in damping at around 22 °C. Δf showed the corresponding antisymmetric pattern.

Coupling between modes has been known for a long time and was given the name "activity dips". The term refers to the fact the resonator does not perform well (is damped) in the respective temperature range. People try to avoid activity dips. The FEM study from Ref. [40] (Sect. 7.8) had the prediction of activity dips as its practical motivation.

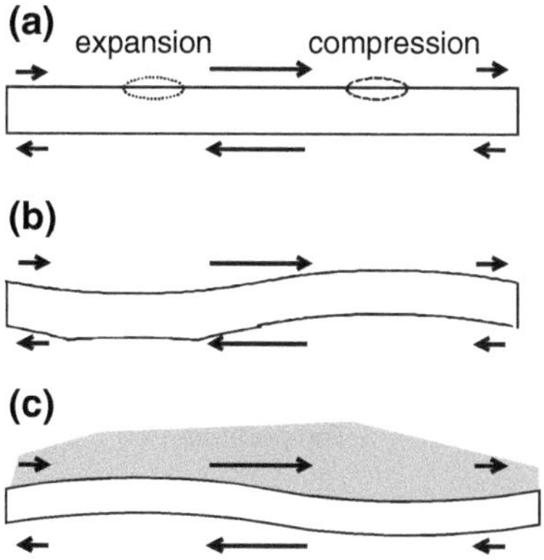

Fig. 7.6 a–c When a parallel plate experiences a hypothetical pure thickness-shear deformation with in-plane gradients, volume is not conserved at those places, were the gradient is large. The pure thickness-shear mode therefore is not realized; the plate bends. If the crystal is immersed in a liquid, volume conservation applies to the liquid above the resonator, as well. The liquid exerts a normal pressure onto the resonator surface and the plate may actually bend the other way (**c**)

7.6 Flexural Deformations, Compressional Waves

Flexural deformations are a severe problem for the operation of the QCM in liquids [31–34]. What is worst about them is, that the details are not known. With compressional waves, one is never sure how large the amplitudes are and to what extent they affect the measurement.

Figure 7.6 sketches the mechanism by which flexural deformations come about. A gradient in amplitude along the upper surface compresses the material at the surface (dashed ellipse in Fig. 7.6a); the plate avoids compression by bending upwards. The motion of the surface then contains a normal component; the plate emits a compressional wave. The problem is that compressional waves propagate in air and liquids (shear waves do not). They can be reflected at the opposing wall of the cell and return to the crystal. In Sect. 1.4 it was argued that surface-specificity was among the principal advantages of the QCM. Compressional waves spoil the surface-specificity.

For plates immersed in a liquid, there is a twist to the volume-conservation argument above: Volume conservation applies both to the crystal and to the liquid. The plate wants to bend upwards, but the liquid at the same time wants to expand vertically and therefore pushes the plate downward. As shown in Ref. [23], the liquid wins. In liquids, the plate actually bends downwards (Fig. 7.6c). Friedt and co-workers estimate the amplitude of the vertical displacement to be of the order of 1/10th of the corresponding tangential amplitude [35].

If one wants to view it that way, compressional waves can give rise to another aesthetically pleasing example of coupled resonances. Figure 7.7 shows an example (see also Refs. [33, 34]). The crystal and the opposing wall of the cell

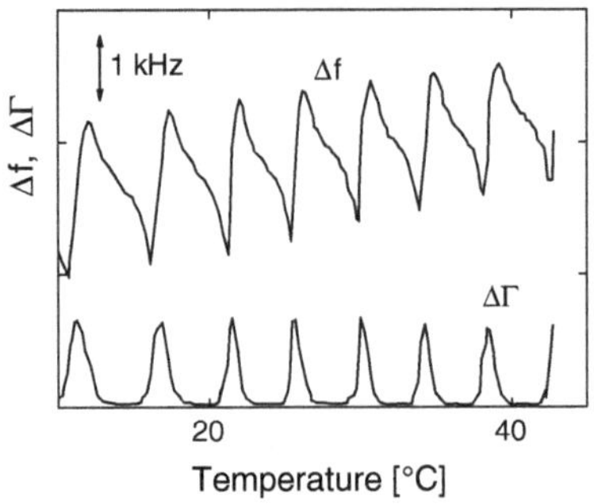

Fig. 7.7 Temperature dependence of Δf and $\Delta\Gamma$ of a resonator in a liquid, where the crystal surface was parallel to the opposing cell wall. There are standing compressional waves inside the cell, the wavelength of which changes with temperature. One observes a sequence of coupled resonances (Sect. 4.6.3)

formed a cavity for conventional ultrasound. The cavity resonates whenever the distance between the resonator and the opposing wall (or some other feature of the cell giving rise to standing waves) is a multiple of the wavelength of compressional sound. The wavelength depends on temperature. Of course, what is shown in Fig. 7.7, is the engineer's nightmare. In order to not let such a thing happen, one should design the cell such that the resonator surface is inclined relative to the opposing wall. This recommendation was actually formulated as early as 1987 by Eggers and Funk in the same paper, where they propose complex resonance frequencies [36]. Eggers and Funk placed a "spoiler" (an irregular piece of Teflon) in front of the crystal in order to deflect the compressional waves. Few researchers have reported on temperature sweeps employing liquid cells. It is always difficult to say why certain experiments were not done (better: not reported), but one may guess that compressional waves were found to be a problem. When changing temperature, one changes the pattern of standing compressional waves.

Compressional-wave effects can never be eliminated in measurements of viscosity because one compares frequencies measured in air and measured in the liquid phase. Compressional waves are only of influence in the second measurement. With regard to adsorption experiments in the liquid, one can hope that the effects of compressional waves are constant, regardless of whether or not there is an adsorbed film. They then disappear from the frequency shift. Probably, the matter should be phrased more optimistically. If the hope expressed above was not justified, the QCM would not be used successfully at so many places.

Compressional waves are not quite as detrimental to QCM experiments in liquids as one might think because the small load approximation does not apply to compressional waves. If the small load approximation would apply to the stress exerted by the compressional wave, the associated frequency shift might well be larger than the one originating from the shear wave. This follows from the fact that the compressional-wave impedance is larger than the shear-wave impedance. The matter is discussed in detail in Sect. 7.6. Roughly speaking, the liquid resists compression about as strongly as the plate itself. The liquid itself much reduces the amplitude of the flexural contribution.

A final note: Flexural modes can also occur when using quartz resonators, which do employ energy trapping, that is, quartz resonators with electrodes covering the entire surface. This can be checked by driving such resonators at high amplitude and searching for steady streaming as described in Sect. 13.2. One does find steady streaming, which leads to the conclusion that these resonators do not vibrate in the pure thickness shear mode. For a numerical treatment of the problem see Ref. [37].

7.7 Energy Trapping Increases the Resonance Frequency

To understand why energy trapping affects the resonance frequency, write the resonance frequency in the following way [38].

$$\frac{\omega_r^2}{c^2} \approx k_z^2 + k_{\perp,eff}^2 = \left(\frac{n\pi}{d_q}\right)^2 + k_{\perp,eff}^2 \tag{7.7.1}$$

The first term is the contribution of the plane wave. The second term accounts for the strain in the plane of the resonator. It was written as $k_{\perp,eff}^2$ although the in-plane strain is not a plane wave. The magnitude of $k_{\perp,eff}^2$ can only be estimated, but it is clear that it must of the order of σ_G^{-2}, where σ_G is the width of the amplitude distribution. σ_G varies a bit between overtones, but we approximate it as constant for the sake of the following simple argument. One can rearrange Eq. 7.7.1 as (Fig. 7.8)

$$\frac{\omega_r}{n} \approx c\left(\frac{\pi}{d_q} + \frac{k_{\perp,eff}^2}{n^2}\right)^{1/2} \tag{7.7.2}$$

Equation 7.7.2 shows that energy trapping increases f_r. The fractional contribution of $k_{\perp,eff}^2$ to f_r decreases with increasing n because k_z^2 scales as n^2, while $k_{\perp,eff}^2$ remains about constant. Because of energy trapping, the value of f_0 is not strictly equal to $c_q/(2d_q)$, as for the parallel plate (Eq. 5.5.13). Using $d_q \approx 330$ μm and $k_{\perp,eff} \approx 2\pi/(3$ mm), the difference is estimated as 0.5 %. The relation $f_r \approx nc_q/(2d_q)$ becomes progressively more accurate with increasing overtone order. In Eq. 4.3.17, the parameter m_q (the mass per unit area of the crystal) was replaced by $Z_q/(2f_0)$. A better approximation would have been

$$m_q = \lim_{n\to\infty} \frac{nZ_q}{2f_n} \tag{7.7.3}$$

with f_n the n-th resonance frequency.

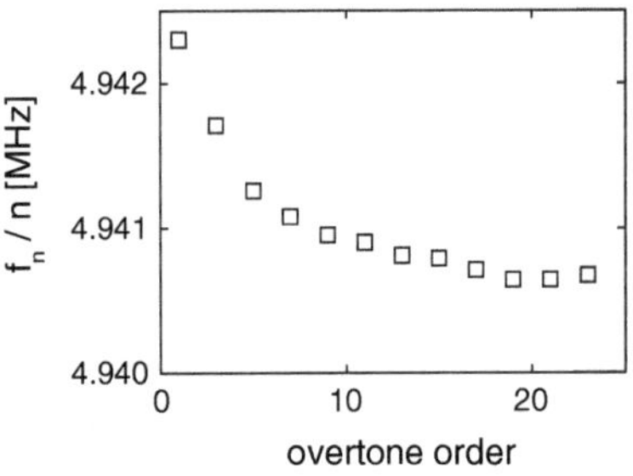

Fig. 7.8 Normalized resonance frequency, f_r/n, versus overtone order for a typical crystal. The limiting value at high frequencies is a good choice for the parameter f_0. The reasons for f_r/n not being constant are piezoelectric stiffening and energy trapping. Data kindly provided by I. Reviakine

There is a second reason why f_n is not equal to nf_0 (with n an integer), which is piezoelectric stiffening. The term producing the difference is the element $-\phi^2/(i\omega C_0)$ in the Mason circuit (Fig. 4.9), which softens the crystal. The influence of piezoelectric stiffening also decreases with increasing n. This can be seen from taking the limit of $n \to \infty$ in Eq. 4.5.22. In this limit, C_1^- approaches C_1.C_1^- does account for piezoelectric stiffening, while C_1 does not. The effect of piezoelectric stiffening is opposite to the effect of energy trapping: It lowers f_r. The statement that m_q is best approximated by the high-n limit of $nZ_q/(2f_n)$ (rather than $Z_q/(2f_0)$) also holds for piezoelectric stiffening. The higher the overtone order, the more accurate is the plane-wave model from Sect. 5.5.

Given that energy trapping changes the resonance frequency, one might suspect that it also affects the mass sensitivity constant in the Sauerbrey equation (see Eq. 8.1.3). Going through the calculation, one finds that such an influence exists, but that it is negligible in magnitude.

7.8 Modeling Resonators in 3D

The ultimate goal of a realistic model would be to reproduce the vibration pattern and the corresponding frequencies with and without the load in three dimensions. There is a considerable body of literature on 3D modeling of acoustic resonators using analytical theory. The work started in the late 1950s [6] and continued with steady progress into the late 1980s [5, 7–10, 39]. Tiersten, in particular, dedicated much effort to the subject and summarized his work in a book in 1995 [10]. Among the problems driving these studies (and solved by them) was the optimum choice for the shape and the thickness of the electrodes. Evidently, the analytical studies must make simplifying assumptions; these often (but not always) follow the lines of the paraxial approximation. Interestingly, the problem of optimum electrode shape has been picked up recently. Two groups have shown that elliptical (rather than circular) electrodes show an improvement with regard to suppression of anharmonic side bands [19, 20].

Today, numerical techniques and the Finite Element Method (FEM), in particular, are gradually replacing the analytical treatments [40–43]. Finite element calculations are more flexible with regard to geometry. Similar code can be applied to other resonators shapes. Flexibility is paid for with limits in the numerical accuracy (for instance originating from the finite mesh size in FEM simulations). At the time of writing, setting up a realistic Finite Element Model of a resonator requires expertise [44].

Given these challenges, it is worthwhile to point out that the approximations underlying the SLA are of help in 3D-modeling. The SLA provides for an interface between numerical calculations on different spatial scales and, also, between calculations being entirely elastic and entirely electrostatic in nature.

- Following the SLA, the sample's influence onto the resonance can be entirely subsumed under the stress at the resonator/sample interface, where the latter is evaluated at the reference frequency. The stress might be calculated analytically or numerically. Even if the traction at the resonator surface is calculated numerically (cf. Chap. 12), this calculation will usually be done separately from the determination of resonator's vibration pattern. It needs to be a separate calculation because the spatial scales are different. The sample will be structured on the nanoscale. On the millimeter scale, typical adsorbates to a QCM surface look homogeneous. If the sample is structured on the nanoscale only, one can calculate the stress using a simulation volume of a few cubic microns. The stress averaged over the bottom surface of this small cell can be assumed to be the same as the stress averaged over the entire resonator surface. This simplification is needed for the computation. A single FEM code cannot cover the nanoscale (relevant for the sample) and the millimeter scale (relevant for the resonator) at the same time. The SLA links the different scales.
  There is an assumption in the discussion above the concerning liquid cells, which is that the geometry of the chamber on the macro-scale neither enters the calculation of the vibration pattern nor the calculation of the interfacial stress. The resonator was assumed to only interact with its immediate environment. As emphasized in Sect. 7.6, compressional waves violate this assumption. If these return from the opposite wall of the chamber, they cause coupled resonances and thereby strongly affect the frequency. In principle, the entire cell can be included in a comprehensive model [45], but the geometry of the cell must then be carefully controlled. Life is easier if the liquid can be thought of as being semi-infinite.
- If the SLA holds, effects of piezoelectric stiffening can treated in a separate calculation, solving an entirely electrostatic problem (Sect. 6.1.7). The coupled equations are not needed. One first calculates the resonator's mode of vibration, ignoring piezoelectric stiffening. This calculation is not simple in the narrow sense of the word, but it is still simpler than the calculation with piezoelectricity included. In a second step, one calculates the electric displacement field, $\hat{\mathbf{D}}(\mathbf{r})$,

making use of Maxwell's first law. Piezoelectricity and the crystal's deformation as calculated in the first step enter the second calculation as a source of polarization. At this level of approximation, the source is fixed; it does not dependent on the electric displacement field calculated in the second step. Once $\hat{\mathbf{D}}(\mathbf{r})$ is known, the induced frequency shift follows from the SLA (Sect. 6.1.7).

All benefits acknowledged, the SLA cannot solve the problem of finding the vibration pattern in the unloaded state. This problem persists. It would be highly desirable to have user-friendly software available, which solves this problem realistically and on a routine basis.

Glossary

Variable	Definition (Comments)
A	Effective area of the resonator plate
$\tilde{c}$	Speed of propagation
C_1	Motional capacitance
$\bar{C}_1$	Motional capacitance of the 4-element circuit (Piezoelectric stiffening and the load have been taken into account)
$\hat{D}$	Electric displacement
d_f	Film thickness
d_q	Thickness of the resonator
d_{26}	Piezoelectric strain coefficient ($d_{26} = 3.1 \times 10^{-12}$ m/V for AT-cut quartz)
e_{26}	Piezoelectric stress coefficient ($e_{26} = 9.65 \times 10^{-2}$ C/m^2 for AT-cut quartz)
f	Frequency
f	As an index: *f*ilm
f_n	Resonance frequency at overtone order n
f_r	Resonance frequency
f_0	Resonance frequency at the fundamental ($f_0 = Z_q/(2m_q) = Z_q/(2\rho_q d_q)$)
G_{el}	Electric conductance
G_q	Shear modulus of AT-cut quartz ($G_q \approx 29 \times 10^9$ Pa)
$\hat{I}$	Electric current
k	Wavenumber

mot	As an index: *mot*ional
m_f	Mass per unit area of a film
m_q	Mass per unit area of the resonator ($m_q = \rho_q d_q = Z_q/(2f_0)$)
n	Overtone order
PP	As an index: *P*arallel *P*late
q	As an index: *q*uartz resonator
Q	Quality factor
r	Distance to the axis of a beam
r	Position
ref	As an index: *ref*erence state of a crystal in the absence of a load
R_1	Motional resistance
S	As an index: *S*urface
$\hat{u}$	Displacement, more general, a field variable obeying the wave equation
$\hat{u}_c$	Amplitude in the center of a Gaussian amplitude distribution
$\hat{u}_E$	Slowly varying envelope
$\hat{U}$	Voltage
$\hat{\mathrm{v}}$	Velocity ($\hat{\mathrm{v}} = \mathrm{i}\omega\hat{u}$)
z	Spatial coordinate along the surface normal
Z_q	Acoustic wave impedance of AT-cut quartz ($Z_q = 8.8 \times 10^6$ kg m^{-2} s^{-1})
Γ	Imaginary part of a resonance frequency
δ	Depth of penetration of a shear wave
δ_L	Loss angle ($\tan(\delta_L) = G''/G' = J''/J'$)
Δ	As a prefix: A shift induced by the presence of the sample
ε	A small quantity (In Taylor expansions)
φ	Azimuthal angle
ϕ	Factor converting between mechanical and electric quantities in the Mason circuit ($\phi = Ae_{26}/d_q$)
λ	Wavelength
σ_G	Standard deviation of a Gaussian distribution
ρ	Density
ω	Angular frequency

References

1. Saleh, B.E.A., Teich, M.C.: Fundamentals of photonics. Wiley, New York (2007)
2. Shockley, W., Curran, D.E., Koneval, D.J.: Trapped-energy modes in quartz filter crystals. J. Acoust. Soc. Am. **41**(4P2), 981–993(1967)
3. Bechmann, R.: Single response thickness-shear mode resonators using circular bevelled plates. J. Sci. Instrum. **29**(3), 73–76 (1952)
4. Josse, F., Lee, Y., Martin, S.J., Cernosek, R.W.: Analysis of the radial dependence of mass sensitivity for modified-electrode quartz crystal resonators. Anal. Chem. **70**(2), 237–247 (1998)
5. Mindlin, R.D.: Optimum sizes and shapes of electrodes for quartz resonators. J. Acoust. Soc. Am. **43**(6), 1329–1331 (1968)
6. Bechmann, R.: Phys. Rev. 100, 1060 (1958)
7. Tiersten, H.F., Smythe, R.C.: Analysis of contoured crystal resonators operating in overtones of coupled thickness shear and thickness twist. J. Acoust. Soc. Am. **65**(6), 1455–1460 (1979)
8. Tiersten, H.F., Stevens, D.S.: An analysis of nonlinear resonance in contoured-quartz crystal resonators. J. Acoust. Soc. Am. **80**(4), 1122–1132 (1986)
9. Stevens, D.S., Tiersten, H.F.: An analysis of doubly rotated quartz resonators utilizing essentially thickness modes with transverse variation. J. Acoust. Soc. Am. **79**(6), 1811–1826 (1986)
10. Tiersten, H.F.: Linear Piezoelectric Plate Vibrations: Elements of the Linear Theory of Piezoelectricity and the Vibrations of Piezoelectric Plates. Springer, Heidelberg (1995)
11. Watanabe, Y., Shikama, Y., Goka, S., Sato, T., Sekimoto, H.: Mode shape measurement of piezoelectric resonators using image processing technique. Jpn. J. Appl. Phys. Part 1 Regul. Pap. Short Notes Rev. Pap. **40**(5B), 3572–3574 (2001)
12. Watanabe, Y., Tominaga, T., Sato, T., Goka, S., Sekimoto, H.: Visualization of mode patterns of piezoelectric resonators using correlation filter. Jpn. J. Appl. Phys. Part 1 Regul. Pap. Short Notes Rev. Pap. **41**(5B), 3313–3315 (2002)
13. Ishizaki, A.; Sekimoto, H.; Watanabe, Y.: Three-dimensional analysis of spurious vibrations of rectangular AT-cut quartz plates. Jpn. J. Appl. Phys. Part 1 Regul. Pap. Short Notes Rev. Pap **36**(3A), 1194–1200 (1997)
14. Spencer, W.J.: Physical Acoustics, Principles and Methods. In: Mason, W.P. (ed.) Physical Acoustics, Principles, and Methods vol. 5, pp 111–161. Academic Press, New York (1968)
15. Haruta, K., Spencer, W.J.: Proc. Annu. Freq. Control Symp. **30**, 92 (1966)
16. Bahadur, H., Parshad, R.: Acoustic vibrational modes in quartz crystals: their frequency, amplitude, and shape determination. In Mason, W.P. (ed.) Physical Acoustics, Principles and Methods vol. 16, pp. 37–171. Academic Press, New York (1982)
17. Sauerbrey, G.: Proc. Annu. Freq. Control Symp. **17**, 63 (1967)
18. Goka, S., Okabe, K., Watanabe, Y., Sekimoto, H.: Multimode quartz crystal microbalance. Jpn. J. Appl. Phys. Part 1 Regul. Pap. Short Notes Rev. Pap. **39**(5B), 3073–3075 (2000)
19. Iwata, H., Hirama, K.: Suppression of inharmonic modes using elliptical periphery electrodes in very high frequency fundamental AT-cut resonators. J. Appl. Phys. Part 1 Regul. Pap. Short Notes Rev. Pap. **40**(5B), 3668–3671 (2001)
20. Ma, T.F., Zhang, C., Jiang, X.N., Feng, G.P.: Thickness shear mode quartz crystal resonators with optimized elliptical electrodes. Chin. Phys. B **20**(4), 047701 (2011)
21. Hess, C., Borgwarth, K., Heinze, J.: Integration of an electrochemical quartz crystal microbalance into a scanning electrochemical microscope for mechanistic studies of surface patterning reactions. Electrochim. Acta **45**(22–23), 3725–3736 (2000)
22. Edvardsson, M., Zhdanov, V.P., Hook, F.: Controlled radial distribution of nanoscale vesicles during binding to an oscillating QCM surface. Small **3**(4), 585–589 (2007)
23. König, R., Langhoff, A., Johannsmann, D.: Steady flows above a quartz crystal resonator driven at elevated amplitude. Phys. Rev. E **89**(4), (2014)

24. Flanigan, C.M., Desai, M., Shull, K.R.: Contact mechanics studies with the quartz crystal microbalance. Langmuir **16**(25), 9825–9829 (2000)
25. Herrscher, M., Ziegler, C., Johannsmann, D.: Shifts of frequency and bandwidth of quartz crystal resonators coated with samples of finite lateral size. J. Appl. Phys. **101**(11), 114909 (2007)
26. Martin, B.A., Hager, H.E.: Velocity profile on quartz crystals oscillating in liquids. J. Appl. Phys. **65**(7), 2630–2635 (1989)
27. Inoue, D., Machida, S., Taniguchi, J., Suzuki, M., Ishikawa, M., Miura, K.: Dynamical frictional force of nanoscale sliding. Phys. Rev. B **86**(11), 4 (2012)
28. Johannsmann, D., Heim, L.O.: A simple equation predicting the amplitude of motion of quartz crystal resonators. J. Appl. Phys. **100**(9), 094505 (2006)
29. Bottom, V.E.: Introduction to Quartz Crystal Unit Design. Van Nostrand Reinhold, New York (1982)
30. Borovsky, B., Mason, B.L., Krim, J.: Scanning tunneling microscope measurements of the amplitude of vibration of a quartz crystal oscillator. J. Appl. Phys. **88**(7), 4017–4021 (2000)
31. Spencer, W.J., Hunt, R.M.: Coupled thickness shear and flexure displacements in rectangular at quartz plates. J. Acoust. Soc. Am. **39**(5P1), 929–935 (1966)
32. Tessier, L., Patat, F., Schmitt, N., Feuillard, G., Thompson, M.: Effect of the generation of compressional waves on the response of the thickness-shear mode acoustic-wave sensor in liquids. Anal. Chem. **66**(21), 3569–3574 (1994)
33. Lin, Z.X., Ward, M.D.: The role of longitudinal-waves in quartz-crystal microbalance applications in liquids. Anal. Chem. **67**(4), 685–693 (1995)
34. Schneider, T.W., Martin, S.J.: Influence of compressional wave generation on thickness-shear mode resonator response in a fluid. Anal. Chem. **67**(18), 3324–3335 (1995)
35. Friedt, J.M., Choi, K.H., Francis, L., Campitelli, A.: Simultaneous atomic force microscope and quartz crystal microbalance measurements: Interactions and displacement field of a quartz crystal microbalance. Jpn. J. Appl. Phys. Part 1 Regul. Pap. Short Notes Rev. Pap. **41**(6A), 3974–3977 (2002)
36. Eggers, F., Funck, T.: Method for measurement of shear-wave impedance in the Mhz region for liquid samples of approximately 1 Ml. J. Phys. E-Sci. Instrum. **20**(5), 523–530 (1987)
37. http://www.comsol.com/model/download/177395/models.mems.thickness_shear_quartz_oscillator.pdf. Accessed 20 Feb 2014
38. Reviakine, I., Morozov, A.N., Rossetti, F.F.: Effects of finite crystal size in the quartz crystal microbalance with dissipation measurement system: Implications for data analysis. J. Appl. Phys. **95**(12), 7712–7716 (2004)
39. Tiersten, H.F.: Perturbation-theory for linear electroelastic equations for small fields superposed on a bias. J. Acoust. Soc. Am. **64**(3), 832–837 (1978)
40. Yang, L., Vitchev, N., Yu, Z.P.: Modal analysis of practical quartz resonators using finite element method. IEEE Trans. Ultrason. Ferroelectr. Freq. Control **57**(2), 292–298 (2010)
41. Lerch, R.: Simulation of piezoelectric devices by 2-dimensional and 3-dimensional finite-elements. IEEE Trans. Ultrason. Ferroelectr. Freq. Control **37**(3), 233–247 (1990)
42. Wang, J., Yu, J.D., Yong, Y.K., Imai, T.: A new theory for electroded piezoelectric plates and its finite element application for the forced vibrations of quartz crystal resonators. Int. J. Solids Struct. **37**(40), 5653–5673 (2000)
43. Wang, J., Yong, Y.K., Imai, T.: Finite element analysis of the piezoelectric vibrations of quartz plate resonators with higher-order plate theory. Int. J. Solids Struct. **36**(15), 2303–2319 (1999)
44. http://www.comsol.com/model/download/177395/models.mems.thickness_shear_quartz_oscillator.pdf. Accessed 20 Feb 2014
45. Reichel, E.K., Riesch, C., Keplinger, F., Jakoby, B.: Modeling of the fluid-structure interaction in a fluidic sensor cell. Sens. Actuators A Phys. **156**(1), 222–228 (2009)

Chapter 8
Gravimetric Sensing

Abstract The advanced QCMs provide information beyond gravimetry, but that is not to say that gravimetry was obsolete. Gravimetry is based on the Sauerbrey equation, derived in this chapter from the small load approximation. Strategies are discussed to improve the limit of detection. At least as important as sensitivity is specificity, meaning the ability to distinguish between different analytes. Specificity is achieved with receptor layers.

8.1 The Sauerbrey Equation and Its Implications

In Sect. 1.3, the Sauerbrey result was briefly explained by reasoning that a film makes the entire resonator thicker and therefore lowers the resonance frequency. Without proof, it was stated Δf that was proportional to the mass of the film, rather than its thickness. We are now in a position to provide that proof and with the SLA at hand, the proof actually is simple. The Sauerbrey result applies to all rigid samples; it is by no means limited to thin films. The sample might be a thin film, but tightly adsorbed nanoparticles behave in the same way. As long as the sample's deformation is negligible, the geometry is irrelevant.

For a rigid sample, the interfacial traction is the product of areal mass density and acceleration. One has $\hat{\sigma}_S = -\omega^2 \hat{u}_S m_f$ with $\hat{u}_S$ the amplitude and m_f the mass per unit area (possibly area-averaged). The index f (for "film") was used, although the sample might not be a film. With $\hat{\sigma}_S = -\omega^2 \hat{u}_S m_f$ the load impedance comes out to be

$$\tilde{Z}_L = \frac{\hat{\sigma}_S}{\hat{v}_S} = \frac{-\omega^2 \hat{u}_S m_f}{\mathrm{i}\omega \hat{u}_S} = \mathrm{i}\omega\, m_f \tag{8.1.1}$$

Inserting this load into the SLA, one finds

$$\frac{\Delta \tilde{f}}{f_0} \approx \frac{\mathrm{i}}{\pi Z_q} \mathrm{i}\omega m_f = -\frac{2nf_0}{Z_q} m_f = -n\frac{m_f}{m_q} \tag{8.1.2}$$

D. Johannsmann, *The Quartz Crystal Microbalance in Soft Matter Research*, Soft and Biological Matter, DOI: 10.1007/978-3-319-07836-6_8,

The relations $\omega = 2\pi n f_0$ (n the overtone order) and $m_q = Z_q/(2f_0)$ were used in steps 2 and 3, respectively. Equation 8.1.2 is the famous Sauerbrey equation [1–3]. The Sauerbrey equation is the quantitative basis for the use of bulk acoustic wave resonators as mass-sensitive detectors [4].

Clearly, the imaginary part in Eq. 8.1.2 is zero on the right-hand side. Sauerbrey layers do not increase the bandwidth. This is the most important check on whether or not the layer under study may be analyzed with the Sauerbrey equation. If $|\Delta\Gamma| \ll |\Delta f|$, one is on a good way with Eq. 8.1.2.

Equation 8.1.2 is often rearranged as

$$m_f = -\frac{Z_q}{2f_0^2}\frac{\Delta f}{n} = -\frac{\sqrt{\rho_q G_q}}{2f_0^2}\frac{\Delta f}{n} = -C\frac{\Delta f}{n} \tag{1.1.1}$$

What is called G_q here (the shear modulus of the crystal), often carries the name μ_q in the literature. C is the mass-sensitivity constant. Its value for a 5 MHz crystal is 17.7 ng cm^{-2} Hz^{-1}. (Conversion: With a density of 1 g/cm^3, 100 ng/cm^2 correspond to a thickness of 1 nm. A 1–nm film shifts the frequency by 6 Hz.)

Again: m_f is the *mass per unit area* (also: "areal mass density"). One might convert it to a mass in units of kg, in principle, but in a sensing context, the mass per unit area is the more meaningful parameter. In sensing, one usually is interested in the concentration of the analyte. The higher the concentration, the higher the coverage of the surface with the analyte, where coverage is proportional to mass per unit area. Alternatively, the concentration might be known, but the equilibrium binding constant might be in question [5, 6]. This will be the case if the sensor surface is functionalized with a receptor layer and if the interaction of the respective recognition element with different molecules is the target of investigation. Again, the higher the (bio-)affinity of the target molecule to the receptor, the larger the adsorbed mass per unit area. Measuring a mass only makes sense if a single molecule (or a single nanoparticle) is adsorbed and if the mass of this object is to be determined. This amounts to mass spectroscopy. Using piezoelectric resonators as mass spectrometers is in reach for the modern nanoelectromechanical devices [7], but not for the standard QCM.

The mass per unit area as derived with the Sauerbrey equation is the "Sauerbrey mass". If the sample is a film with known density, ρ_f, the film thickness, d_f, can be derived from Eq. 8.1.2 as

$$d_f = -\frac{1}{\rho_f}\frac{Z_q}{2f_0^2}\frac{\Delta f}{n} \tag{8.1.4}$$

d_f is also called "Sauerbrey thickness". Applied to films in air, d_f is a fair guess for the geometric thickness. The Sauerbrey relation can also be applied in the liquid phase, but interpreting the Sauerbrey thickness of films immersed in a liquid requires some caution (Sect. 10.3).

Equation 8.1.4 predicts Δf to scale as $-n$ for Sauerbrey films. This scaling behavior is only approximately obeyed in experiment for three reasons. Firstly, the finite compliance of the electrodes leads to a slope in a plot of $\Delta f/n$ versus n^2 (4th term in Eq. 6.2.15). This effect can be avoided with aluminum electrodes because these have a shear-wave impedance similar to AT-cut quartz. Secondly, the flexural contributions lead to a modal mass being different from Am_q. The matter is discussed in Sect. 6.1.4. Finally, there can be effects of piezoelectric stiffening, which may change upon deposition of a thin film. QCM-based gravimetry has excellent precision, but the accuracy unfortunately stays behind precision by quite a bit, even for measurements in the vapor phase.

With a short-time repeatability in frequency of the order of 0.1 Hz, the noise-equivalent mass is 1.7 ng/cm^2 (for a 5 MHz resonator). With a density of 1 g/cm^3 this translates to a noise-equivalent thickness of 17 pm. The QCM has sub-monolayer sensitivity, but the sensitivity nevertheless leaves room for further wishes. In order to quantify adsorption, one hopes for a signal-to-noise ratio around 100 at monolayer coverage. This is difficult to achieve for small molecules.

Can the limit of detection (the LOD) be improved? There are different options (none of them without problems), which can roughly be grouped into two categories. Firstly, one can increase the number of adsorption sites per unit area with a porous capture layer [8] or a soft film capable of swelling in the vapor phase [9]. Such measures are usually part of the design of a receptor layer and we discuss them in that context further below. Secondly, one can try to improve the sensitivity and the noise floor of the device.

Equation 8.1.2 shows that the mass-sensitivity constant scales as f_0^{-2}. Note, however, that an improved mass sensitivity constant does not directly translate to an improved noise-equivalent mass because the scatter on the frequency readings also increases with frequency. The scatter is inversely proportional to the Q-factor; one needs to optimize the "Qf-product" (more precisely the "Qf_0-product"). To see this, invert the Sauerbrey equation as

$$m_f = \frac{Z_q}{2}\frac{1}{f_0}\frac{\Delta f}{f} \tag{8.1.5}$$

The minimum detectable frequency shift is some fraction of the bandwidth. Call this fraction α (that is: $\delta(\Delta f) = \alpha\Gamma$ with δ the uncertainty). The smaller the bandwidth, the more precise the reading of the frequency. The detection limit in mass, $\delta(m_f)$, is

$$\delta(m_f) = \frac{Z_q}{2}\frac{1}{f_0}\frac{\delta(\Delta f)}{f} = \frac{Z_q}{2}\frac{1}{f_0}\frac{\alpha\Gamma}{f} = \frac{\alpha Z_q}{4}\frac{1}{f_0 Q} \tag{8.1.6}$$

Clearly the detection limit scales as $(Qf_0)^{-1}$. In air, the Q-factor is dominated by the resonator's loss tangent (Eq. 4.3.15), which linearly increases with frequency ($\delta_L = \omega\eta_q/G_q'$, Eq. 4.5.21). As a consequence, the detection limit decreases linearly with f_0, not quadratically as the Sauerbrey equation would suggest. In liquids, it is even less advantageous to increase the frequency than in air. In liquids, the bandwidth increases as $f_0^{3/2}$ (Sect. 9.1) and the LOD therefore scales as $f_0^{-1/2}$. Note: This argument applies to bulk acoustic wave (BAW) resonators. For acoustic delay lines (Sect. 15.4) the matter is more complicated, but increased frequency still translates to an improved limit of detection.

Clearly, the LOD can be improved somewhat with thinner plates, but making the plates thinner entails other problems. The most severe among them is that the resonance frequency responds to bending; this effect becomes stronger with thinner plates. Bending can, for instance, be caused by slight fluctuations in the pressure of liquid [10]. The Okahata group has implemented a scheme for dealing with pressure fluctuations, where they make sure that both sides of the crystal are exposed to a liquid and that both liquids are under the same hydrostatic pressure [12]. This scheme reduces the fluctuations in resonance frequency.

There are QCMs employing resonators with a base frequency higher than 10 MHz [11–13]. There even are bulk acoustic wave resonators with fundamental frequencies higher than 100 MHz. These are the FBARs [14–16] and the "inverted mesa resonators" [17]. For more comments on the FBAR see Sect. 15.6. Inverted mesa resonators are prepared by thinning the central portion of a quartz plate with reactive ion etching (RIE). Since these resonators have a rigid frame, they can be handled and mounted easily. Otherwise, such thin crystals would be too fragile to be of practical use. Unfortunately, the increased sensitivity can only be exploited in air or vacuum [18]. When used in liquids, inverted mesa resonators suffer from the dependence of the resonance frequency on bending [19].

Ogi employs high-frequency plates ($f_0 = 170$ MHz) in a microfluidic environment using a technique called "RAMNE-Q". Thin sheets of quartz (not coated with electrodes) are integrated into a microfluidic cell and exposed to the liquid on both sides [20]. Since the resonator is entirely immersed in the liquid, effects of hydrostatic pressure are largely absent. The resonators are held in place by small, soft pads of an elastomer. No energy trapping is employed; there are not even electrodes evaporated onto the plate. The electrical excitation occurs with external electrodes. The use of external electrodes entails a loss in sensitivity because of a reduced coupling efficiency. Also, there may be problems related to piezoelectric stiffening. The electric field emanating from the vibrating crystal permeates the liquid and the liquid's electric properties have an influence on the resonance frequency (Chap. 5). The Q-factor is 1,500, which comes close to the value expected for a parallel plate in water at this frequency.

In order to avoid the problems related to bending, one can resort to guided-wave instruments (Sect. 15.4). Here, a thin resonant structure (more precisely, a thin structure guiding an acoustic wave) is placed onto a supporting substrate, which prevents bending. With surface acoustic wave devices, there is not even the need of a guiding layer. The surface itself guides the wave. The frequencies of the guided-wave devices are in the range of a few hundred MHz. Similarly to the bulk acoustic wave devices, the limit of detection improves with increasing frequency. In particular, the limit of detection (LOD) achieved with surface acoustic wave (SAW) devices in vacuum is better than the LOD of a QCM. SAW devices (Sect.

15.4) are employed in some of the commercially available electronic noses [21, 22]. The matured SAW devices use surface corrugation waves and can therefore only be employed in gas sensing. When it comes to small molecules in the gas phase, the commercial acoustic mass-sensitive devices (of what-ever kind) are still surpassed in the LOD by metal oxide detectors [21]. The latter respond to reducing gases with a change of resistivity. They have low specificity and low stability, but with regard to simplicity and LOD, they are difficult to beat.

MEMS resonators and NEMS resonators (Sect. 15.5) also work at frequencies above 100 MHz and have reached spectacular LODs [23]. (Note: For resonators with a geometry other than the parallel plate, the Sauerbrey equation does not apply. These usually need calibration). At the time of writing, these resonators have not turned into commercial products. MEMS-based mass-sensitive devices are also operated in liquids [24], but the damping is a severe problem.

The upper limit of the dynamic range of the standard QCM depends on the softness of the sample. One needs to stay away from the film resonance (Fig. 10.2, Sect. 10.1). A film resonance occurs when the film thickness is around a quarter of the wavelength of sound. Of course one might attempt to live with the film resonance and analyze the data using equations Eq. 10.1.7 (in the gas phase) or Eq. 10.2.3 (in the liquid phase). Experience has shown that quantitative agreement between the experiment and these models is often poor. The maximum thickness (staying below the film resonance) can be a few microns for rigid films in the dry, but is closer to a few hundred nanometers in the liquid because these layers tend to be swollen in the liquid (therefore being soft, meaning that the wavelength of sound is small). For thickness determination on soft, thick samples, one can resort to resonators with lower frequency, such as the torsional resonators [25]. A potential application would be monitoring of biofilm growth. Biofilms are soft and thick to the extent that the standard QCM determines the sample's viscoelastic properties at the substrate-film interface rather than the film thickness (Sect. 12.6). The Sauerbrey thickness derived with kHz resonators comes closer to the geometrical thickness. Note: Dry films with a thickness in the micron range can also be weighed with a high-quality conventional balance. Thickness determination with kHz resonators only makes sense in the liquid phase.

The section closes with some remarks on a competing technique widely used for similar purposes, which is surface plasmon resonance (SPR) spectroscopy [26]. SPR spectroscopy and acoustic mass sensing both are label-free techniques in the sense that no tags (such as fluorescent groups or radio markers) would be needed [27]. SPR spectroscopy has its own set of problems, the most important being that the adsorbed mass can only be quantitatively determined if the refractive index of the layer is known. The result of an SPR measurement is reported in terms of "refractive index units" (RIU or μRIU [28]) rather than film thickness for that reason. The conversion from RIU to adsorbed mass or layer thickness requires knowledge about the material's refractive index increment, dn/dc. SPR spectroscopy usually is superior to the QCM with regard to sensitivity and baseline stability. Also, there is an imaging mode [29–32], allowing to acquire kinetic adsorption curves on multiple channels simultaneously. A further important difference between SPR spectroscopy and the QCM is that the SPR spectroscopy is less sensitive to swelling of the film in the ambient liquid than the QCM. The output generated by SPR spectroscopy is roughly proportional to the adsorbed mass, while the QCM determines a quantity closer to the geometric layer thickness (Sect. 16.2). QCM-based instrumentation can be simpler than SPR spectroscopy. There is no need for an optical beam. The averaging area is larger than a laser spot. This is an advantage for heterogeneous samples. There is no simple lesson from this comparison, other than that one would wish to have both techniques at one's disposal.

8.2 Receptor Layers

A second performance parameter of a sensor equally important as the limit of detection is *specificity*. One wants the instrument to only respond to the molecule of interest with little cross-sensitivity to contaminants. Specificity usually is achieved with a receptor layer. (It can also be achieved with chromatographic separation upstream of the sensor.) Receptor layers are briefly discussed in the following, although they are a topic separate from the physics of the QCM. Most sensors contain a "transducer", which is separate from the "receptor" [5]. The transducer converts a chemical parameter (here: the adsorbed amount) to some physical signal (here: the frequency shift). Transducers are designed by physicists or engineers. The receptor layer (also: capture layer, recognition layer) ensures that the transducer mostly responds to the molecule of interest. Receptor layers usually are designed by chemists.

An important performance criterion for receptor layers is small nonspecific adsorption. Nonspecific adsorption is less common in liquids than in air, but it is always a problem. In liquids, an adsorbed molecule replaces solvent molecules. If the adsorption energy of a contaminant is about as large as the adsorption energy of the solvent to the same surface, the driving force for adsorption of the contaminant is weak. In air, most organic molecules (even those with a high vapor pressure) tend to adsorb to solid surfaces because of van-der-Waals attraction. Particularly problematic are metal surfaces because of their large polarizibility. Fluorinated surfaces lead to low unspecific adsorption because of the small polarizability [33].

Related to unspecific adsorption is the issue of reversibility and regeneration of the sensor after exposure. Selectivity can be in conflict with reversible adsorption. If two chemical groups recognize each other well, they tend to adhere to each other strongly. The sensor then must be disposed after use or regenerated in some way. Devices which cannot be regenerated are sometimes called "assays" rather than "sensors".

For gas sensing, much experience has been gathered with receptor layers being polymer films capable of swelling the vapor phase [9, 34]. Similar films are in use as stationary phases in gas chromatography [35]. A class of polymers commonly employed are the siloxanes, which have a glass temperature much below room temperature. Relying on a film rather than a solid surface functionalized with receptor groups [36] has two advantages. Firstly, one gains in sensitivity because the mass uptake in a film is higher than the mass adsorbed to a surface. Equally important is that slightly crosslinked, soft polymer films are liquid-like on the molecular scale and therefore equilibrate with the vapor phase [37]. The sorption isotherm reflects the chemical potential of the analyte in both phases. Soft polymer networks do have some memory, meaning that equilibration cannot be taken for granted, but equilibrium thermodynamics is still a good starting point for analysis. Partition coefficients have been determined and tabulated for numerous pairs of polymers and vapors. These predict the distribution of molecules between the vapor phase and the film [9]. With known partition coefficients, the partial pressures of the analytes can be inferred from the mass uptake of the films.

A drawback with polymer films is their broad spectrum of relaxation times, which usually involves memory. There is a scheme to achieve liquid-like behavior of a receptor layer on the local scale in conjunction with macro-scale mechanical integrity, which relies on a mesoscale rigid skeleton rather than a polymer film. One can coat the resonator with a porous structure and fill the pores with a liquid. Mesoporous layers of titania [38], alumina [37], or gold [8] have been used. The liquid takes up the analyte according to its concentration in the vapor phase and the partition coefficient. Ionic liquids are particularly suitable as the liquid because they have low vapor pressure and therefore do not evaporate themselves. With other liquids, the vapor pressure needs to be controlled since they might evaporate themselves.

The information content provided by a sensor system can always be enhanced by use of an array, that is, by a number of different sensors coated such that they respond differently to different analytes. The human nose can be viewed as such an array. Making use of the combined information from many different olfactory receptors, humans can distinguish between a few thousand types of odors. With an array at hand, one can tolerate the fact that any given detector responds to more than one type of analyte. As long as two detectors respond to two gases differently, one can infer both concentrations separately from the combination of the two readings. The procedure is also called "multivariate analysis" [5]. In order for this scheme to work reliably, all sensors must respond differently to the different molecules. In mathematical language, one channel's response should be "orthogonal" to the response of all other channels [9, 39, 40]. If orthogonality is poor, the errors in the derived concentrations are correlated. In gas sensing, the matter has been studied in quite some detail [9, 39, 40] and it turned out to be difficult to find receptor films generating a vector space of sensor outputs with more than five mutually orthogonal dimensions (where the number of five can be debated). This can be phrased differently. For a given molecule, its solubility in a wide variety of polymer matrices can be predicted rather well with five parameters only (among them the polarity, the polarizibility, and the ability to participate in hydrogen bonding). If two molecules are similar in all these parameters, they will be recognized as belonging to the same species. The rules underlying this statement carry the name "linear solvation energy relationships (LSER)" [41]. LSERs were developed in the context of gas-liquid chromatography. Actually, gas-liquid chromatography inserted into a sensor system upstream of a gravimetric detector (or some other detector) is an alternative to a sensor array. When replacing a sensor array by chromatographic separation, one exploits similar physical principles [42].

8.3 Remarks on Gravimetric Sensing in the Liquid Phase

The QCM immersed in a liquid is discussed in generality later in the book. The following comments concern experiments, where the main interest is in the determination of adsorbed mass. Liquid-phase gravimetry is less mature than gas-phase gravimetry, but it is very visible in the literature. The survey by Speight and Cooper [43] counts around 80 publications on protein adsorption in 2010 alone. Here is a list of considerations:

- Effects of mass loading and liquid loading (that is, of variable viscosity) are only additive in the limit of small film thickness (Eq. 10.2.5, Ref. [44]).
- The Sauerbrey equation only holds if the film is much stiffer than the bulk. Viscoelastic effects are more pronounced in the liquid than in the vapor phase. In the vapor phase, the viscoelastic correction scales as m_f^2 (Eq. 10.1.9), while the correction is independent of film thickness in the liquid phase (Eq. 10.2.6). Also, films in liquids are often softer than in the dry because of swelling. Viscoelastic effects can be interesting, but if ignored or poorly understood, they produce artifacts.
- Viscoelastic effects increase in importance with increasing frequency because the penetration depth of the shear wave decreases. At frequencies above 100 MHz, soft bioadsorbates can easily reach a thickness comparable to δ. When the sample is not thin compared to the wavelength of sound (or to δ), the Sauerbrey equation does not apply and the proportionality between adsorbed mass and frequency shift is lost.
- The QCM measures the total mass of the film swollen in the ambient medium [45, 46]. This includes trapped liquid. Swelling can help because it enhances the signal. At the same time swelling complicates the interpretation (Sect. 10.3) [47].
- From a measurement of $\Delta f(n)$ and $\Delta\Gamma(n)$ one can infer whether the film is rigid or soft. Softness is to be distinguished from swelling in the liquid phase. The film may be much more rigid than the bulk, but still be slightly swollen. The degree of swelling can be inferred from the comparison of wet and dry thickness.
- One can always avoid these problems by drying the resonator after dipping it into the solution to be analyzed. This is the "dip-and-dry approach" [48].

- Generally speaking, coating the QCM surface with a receptor layer is easy. There are numerous recipes to construct intricate layer systems and sensing platforms, reviewed by Guilbault in Ref. [49].
- In liquids, a specific interaction between a suitably functionalized surface and an analyte molecule is more easily achieved than in the vapor phase. The most impressive example is the antigen/antibody interaction [50]. QCMs making use of antibodies have acquired a separate name, which is "piezoelectric immunosensors". "Piezoelectric" is often abbreviated as Pz. A certain drawback of immunosensors is that recognition occurs irreversibly because of the high binding constant.
- Polymer matrices can contain pockets fitting to certain analytes. These are created by a process called "molecular imprinting" [51, 52]. Receptor layers based molecular imprinted polymers (MIPS) have also been used on a QCM [53, 54].
- In studies involving specific recognition, the recognition itself often is the research target (as opposed to the concentration of the molecule in question). Determining binding constants of numerous pairs of receptors and candidates for novel drugs is of central importance to the pharmaceutical industry. The field also carries the name biological interaction analysis (BIA) [55].
- In determining the degree of adsorption of an analyte to a receptor layer, one often relies on kinetic measurements. The instrument records the uptake of the analyte versus time, as well as the release of the analyte upon purging the cell with buffer. The fits yield the rate constants k_{on} and k_{off}. The protocol leading to such kinetic data traces is called flow injection analysis (FIA).
- QCM-gravimetry in the liquid phase suffers from the instrumental noise floor when the molecules under study are small. Typical "small" molecules would be candidate molecules for novel drugs. "Large" molecules would be proteins. Various schemes of amplification have been devised, reviewed in Ref. [56]. These amount to linking the recognition event to the adsorption of a large object (or to its replacement by large objects, in the case of a competitive assay). A large object might be a protein or a nano particle. When particles are used, one runs into the problem that these might not be coupled to the resonator surface tightly enough. If a large particle is loosely bound, it can *increase* the resonance frequency (Sect. 11.3, Ref. [57]). This is to be avoided for gravimetric sensing.

An example of a functional multilayer system and its formation process being monitored with a QCM is given in Fig. 8.1. Such complicated processes have ample opportunities for things to go wrong, on the one hand, but they also provide for options to improve specificity, to gain additional information by variation of the protocol, and to give bioanalytical chemistry a role in process.

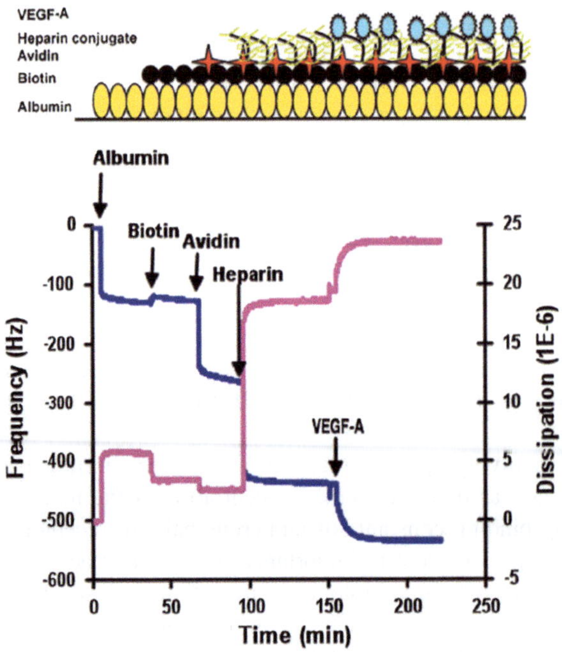

Fig. 8.1 Quartz crystal microbalance with dissipation monitoring analysis of growth factor binding to an immobilized heparin conjugate. The polystyrene surface, used as a proxy for an islet surface, was initially coated with albumin to create a matrix to which biotin could bind. Biotin binding was followed by sequential addition of avidin, heparin conjugate, and finally vascular endothelial growth factor-A (VEGF-A). For each step at which material was added to the sensor, a corresponding dampening of the frequency was obtained (*f*, *blue line*). The corresponding change in the dissipation factor, ΔD, is indicated by purple line. Binding of VEGF-A to the final heparin surface produced a distinct change in both Δf and ΔD. Reprinted with permission from Ref. [58]. Copyright 2010 Mary Ann Liebert Inc.

Glossary

Variable	**Definition (Comments)**
C	Mass-sensitivity constant
d	Thickness of a layer
f	Frequency
f	As an index: *f*ilm
$\tilde{f}_r$	Resonance frequency
f_0	Resonance frequency at the fundamental ($f_0 = Z_q/(2m_q) = Z_q/(2\rho_q d_q)$)

G_q	Shear modulus of AT-cut quartz ($G_q \approx 29 \times 10^9$ Pa, often called μ_q in the literature.)
m	Mass per unit area
n	Overtone order
q	As an index: *q*uartz resonator
Q	Q-factor ($Q = 1/D = f_r/(2\Gamma)$)
ref	As an index: *ref*erence state of a crystal in the absence of a load
S	As an index: *S*urface
t	Time
$\hat{u}$	(Tangential) displacement
$\hat{v}$	Velocity ($\hat{v} = i\omega\hat{u}$)
$\tilde{Z}_L$	Load impedance
Z_q	Acoustic wave impedance of AT-cut quartz ($Z_q = 8.8 \times 10^6$ kg m^{-2} s^{-1})
Γ	Imaginary part of a resonance frequency (Half-bandwidth at half-height of a resonance)
δ	As a prefix: uncertainty, scatter
Δ	As a prefix: A shift induced by the presence of the sample
ε	A small quantity (in Taylor expansions)
ρ	Density
ρ_q	Density of crystalline quartz ($\rho_q = 2.65$ g/cm^3)
$\hat{\sigma}$	(Tangential) stress (also called traction when exerted at a surface)
ω	Angular frequency

References

1. Sauerbrey, G.: Verwendung von Schwingquarzen zur Wägung dünner Schichten und zur Mikrowägung. Zeitschrift für Physik **155**(2), 206–222 (1959)
2. Stockbridge, C.D.: Resonance frequency versus mass added to quartz crystals In: Behrndt, K (ed.) Vacuum Microbalance Techniques, vol. 5, 4th edn. Plenum Press, New York (1966)
3. Lu, C., Czanderna, A.W.: Applications of Piezoelectric Quartz Crystal Microbalances. Elsevier, Amsterdam (1984)
4. King, W.H.: Piezoelectric sorption detector. Anal. Chem. **36**(9), 1735 (1964)
5. Janata, J.: Principles of Chemical Sensors. Springer, New York (2009)
6. Cheng, C.I., Chang, Y.P., Chu, Y.H.: Biomolecular interactions and tools for their recognition: focus on the quartz crystal microbalance and its diverse surface chemistries and applications. Chem. Soc. Rev. **41**(5), 1947–1971 (2012)

7. Hanay, M.S., Kelber, S., Naik, A.K., Chi, D., Hentz, S., Bullard, E.C., Colinet, E., Duraffourg, L., Roukes, M.L.: Single-protein nanomechanical mass spectrometry in real time. Nat. Nanotechnol. **7**(9), 602–608 (2012)
8. van Noort, D., Rani, R., Mandenius, C.F.: Improving the sensitivity of a quartz crystal microbalance for biosensing by using porous gold. Mikrochim. Acta **136**(1–2), 49–53 (2001)
9. Grate, J.W.: Acoustic wave microsensor arrays for vapor sensing. Chem. Rev. **100**(7), 2627–2647 (2000)
10. Zimmermann, B., Lucklum, R., Hauptmann, P., Rabe, J., Büttgenbach, S.: Electrical characterisation of high-frequency thickness-shear-mode resonators by impedance analysis. Sens. Actuators B-Chemical **76**(1–3), 47–57 (2001)
11. Lin, Z.X., Yip, C.M., Joseph, I.S., Ward, M.D.: Operation of an ultrasensitive 30-Mhz quartz-crystal microbalance in liquids. Anal. Chem. **65**(11), 1546–1551 (1993)
12. Sota, H., Yoshimine, H., Whittier, R.F., Gotoh, M., Shinohara, Y., Hasegawa, Y., Okahata, Y.: A versatile planar QCM-based sensor design for nonlabeling biomolecule detection. Anal. Chem. **74**(15), 3592–3598 (2002)
13. http://www.initium2000.com/en/. Accessed 18 May 2013
14. For an overview see Piazza, G., Felmetsger, V., Muralt, P., Olsson, R.H., Ruby, R.: Piezoelectric aluminum nitride thin films for microelectromechanical systems. MRS Bull. **37**(11), 1051–1061 (2012)
15. Wingqvist, G.: AlN-based sputter-deposited shear mode thin film bulk acoustic resonator (FBAR) for biosensor applications—A review. Surf. Coat. Technol. **205**(5), 1279–1286 (2010)
16. Wingqvist, G., Bjurstrom, J., Liljeholm, L., Yantchev, V., Katardjiev, I.: Shear mode AlN thin film electro-acoustic resonant sensor operation in viscous media. Sens. Actuators B-Chemical **123**(1), 466–473 (2007)
17. Birch, J., Marriott, S.P.: Appraisal of the inverted-mesa at-cut quartz resonator for achieving low-inductance high-Q single-response crystal units. Electron. Lett. **15**(20), 641–643 (1979)
18. Kreutz, C., Lorgen, J., Graewe, B., Bargon, J., Yoshida, M., Fresco, Z.M., Frechet, J.M.J.: High frequency quartz micro balances: a promising path to enhanced sensitivity of gravimetric sensors. Sensors **6**(4), 335–340 (2006)
19. Zimmermann, B., Lucklum, R., Hauptmann, P., Rabe, J., Büttgenbach, S.: Electrical characterisation of high-frequency thickness-shear-mode resonators by impedance analysis. Sen. Actuators B-Chemical **76**(1–3), 47–57 (2001)
20. Kato, F., Ogi, H., Yanagida, T., Nishikawa, S., Hirao, M., Nishiyama, M.: Resonance acoustic microbalance with naked-embedded quartz (RAMNE-Q) biosensor fabricated by microelectromechanical-system process. Biosens. Bioelectron. **33**(1), 139–145 (2012)
21. Rock, F., Barsan, N., Weimar, U.: Electronic nose: current status and future trends. Chem. Rev. **108**(2), 705–725 (2008)
22. http://us.msasafety.com/CBRNE-Detectors/CBRNE-Detectors/HAZMATCAD%26reg%3B-and-HAZMATCAD%26reg%3B-Plus/p/000400000200001000. Accessed 11 May 2013
23. Fanget, S., Hentz, S., Puget, P., Arcamone, J., Matheron, M., Colinet, E., Andreucci, P., Duraffourg, L., Myers, E., Roukes, M.L.: Gas sensors based on gravimetric detection-A review. Sens. Actuators B-Chemical **160**(1), 804–821 (2011)
24. Arlett, J.L., Myers, E.B., Roukes, M.L.: Comparative advantages of mechanical biosensors. Nat. Nanotechnol. **6**(4), 203–215 (2011)
25. Bucking, W., Du, B., Turshatov, A., Konig, A.M., Reviakine, I., Bode, B., Johannsmann, D.: Quartz crystal microbalance based on torsional piezoelectric resonators. Rev. Sci. Instrum. **78**(7), 074903 (2007)
26. Homola, J.: Present and future of surface plasmon resonance biosensors. Anal. Bioanal. Chem. **377**(3), 528–539 (2003)
27. Cooper, M.A.: Label-free screening of bio-molecular interactions. Anal. Bioanal. Chem. **377**(5), 834–842 (2003)
28. Rich, R.L., Myszka, D.G.: Advances in surface plasmon resonance biosensor analysis. Curr. Opin. Biotechnol. **11**(1), 54–61 (2000)

29. Rothenhäusler, B., Knoll, W.: Surface-plasmon microscopy. Nature **332**(6165), 615–617 (1988)
30. Yeatman, E., Ash, E.A.: Surface-plasmon microscopy. Electron. Lett. **23**(20), 1091–1092 (1987)
31. Nelson, B.P., Grimsrud, T.E., Liles, M.R., Goodman, R.M., Corn, R.M.: Surface plasmon resonance imaging measurements of DNA and RNA hybridization adsorption onto DNA microarrays. Anal. Chem. **73**(1), 1–7 (2001)
32. http://www.gwcinstruments.com/gwcSPRimager.htm. Accessed 2 Jan 2014
33. Brutschy, M., Lubczyk, D., Muellen, K., Waldvogel, S.R.: Surface pretreatment boosts the performance of supramolecular affinity materials on quartz crystal microbalances for sensor applications. Anal. Chem. **85**(21), 10526–10530 (2013)
34. Schramm, U., Roesky, C.E.O., Winter, S., Rechenbach, T., Boeker, P., Lammers, P.S., Weber, E., Bargon, J.: Temperature dependence of an ammonia sensor in humid air based on a cryptophane-coated quartz microbalance. Sens. Actuators B-Chemical **57**(1–3), 233–237 (1999)
35. Harris, D. C.: Gas chromatography. In: Quantitative Chemical Analysis, pp 675–712, 5th edn. W. H. Freeman and Company, New York (1999)
36. Lubczyk, D., Siering, C., Lorgen, J., Shifrina, Z.B., Mullen, M., Waldvogel, S.R.: Simple and sensitive online detection of triacetone triperoxide explosive. Sens. Actuators B-Chemical **143**(2), 561–566 (2010)
37. Goubaidoulline, I., Vidrich, G., Johannsmann, D.: Organic vapor sensing with ionic liquids entrapped in alumina nanopores on quartz crystal resonators. Anal. Chem. **77**(2), 615–619 (2005)
38. Schön, P., Michalek, R., Walder, L.: Liquid density response of a quartz crystal microbalance modified with mesoporous titanium dioxide. Anal. Chem. **71**, 3305 (1999)
39. Zellers, E.T., Batterman, S.A., Han, M.W., Patrash, S.J.: Optimal coating selection for the analysis of organic vapor mixtures with polymer-coated surface-acoustic-wave sensor arrays. Anal. Chem. **67**(6), 1092–1106 (1995)
40. Park, J., Groves, W.A., Zellers, E.T.: Vapor recognition with small arrays of polymer-coated microsensors. A comprehensive analysis. Anal. Chem. **71**(17), 3877–3886 (1999)
41. Hierlemann, A., Zellers, E.T., Ricco, A.J.: Use of linear salvation energy relationships for modeling responses from polymer-coated acoustic-wave vapor sensors. Anal. Chem. **73**(14), 3458–3466 (2001)
42. Lewis, P.R., Manginell, R.P., Adkins, D.R., Kottenstette, R.J., Wheeler, D.R., Sokolowski, S.S., Trudell, D.E., Byrnes, J.E., Okandan, M., Bauer, J.M., Manley, R.G., Frye-Mason, G.C.: Recent advancements in the gas-phase MicroChemLab. IEEE Sens. J. **6**(3), 784–795 (2006)
43. Speight, R.E., Cooper, M.A.: A survey of the 2010 quartz crystal microbalance literature. J. Mol. Recognit. **25**(9), 451–473 (2012)
44. Martin, S.J., Granstaff, V.E., Frye, G.C.: Characterization of a quartz crystal microbalance with simultaneous mass and liquid loading. Anal. Chem. **63**(20), 2272–2281 (1991)
45. Plunkett, M.A., Wang, Z.H., Rutland, M.W., Johannsmann, D.: Adsorption of pNIPAM layers on hydrophobic gold surfaces, measured in situ by QCM and SPR. Langmuir **19**(17), 6837–6844 (2003)
46. Bingen, P., Wang, G., Steinmetz, N.F., Rodahl, M., Richter, R.P.: Solvation effects in the quartz crystal microbalance with dissipation monitoring response to biomolecular adsorption. A phenomenological approach. Anal. Chem. **80**(23), 8880–8890 (2008)
47. Muratsugu, M., Ohta, F., Miya, Y., Hosokawa, T., Kurosawa, S., Kamo, N., Ikeda, H.: Quartz-crystal microbalance for the detection of microgram quantities of human serum-albumin—relationship between the frequency change and the mass of protein adsorbed. Anal. Chem. **65**(20), 2933–2937 (1993)
48. Shons, A., Dorman, F., Najarian, J.: An immunospecific microbalance. J. Biomed. Mater. Res. **6**, 565 (1972)
49. Vaughan, R. D., Guilbault, G. G.: Piezoelectric immunosensors. In: Steinem, C., Janshoff, A. (eds.) Piezoelectric Sensors. Springer, Berlin (2007)

50. Ebato, H., Gentry, C.A., Herron, J.N., Muller, W., Okahata, Y., Ringsdorf, H., Suci, P.A.: Investigation of specific binding of antifluorescyl antibody and fab to fluorescein lipids in langmuir-blodgett deposited films using quartz-crystal microbalance methodology. Anal. Chem. **66**(10), 1683–1689 (1994)
51. Lee, S.-W., Kunitake, T.: Handbook of molecular imprinting: advanced sensor applications. Pan Stanford Publishing, Singapore (2012)
52. Suriyanarayanan, S., Cywinski, P.J., Moro, A.J., Mohr, G.J., Kutner, W.: Chemosensors based on molecularly imprinted polymers. In Mol. Imprinting **325**, 165–265 (2012)
53. Dickert, F.L., Tortschanoff, M., Bulst, W.E., Fischerauer, G.: Molecularly imprinted sensor layers for the detection of polycyclic aromatic hydrocarbons in water. Anal. Chem. **71**(20), 4559–4563 (1999)
54. Reddy, S.M., Phan, Q.T., El-Sharif, H., Govada, L., Stevenson, D., Chayen, N.E.: Protein crystallization and biosensor applications of hydrogel-based molecularly imprinted polymers. Biomacromolecules **13**(12), 3959–3965 (2012)
55. Cheng, C.I., Chang, Y.P., Chu, Y.H.: Biomolecular interactions and tools for their recognition: focus on the quartz crystal microbalance and its diverse surface chemistries and applications. Chem. Soc. Rev. **41**(5), 1947–1971 (2012)
56. Cabric, S., Sanchez, J., Johansson, U., Larsson, R., Nilsson, B., Korsgren, O., Magnusson, P.U.: Anchoring of vascular endothelial growth factor to surface-immobilized heparin on pancreatic islets: implications for stimulating islet angiogenesis. Tissue Eng. Part A **16**(3), 961–970 (2010)
57. Tai, D.F., Lin, C.Y., Wu, T.Z., Chen, L.K.: Recognition of dengue virus protein using epitope-mediated molecularly imprinted film. Anal. Chem. **77**(16), 5140–5143 (2005)
58. Pomorska, A., Shchukin, D., Hammond, R., Cooper, M.A., Grundmeier, G., Johannsmann, D.: Positive frequency shifts observed upon adsorbing micron-sized solid objects to a quartz crystal microbalance from the liquid phase. Anal. Chem. **82**(6), 2237–2242 (2010)

Chapter 9
Homogeneous Semi-infinite Samples

Abstract The load impedance of a homogeneous, semi-infinite medium in contact with the resonator surface is equal to the material's shear-wave impedance, which leads to the Gordon-Kanazawa-Mason result. For Newtonian liquids the QCM determines the viscosity-density product. If the density is known independently, one can infer the viscosity. The Gordon-Kanazawa-Mason result can be extended to viscoelastic media, in which case the (complex) viscosity is often converted to the complex shear modulus at MHz frequencies. The formulation can be extended to cover nematic liquid crystals, colloidal dispersions, interfaces with shallow surface roughness, and samples, which touch the resonator in the center, only.

9.1 Newtonian Liquids

For the viscoelastic half-space, the load impedance, $\tilde{Z}_L$, is the same as the shear-wave impedance of the medium, $\tilde{Z}$. This sounds simple, but even these simple samples have their intricacies. Consider the Newtonian liquid first. For simple liquids (consisting of small molecules), all relaxations are so fast that viscoelastic dispersion is negligible in the MHz range (Sect. 3.7). For instance, the relaxation time for water according to dielectric spectroscopy is below a nanosecond [1]. As far as the QCM is concerned, water can safely be regarded as a Newtonian liquid. Newtonian liquids have $\eta'' = 0$ and $\eta' = const$, independent of frequency.

For the Newtonian liquid, the SLA predicts

$$\begin{aligned}\frac{\Delta\tilde{f}}{f_0} &= \frac{\mathrm{i}}{\pi Z_q}\tilde{Z}_L = \frac{\mathrm{i}}{\pi Z_q}\tilde{Z}_{liq} = \frac{\mathrm{i}}{\pi Z_q}\sqrt{\mathrm{i}\omega\rho_{liq}\eta_{liq}} \\ &= \frac{-1+\mathrm{i}}{\sqrt{2}\pi Z_q}\sqrt{\omega}\sqrt{\rho_{liq}\eta_{liq}}\end{aligned} \tag{9.1.1}$$

D. Johannsmann, *The Quartz Crystal Microbalance in Soft Matter Research*, Soft and Biological Matter, DOI: 10.1007/978-3-319-07836-6_9,

The index *liq* denotes the bulk medium. η_{liq} is the viscosity, $\tilde{Z}_{liq} = \rho_{\mathrm{liq}}\tilde{c}_{liq} = (\rho_{\mathrm{liq}}\tilde{G}_{liq})^{1/2} = (\mathrm{i}\omega\rho_{liq}\eta_{liq})^{1/2}$ is the shear-wave impedance.

Equation 9.1.1 is the Gordon-Kanazawa-Mason result. Δf and $\Delta\Gamma$ are equal and opposite for Newtonian liquids. They both scale as $n^{1/2}$. As Eq. 9.1.1 shows, the QCM determines the viscosity-density product, not the viscosity itself. However, the density is often known. Conversion from the viscosity-density product to viscosity rarely is a problem. (If desired, the density of a liquid can be determined separately with a rigid mesoporous layer on the quartz surface, which takes up the liquid and thereby decreases the resonance frequency with no corresponding increase in damping [2, 3]. However, there are easier ways to determine a liquid's density; this scheme has little practical relevance). With known density, the viscosity is derived as

$$\eta_{liq} = \frac{\pi Z_q^2}{\rho_{liq} f} \frac{\Delta f^2}{f_0^2} = \frac{\pi Z_q^2}{\rho_{liq} f} \frac{\Delta \Gamma^2}{f_0^2} \tag{9.1.2}$$

Importantly, the QCM only probes the region close to the interface. The displacement pattern is of the form

$$\begin{aligned} \hat{u}(z,t) &= \mathrm{Re}\left(\hat{u}_S \exp\left(\mathrm{i}\left(\omega t - \tilde{k}z\right)\right)\right) \\ &= \hat{u}_S \cos(\omega t - k'z) \exp(-k''z) \\ &= \hat{u}_S \cos\left(\omega t - \frac{z}{\delta}\right) \exp\left(-\frac{z}{\delta}\right) \end{aligned} \tag{9.1.3}$$

$\hat{u}_S$ was assumed as real in line 2. In line 3, the complex wavenumber was converted to an inverse penetration depth using $k = \omega/\tilde{c} = \omega(\rho/(\mathrm{i}\omega\eta))^{1/2} = (1 + \mathrm{i})/\delta$. The penetration depth is $\delta = (2\eta/(\omega\rho))^{1/2}$. In water, the penetration depth at 5 MHz is 250 nm. It scales as $n^{-1/2}$. The displacement pattern is shown in Fig. 9.1.

The presence of a gas (such as air) does have an influence on the resonance frequency, as can be noticed when operating a QCM in a vacuum chamber [4]. These effects cannot be explained with the Gordon-Kanazawa-Mason result. The viscous drag exerted by air is small, as can be seen by inserting the density of 1.2×10^{-3} g/cm^3 and the viscosity of 18 μPa s into Eq. 9.1.1. Larger than the viscous drag are the effects of compressional waves and a slight dependence of the crystal's elastic stiffness on hydrostatic pressure. The influences of a gas phase on $\Delta\tilde{f}$ is neglected in the following.

Jakoby has reviewed the use of acoustic devices for the determination of viscosity [5]. He also covers micromachined sensors with geometries other than the parallel-plate. A commercial instrument (the ViSmart™ Sensor) is available from Vectron. Acoustic waveguides can also be used to determine a viscosity. The principal benefit is the small sample volume. A number of different effects interfere with acoustic measurement of the viscosity. Among them are nanobubbles, roughness, slip, compressional waves, and adsorbed contaminants. The reproducibility of viscosity measurements between different QCM crystals is in the

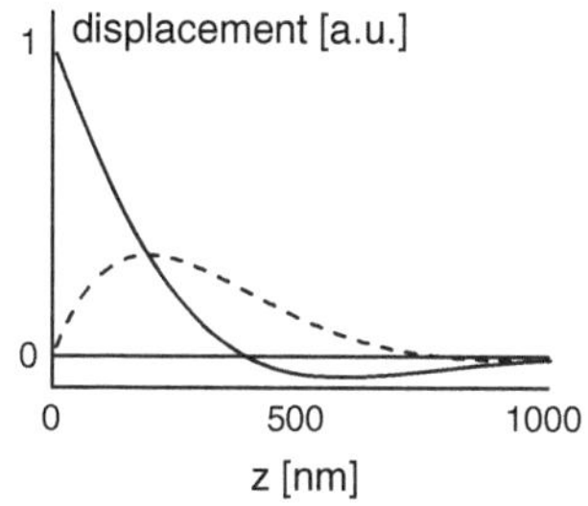

Fig. 9.1 Displacement field of a shear wave in a Newtonian liquid according to Eq. 9.1.3. The wave decays within a few hundred nanometers from the surface. The *full* and the *dashed curve* show the displacement at $t = 0$ and $t = T/4$ with T the oscillation period

range of ~10 %. (At least this is so in the author's laboratory.) Also, remember that the MHz viscosity can be different from the viscosity in steady shear in case the sample is not a simple liquid. Such differences can be expected for engine oil. (They are absent for water). What the QCM determines, may not be, what the user is interested in. kHz resonators (torsional resonators [6], in particular, see Sect. 15.3) come closer to most technical applications in this regard. Viscosity as a sensing target of the QCM is less relevant than microweighing.

For viscoelastic semi-infinite media, Eq. 9.1.1 is generalized as

$$\frac{\Delta\tilde{f}}{f_0} = \frac{-1+\mathrm{i}}{\sqrt{2}\pi Z_q}\sqrt{\rho_{liq}\omega\left(\eta'_{liq} - \mathrm{i}\eta''_{liq}\right)} = \frac{\mathrm{i}}{\pi Z_q}\sqrt{\rho_{liq}\left(G'_{liq} + \mathrm{i}G''_{liq}\right)} \quad (9.1.4)$$

The complex viscosity is derived from Δf and $\Delta\Gamma$ as

$$\begin{aligned} \eta'_{liq} &= \frac{G''_{liq}}{\omega} = -\frac{\pi Z_q^2}{\rho_{liq} f_r}\frac{\Delta f \Delta\Gamma}{f_0^2} && \text{(a)} \\ \eta''_{liq} &= \frac{G'_{liq}}{\omega} = \frac{1}{2}\frac{\pi Z_q^2}{\rho_{liq} f_r}\frac{\left(\Delta\Gamma^2 - \Delta f^2\right)}{f_0^2} && \text{(b)} \end{aligned} \quad (9.1.5)$$

Using the shear modulus ($\tilde{G}_{liq} = \mathrm{i}\omega\tilde{\eta}_{liq}$) and complex variables, this result can be formulated more compact as

$$\tilde{G}_{liq}\rho_{liq} = -\left(\pi Z_q \frac{\Delta\tilde{f}}{f_0}\right)^2 \quad (9.1.6)$$

The density, ρ_{liq}, was moved to the left-hand side to emphasize that the QCM determines the product of $\tilde{G}_{liq}$ and ρ_{liq}. Note: $\tilde{G}_{liq}$ is expected to differ between overtones. A viscoelastic spectrum (in a frequency range of about a decade) can be explicitly obtained.

The depth of penetration of the shear wave in viscoelastic media is the inverse of k'', which is given as

$$k''^{-1} = -\left(\mathrm{Im}\left(\frac{\omega}{\tilde{c}}\right)\right)^{-1} = -\frac{1}{\omega\sqrt{\rho_{liq}}}\left(\mathrm{Im}\left(\frac{1}{\sqrt{\tilde{G}_{liq}}}\right)\right)^{-1} = -\frac{1}{\omega\sqrt{\rho_{liq}}}\frac{\left|\tilde{G}_{liq}\right|}{\mathrm{Im}\left(\sqrt{\tilde{G}_{liq}}\right)} \tag{9.1.7}$$

Contrary to intuition, Γ depends on a material's storage modulus more than on its loss modulus. More specifically, if one keeps the absolute value of the complex viscosity constant ($|\eta'_{liq} - i\eta''_{liq}| \approx$ *const.*) and decreases the loss angle (making the medium more elastic) the bandwidth *increases*. To see this, set G''_{liq} to zero in Eq. 9.1.4. The right-hand side then turns imaginary. Δf is zero and $\Delta\Gamma$ is proportional to $G'^{1/2}_{liq}$. This behavior follows from the fact that the more elastic material transports the wave over a longer distance (has a larger depth of penetration). Even though the local rate of dissipation is lowered when making the material more elastic, the rate by which energy is withdrawn from the crystal (which is, what Γ measures) increases because the increase in volume permeated by the shear wave outweighs the decrease in the local rate of dissipation.

The shear modulus as determined with Eq. 9.1.6 is the shear modulus *at the respective frequency*. It must not be confused with the low frequency shear stiffness. For polymers, this statements is sometimes phrased in the form: *"The QCM operates at* –20 °C". Of course the QCM does not operate at –20 °C, but a large class of amorphous polymers appear to the QCM, as though they had been cooled to somewhere below 0 °C. This is the consequence of time-temperature superposition (Sect. 3.7). Time-temperature superposition states that a measurement at high frequency can be equivalent to a measurement at lower frequency and lower temperature. There is a "shift factor", a_T, which expresses this equivalence. The shift factors are in a similar range for many polymers. Seven decades in frequency (the ratio of 10 MHz and 1 Hz) translate to a shift in temperature of a few tens of degrees. The details vary, but one can still think of a QCM measurement on a polymer as providing the shear modulus of this material at a temperature, which is lower than the actual temperature of the measurement.

9.2 Nematic Liquid Crystals

In nematic liquid crystals (LCs) the relation between shear stress and shear rate depends on the orientation of the preferred direction (the "director") relative to the flow direction and the gradient direction. Consequently, Δf and $\Delta\Gamma$ depend on the orientation of the director relative to the sensor surface. The orientation at the surface is controlled by a process called "orientational anchoring" [7]. The preferred direction depends on the surface topography and the organization of the molecules at the surface. If the material is aligned parallel to the surface, the

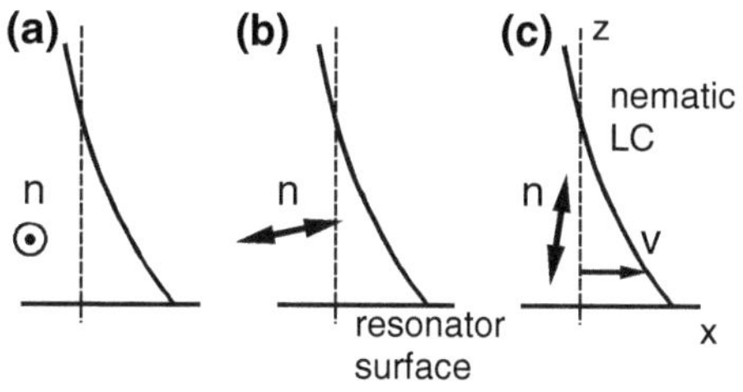

Fig. 9.2 Three different orientations of the nematic director relative to the crystal surface. The effective viscosities in configuration **b** and **c** are the same [12]. They are different from the configuration shown in **a**, where the preferred direction of the liquid crystal is along the vorticity direction

in-plane orientation can be selected by rubbing the surface. This is what the LC–display industry does; orientational anchoring is of central importance for displays.

Orientational anchoring can also be used as an amplification scheme in sensing [8]. If proteins adsorb to a surface, they change the local chemistry and topography to the extent that they can be visualized optically by bringing the respective surface into contact with a nematic LC and imaging the sample with a polarizing microscope. Changes in orientational anchoring are also (easily) detected with a QCM, based on the dependence of the apparent viscosity on LC–orientation [9, 10].

On paper, the flow behavior of nematics has been all worked out. The formalism carries the name "nematodynamics" and is covered in, for instance, the book by deGennes and Prost [11]. The flow behavior of nematics is specified by no less than five different viscosities, all of which are complex functions of frequency and temperature [11]. The formalism is elegant in many ways, but it is a bit of an academic exercise because of the unknown viscosities. One can measure some of them, but it is difficult to devise experiments, which would be able to falsify the model as such.

Interestingly, there is one such test, addressing a special aspect: The test amounts to comparing the high-frequency shear-wave impedances of two nematic samples aligned along x and z in Fig. 9.2. They are predicted to be the same and the theory can be tested in this regard.

At some point in the theory, the number of free parameters reduces by one because of the so-called Onsager relations. It appears to be less than certain that the Onsager relations actually do apply [12]. If they do, the "Rapini equality" holds, which (translated to the QCM-experiment) says that the frequency shift induced by a nematic LC does not depend on whether the director is along the displacement direction or along the gradient direction (panels b and c in Fig. 9.2). The Onsager relations are part of non-equilibrium thermodynamics. They cannot be explained in a few lines, but it is safe to say that they are more difficult than they look at first glance. The book by deGennes and Prost at this point contains a footnote, saying: "The experts in classical mechanics (Ericksen, Truesdell) consider that the use of the Onsager relations in hydrodynamics requires special caution. The Rapini equality, which does depend on the validity of these relations, provides a direct experimental

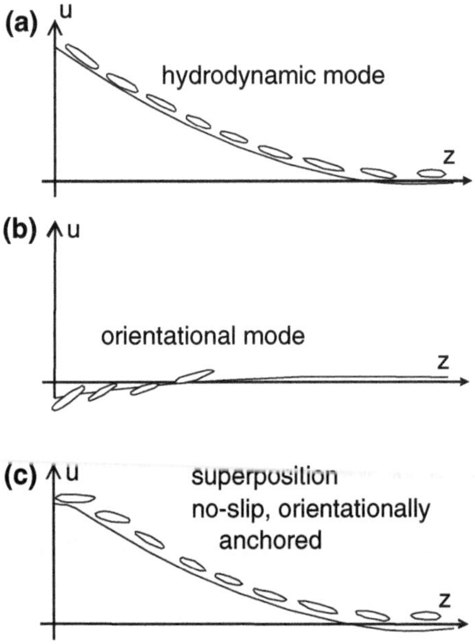

Fig. 9.3 At a solid surface, a nematic liquid crystal is pinned both with regard to position ("no-slip condition") and to orientation ("orientational anchoring"). Displacement and orientation are separate dynamical variables, giving rise to two separate types of decaying waves. They both combine displacement and rotation. For the hydrodynamic mode (**a**), the orientation mostly follows the shear gradient. For the orientational mode (**b**), the displacement is small and the orientation substantially deviates from the shear gradient. Inserting materials parameters, one finds that the depth of penetration of the orientational mode is smaller than that of the hydrodynamic mode. A superposition of both modes (**c**) is needed to fulfill the boundary conditions on orientation and displacement

check on this point". Again, the Rapini equality can be tested rather easily. The two effective viscosity from Fig. 9.2b, c might be unequal. The experiment was undertaken by Kiry and Martinoty in 1977 on the material 4-*n*-pentyl-4′cyanobiphenyl (5CB) [13, 14], using shear-wave reflectometry. Kiry and Martinoty applied the instrument described in Sect. 15.1. As shown in Sect. 4.2, ultrasonic reflectometry is equivalent to a QCM experiment. The author had his own go at the problem [9]. Part of the motivation was to show that shear-wave reflectometry and QCM-experiments are indeed equivalent. However, agreement was not achieved in this particular experiment; the two acoustic impedances were found to be unequal. There is a caveat though: Tilted orientation is not covered in Ref. [13] and tilt may have been present in the author's experiments. Tilted orientation leads to a coupling between shear waves and compressional waves, which may explain the large loads found for the inclined state.

On the conceptual side, nematics on a QCM are intriguing samples because orientation is not trivially coupled to deformation. In addition to the displacement field, there is the field of mean molecular orientation (the "director"). Orientation is a dynamical variable on its own. The displacement above the resonator surface is not a decaying cosine as in simple liquids, but rather a superposition of two such functions ("modes", Fig. 9.3). One mode is predominantly translational ("hydrodynamic" in Ref. [14]), while the other one is predominantly orientational. The orientational mode primarily exerts a torque at the interface. Both modes are needed to fulfill the boundary conditions with regard to orientation and displacement at the same time (Fig. 9.3c). While this sounds interesting, it turns out that the stress exerted by the orientational mode is small. The peculiarities of nematics under shear excitation have not translated to a practical sensing scheme.

9.3 Colloidal Dispersions

Colloidal dispersions consist of small particles embedded in a liquid. The particles are colloidally stable; they neither aggregate nor sediment. Colloidal dispersions are of outstanding importance in the chemical and the pharmaceutical industry. Biofluids also contain numerous dispersed particles. Particles tend to adsorb to surfaces and particle adsorption is intensely investigated with the QCM [15–19]. Here the question is a different one: Suppose that particle adsorption is prevented somehow: Can one learn anything about the bulk of the dispersion by determining its effective viscosity at MHz frequencies? "Effective" here means "as calculated from $\Delta\tilde{f}$ with Eq. 9.1.5." The question relates to a topic frequently recurring in the modeling process: Consider a heterogeneous medium with some random structure and let the scale of heterogeneity be below the wavelength of shear sound: Can this medium be described by some kind of effective medium theory? What information about the medium is contained in the effective parameters?

Effective medium theories have a long tradition in colloid science [20]. A famous example is the Einstein-formula for the viscosity of a colloidal dispersion. Einstein predicts the macroscopic viscosity of colloidal dispersion to be $\eta_{eff} = \eta_{liq}\,(1 + 5/2\varphi)$ with φ the particle volume fraction and η_{liq} the viscosity of the liquid phase. The particle size does not enter. The Einstein relation holds for non-interacting spherical particles at low volume fraction and low frequency. The effective viscosity at higher solids content is a complicated matter. In particular, it depends on the interaction between particles. Conversely, one can learn about the interaction between particles by measuring the macroscopic complex shear modulus as a function of particle volume fraction and frequency.

In this context, there is specific advantage of high-frequency rheology. One of the characteristic time scales in colloid science is the time needed for a rearrangement of the local cages around a particle. This time is given by d_P^2/D with d_P^2 the interparticle distance (often approximated by the particle diameter) and D the particle diffusivity. If the measurement of the viscosity occurs on a time scale faster than d_P^2/D, this can make data analysis easier. In particular, there is a relation between the curvature of the interparticle potential at the mid-plane between two particles, on the one hand, and the high-frequency shear modulus, on the other [21].

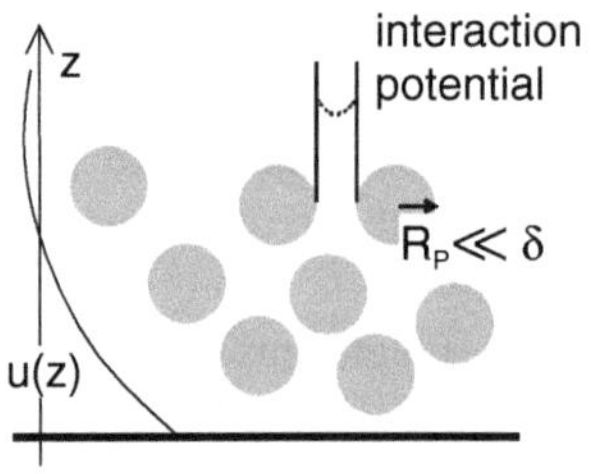

Fig. 9.4 When exposed to a colloidal dispersion of non-adsorbing spheres with a size of less than the penetration depth, the QCM determines an effective shear modulus, to be interpreted in the frame of an effective medium theory. The effective shear modulus reflects the interactions between the particles

Hydrodynamic interactions complicate the picture [22]. Still, one can assess the charge on the particle surfaces [23].

In the past, "high-frequency rheology" has mostly meant rheology in the range of many kHz, as opposed to a few MHz [24–27]. Can one expand the frequency range at the upper end and learn even more from an experiment as sketched in Fig. 9.4? A first problem with such experiments is that one must be sure to indeed avoid particle adsorption. The higher the frequency, the more sensitive is the device to adsorbed mass. Further complicating the situation, there are characteristic times other than the time needed for rearrangement of local cages. These are the "hydrodynamic time scale" [28], τ_{hyd}, given as $d_P^2\rho_{\mathrm{liq}}/\eta_{liq}$ and the "momentum relaxation time", τ_{MR}, given as $(\rho_P - \rho_{liq})R_P^2/\eta_{liq}$ (numerical factors were ignored, see also Sect. 11.6). Both times must be smaller than the period of oscillation if high-frequency rheology in the sense of Ref. [22] shall be of use. Consider τ_{hyd} first. Effective medium theories only hold as long as all structural heterogeneities occur on a scale below δ. This condition is more difficult to fulfill, if one goes up in frequency. τ_{hyd} is the inverse frequency, at which d_p is comparable to δ. The penetration depth is 250 nm at 5 MHz in water and it decreases with increasing overtone order. Polymer dispersions have particle sizes in the range of 100 nm, which is awkwardly close to the penetration depth. Gold nanoparticles and nanoclays are smaller (around 10 nm, possible a few tens of nanometers), but a similar argument applies to the distance between particles, d_P. d_P should also be smaller than δ.

The momentum relaxation time is related to inertial forces. Typical colloidal particles are too small to feel inertia. However, neglecting inertia is questionable at MHz frequencies because the forces of inertia scale as ω^2 [23]. To assess whether inertia makes a significant contribution to the shear-wave impedance, one compares the inertial force to the drag force. The two can be estimated as

$$\begin{aligned}
\hat{F}_{drag} &\approx \xi_P \hat{\mathrm{v}} = 6\pi\eta_{liq} R_P (i\omega \hat{u}_S) \\
\hat{F}_{inertia} &\approx \frac{4\pi}{3}\left(\rho_P - \rho_{liq}\right) R_P^3 \frac{\mathrm{d}\hat{\mathrm{v}}}{\mathrm{d}t} = \frac{4\pi}{3}\left(\rho_P - \rho_{liq}\right) R_P^3 \left(-\omega^2 \hat{u}_S\right)
\end{aligned} \tag{9.3.1}$$

R_P is the particle radius. The first relation is Stokes' law. Requiring that the two forces be comparable in magnitude ($|\hat{F}_{drag}| \approx |\hat{F}_{inertia}|$ at $\omega = \omega_c$) leads to a characteristic frequency of

$$\omega_c \approx \frac{9}{2} \frac{\eta_{liq}}{\left(\rho_P - \rho_{liq}\right) R_P^2} \tag{9.3.2}$$

The momentum relaxation time is the inverse of ω_c (again: numerical factors aside). Using $\eta \approx 1$ mPa s, $R_P \approx 100$ nm, and $\Delta\rho \approx 0.1$ g/cm^3 one arrives at $\omega_c/(2\pi) \approx 700$ MHz. Inertial effects are small, but not necessarily irrelevant. Inertial effects in colloids have a this point not been explored with the QCM, but such studies should be possible. (Inertial effects in colloids *have* been observed with an optical technique called nanorheology, see Ref. [29]). It was stated in Sect. 3.7 that $G(t)$ can never be negative and that $G'(\omega)$ monotonically increases with frequency because inertial forces are irrelevant in viscoelastic media. Conventional viscoelastic materials respond to stress with relaxation processes only; there are no resonances. This statement needs to be revisited, if the frequency of excitation is beyond ω_c.

9.4 Elastomers Contacting the Resonator in the Center only

The near-surface viscoelastic properties of elastomers are of outstanding practical importance [30–32]. The QCM can make a contribution to this research because it probes the material's viscoelastic properties within in the penetration depth, only. Unfortunately, a QCM-based measurement of G' and G'' of a polymer melt or a typical elastomer is difficult. Most polymers are so viscous that they overdamp the resonance. (The viscosity is complex for polymer melts, but that does not change the previous statement.) The resonances become so broad that they cannot be fitted reliably. Large drive levels, careful calibration, and slow frequency sweeps with good signal-to-noise ratio help, but the matter remains difficult.

If fitting the broad resonances turns out to be impossible, one can confine the area of contact to a small spot in the center of the plate. The damping then is much reduced. This mode of measurement was first demonstrated by Flanigan et al. who combined the QCM with a JKR apparatus (Refs. [32, 33], see also Sect. 16.5). Because of the finite contact area, Eq. 4.6.1 must be modified as

$$\frac{\Delta\tilde{f}}{f_0} = \frac{i}{\pi Z_q} K_A(A_c) \frac{A_c}{A} \tilde{Z}_L \tag{9.4.1}$$

A_c is the contact area and K_A is a sensitivity factor, accounting for the nontrivial amplitude distribution over the area of the crystal. Equation 9.4.1 assumes plane waves, that is, a contact area much larger than the decay length of the shear wave. Comparison with Eq. 6.1.31 shows that the combination of prefactors K_A A_c/A is equal to

$$K_A(A_c)\frac{A_c}{A} = \frac{\int\limits_{\textit{contact area}} \left|\hat{u}_S^2(\mathbf{r}_S)\right| \mathrm{d}^2\mathbf{r}_S}{\int\limits_{\substack{\textit{resonator}\\ \textit{surface}}} \left|\hat{u}_S^2(\mathbf{r}_S)\right| \mathrm{d}^2\mathbf{r}_S} \tag{9.4.2}$$

û*S* is the amplitude of motion at the surface and r*S* is a location on the resonator surface. The prefactors in Eq. 9.4.1 can be determined by a calibration with a substance of known shear-wave impedance. One deposits droplets of variable radius on the resonator surface and determines $K_A(A_c)$ from the induced complex frequency shift. Note: Since the efficiency of energy trapping depends on overtone order, the calibration has to be repeated for every overtone.

There is a complication: By contacting the resonator in the center only, one *changes* the degree of energy trapping and thereby changes the resonance frequency (Sect. 7.7) [34]. There is a contribution to Δf originating from enhanced energy trapping. This contribution is not covered by Eq. 9.4.1. The effect can be noticed in experiment because energy trapping affects frequency more than bandwidth and the ratio of Δf and $\Delta\Gamma$ therefore depends on contact area. This should not be the case, following Eq. 9.4.1. The dependence of $\Delta\Gamma/(-\Delta f)$ on contact area has been analyzed in Ref. [35]. The solution to this problem consists in doing all experiments at variable contact size and extrapolating to vanishing contact area. Arguably, this complication mitigates the experiment's beauty and simplicity. On the other hand, the experiments principal virtue is unchanged: The instrument determines the sample's shear modulus close to the interface. (Remember: It is the MHz shear modulus.)

9.5 Solid-Liquid Interfaces with Shallow Roughness

An effect of surface roughness on the resonance frequency is plausible from intuition. An early experiment is reported in Ref. [36]. Roughness is expected to lower the resonance frequency because of the trapped liquid. The Tel-Aviv group has put forward a more quantitative mathematical description [37]. These equations also cover the half-bandwidth and the overtone dependence of Δf and $\Delta\Gamma$.

There are two separate models. A surface, where the vertical scale of roughness is large compared to the lateral scale ($h_r \gg l_r$), is treated like a porous layer. The Brinkman equation is employed, which amounts to an effective medium theory. See Ref. [37] for more details. If the aspect ratio is low (that is, if the height of the features is much less than their width, Fig. 9.5), effects of roughness can be captured by a small modification to the plane-wave picture. The displacement pattern mostly is a decaying shear wave. Superimposed onto the shear wave is a

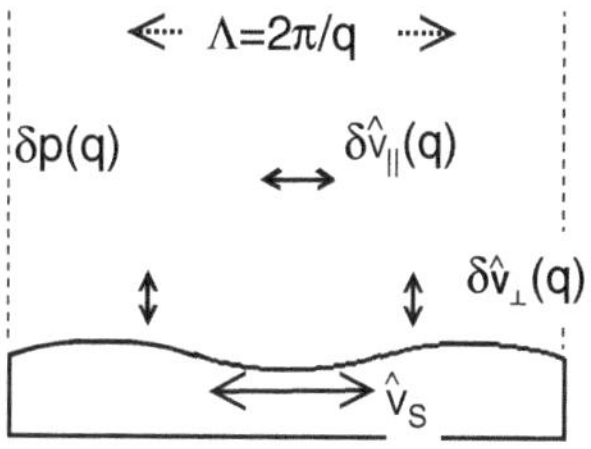

Fig. 9.5 Shallow roughness can be accounted for with a small-amplitude flow field superimposed onto the shear wave. The small added flow field is characterized by a vertical component, $\delta\hat{v}_\perp(q)$, a tangential component, $\delta\hat{v}_\parallel(q)$, and a pressure, $\delta p(q)$. The resulting surface traction is treated as an additive contribution to $\hat{\sigma}_S$. Since the contribution is additive, one can decompose the surface topography $\delta z(x, y)$ into its Fourier components (each component characterized by its wave vector, q) and solve the incompressible Stokes problem (see Sect. 9.1) for every Fourier component separately

small-amplitude flow field, ensuring continuity of stress and displacement everywhere at the corrugated surface. The stress resulting from this second displacement field is approximated as being additive. Because the (small) stress is additive, one may Fourier-decompose the surface corrugation and solve the equations of hydrodynamics for each Fourier component (characterized by its wave vector, q) separately. The total roughness-induced stress is the sum of the stresses induced by the Fourier components. The frequency shift resulting from shallow roughness can be written down in analytical form. Again, the feature height (the root-means-square value of the height distribution), h_r, is assumed to be much less than the feature width, l_r. Also, the feature height must be less than the penetration depth of the shear wave δ.

For the general case, the power spectral density of the roughness distribution must be known. As always, assumptions can be made. For instance, one might assume the spectral density to follow a Gaussian. Two limiting cases have to be distinguished. If the lateral scale is much less than δ ($l_r \ll \delta$) the result is:

$$\begin{aligned}\frac{\Delta f}{f_0} &\approx \frac{-1}{\pi Z_q}\sqrt{\frac{\rho_{liq}\omega\eta_{liq}}{2}}\left(1+3\sqrt{\pi}\frac{h_r^2}{l_r\delta}-2\frac{h_r^2}{\delta^2}\right)\\ \frac{\Delta\Gamma}{f_0} &= \frac{1}{\pi Z_q}\sqrt{\frac{\rho_{liq}\omega\eta_{liq}}{2}}\left(1+2\frac{h_r^2}{\delta^2}\right)\end{aligned} \tag{9.5.1}$$

For the opposite case of $l_r \gg \delta$, one has

$$\begin{aligned}\frac{\Delta f}{f_0} &\approx \frac{-1}{\pi Z_q}\sqrt{\frac{\rho_{liq}\omega\eta_{liq}}{2}}\left(1+2\frac{h_r^2}{l_r^2}+\sqrt{\pi}\frac{h_r^2}{l_r\delta}\right)\\ \frac{\Delta\Gamma}{f_0} &= \frac{1}{\pi Z_q}\sqrt{\frac{\rho_{liq}\omega\eta_{liq}}{2}}\left(1+2\frac{h_r^2}{l_r^2}\right)\end{aligned} \tag{9.5.2}$$

Again, both relations assume $h_r \ll l_r$ and $h_r \ll \delta$. With $h_r = 0$, Eqs. 9.5.1 and 9.5.2 reduce to the Gordon-Kanazawa-Mason result. Clearly, roughness effects are quadratic in h_r. Note: In Ref. [37], Γ is the *full*-width at half-height, not the half-width at half-height, as in this book. Also, Ref. [37] contains a misprint in Eqs. 2 and 3: f^2 should be $f_0 f$. This can be seen from setting $h_r = 0$, in which case Eqs. 2 and 3 must reduce to the Gordon-Kanazawa-Mason result.

Roughness certainly is relevant to experiment. For illustration, use Eq. 9.5.1 and insert values typical for a gold surface ($h_r \approx 3$ nm and $l_r \approx 10$ nm): With $\omega = 2\pi \times 5$ MHz, $\rho_{liq} = 1$ g/cm^3, and $\eta_{liq} = 1$ mPa s, one arrives at a roughness-induced frequency shift of 13.5 Hz. The roughness-induced shift in half-bandwidth is 0.2 Hz. The shift in frequency is larger than the shift in half-bandwidth because a rough surface traps the liquid in the valleys. A value of 13.5 Hz for the roughness-induced frequency shift definitely is relevant to the experiment. If, for instance, an adsorbate smoothens the surface, this will have an effect on the frequency. Large roughness is often observed in electrodeposition experiments. An experimental study of roughness effects, which compares experimental values of Δf and $\Delta\Gamma$ to the predictions from the Tel-Aviv group, is reported in Ref. [38].

Glossary

Variable	Definition (Comments)
a_T	Shift factor (Sect. 3.7)
A	(Effective) area of the resonator plate (Sect. 7.4)
c	Speed of (shear) sound ($\tilde{c} = (\tilde{G}/\rho)^{1/2}$)
d_P	Interparticle distance (Sect. 9.3)
D	Diffusivity (Sect. 9.3, do not confuse with the dissipation factor ($1/Q$))
eff	As an index: *eff*ective, mostly used in the context of an effective medium
f	Frequency
f_0	Resonance frequency at the fundamental ($f_0 = Z_q/(2m_q) = Z_q/(2\rho_q d_q)$)
f_r	Resonance frequency
$\hat{F}_{inertia}$	An inertial force (Sect. 9.3)
$\tilde{G}$	Shear modulus
K_A	A sensitivity factor (Sect. 9.4, taking care of an amplitude distribution)
h_r	Characteristic vertical scale of roughness (Sect. 9.5)
$\tilde{k}$	Wavenumber ($\tilde{k} = \omega/\tilde{c}$)

liq	As an index: *liq*uid
l_r	Characteristic horizontal scale of roughness (Sect. 9.5)
M	Mass
n	Overtone order
p	Pressure (Sect. 9.5)
q	Wave vector (Sect. 9.5)
P	As an index: *P*article
$\mathbf{r}_S$	A position on the resonator surface
R_P	Particle radius
S	As an index: *S*urface
t	Time
$\hat{u}$	(Tangential) displacement
$\hat{\mathrm{v}}$	Velocity
z	Spatial coordinate perpendicular to the surface
$\tilde{Z}_{liq}$	Acoustic wave impedance of a liquid ($\tilde{Z}_{liq} = (\mathrm{i}\omega\rho_{liq}\eta_{liq})^{1/2}$)
$\tilde{Z}_L$	Load impedance
Z_q	Acoustic wave impedance of AT-cut quartz ($Z_q = 8.8 \times 10^6$ kg $\mathrm{m}^{-2}\mathrm{s}^{-1}$)
Γ	Imaginary part of a resonance frequency
δ	As a prefix: a small quantity (Fig. 9.5)
δ	Penetration depth of a shear wave (Newtonian liquids: $\delta = (2\eta_{liq}/(\rho_{liq}\omega))^{1/2}$)
Δ	As a prefix: A shift induced by the presence of the sample
φ	Particle volume fraction
$\tilde{\eta}_{liq}$ η_{liq}	Viscosity
ρ	Density
τ_{hyd}	Hydrodynamic time scale (Sect. 9.3)
τ_{MR}	Momentum relaxation time (Sect. 9.3)
ξ	Drag coefficient
ω	Angular frequency
ω_c	A critical frequency, above which inertial effects are noticeable (Sect. 9.3)

References

1. Nandi, N., Bhattacharyya, K., Bagchi, B.: Dielectric relaxation and solvation dynamics of water in complex chemical and biological systems. Chem. Rev. **100**(6), 2013–2045 (2000)
2. Goubaidoulline, C., Reuber, J., Merz, F., Johannsmann, D.: Simultaneous determination of density and viscosity of liquids based on quartz-crystal resonators covered with nanoporous alumina. J. Appl. Phys. **98**(1), 014305 (2005)
3. Schön, P., Michalek, R., Walder, L.: Liquid density response of a quartz crystal microbalance modified with mesoporous titanium dioxide. Anal. Chem. **71**, 3305 (1999)
4. Stockbridge, C.D.: Effects of gas pressure on quartz crystal microbalances. In: Behrndt, K.H. (ed.) Vacuum Microbalance Techniques, vol. 5. Plenum Press, New York (1966)
5. Jakoby, B., Beigelbeck, R., Keplinger, F., Lucklum, F., Niedermayer, A., Reichel, E.K., Riesch, C., Voglhuber-Brunnmaier, T., Weiss, B.: Miniaturized sensors for the viscosity and density of liquids-performance and issues. IEEE Trans. Ultrason. Ferroelectr. Freq. Control **57**(1), 111–120 (2010)
6. http://www.flucon.de/, Accessed 28 Mar 2013
7. Jerome, B.: Surface effects and anchoring in liquid-crystals. Rep. Prog. Phys. **54**(3), 391–451 (1991)
8. Gupta, V.K., Skaife, J.J., Dubrovsky, T.B., Abbott, N.L.: Optical amplification of ligand-receptor binding using liquid crystals. Science **279**(5359), 2077–2080 (1998)
9. Domack, A., Johannsmann, D.: High frequency effective viscosities of nematic liquid crystals with tilted orientation. Appl. Phys. Lett. **80**(25), 4750–4752 (2002)
10. Muramatsu, H., Iwasaki, F.: Monitoring of the changes of dynamic viscoelastic properties of liquid crystals during the orientation with voltage pulses using a quartz crystal resonator. Mol. Cryst. Liq. Cryst. Sci. Technol. Sect. A **258**, 153–162 (1995)
11. deGennes, J.P., Prost, J.: The Physics of Liquid Crystals. Oxford University Press, Oxford (1993) (Ch. 5.2.2)
12. Parodi, O.: Stress tensor for a nematic liquid crystal. J. Phys. **31**(7), 581–584 (1970)
13. Kiry, F., Martinoty, P.: Ultrasonic investigation of anisotropic viscosities in a nematic liquid-crystal. J. Phys. **38**(2), 153–157 (1977)
14. Martinoty, P., Candau, S.: Determination of viscosity coefficients of a nematic liquid crystal using a shear waves reflectance technique. Mol. Cryst. Liq. Cryst. **14**(3–4), 243 (1971)
15. Berglin, M., Olsson, A., Elwing, H.: The interaction between model biomaterial coatings and nylon microparticles as measured with a quartz crystal microbalance with dissipation monitoring. Macromol. Biosci. **8**(5), 410–416 (2008)
16. Fatisson, J., Domingos, R.F., Wilkinson, K.J., Tufenkji, N.: Deposition of TiO_2 nanoparticles onto silica measured using a quartz crystal microbalance with dissipation monitoring. Langmuir **25**(11), 6062–6069 (2009)
17. Olofsson, A.C., Hermansson, M., Elwing, H.: Use of a quartz crystal microbalance to investigate the antiadhesive potential of N-acetyl-L-cysteine. Appl. Environ. Microbiol. **71**(5), 2705–2712 (2005)
18. Molino, P.J., Hodson, O.A., Quinn, J.F., Wetherbee, R.: The quartz crystal microbalance: a new tool for the investigation of the bioadhesion of diatoms to surfaces of differing surface energies. Langmuir **24**(13), 6730–6737 (2008)
19. Poitras, C., Fatisson, J., Tufenkji, N.: Real-time microgravimetric quantification of *Cryptosporidium parvum* in the presence of potential interferents. Water Res. **43**(10), 2631–2638 (2009)
20. Russel, W.B., Schowalter, D.A.S., Schowalter, W.R.: Colloidal Dispersions. Cambridge University Press, Cambridge (1991)
21. Buscall, R., Goodwin, J.W., Hawkins, M.W., Ottewill, R.H.: Viscoelastic properties of concentrated lattices 2 theoretical-analysis. J. Chem. Soc. Faraday Trans. I **78**, 2889–2899 (1982)

22. Bergenholtz, J., Willenbacher, N., Wagner, N.J., Morrison, B., van den Ende, D., Mellema, J.: Colloidal charge determination in concentrated liquid dispersions using torsional resonance oscillation. J. Colloid Interface Sci. **202**(2), 430–440 (1998)
23. Wagner, N.J.: The high-frequency shear modulus of colloidal suspensions and the effects of hydrodynamic interactions. J. Colloid Interface Sci. **161**(1), 169–181 (1993)
24. Huang, J.S., Varadaraj, R.: Colloid and interface science in the oil industry. Curr. Opin. Colloid Interface Sci. **1**(4), 535–539 (1996)
25. Bell, J., Kohler, T., Woermann, D.: Change of the resonance frequency of a quartz crystal microbalance in contact with an aqueous dispersion of solid particles. Berichte Der Bunsen-Gesellschaft-Phys. Chem. Chem. Phys. **101**(6), 879–883 (1997)
26. Lionberger, R.A., Russel, W.B.: High-frequency modulus of hard-sphere colloids. J. Rheol. **38**(6), 1885–1908 (1994)
27. Wagner, N.J.: The High-frequency shear modulus of colloidal suspensions and the effects of hydrodynamic interactions. J. Colloid Interface Sci. **161**(1), 169–181 (1993)
28. Kao, M.H., Yodh, A.G., Pine, D.J.: Observation of Brownian-motion on the time scale of hydrodynamic interactions. Phys. Rev. Lett. **70**(2), 242–245 (1993)
29. Atakhorrami, M., Mizuno, D., Koenderink, G.H., Liverpool, T.B., MacKintosh, F.C., Schmidt, C.F.: Short-time inertial response of viscoelastic fluids measured with Brownian motion and with active probes. Phys. Rev. E **77**(6), 13 (2008)
30. Persson, B.N.J.: Theory of rubber friction and contact mechanics. J. Chem. Phys. **115**(8), 3840–3861 (2001)
31. Baumberger, T., Caroli, C., Ronsin, O.: Self-healing slip pulses along a gel/glass interface. Phys. Rev. Lett. **88**(7), 075509 (2002)
32. Shull, K.R.: Contact mechanics and the adhesion of soft solids. Mater. Sci. Eng. R **36**(1), 1–45 (2002)
33. Flanigan, C.M., Desai, M., Shull, K.R.: Contact mechanics studies with the quartz crystal microbalance. Langmuir **16**(25), 9825–9829 (2000)
34. Efimov, I., Hillman, A.R., Schultze, J.W.: Sensitivity variation of the electrochemical quartz crystal microbalance in response to energy trapping. Electrochim. Acta **51**(12), 2572–2577 (2006)
35. Herrscher, M., Ziegler, C., Johannsmann, D.: Shifts of frequency and bandwidth of quartz crystal resonators coated with samples of finite lateral size. J. Appl. Phys. **101**(11), 114909 (2007)
36. Martin, S.J., Frye, G.C., Ricco, A.J., Senturia, S.D.: Effect of surface-roughness on the response of thickness-shear mode resonators in liquids. Anal. Chem. **65**(20), 2910–2922 (1993)
37. Daikhin, L., Gileadi, E., Katz, G., Tsionsky, V., Urbakh, M., Zagidulin, D.: Influence of roughness on the admittance of the quartz crystal microbalance immersed in liquids. Anal. Chem. **74**(3), 554–561 (2002)
38. Bund, A.M., Schneider, M.: Characterization of the viscoelasticity and the surface roughness of electrochemically prepared conducting polymer films by impedance measurements at quartz crystals. J. Electrochem. Soc. **149**, 331 (2002)

Chapter 10
Stratified Layer Systems

Abstract Samples, which are homogeneous in the resonator plane, can be modeled as acoustic multilayers. The deformation pattern is a plane wave. Thin films exposed to air behave as predicted by Sauerbrey. For somewhat thicker films, there is a viscoelastic correction scaling as the square of the film's mass. For films exposed to a liquid, the viscoelastic correction is independent of thickness. If the layer is soft, the correction can be substantial, even for molecularly thin films. Under certain conditions, the film's elastic compliance, J_f', can be calculation from the ratio of $\Delta\Gamma$ and ($-\Delta f$). Thick films display a film resonance.

10.1 Viscoelastic Film in Air

The geometry of the film in air is shown in Fig. 10.1. The acoustic wave is reflected at the outer edge of the film and returns to the resonator surface. The stress at the resonator surface is

$$\begin{aligned}\hat{\sigma}_S &= -\tilde{G}_f \left.\frac{\mathrm{d}\hat{u}(z)}{\mathrm{d}z}\right|_{z=d_q} = -\tilde{G}_f\left(-\mathrm{i}\tilde{k}_f\right)\left(\hat{u}_f^- - \hat{u}_f^+\right) \\ &= \mathrm{i}\tilde{G}_f\frac{\omega}{\tilde{c}_f}\left(\hat{u}_f^- - \hat{u}_f^+\right) \\ &= \mathrm{i}\omega\tilde{Z}_f\left(\hat{u}_f^- - \hat{u}_f^+\right)\end{aligned} \tag{10.1.1}$$

Note the minus sign before $\tilde{G}_f$ in line 1. It occurs because $\hat{\sigma}_S$ is exerted by the resonator onto the sample. For a film in air one has $\tilde{Z}_{bulk} \approx 0$ and the reflectivity at the film-air interface is $\tilde{r}_{f,bulk} = (\tilde{Z}_f - \tilde{Z}_{bulk})/(\tilde{Z}_f + \tilde{Z}_{bulk}) = 1$. $\hat{u}_f^+$ then is related to $\hat{u}_f^-$ by

$$\begin{aligned}\hat{u}_f^+ &= \exp\left(-\mathrm{i}\tilde{k}_f d_f\right)\tilde{r}_{f,bulk}\exp\left(-\mathrm{i}\tilde{k}_f d_f\right)\hat{u}_f^- \\ &= \exp\left(-2\mathrm{i}\tilde{k}_f d_f\right)\hat{u}_f^-\end{aligned} \tag{10.1.2}$$

D. Johannsmann, *The Quartz Crystal Microbalance in Soft Matter Research*, Soft and Biological Matter, DOI: 10.1007/978-3-319-07836-6_10,

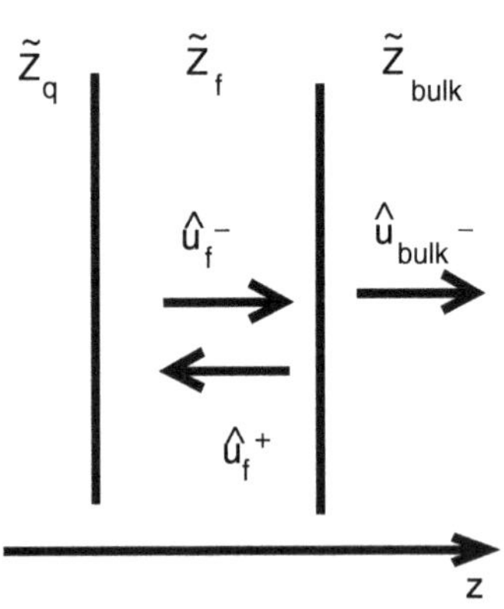

Fig. 10.1 Layer system and definition of variables for the single film. If the bulk is a vapor phase, its acoustic wave impedance is small

The load impedance is

$$\begin{aligned}\tilde{Z}_L &= \frac{\hat{\sigma}_S}{\hat{v}_S} = \frac{\mathrm{i}\omega\,\tilde{Z}_f\left(\hat{u}_f^- - \hat{u}_f^+\right)}{\mathrm{i}\omega\left(\hat{u}_f^- + \hat{u}_f^+\right)} = \tilde{Z}_f\frac{1-\exp\left(-2\mathrm{i}\tilde{k}_f d_f\right)}{1+\exp\left(-2\mathrm{i}\tilde{k}_f d_f\right)}\\ &= \tilde{Z}_f\frac{\exp\left(\mathrm{i}\tilde{k}_f d_f\right)-\exp\left(-\mathrm{i}\tilde{k}_f d_f\right)}{\exp\left(\mathrm{i}\tilde{k}_f d_f\right)+\exp\left(\mathrm{i}\tilde{k}_f d_f\right)}\\ &= \mathrm{i}\tilde{Z}_f\tan\left(\tilde{k}_f d_f\right)\end{aligned} \tag{10.1.3}$$

The relation $\tan(x) = -\mathrm{i}(\exp(\mathrm{i}x) - \exp(-\mathrm{i}x))/(\exp(\mathrm{i}x) + \exp(-\mathrm{i}x))$ was used in line 3. The frequency shift is

$$\frac{\Delta\tilde{f}}{f_0} = \frac{-1}{\pi Z_q}\tilde{Z}_f\tan\left(\tilde{k}_f d_f\right) \tag{10.1.4}$$

In the derivation of Eq. 10.1.4 we have applied the SLA. The corresponding equation derived without invoking the SLA (but still neglecting piezoelectric stiffening) is

$$\tilde{Z}_q\tan\left(\tilde{k}_q d_q\right) = -\tilde{Z}_f\tan\left(\tilde{k}_f d_f\right) \tag{10.1.5}$$

Equation 10.1.5 follows from Eqs. 10.1.3 and 4.6.8. Using $\tilde{k}_{q,ref} \approx \tilde{k}_{q,OC} = n\pi/d_q$ as well as $\tan(n\,\pi + x) = \tan(x)$ for integer n, this amounts to

$$\tan\left(\left(\Delta\tilde{k} + \tilde{k}_{ref}\right)d_q\right) = \tan\left(\Delta\tilde{k}d_q\right) = \tan\left(\frac{2\pi d_q}{\tilde{c}_q}\Delta\tilde{f}\right) = -\frac{\tilde{Z}_f}{\tilde{Z}_q}\tan\left(\frac{2\pi d_f}{\tilde{c}_f}\tilde{f}\right) \tag{10.1.6}$$

Equation 10.1.6 is usually ascribed to Lu and Lewis (Ref. [1], for the extension to layers with finite viscosity see Ref. [2]). In the context of film-thickness monitors,

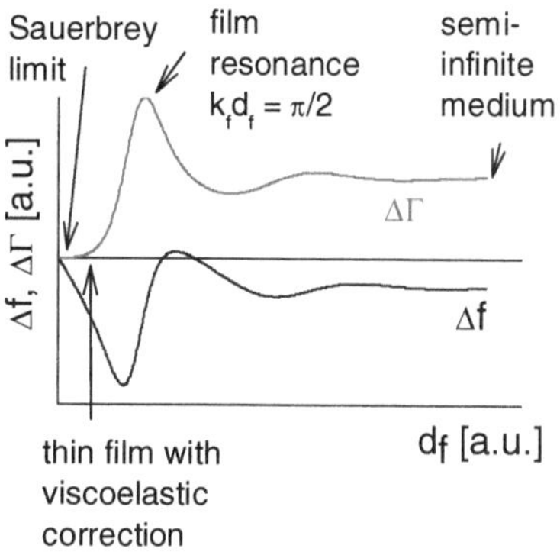

Fig. 10.2 A plot of Δf and ΔΓ versus film thickness according to Eq. 10.1.4. The loss angle of the film is δL = 40°

the use of Eq. 10.1.6 is called "Z-match method" [3]. The user provides the wave impedance of the coating to be deposited. Properly accounting for a nontrivial Z-ratio ($Z_f/Z_q \neq 1$), the instrument calculates a value for the deposited mass, which is more accurate than the Sauerbrey mass. Tabulated values of Z_f for some common coating materials can be found in Ref. [4].

It is instructive to rewrite Eq. 10.1.4 as

$$\begin{aligned}\frac{\Delta\tilde{f}}{f_0} &= \frac{-1}{\pi\, Z_q}\tilde{Z}_f \tan\left(\tilde{k}_f d_f\right)\\ &= \frac{-\tilde{Z}_f}{\pi\, Z_q}\tan\left(\frac{\omega}{\tilde{c}_f}d_f\right) = \frac{-\tilde{Z}_f}{\pi\, Z_q}\tan\left(\frac{\omega\sqrt{\rho_f}}{\sqrt{\tilde{G}_f}}d_f\right) = \frac{-\tilde{Z}_f}{\pi\, Z_q}\tan\left(\frac{\omega}{\tilde{Z}_f}\rho_f d_f\right)\\ &= \frac{-\tilde{Z}_f}{\pi Z_q}\tan\left(\frac{\omega}{\tilde{Z}_f}m_f\right)\end{aligned} \quad (10.1.7)$$

Equation 10.1.7 shows that the acoustic properties of a film are fully specified by two parameters, which are its acoustic wave impedance, $\tilde{Z}_f$, and its mass per unit area, m_f. (Strictly speaking, there are three parameters, which are Z_f', Z_f'', and m_f.) Equation 10.1.4 misleadingly suggests that one might be able to separately determine the film thickness and the density of the film. Independent information is needed to separate thickness from mass.

Figure 10.2 shows Δf and $\Delta\Gamma$ as predicted by Eq. 10.1.4 versus film thickness. The arrows indicate four different regimes, which are the Sauerbrey regime at very low thickness, the "thin-film regime", in which the viscoelastic correction is small compared to the Sauerbrey-term, the film resonance(s), [5, 6], and the limit of infinite thickness. In the limit of $d_f \rightarrow 0$, one expands the tangent as $\tan(x) \approx x$ and recovers the Sauerbrey equation. In the limit of $d_f \rightarrow \infty$, the complex tangent approaches $-i$. (Remember that $k'' > 0$ and $\tilde{k} = k' - ik''$, Eq. 4.1.6). Equation 10.1.4 then approaches the Gordon-Kanazawa-Mason result (Eq. 9.1.5).

The maximum in $\Delta\Gamma(d_f)$ occurring at $k_f d_f = \pi/2$ is called "film resonance". At the film resonance, Δf *in*creases with increasing thickness. Δf can even turn positive, if the film resonance is sharp enough. The fact that the film resonance is a coupled resonance in the sense of Sect. 4.6.3 becomes clear upon approximating $\tan(\tilde{k}_f d_f)$ by $(\tilde{k}_f d_f - \pi/2)^{-1}$ in the vicinity of $\tilde{k}_f d_f = \pi/2$:

$$\begin{aligned}\frac{\Delta \tilde{f}}{f_0} &= \frac{-1}{\pi\, Z_q}\tilde{Z}_f \tan\left(\tilde{k}_f d_f\right)\\ &\approx \frac{-1}{\pi\, Z_q}\tilde{Z}_f \frac{1}{\tilde{k}_f d_f - \frac{\pi}{2}}\\ &= \frac{-\tilde{Z}_f}{\pi Z_q}\frac{2}{\pi}\frac{1}{\frac{\pi \tilde{k}_f d_f}{2} - 1}\\ &= \frac{\tilde{Z}_f}{\pi Z_q}\frac{2}{\pi}\frac{\tilde{\omega}_{FR}}{\tilde{\omega}_{FR} - \omega}\end{aligned} \tag{10.1.8}$$

The index *FR* denotes the *F*ilm *R*esonance. The film's resonance frequency, ω_{FR}, is given by the condition that $\tilde{\omega}_{FR}/\tilde{c}_f d_f = \tilde{k}_f d_f = \pi/2$. Equation 10.1.8 has the same structure as Eq. 4.6.24.

Figure 10.3a sketches why the film itself forms a resonator at $k_f d_f = \pi/2$. At $k_f d_f = \pi/2$, the film thickness is one quarter of the wavelength of sound. The displacement pattern is a standing wave with an antinode at the surface and a node at the interface with the resonator. Because there is a node at the substrate-film interface, a small amplitude at the resonator surface translates to a large amplitude at the film-air interface, which is characteristic of a resonance. The film can be viewed as a viscoelastic plate rigidly clamped on one side. The film is not strictly clamped because the resonator has a finite elastic compliance. However, since most polymer films are much softer than quartz, one can approximate the film as being clamped. If the film is not much softer than quartz, Eq. 10.1.4 must be replaced by Eq. 10.1.6.

If the film has little internal dissipation, the film resonance amounts to a sharp peak in a plot $\Delta\Gamma$ versus d_f. Note, however, that the SLA does not apply under these conditions. Δf and $\Delta\Gamma$ plotted versus *n* only form a resonance curve of their own, if the damping factor of the coupled resonance is large enough (cf. Eq. 4.6.26). For samples with low internal dissipation, one has two separate modes around $k_f d_f = \pi/2$ [7]. Two separate modes have been seen with Langmuir-Blodgett films, sequentially deposited on the resonator surface [8]. One also can find the resonance frequency to discontinuously jump upwards in swelling experiments. In this case the instrument at some point looses the lower branch because the respective resonance becomes to weak. It then searches a resonance and finds the upper branch. Only sufficiently viscous films follow Eq. 10.1.4 and Fig. 10.2.

Figure 10.3b shows the same data as Fig. 10.2, plotted in polar form ($\Delta\Gamma$ versus Δf). Clearly there is a spiral, characteristic for the damped coupled resonance.

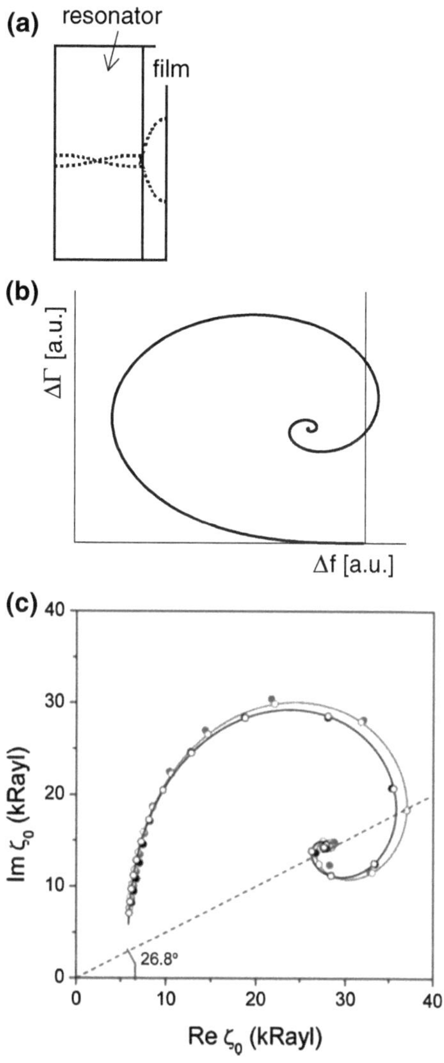

Fig. 10.3 **a** At the film resonance, the film thickness equals a quarter of the wavelength of sound. The displacement pattern is a standing wave with an antinode at the film-air interface and a node at the film-substrate interface. A small amplitude at the resonator surface thereby translates to a large amplitude at the film-air interface. **b** The simulated data from Fig. 10.2 shown in polar form. One finds the characteristic spiral. **c** Experimental results showing a film resonance. The film was formed from polyelectrolyte multilayers [48]. The layers were deposited sequentially and QCM measurements were undertaken at various stages during the build-up. The authors converted Δf and $\Delta\Gamma$ to a load impedance making use of Eq. 4.6.1. Because Eq. 4.6.1 contains the imaginary unit, panels b and c differ by a 90°-rotation. 1 Rayl corresponds 1 kg m^{-2} s^{-1} (Sect. 4.2). Panel c is reprinted with permission from Ref. [9]. Copyright 2007 American Chemical Society

Contrasting to the case of a sphere on a plate (Sects. 4.6.3 and 11.4), the spiral can make more than one turn. The subsequent turns are the higher overtones of the film resonance. An experimental example, which follows Eq. 10.1.4 particularly well, is shown in Fig. 10.3c [9]. Salomaki and Kankare display their results in polar form and they convert Δf and $\Delta\Gamma$ to $\tilde{Z}_L$ using Eq. 4.6.1. The authors use the metric Rayl as the unit of the impedance (cf. Sect. 4.2). The metric Rayl is equal to 1 kg m^{-2} s^{-1}. The agreement between Fig. 10.3c and the model is quite remarkable. Figure 10.3c is a rare example, where the higher overtones of the film resonance are seen. The spiral makes 1.5 turns. (The "higher film resonances" have $k_f d_f = m\,\pi/2$ with m an odd integer).

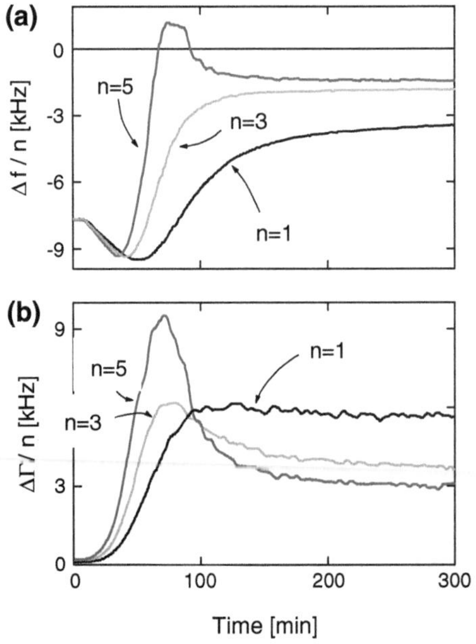

Fig. 10.4 Experimental example of a film resonance. **a** Normalized frequency shift. **b** Normalized shift of half bandwidth. A poly-isobutylene film was exposed to toluene vapor. As the solvent uptake proceeds, both the film thickness and the film's softness increase. The softening actually is more important than the increase in thickness. Softening lowers the wavelength of shear sound. The highest overtone ($n = 5$) goes into resonance first because the corresponding wavelength is shorter than on $n = 3$ and $n = 1$. Quantitative agreement with Eq. 10.1.7 is poor. There are different potential reasons (among them the gradient of softness along the z-axis). Less-than-perfect agreement between data and theory at conditions close the film resonance is often observed. Adapted from Ref. [10]

Figure 10.4 shows a second experimental example of a film resonance [10]. In this case, a polymer film was swollen in solvent vapor. Upon exposing the film to the vapor phase, it becomes thicker and softer. The highest overtone shown ($n = 5$ at 25 MHz) reaches the film resonance first because it has the shortest wavelength. The third overtone (15 MH) follows later. For the fundamental, the film resonance is barely reached.

The film resonance experimentally confirms the multilayer formalism, but it is of limited practical relevance, otherwise. The thin-film limit is much more important. Many samples show small viscoelastic effects without showing the film resonance itself. In the "thin film regime", one expands tan $(\tilde{k}_f d_f)$ from Eq. 10.1.4 to third order in $\tilde{k}_f d_f (\tan(\varepsilon) \approx \varepsilon + \varepsilon^3/3)$, which results in

$$\begin{aligned}\frac{\Delta\tilde{f}}{f_0} &= \frac{-1}{\pi Z_q}\tilde{Z}_f \tan\left(\tilde{k}_f d_f\right) \approx \frac{-1}{\pi Z_q}\tilde{Z}_f\left(\tilde{k}_f d_f + \frac{1}{3}\left(\tilde{k}_f d_f\right)^3\right) \\ &= \frac{-1}{\pi Z_q}\omega\, m_f\left[1 + \frac{(n\pi)^2}{3}\frac{\tilde{J}_f}{\rho_f}Z_q^2\left(\frac{m_f}{m_q}\right)^2\right] \\ &= \frac{-1}{\pi Z_q}\omega\, m_f\left[1 + \frac{(n\pi)^2}{3}\frac{Z_q^2}{\tilde{Z}_f^2}\left(\frac{m_f}{m_q}\right)^2\right]\end{aligned} \tag{10.1.9}$$

The relations $\tilde{k}_f = \omega/\tilde{c}_f = \omega(\rho\tilde{J}_f)^{1/2}$ and $\omega = 2\pi n f_0 = \pi n Z_q/m_q$ were used in line 2. The relation $\tilde{Z}_f = (\rho_f/\tilde{J}_f)^{1/2}$ was used in line 3.

At this point we recall three results from Sects. 6.1.4 and 6.2:

(a) Revisiting the approximations inherent to the SLA, one finds that the term $(\tilde{J}_f/(\rho_f)Z_q^2$ needs to be replaced by $(\tilde{J}_f/(\rho_f)Z_q^2 - 1$:

$$\begin{aligned}\frac{\Delta\tilde{f}}{f_0} &\approx \frac{-\omega\, m_f}{\pi Z_q}\left[1 + \frac{(n\pi)^2}{3}\left(\frac{Z_q^2}{\tilde{Z}_f^2} - 1\right)\left(\frac{m_f}{m_q}\right)^2\right] \\ &= \frac{-\omega\, m_f}{\pi Z_q}\left[1 + \frac{(n\pi)^2}{3}\left(\frac{\tilde{J}_f}{\rho_f}Z_q^2 - 1\right)\left(\frac{m_f}{m_q}\right)^2\right]\end{aligned} \tag{10.1.10}$$

The difference between Eqs. 10.1.9 and 6.2.8 matters when the film's acoustic wave impedance is comparable to Z_q.

(b) The parameter m_q in the denominator is half the plate's mass within the parallel plate model, but it turns into the "modal mass" when accounting for deviations from that model (Sect. 6.1.4). The true modal mass as a function of overtone order can be determined with thin rigid films. An overtone dependence of $\Delta f/n$ can only be attributed to viscoelastic effects after the modal mass has been corrected for.

(c) A dependence of $\Delta f/n$ on n^2 can also result from electrode effects. See the text around Eq. 6.2.15 for details.

For small m_f/m_q (neglecting the second term in square brackets), Eq. 10.1.10 reduces to the Sauerbrey equation. For slightly thicker films, there is a viscoelastic correction proportional to n^2, m_f^2, and $\tilde{J}_f$. Importantly, the viscoelastic contribution is large for soft films (with large $\tilde{J}_f$). The non-gravimetric QCM plays out its strength when the sample is softer than quartz. Viscoelastic effects disappear in the thin-film limit because the film's own inertia is the only source of shear stress; the sample is not clamped from the other side. It is difficult to determine the softness of a film thinner than about 10 nm and it is quite impossible to determine the softness of a monomolecular adsorbate. This contrasts to the situation in liquids. A liquid exerts a stress onto the film from the outer edge and thereby makes viscoelastic effects visible even for adsorbed monolayers.

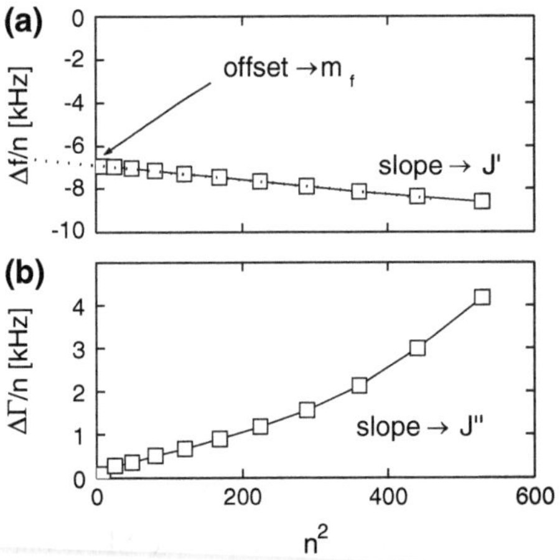

Fig. 10.5 Plot of $\Delta f/n$ (**a**) and $\Delta\Gamma/n$ (**b**) versus n^2 as suggested by Eq. 10.1.10. The sample was a spin-cast film of poly-isobutylene. The thickness as derived from the offset in panel a is about 1.6 μm. In this particular example, the slope $d(\Delta f/n)/d(n^2)$ is about constant, indicating a weak dependence of J'_f on frequency. The frequency dependence of J''_f is more evident (panel b). Quantitative evaluation leads to shear moduli in the range of 100 MPa. Such high values appear surprising at first glance because poly-isobutylene is a soft polymer. The high value is explained by viscoelastic dispersion and the high frequency of the measurement (cf. Fig. 3.3). Adapted from Ref. [48]

A plot adapted to the structure of Eq. 10.1.10 has n^2 along the x-axis and the normalized shifts of frequency and half-bandwidth ($\Delta f/n$) and $\Delta\Gamma/n$) along y. An example is shown in Fig. 10.5. If $\tilde{J}(\omega)$ were constant, straight lines would result and the compliances (elastic and viscous) would be proportional to the slopes. However, $\tilde{J}(\omega)$ usually does depend on frequency. For J'' this is not a problem because $J''(\omega)$ can be explicitly derived from the $\Delta\Gamma/n$. However, there is a problem for the shear compliance because one needs to extrapolate $\Delta f(n)$ to $n = 0$ in order to determine the mass, m_f. An approximate law for $J'(\omega)$ must be assumed in order to do the extrapolation. Typical would be a power law as in Eq. 10.4.1.

For thin films in air, $\Delta\Gamma$ can be converted to J'' using [11]

$$\Delta\Gamma \approx \frac{8}{3\rho_f Z_q} f_0^4 m_f^3 n^3 \pi^2 J'' = \frac{8 f_0^4 n^3 \pi^2}{3 Z_q} \rho_f^2 d_f^3 J'' \qquad (10.1.11)$$

m_f can be estimated from the Sauerbrey equation ($m_f \approx -m_q \Delta f/f_r$), resulting in

$$\frac{J''_f}{\rho_f} \approx \frac{3}{(n\pi)^2} \frac{1}{Z_q^2} \left(\frac{f}{\Delta f}\right)^2 \left(\frac{\Delta\Gamma}{-\Delta f}\right) \qquad (10.1.12)$$

10.2 Viscoelastic Film in a Liquid

Start the analysis from Fig. 10.1, but let the bulk medium be a liquid rather than air. Equation 10.1.1 still applies, but the reflectivity at the outer interface now is $\tilde{r}_{f,bulk} = (\tilde{Z}_f - \tilde{Z}_{liq})/(\tilde{Z}_f + \tilde{Z}_{liq})$. The amplitude of the reflected wave at the resonator surface, $\hat{u}^+$, is given as

$$\hat{u}^+ = \exp(-2i\tilde{k}_f d_f)\frac{\tilde{Z}_f - \tilde{Z}_{liq}}{\tilde{Z}_f + \tilde{Z}_{liq}}\hat{u}^- \tag{10.2.1}$$

Inserting Eq. 10.2.1 into Eq. 10.1.1 leads to

$$\begin{aligned}
\tilde{Z}_L &= \tilde{Z}_f \frac{1 - \exp(-2\mathrm{i}\tilde{k}_f d_f)\frac{\tilde{Z}_f - \tilde{Z}_{liq}}{\tilde{Z}_f + \tilde{Z}_{liq}}}{1 + \exp(-2\mathrm{i}\tilde{k}_f d_f)\frac{\tilde{Z}_f - \tilde{Z}_{liq}}{\tilde{Z}_f + \tilde{Z}_{liq}}} \\
&= \tilde{Z}_f \frac{\tilde{Z}_f(\exp(\mathrm{i}\tilde{k}_f d_f) - \exp(-\mathrm{i}\tilde{k}_f d_f)) + \tilde{Z}_{liq}(\exp(\mathrm{i}\tilde{k}_f d_f) + \exp(-\mathrm{i}\tilde{k}_f d_f))}{\tilde{Z}_f(\exp(\mathrm{i}\tilde{k}_f d_f) + \exp(-\mathrm{i}\tilde{k}_f d_f)) + \tilde{Z}_{liq}(\exp(\mathrm{i}\tilde{k}_f d_f) - \exp(-\mathrm{i}\tilde{k}_f d_f))} \\
&= \tilde{Z}_f \frac{\tilde{Z}_f 2\mathrm{i}\sin(\tilde{k}_f d_f) + \tilde{Z}_{liq} 2\cos(\tilde{k}_f d_f)}{\tilde{Z}_f 2\cos(\tilde{k}_f d_f) + \tilde{Z}_{liq} 2\mathrm{i}\sin(\tilde{k}_f d_f)}
\end{aligned} \tag{10.2.2}$$

Upon dividing numerator and denominator by $2\cos(\tilde{k}_f d_f)$ and inserting the result into the SLA, one finds [12, 13],

$$\frac{\Delta\tilde{f}}{f_0} = \frac{-\tilde{Z}_f}{\pi Z_q}\frac{\tilde{Z}_f\tan(\tilde{k}_f d_f) - \mathrm{i}\tilde{Z}_{liq}}{\tilde{Z}_f + \mathrm{i}\tilde{Z}_{liq}\tan(\tilde{k}_f d_f)} \tag{10.2.3}$$

As in Eq. 10.1.7, one can express $\tilde{k}_f d_f$ as $(\omega/\tilde{Z}_f)d_f$, which shows that the thickness and the mass per unit area of a film cannot be derived independently. With regard to shear waves, the properties of a film are fully specified by the parameters $\tilde{Z}_f$ and m_f.

Similar to the film in air, the Gordon-Kanazawa-Mason result is recovered in the limit of $d_f \to \infty$. At intermediate thickness, there again is a film resonance. The film resonance has been seen in liquids; Fig. 10.6 shows an example. The importance of the film resonance in liquid sensing is limited and the discussion therefore continues with the thin-film limit.

Expanding Eq. 10.2.3 to first order in d_f, one finds

$$\frac{\Delta\tilde{f}}{f_0} \approx \frac{\mathrm{i}}{\pi Z_q}\left[\tilde{Z}_{liq} + \mathrm{i}\tilde{Z}_f\tilde{k}_f d_f\left(1 - \frac{\tilde{Z}_{liq}^2}{\tilde{Z}_f^2}\right)\right] \tag{10.2.4}$$

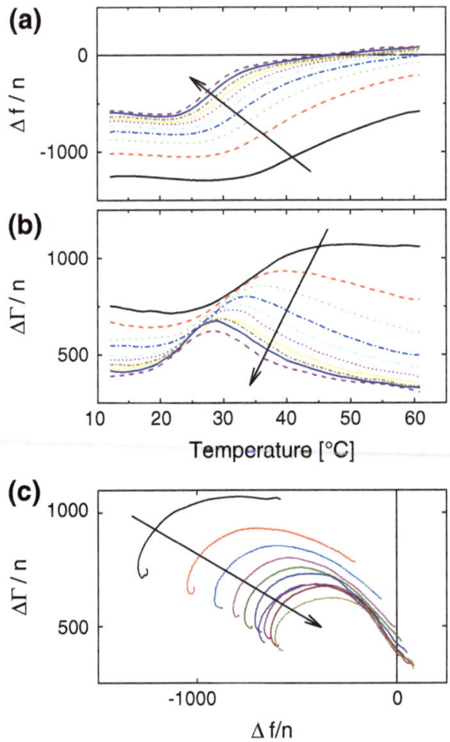

Fig. 10.6 Example of a film resonance observed in the liquid phase. **a** Normalized frequency shift. **b** Normalized shift of half bandwidth. **c** Polar diagram. Slightly simplifying the matter, the film resonance corresponds to a situation where the argument of the tangent in Eq. 10.2.3 is close to $\pi/2$. The experiment shown here was actually conducted on a polymer brush, rather than a film. The polymer brush has a segment density profile $\varphi(z)$, which gradually approaches zero at the edge; there is no sharp interface between the polymer and the bulk liquid. Even this sample shows a film resonance. The thickness of the brush was varied, using temperature. The solvent quality of this polymer-solvent pair (polystyrene-cyclohexane) improves with temperature. As temperature increases, the brush swells. Swelling increases the thickness and at the same time softens the material because the more dilute polymer has weaker interactions between neighboring chains. Arrows indicate increasing overtone order. Adapted from Ref. [12]

The two terms can be attributed to the bulk viscosity and to the adsorbed mass. For a sufficiently stiff film $(|\tilde{Z}_f^2| \gg |\tilde{Z}_{liq}^2|)^2$ the second term in square brackets is the Sauerbrey term:

$$\begin{aligned}\frac{\Delta\tilde{f}}{f_0} &\approx \frac{\mathrm{i}}{\pi Z_q}\left(\tilde{Z}_{liq} + \mathrm{i}\tilde{Z}_f\tilde{k}_f d_f\right) = \frac{\mathrm{i}}{\pi Z_q}\left(\sqrt{\mathrm{i}\omega\rho_{liq}\tilde{\eta}_{liq}} + \mathrm{i}\omega m_f\right)\\ &= \frac{-1+\mathrm{i}}{\sqrt{2}\pi Z_q}\sqrt{\omega\rho_{liq}\tilde{\eta}_{liq}} - \frac{2nf_0}{Z_q} m_f\end{aligned} \tag{10.2.5}$$

The frequency shift is a sum of a viscous contribution (following the Gordon-Kanazawa-Mason equation) and an inertial contribution (following Sauerbrey). For stiff films the QCM works as a gravimetric sensor even in the liquid phase. Additivity of the two contributions was found experimentally in Ref. [14].

The way in which viscoelasticity enters the equations of the thin-film limit deserves a closer look. The frequency shift is usually determined with respect to a reference state, where the quartz is already immersed in liquid. Referencing the measurement to the bare quartz in the liquid, one obtains [13–16]

$$\begin{aligned}\frac{\Delta\tilde{f}}{f_0} &\approx \frac{i}{\pi Z_q}\left(\tilde{Z}_{tot}-\tilde{Z}_{liq}\right)=\frac{-1}{\pi Z_q}\tilde{Z}_f\tilde{k}_f d_f\left[1-\frac{\tilde{Z}_{liq}^2}{\tilde{Z}_f^2}\right]\\ &=\frac{-\omega m_f}{\pi Z_q}\left[1-\frac{\tilde{Z}_{liq}^2}{\tilde{Z}_f^2}\right]=\frac{-\omega m_f}{\pi Z_q}\left[1-\frac{\tilde{J}_f(\omega)}{\rho_f}i\omega\rho_{liq}\eta_{liq}\right]\\ &=\frac{-\omega m_f}{\pi Z_q}\left[1-2\pi\, in\frac{\tilde{J}_f(\omega)}{\rho_f}f_0\rho_{liq}\eta_{liq}\right]\end{aligned} \tag{10.2.6}$$

The prefactor has the same form as the right-hand side of the Sauerbrey equation; the term in square brackets is the viscoelastic correction. As opposed to experiments air, softness *in*creases the resonance frequency. It decreases the apparent Sauerbrey mass (the mass obtained when naively analyzing experimental data with the Sauerbrey equation).

An experimental example is shown in Fig. 10.7. These data were obtained on a layer of adsorbed vesicles. The details are unessential. Of course vesicles are discrete objects, but the layer appears like a soft film to the QCM. Characteristically for a soft film in a liquid, the plot of $\Delta f/n$ versus n slopes upwards. The larger the slope, the larger is the sample's viscous (!) compliance. Note: When a film is immersed in a liquid, a finite *elastic* compliance mostly affects the *bandwidth*, while a finite *viscous* compliance mostly affects the *frequency*. It is the other way round in air.

Contrasting to the dry state, the viscoelastic correction is independent of film thickness in a liquid environment. For films in air the viscoelastic correction scales as the square of the mass (Eq. 10.1.9) because the film surface is stress-free in air. The film only shears under its own weight. This is different in liquids because the liquids exerts a stress onto the film. The viscoelastic correction may be sizeable, even for molecularly thin films.

Equation 10.2.6 predicts $-\Delta f$ to increase with overtone order, as shown in Fig. 10.7. In experiment, one can also find $-\Delta f$ decreasing with n. The reasons are unclear; roughness may play a role.

Because Eq. 10.2.6 is linear in mass, it also holds in an integral sense [17];

$$\frac{\Delta\tilde{f}}{f_0}\approx-\frac{-\omega}{\pi Z_q}\int_0^\infty\left(\frac{\tilde{Z}^2(z)-\tilde{Z}_{liq}^2}{\tilde{Z}^2(z)}\right)\rho(z)dz\approx-\frac{\rho_{liq}\omega}{\pi Z_q}\int_0^\infty\left(\frac{\tilde{G}(z)-\tilde{G}_{liq}}{\tilde{G}(z)}\right)dz \tag{10.2.7}$$

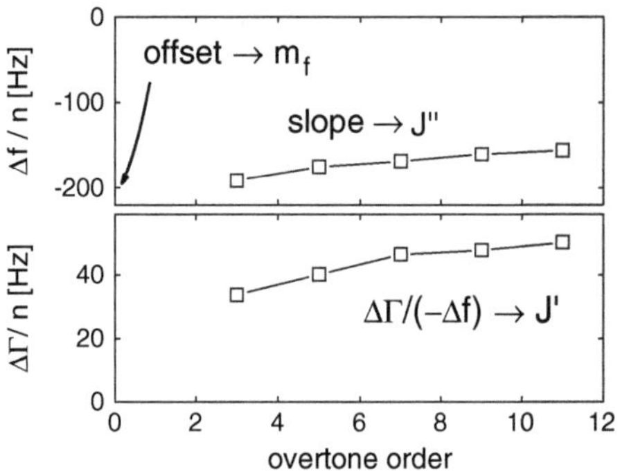

Fig. 10.7 Plot of normalized shifts of frequency and half-bandwidth versus overtone order as suggested by Eq. 10.2.6. The sample was a layer of vesicles. Details of the sample preparation are unessential. Characteristically, the slope of $\Delta f/n$ versus n is positive. If the slope were constant over the entire data range, this would indicate an elastic compliance independent of frequency. Given that the sample consists of discrete adsorbed objects, the derived values of elastic and viscous compliance must be viewed as effective values. Data kindly provided by Ilya Reviakine

In step 2, $\rho(z)$ was assumed to be about constant and equal to ρ_{liq}. Equation 10.2.7 will be central importance in Sects. 10.6 and 16.2.

Two notes concerning the literature:

- The Voigt model from Ref. [18] is equivalent to the acoustic multilayer formalism. For the proof see the appendix of Ref. [13]. Although not readily apparent, the "Voigt model" also makes use of the SLA. At the end of the derivation, Voinova et al. invoke an approximation from a previous paper of the group (Ref. [19]), which essentially is the SLA.
- The "missing-mass effect" from Ref. [15] is equivalent to the viscoelastic correction in Eq. 10.2.6 (Ref. [11]).

10.3 Equivalent Mass and Equivalent Thickness

The interpretation of QCM measurements in liquids is somewhat of a challenge. Practitioners often just apply the Sauerbrey equation (Eq. 8.1.2) to their data and call the resulting areal mass density the "Sauerbrey mass". The corresponding thickness is the "Sauerbrey thickness". The Sauerbrey thickness can certainly serve to compare different experiments, but it must not be naively identified with the geometric thickness. Here is a list of worthwhile considerations:

(a) The QCM always measures a film's areal mass density, never its geometric thickness. The conversion from areal mass density to thickness requires the physical density as independent input.

(b) One can account for the viscoelastic correction by fitting the experimental sets of $\{\Delta f(n)\}$and $\{\Delta\Gamma(n)\}$, with in Eq. 10.2.6. The parameter m_f then is one of the fit parameters. The fits are not trivial though, there sometimes is a question with regard to uniqueness.
(c) If full-fledged fitting shall be avoided, one can extrapolate the function $\Delta f(n)$ to $n = 0$. At $n = 0$, the viscoelastic correction disappears.
(d) If the correction factor is significantly different from unity, it also affects the bandwidth and makes $-\Delta f/n$ depend on overtone order. If, conversely, such effects are absent (if $\Delta\Gamma \ll (-\Delta f)$ and $\Delta f/n \approx const$), one may assume $1 - \left|\tilde{Z}_{liq}^2/\tilde{Z}_f^2\right| \approx 1$. The Sauerbrey equation then is trustworthy.
(e) If the viscoelastic correction is insignificant as discussed in (c), this does not imply that the film would not be swollen in the ambient liquid. It only implies that the (swollen) film is much more rigid than the ambient liquid. The amount of swelling can only be inferred from the comparison of wet and dry thickness. QCM data taken on the wet sample alone do not allow to infer the degree of swelling (cf. Sect. 16.2).
(f) Soft samples often have a fuzzy interface with the ambient liquid. Fuzzy interfaces usually are soft interfaces. They will induce significant viscoelastic effects ($\Delta\Gamma > 0$, $\Delta f/n$ a function of n). In the absence of such effects, one may conclude that the border of the film with the liquid is sharp.
(g) The degree of swelling of a film in aqueous solution can be inferred from H_2O/D_2O exchange experiments, see Ref. [20].

10.4 Determination of Viscoelastic Constants

In principle, both J_f' and J_f'' can be extracted from the n-dependence of Δf and $\Delta\Gamma$. A few difficulties must be kept in mind, though. Firstly and importantly, J_f' and J_f'' themselves depend on n. Secondly, J_f' and J_f'' cannot be derived explicitly because the mass is not a priori known. Finally, films in a liquid environment can show an overtone dependence of $\Delta f/n$ because of roughness (Sects. 9.5 and 12.2). (And of course there is always the possibility of compressional waves having an influence.)

Rheological spectra usually are smooth. They are displayed on logarithmic scales (Sect. 3.7). Since the experimentally accessible range of frequencies is only about a decade, it is fair to approximate the frequency dependence of J_f' and J_f'' by a power law with power law exponents β' and β''

$$\begin{aligned} J_f'(f) &\approx J_f'\left(f_{ref}\right)\left(\frac{f}{f_{ref}}\right)^{\beta'} \\ J_f''(f) &\approx J_f''\left(f_{ref}\right)\left(\frac{f}{f_{ref}}\right)^{\beta''} \end{aligned} \qquad (10.4.1)$$

Note: The index "*ref*" here denotes a frequency somewhere in the middle of the frequency range of the QCM (as opposed to a reference state of the crystal without a sample). Power laws analogous to Eq. 10.4.1 can be formulated for other choices of viscoelastic parameters (G'_f and G''_f or G'_f and η'_f). Unfortunately, power laws in J'_f and J''_f do not turn into power laws in G'_f and G''_f when transformed with Eq. 3.7.4. A set of power laws in J'_f and J''_f therefore is not strictly equivalent to a similar set in G'_f and G''_f.

There are limits on β' and β''. If expressing viscoelasticity in terms of compliance (J'_f and J''_f as in Eq. 10.4.1), one has $-2<\beta'<0$ and $-1<\beta''<1$. If moduli are used (G'_f and G''_f), one has $0<\beta'<2$ and $-1<\beta''<1$. Users of the QCM-D often express viscoelasticity in terms of $\mu = G'_f$ and $\eta = G''_f/\omega$. The power law exponent for η is between -2 and 0.

10.5 The ΔΓ/Δf-Ratio and a Film's Elastic Compliance

The term in square brackets in Eq. 10.2.6 is the viscoelastic correction to the Sauerbrey equation. The film thickness only enters as a prefactor. In the limit of thin films, the viscoelastic correction is independent of film thickness and the thickness can be eliminated from the equations by taking the ratio $\Delta\Gamma$ to ($-\Delta f$) (the "$\Delta\Gamma/\Delta f$-ratio"). The $\Delta\Gamma/\Delta f$-ratio is an intrinsic property of the adsorbed material. It is related to the material's density and viscoelastic constants as

$$\frac{\Delta\Gamma}{-\Delta f} = \frac{-\mathrm{Im}\left(1 - \tilde{Z}_{liq}^2/\tilde{Z}_f^2\right)}{\mathrm{Re}\left(1 - \tilde{Z}_{liq}^2/\tilde{Z}_f^2\right)} \tag{10.5.1}$$

It turns out that the $\Delta\Gamma/\Delta f$-ratio mostly is a measure of a film's *softness*. To see this more quantitatively, we make further approximations. Assume the liquid to be Newtonian. The shear-wave impedance of this liquid is $\tilde{Z}_{liq} = (i\omega\rho_{liq}\eta_{liq})^{1/2}$, which leads to

$$\frac{\Delta\Gamma}{-\Delta f} = \frac{-\mathrm{Im}\left(1 - i\omega\rho_{liq}\eta_{liq}\tilde{J}_f/\rho_f\right)}{\mathrm{Re}\left(1 - i\omega\rho_{liq}\eta_{liq}\tilde{J}_f/\rho_f\right)} = \frac{\omega\rho_{liq}\eta_{liq}J'_f}{\rho_f - \omega\rho_{liq}\eta_{liq}J''_f} \tag{10.5.2}$$

The densities in soft-matter experiments often are similar. Approximating ρ_f by ρ_{liq} yields

$$\frac{\Delta\Gamma}{-\Delta f} \approx \frac{\omega\eta_{liq}J'_f}{1 - \omega\eta_{liq}J''_f} \tag{10.5.3}$$

Most films of interest are substantially more rigid than the ambient liquid. Even films, which are soft at low frequencies, often appear as rigid in the MHz range. The denominator in Eq. 10.5.3 can be rewritten as $1 - J''_f / J''_{liq}$ where $J''_{liq} = (\omega\eta_{liq})^{-1}$ is the viscous compliance of the liquid. If the viscous compliance of the film is much smaller than the viscous compliance of the liquid (given as $(\omega\eta_{liq})^{-1}$), the denominator is close to unity and one has [17]

$$\frac{\Delta\Gamma}{-\Delta f} \approx \omega\eta_{liq} J'_f = 2\pi n f_0 \eta_{liq} J'_f \tag{10.5.4}$$

This is one of the cases, where the shear compliance as a parameter quantifying viscoelastic response is more convenient than the shear modulus. One can easily determine J'_f from experiment, but not J''_f. Because the conversion to shear modulus requires knowledge of both J'_f and J''_f (Eq. 3.7.4), neither G'_f nor G''_f can be derived (but J'_f can).

If the density of the film and the liquid are unequal, one can still neglect the second term in the denominator in Eq. 10.5.3, which leads to

$$\frac{\Delta\Gamma}{-\Delta f} \approx \frac{\rho_{liq}}{\rho_f} \eta_{liq} \omega J'_f = \frac{\rho_{liq}}{\rho_f} 2\pi n f_0 \eta_{liq} J'_f = \mathrm{Re}\left(\frac{\left|Z^2_{liq}\right|}{\tilde{Z}^2_f}\right) \tag{10.5.5}$$

Again: Eq. 10.5.1 only holds for thin films. The detailed calculations show that Eqs. 10.5.4 and 10.5.5 should only be applied if the thickness is a few nanometers at most.

The Gizeli group has carried this concept even further [21, 22]. They take the ratio $\Delta D/(-\Delta F)$ (D the dissipation factor and $\Delta F = \Delta f/n$ the normalized frequency shift) and call this parameter the "acoustic ratio". While the acoustic ratio clearly is equivalent to what is called $\Delta\Gamma/\Delta f$ above, the meaning given to it by the Gizeli group is different. These workers attached DNA strands to the QCM surface. They find that acoustic ratio to be constant up to a certain limit in coverage. From the fact that $\Delta D/(-\Delta F)$ is independent of coverage, they conclude that the acoustic ratio is a property of the *molecules* under investigation (as opposed to a property of the layer). Evidently, this argument only holds in the range where $\Delta D/(-\Delta F)$ is indeed independent of ΔF. The acoustic ratio was found to depend on the length of the DNA strands adsorbed. It increased with increasing length. In the discussion, the group draws an analogy to the intrinsic viscosity of a dilute solution, $[\eta]$, which is defined as $[\eta] = (\eta - \eta_0)/(\eta_0 c)$, to be evaluated in the limit of low c. η_0 is the viscosity of the pure solvent, c is the concentration of the solute. In the limit of low concentration, intermolecular interactions do not affect the viscosity and $[\eta]$ is a property of the solute/solvent pair.

Two more remarks:

- From a mathematical point of view, it is not necessarily clear that the $\Delta\Gamma/\Delta f$-ratio should be proportional to the intrinsic viscosity of the adsorbate molecules. The Gizeli group has established experimentally that there is a correlation between the $\Delta\Gamma/\Delta f$-ratio and the intrinsic viscosity of the same molecules in the bulk. In principle, the $\Delta\Gamma/\Delta f$-ratio might relate to other molecular parameters, as well. That would not invalidate the argument as such. Whatever the parameter is, it would still be a property of the solute/solvent pair.
- A $\Delta\Gamma/\Delta f$-ratio independent of $-\Delta f$ is not always observed. Figure 12.8 shows counter examples. The $\Delta\Gamma/\Delta f$-ratio only is a property of the molecules involved, as long as it is independent of coverage.

10.6 Multilayers and Continuous Viscoelastic Profiles

For samples consisting of multiple layers, the explicit equations predicting $\Delta\tilde{f}$ become long and complicated. For two layers in a liquid, the equation still fits into one line, which is [6]:

$$\frac{\Delta\tilde{f}}{f_0} = \frac{-\tilde{Z}_1\,\tilde{Z}_2\left(\tilde{Z}_1\tan\left(\tilde{k}_1 d_1\right)+\tilde{Z}_2\tan\left(\tilde{k}_2 d_2\right)\right)+\mathrm{i}\tilde{Z}_{liq}\left(\tilde{Z}_1\tan\left(\tilde{k}_2 d_2\right)\tan\left(\tilde{k}_1 d_1\right)-\tilde{Z}_2\right)}{\pi Z_q\,\tilde{Z}_2\left(\tilde{Z}_1-\tilde{Z}_2\tan\left(\tilde{k}_2 d_2\right)\,\tan\left(\tilde{k}_1 d_1\right)\right)+\mathrm{i}\tilde{Z}_{liq}\left(\tilde{Z}_1\tan\left(\tilde{k}_2 d_2\right)+\tilde{Z}_2\tan\left(\tilde{k}_1 d_1\right)\right)} \tag{10.6.1}$$

The indices 1, 2, and *liq* denote the two films and the liquid, respectively. The author has made a software package for the analysis of multilayers available on the web [23]. This software calculates Δf for up to 4 layers. Of course the uniqueness of the fit must be checked carefully.

Continuous viscoelastic profiles can be approximated by a sequence of many thin layers, evaluated with the matrix method (Sect. 1.5.3). There is a second method. It is not necessarily difficult, but it requires the use of one of the advanced mathematical software packages like Mathematica or MATLAB. These can solve ordinary differential equations numerically. The respective command is a one-liner. If the functions $\rho(z), G'(z)$, and $G''(z)$ are given, one can calculate the displacement $\hat{u}(z)$ with Mathematica and infer the stress-velocity ratio from $\hat{u}(z=0)$ and $\mathrm{d}\hat{u}/\mathrm{d}z(z=0)$ One has

$$Z_L = \frac{\hat{\sigma}_S}{\hat{v}_S} = \frac{-\tilde{G}\frac{\mathrm{d}}{\mathrm{d}z}\hat{u}(z)\Big|_{z=0}}{\mathrm{i}\omega\hat{u}(z)\big|_{z=0}} \tag{10.6.2}$$

$\hat{u}(z)$ is the solution to the wave equation, which in this case is

$$-\rho(z)\omega^2\hat{u}(z) = \frac{\mathrm{d}\hat{\sigma}}{\mathrm{d}z} = \frac{\mathrm{d}}{\mathrm{d}z}\left(\tilde{G}(z)\frac{\mathrm{d}\hat{u}(z)}{\mathrm{d}z}\right) \tag{10.6.3}$$

Note that the shear modulus must be inside the second derivative in case $\tilde{G}$ itself is a function of z.

Mathematica code implementing this method is provided in Sect. 10.8. (This section also discusses the initial conditions). The input to such a model consists of two steps. First, an assumption is needed on the dependence of the polymer volume fraction on distance, $\varphi(z)$, where φ is the polymer volume fraction. More difficult is the question of how the shear modulus should depend on φ. Remember: The model needs the shear modulus at MHz frequencies. The low frequency shear modulus might be determined with conventional rheology on bulk solutions of the respective material, but such low-frequency moduli are of little help. Input from polymer physics is needed. What is written down in Sect. 10.8 is just a guess.

If the overall thickness of the sample is much below the penetration depth, there are relations between the complex frequency shifts and certain integrals. This limit was called the thin-film limit in Sect. 10.2. Another term used in optics is the "long-wavelength limit" [24]. Applying Eq. 10.2.7 to viscoelastic profiles and assuming $\rho(z) \approx \rho_{liq}$, one finds

$$\begin{aligned}\frac{\Delta f}{f_0} &\approx -\frac{\rho_{liq}\omega}{\pi Z_q}\int_0^\infty \left(1-\omega\eta_{liq}J''(z)\right)\mathrm{d}z\\ \frac{\Delta\Gamma}{f_0} &\approx \frac{\rho_{liq}\omega}{\pi Z_q}\int_0^\infty \omega\eta_{liq}J'(z)\,\mathrm{d}z\end{aligned} \tag{10.6.4}$$

The Sauerbrey thickness is given as

$$d_{app} = -\frac{1}{\rho_{liq}}\frac{Z_q}{2nf_0^2}\Delta f \approx \int_0^\infty \left(1-\omega\eta_{liq}J''(z)\right)\mathrm{d}z \tag{10.6.5}$$

The index *app* denotes the *app*arent thickness. Based on the $\Delta\Gamma/\Delta f$-ratio (Sect. 10.5), one can define an apparent elastic compliance as $J'_{app} = \Delta\Gamma/(-\Delta f)/(\omega\eta_{liq})$. The apparent elastic compliance is related to the viscoelastic profile as

$$J'_{app} = \frac{1}{\omega\eta_{liq}}\left(\frac{\Delta\Gamma}{-\Delta f}\right) \approx \frac{\int_0^\infty J'(z)\,\mathrm{d}z}{\int_0^\infty 1-\omega\eta_{liq}J''(z)\,\mathrm{d}z} \tag{10.6.6}$$

10.7 Slip

Slippage at an interface between a solid and a simple liquid has for a long time been regarded as a misunderstanding. As the textbooks explain, the solid-liquid interface obeys a no-slip condition because the attraction between the small molecules forming the liquid phase and a wall is at least as strong as the mutual attraction between the small molecules [25, 26]. At first glance, there appears to be little reason, why a material should flow more easily close to a wall than in the bulk. On the contrary: Adsorption of molecules to the surface should enhance the local viscosity. There are many examples of solidified layers of liquid molecules next to a solid surface. The opposite phenomenon, a layer of highly mobile molecules next to a wall, was long considered an exception, if not a phantom.

Today there is solid experimental evidence for slip even in simple liquids [27–31]. Different mechanisms have been put forward. For instance, nanobubbles can facilitate a flow. The hydrophobic interaction can also reduce the density of water near a hydrophobic surface, thereby reducing the local viscosity [29].

In passing we mention two other meanings of slip, different from what is discussed below. Firstly, slip can mean "*solid-like sliding*" in the sense of friction and tribology [32]. Some aspects of tribology with relevance to the QCM are discussed in Sect. 12.1. Secondly, slip can denote interfacial shear-thinning in *complex fluids* [33, 34]. A well-known example is tooth paste. When tooth paste is squeezed out of the tube, it experiences plug flow, meaning that bulk of the material remains undeformed and the shear gradient is concentrated close to the wall. In tooth paste, large stress causes structural transformations, which lower the viscosity. On start-up, the pressure-driven flow has a parabolic profile. Soon afterwards, the structural transformation sets in close to the wall, where the shear gradient is largest. There is positive feed-back, eventually leading to strong shear-thinning at the wall and movement without internal shear everywhere else. MHz shear waves are not suited to study shear-thinning in complex fluids. Firstly, the structural transformations are too slow to be induced by MHz excitation. Also, the fluids giving rise to plug flow all are viscous to the extent that they overdamp the resonator's vibration. Further, the stress required to induce shear-thinning is too large to be exerted by a QCM.

We now return to simple liquids. Slip in simple liquids is of practical relevance, whenever the flow occurs on small spatial scales. An obvious example is flow in porous media [35], where the pore size may easily be comparable to the slip length. (For the definition of the slip length see Fig. 10.8). A second example is sedimentation [36], where the slip length must be compared to the particle diameter. Nanobubbles and slip also are of importance in electrochemistry, key words being gas diffusion electrodes [37] and electrophoretic deposition [38].

In simple liquids, the layer of anomalous viscosity (be it decreased or increased) is few nanometers thick, at most. Only molecules in the vicinity of the surface feel the effect, which the surface has on the material's density and dynamics. Unless the viscosity in the denominator of Eq. 10.7.1 is close to zero, the slip length is a

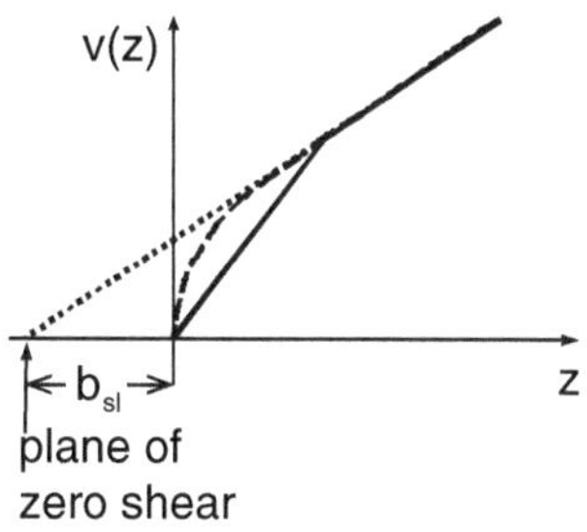

Fig. 10.8 Flow profile above a solid surface with slip. The shear gradient, dv(z)/dz, is inversely proportional to the local viscosity, $\eta(z)$. From Newton's law of action, the shear stress is constant, independent of z. *Solid line* The viscosity is reduced inside a hypothetical discrete layer. *Dashed line* The viscosity $\eta(z)$ varies gradually. If there is slip, the viscosity is smaller near the surface than in the bulk. At the surface, the shear gradient is correspondingly increased. The slip length, b_{sl}, is the distance from the surface to the extrapolated plane of zero shear (*dotted line*)

few nanometers, as well, and the measurement of the slip length by conventional macroscopic techniques is difficult. As has been pointed out by numerous authors, MHz shear waves can be exploited to this end [39–43]. Their penetration depth is around 100 nm. If the slip length is—for instance—1 % of the penetration depth, slip is readily detected.

The continuum model of slip in simple liquids amounts to a layer of decreased viscosity near a surface. The slip length, b_{sl}, is defined by extrapolating the linear portion of the flow profile, v(z), to the plane of zero shear (Fig. 10.8) [44]. Again, the near-surface viscosity may be decreased or increased. It can be increased by a layer of adsorbed molecules. The distance between the surface and the plane of zero shear then (by definition) is the hydrodynamic thickness of the respective layer [45]. The slip length can be viewed as an apparent negative hydrodynamic thickness.

For steady flows, the slip length is given by [29]

$$b_{sl} = \left(\frac{\eta_{liq}}{\eta_{SL}} - 1\right) d_{SL} \tag{10.7.1}$$

d_{SL} is the thickness of the layer with reduced viscosity (the *S*lipping *L*ayer, solid line in Fig. 10.8) and η_{SL} is the viscosity inside this layer. Since 10.7.1 is linear in d_{SL}, it also holds for continuous profiles, $\eta(z)$, (dashed line in Fig. 10.8) in an integral sense:

$$b_{sl} = \int_0^\infty \left(\frac{\eta_{liq}}{\eta(z)} - 1\right) \mathrm{d}z. \tag{10.7.2}$$

The frequency shift induced by a layer with reduced viscosity can be predicted from an adapted version of Eq. 10.2.7, which is [46]:

$$\begin{aligned}\frac{\Delta\tilde{f}}{f_0} &= -\frac{2\omega}{\pi Z_q}\int_0^\infty\left(1-\frac{Z_{liq}^2}{\tilde{Z}^2(z)}\right)\rho(z)\mathrm{d}z\\ &= -\frac{2f}{Z_q}\int_0^\infty\left(1-\frac{\mathrm{i}\omega\rho_{liq}\eta_{liq}}{\mathrm{i}\omega\rho(z)\eta(z)}\right)\rho(z)\mathrm{d}z\\ &= \frac{2f}{Z_q}\rho_{liq}\int_0^\infty\left(\frac{\eta_{liq}}{\eta(z)}-\frac{\rho(z)}{\rho_{liq}}\right)\mathrm{d}z\end{aligned} \tag{10.7.3}$$

If the density of the slipping layer is the same as the density of the bulk liquid, ($\rho(z) \approx \rho_{liq}$), one has

$$\frac{\Delta\tilde{f}}{f_0} \approx \frac{2f\rho_{liq}}{Z_q}\int_0^\infty\left(\frac{\eta_{liq}}{\eta(z)}-1\right)\mathrm{d}z \approx \frac{2f\rho_{liq}}{Z_q}b_{sl} \tag{10.7.4}$$

With constant density, slip looks like a negative Sauerbrey mass, where the slip length is equal to the negative Sauerbrey thickness.

If the densities of the slipping layer and the bulk liquid are unequal, the situation is more complicated. There is an "acoustic slip length", $b_{sl,ac}$, given as

$$b_{sl,ac} = \int_0^\infty\left(\frac{\eta_{liq}}{\eta(z)}-\frac{\rho(z)}{\rho_{liq}}\right)\mathrm{d}z \tag{10.7.5}$$

When slip is caused by a nanobubbles, the slipping layer's average density is less than the density of the bulk. The acoustic slip length then is different from the conventional slip length.

Equations 10.7.2 and 10.7.4 ignore roughness and heterogeneities [47]. Bubbles and nanobubbles do constitute such heterogeneities. Nanobubbles do not necessarily induce slip. Since nanobubbles constitute a laterally structured sample, we defer their discussion to Sect. 12.4.

10.8 Appendix: Modeling Viscoelastic Profiles Using the Wave Equation

In Sect. 10.6 it was stated that the complex frequency shift induced by a sample with continuous $\tilde{G}(z)$ and $\rho(z)$ can be found by solving the wave equation numerically. The calculation amounts to a few lines of code in Mathematica and we provide an example below.

The following assumptions were made to set up the model:

- The polymer volume fraction was chosen as $\varphi = \varphi_0\ 1/2(1 - \tanh(z - z_i)/w))$. φ_0 is the polymer volume fraction at $z = -\infty$, z_i is the point of inflection, and w is the width of the interface.
- The density, ρ, was assumed to be constant and equal to 1 g/cm^3.
- The shear modulus was assumed to obey a Maxwell model, meaning that $\tilde{G}$ is given as $\tilde{G} = \mathrm{i}\omega\eta_{liq} + G_\infty \mathrm{i}\omega\tau/(1 + \mathrm{i}\omega\tau)$. G_∞ is the elastic shear modulus at infinite frequency and τ is a relaxation time. τ was made to depend on polymer volume fraction as $\tau = 0.1$ ns $\times\ 10^3\ \varphi/\varphi_0$. At $z = \infty$, the shear modulus equals the (complex) shear modulus of the bulk liquid.

There is subtlety with regard to the initial condition of the problem, which is the amplitude at infinity (where "infinity" is some value z_{max} far outside the region of interest). Mathematica integrates the wave equation from z_{max} down to the resonator surface. Since the wave equation contains the second derivative, initial conditions are needed for both $\hat{u}$ and $\mathrm{d}\hat{u}/\mathrm{d}z$ at $z = z_{max}$. It turns out that all initial conditions lead to the same frequency shift. Consider $\hat{u}(z_{\mathrm{max}})$ first. Because what is analyzed eventually is the ratio of stress to velocity, $\hat{\sigma}_S/\hat{v}_S$, one is free to pick the amplitude $\hat{u}(z_{\mathrm{max}})$ arbitrarily. Multiplying $\hat{u}(z)$ by some complex number does not change the ratio of $\hat{\sigma}_S/\hat{v}_S$. Somewhat surprisingly, the value of the first derivative of $\hat{u}(z)$ at z_{max} has no influence on $\hat{\sigma}_S/\hat{v}_S$, either. Some value must be chosen in order to get Mathematica running, but all choices lead to the same $\Delta\tilde{f}$. While one might think that $\mathrm{d}\hat{u}(z)/\mathrm{d}z$ must be chosen as $-ik\hat{u}(z)$, any other choice is fine as well. This is so because any other choice will represent a superposition of two waves decaying towards positive and negative z. The second solution (decaying towards negative z) is unphysical because it grows to infinity far away from the resonator surface. However, the presence of this unphysical contribution is no problem if z_{max} is large enough. If z_{max} is far away from the surface, the unphysical part of the solution decays to zero at the resonator surface and has no influence on $\Delta\tilde{f}$.

The code below should not be misunderstood as a realistic model. It is a tool box, allowing users to develop their own model [49][1].

---

[1] A Mathematica file with the content given below is available for download at http://www.pc.tu-clausthal.de/en/forschung/ak-johannsmann/qcm-modellierung/

```
                         (* various parameters, all in SI units *)
phi0 = 0.6;              (* Prefactor of polymer volume fraction*)
zi = 10^-6;              (* Point of inflection *)
w = 10^-7;               (* Width of fuzzy interface *)
Ginf = 10^7;             (* Elastic shear modulus at infinite frequency*)
rho = 10^3;              (* Density, chosen as constant *)
etaliq = 10^-3;          (* Viscosity of the liquid *)
f0 = 5 10^6;             (* Frequency of fundamental*)
om = 2 Pi f0;            (* Radial frequency *)
Zq = 8.8 10^6;           (* Acoustic impedance of AT-cut quartz *)
zmax = 2 10^-6;          (* Upper limit of integration range *)

        (* functional forms of phi(z), tau(phi), and G(tau) *)
phi = phi0 1/2(1-Tanh[(z-zi)/w]);
tau = 10^-10 Power[10, 3phi / phi0];
G = Ginf I om tau/(1+I om tau) + I om etaliq;

        (* Display G'(z) and G"(z)*)
Print["Re[G]"];  Plot[Re[G], {z,0,zmax}];
Print["Im[G]"];  Plot[Im[G], {z,0,zmax}];
        (* The above two lines do not run on the Mathematica
        version available online at www.mathics.net
        The rest of the program should run*)

        (*Below are the essential steps*)
G2 = G /. z -> zmax-z;
        (*revert z-scale, Mathematica integrates forward *)
WaveEquation = D[G2 D[u[z],z],z] = = -rho om^2 u[z];
        (*The differential equation to be solved*)
solution = NDSolve[{WaveEquation, u[0]==1,u'[0]==1},u,{z,0,zmax}];
        (*Calculate the displacement field*)
displacement = u[z] /. solution[[1]] /. z -> zmax-z;
        (*Again revert the z-scale*)
velocity = I om displacement /. z->0;
        (*Calculate the velocity at z = 0*)
stress = -G D[displacement, z] /. z->0;
        (*Calculate the stress at z = 0.
        The minus sign occurs because the stress
        to be inserted into the SLA acts into the +z direction*)
Df = f0 I /(Pi Zq) stress / velocity
        (*Calculate the frequency shift with the SLA*)
```

Glossary

Variable	Definition (Comments)
app	As an index: *app*arent
b_{sl}	Slip length
$b_{sl,ac}$	Acoustic slip length (Eq. 10.7.5)
$\tilde{c}$	Speed of (shear) sound ($\tilde{c} = (\tilde{G}/\rho)^{1/2}$)
d	Thickness of a layer
d_q	Thickness of the resonator ($d_q = m_q/\rho_q = Z_q/(2\rho_q f_0)$)
f	Frequency
f	As an index: *f*ilm
f_0	Resonance frequency at the fundamental ($f_0 = Z_q/(2m_q) = Z_q/(2\rho_q d_q)$)
FR	As an index: *F*ilm *R*esonance
$\tilde{G}$	Shear modulus
G_∞	Limiting storage modulus at high frequency
$\tilde{J}$	Shear compliance ($\tilde{J} = 1/\tilde{G}$)
$\tilde{k}$	Wavenumber ($\tilde{k} = \omega/\tilde{c}$)
liq	As an index: *liq*uid
m	Mass per unit area
m_q	Mass per unit area of the resonator ($m_q = \rho_q d_q = Z_q/(2f_0)$)
n	Overtone order
$\tilde{r}$	Amplitude reflection coefficient (reflectivity, for short)
ref	As an index: *ref*erence state of a crystal in the absence of a load or reference frequency for viscoelastic constants (Eq. 10.4.1)
S	As an index: *S*urface
SL	As an index: *S*lipping *L*ayer
t	Time
$\hat{u}$	(Tangential) displacement
$\hat{v}$	Velocity
w	Width of a fuzzy interface (Sect. 10.8)
z_i	Point of inflection of a segment density profile (Sect. 10.8)

$\tilde{Z}_{liq}$	Shear-wave impedance of a liquid ($\tilde{Z}_{liq} = (i\omega\rho_{liq}\eta_{liq})^{1/2}$)
$\tilde{Z}_L$	Load impedance
z_{max}	Limit of integration range (Sect. 10.8)
Z_q	Acoustic wave impedance of AT-cut quartz ($Z_q = 8.8 \times 10^6$ kg m^{-2} s^{-1})
β', β''	Power law exponents (Eq. 10.4.1)
Γ	Imaginary part of a resonance frequency
δ	Penetration depth of a shear wave (Newtonian liquids: $\delta = (2\eta_{liq}/(\rho_{liq}\omega))^{1/2}$)
Δ	As a prefix: A shift induced by the presence of the sample
φ	Polymer volume fraction (Sect. 10.8)
$\tilde{\eta}, \eta$	Viscosity $\tilde{\eta} = \tilde{G}/(i\omega)$
ρ	Density
$\hat{\sigma}$	(Tangential) stress
$\hat{\sigma}_s$	Tangential stress at the surface, also: "traction"
τ	Relaxation time
ω	Angular frequency

References

1. Lu, C.S., Lewis, O.: Investigation of film-thickness determination by oscillating quartz resonators with large mass load. J. Appl. Phys. **43**(11), 4385 (1972)
2. Crane, R.A., Fischer, G.: Analysis of a quartz crystal microbalance with coatings of finite viscosity. J. Phys. D-Appl. Phys. **12**(12), 2019–2026 (1979)
3. Benes, E.: Improved quartz crystal microbalance technique. J. Appl. Phys. **56**(3), 608–626 (1984)
4. http://www.ecse.rpi.edu/~schubert/Course-Teaching-modules/A040-Parameters-used-in-E-beam-deposition-system.pdf. Accessed 1 June 2013
5. Johannsmann, D., Mathauer, K., Wegner, G., Knoll, W.: Viscoelastic properties of thin-films probed with a quartz-crystal resonator. Phys. Rev. B **46**(12), 7808–7815 (1992)
6. Granstaff, V.E., Martin, S.J.: Characterization of a thickness-shear mode quartz resonator with multiple nonpiezoelectric layers. J. Appl. Phys. **75**(3), 1319–1329 (1994)
7. Martin, S.J., Bandey, H.L., Cernosek, R.W., Hillman, A.R., Brown, M.J.: Equivalent-circuit model for the thickness-shear mode resonator with a viscoelastic film near film resonance. Anal. Chem. **72**(1), 141–149 (2000)
8. Wolff, O.: Private communication
9. Salomaki, M., Kankare, J.: Modeling the growth processes of polyelectrolyte multilayers using a quartz crystal resonator. J. Phys. Chem. B **111**(29), 8509–8519 (2007)

10. Domack, A., Johannsmann, D.: Plastification during sorption of polymeric thin films: a quartz resonator study. J. Appl. Phys. **80**(5), 2599–2604 (1996)
11. Johannsmann, D.: Viscoelastic analysis of organic thin films on quartz resonators. Macromol. Chem. Phys. **200**(3), 501–516 (1999)
12. Domack, A., Prucker, O., Ruhe, J., Johannsmann, D.: Swelling of a polymer brush probed with a quartz crystal resonator. Phys. Rev. E **56**(1), 680–689 (1997)
13. Johannsmann, D.: Viscoelastic, mechanical, and dielectric measurements on complex samples with the quartz crystal microbalance. Phys. Chem. Chem. Phys. **10**(31), 4516–4534 (2008)
14. Martin, S.J., Granstaff, V.E., Frye, G.C.: Characterization of a quartz crystal microbalance with simultaneous mass and liquid loading. Anal. Chem. **63**(20), 2272–2281 (1991)
15. Voinova, M.V., Jonson, M., Kasemo, B.: 'Missing mass' effect in biosensor's QCM applications. Biosens. Bioelectron. **17**(10), 835–841 (2002)
16. Kankare, J.: Sauerbrey equation of quartz crystal microbalance in liquid medium. Langmuir **18**(18), 7092–7094 (2002)
17. Du, B.Y., Johannsmann, D.: Operation of the quartz crystal microbalance in liquids: Derivation of the elastic compliance of a film from the ratio of bandwidth shift and frequency shift. Langmuir **20**(7), 2809–2812 (2004)
18. Voinova, M.V., Rodahl, M., Jonson, M., Kasemo, B.: Viscoelastic acoustic response of layered polymer films at fluid-solid interfaces: continuum mechanics approach. Phys. Scr. **59**(5), 391–396 (1999)
19. Rodahl, M., Kasemo, B.: On the measurement of thin liquid overlayers with the quartz-crystal microbalance. Sens. Actuators A-Phys. **54**(1–3), 448–456 (1996)
20. Craig, V.S.J., Plunkett, M.: Determination of coupled solvent mass in quartz crystal microbalance measurements using deuterated solvents. J. Colloid Interface Sci. **262**(1), 126–129 (2003)
21. Tsortos, A., Papadakis, G., Gizeli, E.: Shear acoustic wave biosensor for detecting DNA intrinsic viscosity and conformation: a study with QCM-D. Biosens. Bioelectron. **24**(4), 836–841 (2008)
22. Papadakis, G., Tsortos, A., Bender, F., Ferapontova, E.E., Gizeli, E.: Direct detection of DNA conformation in hybridization processes. Anal. Chem. **84**(4), 1854–1861 (2012)
23. http://www.pc.tu-clausthal.de/en/forschung/ak-johannsmann/qcm-modellierung/
24. Lekner, J.: Theory of Reflection of Electromagnetic and Particle Waves. Springer, Berlin (1987)
25. Bernoulli, D.: Hydrodynamica (1738). http://en.wikipedia.org/wiki/Hydrodynamica. Accessed 15 June 2014
26. Larson, R.G.: The Structure and Rheology of Complex Fluids. Oxford University Press, New York (1998)
27. Vinogradova, O.I.: Slippage of water over hydrophobic surfaces. Int. J. Miner. Process. **56**(1–4), 31–60 (1999)
28. Thompson, P.A., Troian, S.M.: A general boundary condition for liquid flow at solid surfaces. Nature **389**(6649), 360–362 (1997)
29. Huang, D.M., Sendner, C., Horinek, D., Netz, R.R., Bocquet, L.: Water slippage versus contact angle: a quasiuniversal relationship. Phys. Rev. Lett. **101**(22), 226101 (2008)
30. Barrat, J.L., Bocquet, L.: Large slip effect at a nonwetting fluid-solid interface. Phys. Rev. Lett. **82**(23), 4671–4674 (1999)
31. Neto, C., Evans, D.R., Bonaccurso, E., Butt, H.J., Craig, V.S.J.: Boundary slip in Newtonian liquids: a review of experimental studies. Rep. Prog. Phys. **68**(12), 2859–2897 (2005)
32. Bowden, F.P., Tabor, D.: Friction lubrication and wear—a survey of work during last decade. Br. J. Appl. Phys. **17**(12), 1521–1524 (1966)
33. Tucker, C.L., Moldenaers, P.: Microstructural evolution in polymer blends. Annu. Rev. Fluid Mech. **34**, 177–210 (2002)

34. Barnes, H.A.: A review of the slip (wall depletion) of polymer-solutions, emulsions and particle suspensions in viscometers—its cause, character, and cure. J. Nonnewton. Fluid Mech. **56**(3), 221–251 (1995)
35. Lefevre, B., Saugey, A., Barrat, J.L., Bocquet, L., Charlaix, E., Gobin, P.F., Vigier, G.: Intrusion and extrusion of water in highly hydrophobic mesoporous materials: effect of the pore texture. Colloids Surf. A-Physicochem. Eng. Aspects **241**(1–3), 265–272 (2004)
36. Boehnke, U.C., Remmler, T., Motschmann, H., Wurlitzer, S., Hauwede, J., Fischer, T.M.: Partial air wetting on solvophobic surfaces in polar liquids. J. Colloid Interface Sci. **211**(2), 243–251 (1999)
37. Al-Fetlawi, H., Shah, A.A., Walsh, F.C.: Modelling the effects of oxygen evolution in the all-vanadium redox flow battery. Electrochim. Acta **55**(9), 3192–3205 (2009)
38. Zhitomirsky, I.: Cathodic electrodeposition of ceramic and organoceramic materials. Fundamental aspects. Adv. Colloid Interface Sci. **97**(1–3), 279–317 (2002)
39. Ferrante, F., Kipling, A.L., Thompson, M.: Molecular slip at the solid-liquid interface of an acoustic-wave sensor. J. Appl. Phys. **76**(6), 3448–3462 (1994)
40. McHale, G., Lucklum, R., Newton, M.I., Cowen, J.A.: Influence of viscoelasticity and interfacial slip on acoustic wave sensors. J. Appl. Phys. **88**(12), 7304–7312 (2000)
41. Ellis, J.S., Hayward, G.L.: Interfacial slip on a transverse-shear mode acoustic wave device. J. Appl. Phys. **94**(12), 7856–7867 (2003)
42. Daikhin, L., Gileadi, E., Tsionsky, V., Urbakh, M., Zilberman, G.: Slippage at adsorbate-electrolyte interface. Response of electrochemical quartz crystal microbalance to adsorption. Electrochim. Acta **45**(22–23), 3615–3621 (2000)
43. Zhuang, H., Lu, P., Lim, S.P., Lee, H.P.: Effects of interface slip and viscoelasticity on the dynamic response of droplet quartz crystal microbalances. Anal. Chem. **80**(19), 7347–7353 (2008)
44. Tretheway, D.C., Meinhart, C.D.: Apparent fluid slip at hydrophobic microchannel walls. Phys. Fluids **14**(3), L9–L12 (2002)
45. Klein, J., Kumacheva, E., Perahia, D., Mahalu, D., Warburg, S.: Interfacial sliding of polymer-bearing surfaces. Faraday Discuss. **98**, 173–188 (1994)
46. Urbakh, M., Tsionsky, V.; Gileadi, E.; Daikhin, L.: Probing the solid/liquid interface with the quartz crystal microbalance. In: Steinem, C., Janshoff, A. (eds.) Piezoeletric Sensors. Springer, Heidelberg (2006)
47. Du, B.Y., Goubaidoulline, E., Johannsmann, D.: Effects of laterally heterogeneous slip on the resonance properties of quartz crystals immersed in liquids. Langmuir **20**, 10617–10624 (2004)
48. Decher, G.: Fuzzy nanoassemblies: toward layered polymeric multicomposites. Science **277**(5330), 1232–1237 (1997)
49. Wolff, O., Seydel, E., Johannsmann, D.: Viscoelastic properties of thin films studied with quartz crystal resonators. Faraday Discuss. **107**, 91–104 (1997)

Chapter 11
Point Contacts and Contact Stiffness

Abstract If an external object (such as a sphere) touches the resonator surface across a contact with a diameter much below the wavelength of sound and, also, much below the size of the particle, one can infer the stiffness of the contact from the frequency shift. The situation is particularly transparent for particles, which are heavy enough to be clamped in space by inertia, so that they do not follow MHz motion of the resonator. In this case, the frequency shift is positive and proportional to the contact stiffness. Smaller particles give rise to coupled resonances. Coupled resonances can be viewed as absorption lines in shear-wave spectroscopy.

11.1 General

In Chaps. 9 and 10, all modeling occurred within the plane-wave picture. The samples where either laterally homogeneous or almost lateral homogeneous in the sense that the plane-wave picture could be suitably adapted. For instance, the sample did not cover the entire resonator surface in Sect. 9.4, but the diameter of the contact was still much larger than λ and the displacement pattern therefore was almost a plane wave. We now turn to the opposite limit, where the lateral scale inherent to the sample is much smaller than the wavelength of sound. More precisely, we discuss discrete contacts of small size, which are elastically independent from each other. For such samples, the bulk of the object touching the surface remains undeformed (Fig. 11.1a). It experiences translation and rotation, but no internal deformation. The contact region, on the contrary, does experience deformation. One may integrate the tangential stress over the contact area and obtain a force. Because the contact is small, one can define a displacement vector based on the relative displacement evaluated outside the deformation zone. The ratio between force and displacement is the tangential spring constant (also: "contact

D. Johannsmann, *The Quartz Crystal Microbalance in Soft Matter Research*,
Soft and Biological Matter, DOI: 10.1007/978-3-319-07836-6_11,

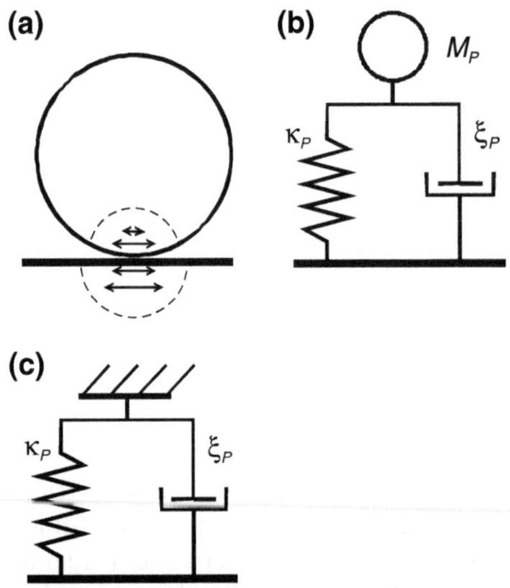

Fig. 11.1 **a** Sketch of a contact between a sphere and the resonator surface. It is essential that the contact is smaller than both the sphere and the wavelength of sound. Only the region close to the contact (*dashed circle*) is deformed. Together with the deformed region of the substrate, it forms a Hookean spring. **b** Mechanical equivalent circuit. Viscous dissipation is covered by the dashpot. **c** If the sphere is heavy enough, it is clamped in space by inertia. The sphere can then be replaced by a rigid wall

stiffness"), $\tilde{\kappa}_P$. Note: The contact stiffness here refers to a *single* contact, as opposed to a multi-contact-interface. It has dimensions of N/m.

A typical experimental configuration is a sphere with radius R_P deposited on the sensor surface. The index *P* denotes a Particle. The external object does not necessarily have to be a sphere, but its surface must be curved at the contact to ensure a small contact size. Particles of arbitrary shape may form multiple contacts or even a flat contact. This must be avoided because it spoils the separation of scales between contact size, on the one hand, and particle size and wavelength, on the other.

The mechanical equivalent circuit is shown in Fig. 11.1b. The contact is depicted as a spring in parallel to a dashpot (a "Voigt element"). For the sake of complex formulation, the spring constant, κ_P, and the drag coefficient, ξ_P, were subsumed under one complex parameter $\tilde{\kappa}_P = \kappa_P + i\omega\xi_P$, in the following. In principle, κ_P and ξ_P may depend on frequency because of viscoelastic dispersion. They are approximated as constant for simplicity. Note that the model is one-dimensional. The real particle may undergo translation or rotation. The link may be sheared or bent (or both). In this regard, the equivalent circuit from Fig. 11.1b is too simple. This problem is treated Sect. 11.5. It will turn out that the equivalent circuit from Fig. 11.1b remains valid, but that the parameters M_P, κ_P, and ξ_P need reinterpretation. Along similar lines: What causes the dissipation quantified by ξ_P is not entirely clear. One often finds a substantial increase in $\Delta\Gamma$ upon deposition of rigid particles onto a rigid sensor surface, which suggests that interfacial processes contribute to the damping.

Earlier in the book, modeling was largely based on plane waves. In the wave picture (Sect. 4.4), point contacts amount to sources of scattering. A thorough discussion of scattering would go beyond the scope of this book. There is one aspect of scattering, though, which is relevant and not easily explained otherwise. Guided by what is known from optics, one might think that even perfectly elastic contacts would damp the resonance. Elastic optical scattering does attenuate optical beams. By analogy, elastic acoustic scattering at the contacts does have an influence on $\Delta\Gamma$. However, the small-load-approximation predicts this influence to be insignificant. The detailed argument is provided in Sect. 11.7. As shown in the Sect. 11.7, the assumptions, which lead to the small-load-approximation, also predict the contribution of elastic scattering to $\Delta\Gamma$ to be insignificant.

The influence of a sphere attached to the resonator onto the frequency shift depends on the ratio between the mass and the contact stiffness (Eq. 11.4.2, Sect. 4.6.3). The limit of small particles rigidly attached to the resonator surface is simple. These particles behave according to Sauerbrey. They increase the mass of the resonator, shifting the resonance frequency downward. (In this limit the separation of scales invoked below is not required. For inertial loads, geometry is irrelevant.) The opposite limit (large particles, soft contacts) is similarly simple, but less widely known. This is the elastic-coupling limit. In the elastic-coupling limit, the QCM essentially measures the contact stiffness, regardless of how large the mass is (Sect. 11.3). Saying that the QCM measures the contact stiffness does not imply a statement on how the contact stiffness relates to the microscopic structure of the contact and the materials parameters. Predicting a contact stiffness from geometry and materials parameters is difficult [1]. Arguable, one only rephrases one's lack of understanding when converting frequency shift to contact stiffness with Eq. 11.3.5. Equation 11.3.5 does not explain why some contacts are stiffer than others, but it translates the result of a QCM experiment to the language of contact mechanics.

The intermediate case (ratio of mass and stiffness in-between the limits of inertial loading and elastic-coupling) is complicated, but of much practical importance. It is discussed in Sects. 11.4 and 11.5. To start out, we invest one section on the background, which is contact mechanics. Some of this material will also be needed in Chap. 13, dealing with nonlinear interactions.

11.2 Some Element of Contact Mechanics

The contact stiffness is quickly defined and also readily extracted from the QCM experiment (Eq. 11.3.5), but its relation to the microscopic structure of the contact region is highly nontrivial. Contact mechanics is an active field of research with a wide range of applications [2–4]. One must keep in mind, though, that the QCM only accesses a small subsection of the many contact mechanics problems. Among the topics not amenable to QCM experiments are steady sliding, steady rolling, and

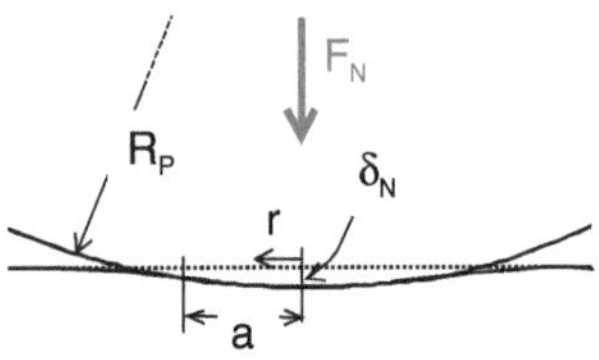

Fig. 11.2 Hertzian contact: definition of variables

sudden impacts. Only those phenomena can be studied, which are caused by small-amplitude periodic excitation. A further constraint follows from the requirement of well-separated spatial scales. One can certainly do experiments employing larger contacts, but the interpretation is difficult. Only those geometries are well understood, where the contact area is small.

The simplest model for predicting forces and displacements at contacts dates from 1882 and was formulated by Heinrich Hertz [5]. The Hertzian contact consists of a sphere with radius R_P pushed against a plate by an external force, F_N (Fig. 11.2). There is no adhesion, no roughness, and no third phase which would exert capillary forces, lubricate the contact, or affect the contact in any other way. Both contacting media are isotropic, characterized by a Young's modulus, E, and a Poisson ratio, ν. According to Hertz, the contact radius, a, depends on the normal force as

$$a = \left(\frac{3}{4}\frac{F_N R_P}{E^*}\right)^{1/3} \tag{11.2.1}$$

E^* is an effective modulus, given as

$$\frac{1}{E^*} = \frac{1-\nu_1^2}{E_1} + \frac{1-\nu_2^2}{E_2} \tag{11.2.2}$$

The indices 1 and 2 label the two media. The compression, δ_N, (Fig. 11.2) is related to the contact radius by

$$\delta_N = \frac{a^2}{R_P} \tag{11.2.3}$$

Combining Eq. 11.2.1 with Eq. 11.2.3, one finds

$$F_N = \frac{4}{3} E^* \delta_N^{3/2} R_P^{1/2} \tag{11.2.4}$$

The differential contact stiffness is given as

$$\frac{dF_N}{d\delta_N} = 2E^* R_P^{1/2} \delta_N^{1/2} = 2E^* \left(\frac{4}{3}\frac{R_P}{E^*}\right)^{1/3} F_N^{1/3} \tag{11.2.5}$$

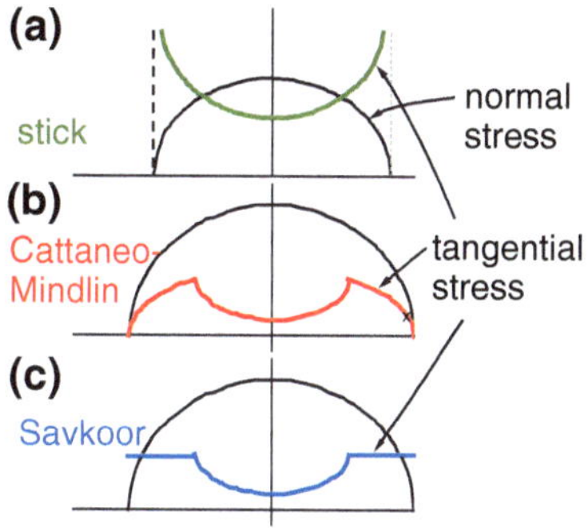

Fig. 11.3 Distribution of normal stress and tangential stress for a Hertzian contact under a tangential load. **a** A hypothetical contact sticking everywhere, even close to the rim, where the tangential stress diverges. **b, c** A contact with partial slip according to Cattaneo-Mindlin [9, 10] and Savkoor [13]

Substituting Eq. 11.2.3 into Eq. 11.2.5 leads to

$$\frac{\mathrm{d}F_N}{\mathrm{d}\delta_N} = 2E^* a \tag{11.2.6}$$

The differential contact stiffness is proportional to the contact radius, not to the contact area. Clearly, the Hertzian contact is not a Hookean spring. The Hertzian contact behaves nonlinearly because the contact stiffness depends on the contact size and because the contact size, in turn, depends on the normal force.

The stress distribution inside the contact is non-uniform. The normal stress as a function of the distance to the center of the contact, r, is given as

$$p(r) = p_0\left(1 - \frac{r^2}{a^2}\right)^{1/2} \tag{11.2.7}$$

$p_0 = 3/2\ F_N/(\pi a^2)$ is the normal stress at the center. $p(r)$ as given by Eq. 11.2.7 is sketched as a solid line in Fig. 11.3a.

Again, the Hertz-model ignores adhesive forces, roughness, and influences of an intervening third phase. One of the many improvements is the JKR model [6]. Including adhesive forces into the model (as the JKR model does), one finds a stress concentration at the edge of the contact. Such stress concentrations are characteristic for contact mechanics and will be needed in Sect. 13.3 to explain QCM experiments. The JKR contact extends the Hertz model by an interfacial energy, γ_S. A nonzero γ_S is caused by short-range attractive forces. Following JKR theory, the contact radius is

$$a^3 = \frac{3R_P}{4E^*}\left(F_N + 6\gamma_S\pi R_P + \sqrt{12\gamma_S\pi R_P F_N + (6\gamma_S\pi R_P)^2}\right) \tag{11.2.8}$$

The distribution of normal stress is

$$p(r) = p_0\left(1 - \frac{r^2}{a^2}\right)^{1/2} + p_1\left(1 - \frac{r^2}{a^2}\right)^{-1/2} \tag{11.2.9}$$

with $p_0 = 2aE^*/(\pi R_P)$ and $p_1 = E^*/\pi(\delta_N/a - a/R_P)$. The second term contains an inverse square-root singularity in tensile stress at the edge of contact. Of course, the JKR model is incomplete in this regard. The tensile stress cannot be infinite; some further refinement of the model is needed. The refinement might start out from recognizing that the edge is not infinitely sharp. Actually, giving contacts rounded shapes at the edge can significantly increase their pull-off force [7]. Rounded edges lower the peak in tensile stress of the adhesive contact and thereby increase the force of detachment. Rounded edges can also help to prevent partial slip and fretting wear [8]. Regardless of how the infinity at the edge is avoided: The JKR contact has a stress concentration close to the edge. Given the large stress, one can expect nonlinear stress-strain relations in this zone. Stress concentrations are a second source of nonlinearities in contact mechanics (in addition to the load-dependence of the contact area).

The behavior of contacts under *tangential* loads is different in many ways from the behavior under normal load. For instance, the Hertzian contact responds to a tangential force with either partial slip or (in the case of large loads) gross slip. When two surfaces slide against each other, some stress is released, but there still is sliding friction, meaning that the sliding portion of the contact contributes to the overall tangential force (Fig. 11.4).

In a first step, consider, a sticking contact between identical materials. The equations of continuum elasticity lead to a distribution of tangential stress given by [3]

$$\sigma(r) = \sigma_1\left(1 - \frac{r^2}{a^2}\right)^{-1/2} \tag{11.2.10}$$

The stress at the center is given as $\sigma_1 = F_x/(2\pi a^2)$ with F_x the tangential force. At this level of approximation, the Hertzian contact under tangential load is a Hookean spring. The total force, F_x, is proportional to the tangential displacement. The tangential stiffness, κ_P, is

$$\kappa_P = 2G^* a \tag{11.2.11}$$

G^* is an effective modulus, given as

$$\frac{1}{G^*} = \frac{2 - \nu_1}{4G_1} + \frac{2 - \nu_2}{4G_2} \tag{11.2.12}$$

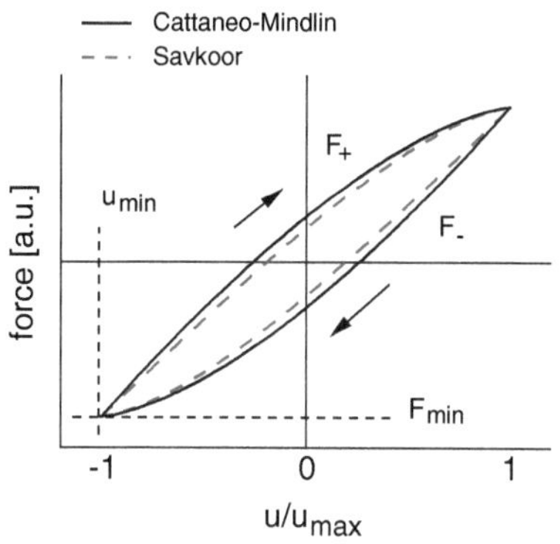

Fig. 11.4 Force-displacement relation under conditions of partial slip following two different models. *Solid black line* $\sigma = \mu p$ in the sliding zone with μ the friction coefficient and p the normal stress (after Cattaneo [9] and Mindlin [10]). *Dashed red line* $\sigma = \tau_0 = \text{const}$ in the sliding zone (after Savkoor [13, 14])

ν is the Poisson ratio and G is the shear modulus. G is related to the Young's modulus, E, by the relation $G = E/(2(1 + \nu))$.

As in the case of the normal load with adhesion (treated with the JKR model), the stress goes to infinity at the edge (thick line in Fig. 11.3a), which is unphysical. The infinity can be avoided by different mechanisms, among them a rounded edge. Differing from pulling problem, the singularity can also be removed by a mechanism called partial slip (also: "microslip"). Cattaneo [9] and Mindlin [10], who independently developed the simplest model of partial slip, make the following assumptions:

(a) Those parts of the contact, where the tangential stress exceeds a critical value, undergo slip.
(b) The tangential stress in the sliding zone is given as μp with p the normal stress and μ a "friction coefficient". μp also is the critical tangential stress of the onset of slip.
(c) The distribution of the normal stress follows the Hertz model (Eq. 11.2.7).

The stress distribution resulting from these assumptions is shown as a thick line in Fig. 11.3b. Clearly, the singularity is removed, which was the goal of the model. While assumptions (b) and (c) can be debated, assumption (a) is not at all special or peculiar. Partial slip is an everyday experience. One can convince oneself of its existence by pushing one's thumb against a surface and gently moving it sideways. In a technical context, partial slip leads to fretting wear [11, 12]. When contacts are subjected to prolonged periodic tangential stress (for instance due to vibrations), they wear off at the edge. Contacts between rough surfaces show microslip and fretting wear, as well. Figure 11.5 shows a photograph from Johnson's book [2]. A steel plate, which had been contacted with a sphere and which had experienced oscillatory tangential loads for some time, showed scratches at the edge of the contact area.

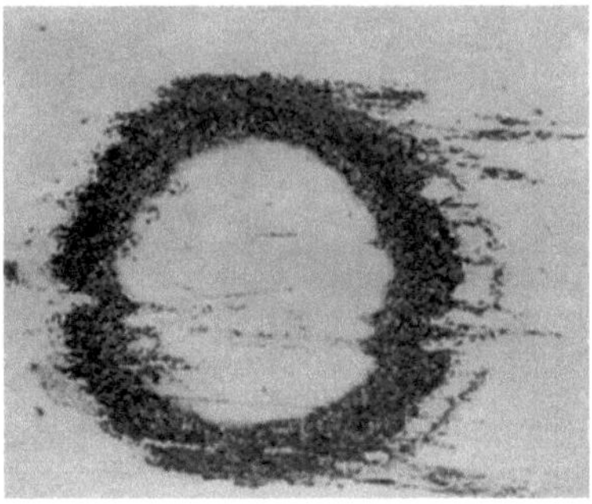

Fig. 11.5 Upon prolonged exposure of a contact to oscillatory tangential stress, a characteristic pattern of wear develops. There a scratches in those areas, which have undergone partial slip (also "microslip"). Reprinted with permission from K.L. Johnson, *Contact Mechanics*. Copyright 1985 Cambridge University Press

Contrasting to the assumption of partial slip as such, assumptions *b* and *c* above are choices, which may or may not be realistic and can be modified. Another choice has been studied in detail by Savkoor, which is $\tau_S = const$ in the slipping zone [13, 14]. The force-displacement relations following from Cattaneo-Mindlin and Savkoor are shown as a solid and a dashed line in Fig. 11.4. They will be needed in the discussion of QCM experiments in Sect. 13.3.

11.3 Large Particles

Contact mechanics experiments with the QCM are most easily analyzed if the particles are so heavy that they are clamped in space by inertia. An estimate of the minimum size needed to achieve inertial clamping is given at the end of this section. The minimum size is a few tens of microns in diameter. (If smaller contacts are to be studied, one can glue these to a backing plate.)

In the equivalent circuit representation, the bulk of the particle is depicted as a wall (Fig. 11.1c). With the mass, M_P, replaced by a rigid support, the force, $\hat{F}$, is proportional to $\tilde{\kappa}_P \hat{u}_S$. One has

$$\hat{F} = \tilde{\kappa}_P \hat{u}_S = (\kappa_P + i\omega\xi_P)\hat{u}_S \tag{11.3.1}$$

$\hat{F}$ has the same sign as $\hat{u}_S$ (even though it is restoring force) because the stress inserted into the SLA is the stress exerted by the resonator onto the sample.

Let N_P be the number of particles per unit area. The load impedance is

$$\tilde{Z}_L = \frac{\langle \hat{\sigma}_S \rangle_{area}}{\hat{v}_S} = N_P \frac{\hat{F}}{i\omega\hat{u}_S} = N_P \frac{\tilde{\kappa}_P}{i\omega} \tag{11.3.2}$$

Angle brackets denote an average. From the SLA one finds:

$$\Delta\tilde{f} = \mathrm{i}\frac{f_0}{\pi\tilde{Z}_q}\tilde{Z}_L = \frac{f_0}{\pi Z_q}\frac{N_P\tilde{\kappa}_P}{\omega} = \frac{1}{2n\pi^2 Z_q}N_P\tilde{\kappa}_P \tag{11.3.3}$$

If the number of spheres is small, one should include a statistical weight into Eq. 11.3.3, accounting for the amplitude distribution (Sect. 4.6.2). One writes:

$$\Delta\tilde{f} = \frac{N_p}{2n\pi^2 Z_q}\frac{\sum\left|\hat{u}_S^2(\mathbf{r}_{S,i})\right|\tilde{\kappa}_{P,i}}{\sum\left|\hat{u}_S^2(\mathbf{r}_{S,i})\right|} \tag{11.3.4}$$

The index i labels the contacts. $u_S(\mathbf{r}_{S,i})$ is the local amplitude of oscillation.

Characteristic for the elastic-loading condition, Eq. 11.3.3 predicts Δf to be positive. Δf is positive because the contact increases the effective stiffness of the composite resonator [15, 16]. This situation is termed "elastic coupling". Elastic coupling contrasts to mass-loading, where the frequency decreases. Remember that a resonance frequency can be expressed as $(\kappa/M)^{1/2}$ with κ a stiffness and M a mass. With inertial loads, resonator's modal mass increases and the frequency decreases, in consequence. If the resonator is elastically coupled to an outside object fixed in space, the stiffness increases, which moves the resonance frequency up.

Equation 11.3.3 can be inverted as

$$\tilde{\kappa}_P = \frac{2n\pi^2 Z_q}{N_P}\Delta\tilde{f} \tag{11.3.5}$$

The fact that the QCM can determine a contact stiffness was first noted by Dybwad in his seminal paper from 1985 [15]. The appendix of Ref. [17] shows that Dybwad's expression reduces to Eq. 11.3.3 in the limit of large spheres.

In Eq. 11.3.5, Δf scales as $1/n$ with n the overtone order, which is called "Dybwad scaling", here. Dybwad scaling contrasts to Sauerbrey scaling ($\Delta f \sim -n$). By checking for the overtone dependence, one can confirm that a positive frequency shift is indeed the consequence of elastic coupling across a small contact [22].

For single-asperity contacts, the established models of contact mechanics (Hertz theory being the simplest one) make predictions for κ_P [3]. For contacts between rough interfaces [18], contacts with intervening soft films [19], or contacts exposed to water vapor [22] the situation is much less clear. For the stiffness of multicontact interfaces (contacts between rough surfaces) see Sect. 11.8.

Measurements of contact stiffnesses based on the detuning of a resonance are also performed with AFM-based instrumentation [21] and, further, with macroscopic equipment [20]. The approach is more generally referred to as the "contact resonance method". When employing a QCM in this role, the contact radius must be much below the wavelength of sound for Eq. 11.3.5 to hold. Otherwise, inertial effects play a role. In principle, a similar condition also applies with to macroscopic devices. The situation there is different, though, because the frequency is lower and the wavelength of sound is larger.

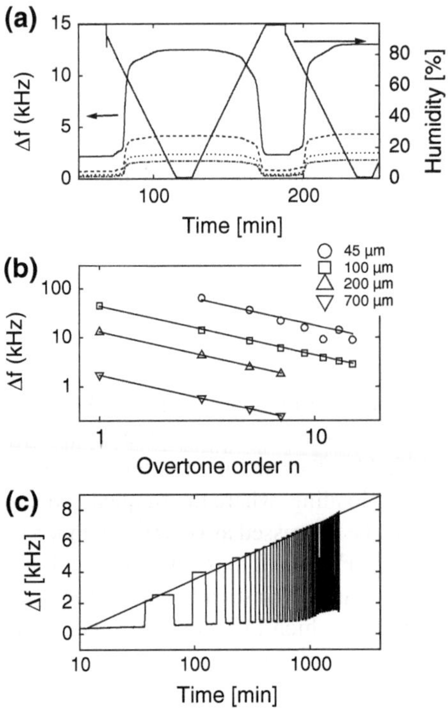

Fig. 11.6 An example, where the QCM was used to determine contact stiffness. The contact was established between glass spheres and the resonator surface. The stiffness strongly depended on ambient humidity. At low humidity, capillary forces pull the spheres towards the plate, thereby stiffening the bonds. Panel **a** shows Δf versus time for various overtones obtained with 200 μm spheres. In the dried-out state, the frequency shift is positive and scales as $1/n$ (**b**). Δf in the dry state increases over prolonged cycling (**c**). This behavior is also referred to as capillary aging. It is one of the reasons why the static friction coefficient increases with time. Adapted from Ref. [22]

Figure 11.6 shows an experimental example [22]. A resonator was densely covered with spheres (R_P between 45 and 700 μm) and the response of the frequency, initially, was not at all spectacular. The frequency did shift slightly, but the effect was poorly reproducible. Also, the frequency shift was not always positive and did not obey Dybwad scaling. In a second step, the sample was exposed to humid air. Again, the frequency changed by a small amount, but these changes were neither impressive, nor were they easily interpreted based on the n-dependence. In a third step, dry air was blown over the sample. At this point, there was a dramatic increase in frequency. Δf on the fundamental mode was more than 10 kHz. In this state, Dybwad scaling was well obeyed (Fig. 11.6b).

In these experiments, the contact stiffness clearly cannot be predicted from the Hertz model (Eq. 11.2.12). When the dry spheres first touch the resonator surface, Δf is much smaller than predicted by Hertz because of roughness. After exposing

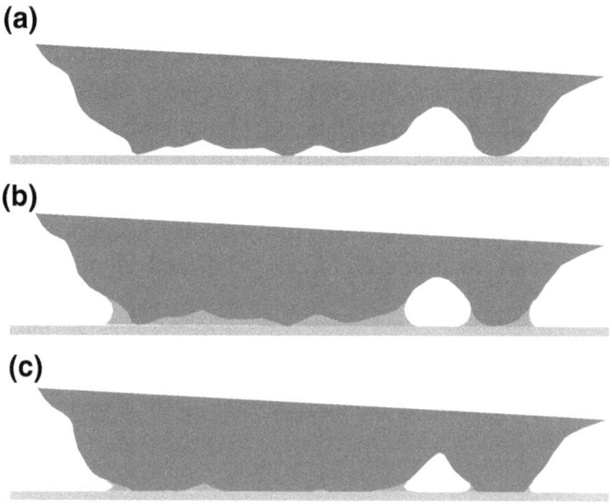

Fig. 11.7 Capillary forces are hysteretic. When the sample is first exposed to humid air, capillary bridges form between asperities, which had been separated in the dry state. The capillary force pulls the asperities together and the asperities remain in contact when the gap dries out. **a** Initial state of a dry contact display roughness on the nanoscale. **b** The same contact at high humidity. **c** The contact of after drying

the contacts to humid air for some time and drying them again, the stiffness is much larger than predicted by Hertz for the reasons sketched in Fig. 11.7. The mechanism is also called the "sandcastle effect" [23, 24]. Initially, the contact is established across a few small asperities, which only exert a negligible tangential force (This is a long-standing experience in QCM manufacturing. Non-electroded quartz blanks can be tested by placing them on a—slightly rough—metal surface and using this metal surface as the lower electrode. The few load-bearing asperities have hardly any influence on the resonance.) Upon admitting humid air to the chamber, there is capillary condensation at the points of contact. However, the capillary force pulling the two surfaces together is weak because the air-water interface has a large curvature, following the Kelvin equation [25]. The capillary pressure (as predicted by the Laplace equation) is correspondingly small. An impressive change in frequency is only observed when the sample is dried *after* having been exposed to humid air. Upon drying, the negative capillary pressure increases and pulls the two surfaces together. In consequence, the contact becomes stiffer than it was originally. The formation of capillary bridges is hysteretic.

The above argument relies on capillarity alone. There also is the possibility that the presence of the water slightly modifies the properties of the glass surface. Impurities might leach out and glue the surfaces together after drying. Chemistry might play a role (here and in the sand castle effect, in general).

As Fig. 11.6c shows, the contact stiffness increases with the time spent in the wet state (Fig. 11.6c). The dependence of Δf on time follows the logarithm, typical for aging phenomena [26]. Contact aging also manifests itself in the increase of the static friction coefficient with the time of rest. The static friction coefficient also increases logarithmically with the rest time [27]. Capillary hysteresis is one of the reasons for contact aging [27]. Another one is asperity creep [28, 29].

The section concludes with a calculation of the sphere size needed a achieve inertial clamping. The particle rests in place if the particle resonance frequency $\tilde{\omega}_P = (\kappa_P/M_P)^{1/2}$ (see text around Eq. 11.4.2 in the next section) is much lower than the frequency of the resonator. For the reasons stated in Sect. 11.5, the parameter M_P is not the mass of the particle itself, but rather 2/5 of this mass ($M_P = (2/5)(4\pi/3)\rho_P R_P^3$). The contact stiffness can be estimated from Hertz theory [3]. The Hertzian contact has a tangential stiffness of $\kappa_P = 2G^*a$ with a the contact radius as predicted by Eq. 11.2.1 and G^* the effective modulus given by Eq. 11.2.12. The normal force (which enters the calculation of a) is $F_N = g\rho_P(4\pi/3)R_P^3$ with $g = 9.81$ m/s^2 the standard acceleration. Combining the above equations and assuming $G_1 \approx G_2 \approx G$ and $\nu_1 \approx \nu_2 \approx \nu \approx 0.2$ yields

$$\omega \approx 1.3 G^{1/3} \rho_P^{-1/3} R_P^{-5/6} g^{1/6} \tag{11.3.6}$$

Requiring that ω_P be smaller than ω by a factor 2 and solving Eq. 11.3.6 for R_P yields

$$R_P \geq 0.6\,\omega^{-6/5} g^{1/5} G^{2/5} \rho_P^{-2/5} \tag{11.3.7}$$

Inserting the values of quartz on both sides of the contact ($\rho_P = 2.65$ g/cm^3, $G \approx 30$ GPa) and further using $\omega = 2\pi$ 5 MHz, one arrives at a sphere radius of about 60 μm (This estimate assumes free spheres. Rolling can be prevented by gluing spheres to a backing plate). Again: Whether or not the sphere is inertially clamped, can be checked experimentally based on Dybwad scaling ($\Delta f \sim 1/n$). If the frequency shift is positive but not proportional to $1/n$, elastic coupling is in question.

11.4 Contacts with Particles of Intermediate Size

Particles, which are too small to be clamped by inertia and too large to be tightly coupled to the crystal, clearly are a more difficult case [30–33]. These give rise to a coupled resonance with a least three free parameters in the model. There are two reasons to describe this case in detail: Firstly, many bio-colloids fall into this class. The particles of interest include bacteria [34–39], diatoms [40], and protozoan oocysts [41]. The prospect of being able to measure the stiffness of a link between a bio-colloid and a solid substrate has attracted much attention. Secondly, a coupled resonance amounts to an absorption line in shear-wave spectroscopy. With a QCM, one can do vibrational spectroscopy on surface-attached colloids.

Particles of intermediate size are described by the equivalent circuit shown in Fig. 11.1b. Applying the mechanical Kirchhoff rules (Sect. 4.7.3) to Fig. 11.1b, one finds the load impedance as

$$\tilde{Z}_L = N_P \left(\left(\frac{\tilde{\kappa}_P}{i\omega} \right)^{-1} + (i\omega\, M_P)^{-1} \right)^{-1}$$
$$= N_P \frac{i\omega\, M_P \tilde{\kappa}_P}{\tilde{\kappa}_P - \omega^2 M_P} \tag{11.4.1}$$

Equation 11.4.1 can be simplified by introducing the particle resonance frequency, $\tilde{\omega}_P$, as

$$\tilde{\omega}_P = 2\pi \tilde{f}_P = \sqrt{\frac{\tilde{\kappa}_P}{M_P}} \tag{11.4.2}$$

A word of caution: $\tilde{\kappa}_P$ and M_P will turn out to be a "modal stiffness" and a "modal mass" later. There may be ambiguities in interpretation. As in Sect. 4.1.3, the undamped particle resonance frequency is defined as

$$\omega_{P,0} = \sqrt{\frac{\kappa_P}{M_P}} \tag{11.4.3}$$

Because of damping, $\omega_{P,0}$ is different from ω_P', where ω_P' is the "damped particle resonance frequency". ω_P' is given as

$$\omega_P' = \sqrt{\left(\frac{\kappa_P}{M_P}\right)^2 - \left(\frac{\xi_P}{2M_P}\right)^2} \tag{11.4.4}$$

Differing from Sect. 4.1.3, the difference between $\omega_{P,0}$ and ω_P', can be substantial because the coupled resonance can be strongly damped.

Assume all contacts to have the same $\tilde{\omega}_P$. Inserting Eq. 11.4.2 into Eq. 11.4.1 and using the SLA, one finds

$$\Delta\tilde{f} = -N_P \frac{f_0}{\pi\, Z_q} \omega\, M_P \frac{\tilde{\omega}_P^2}{\tilde{\omega}_P^2 - \omega^2} = N_P \frac{f_0}{\pi\, Z_q} \omega\, \tilde{\kappa}_P \frac{1}{\omega^2 - \tilde{\omega}_P^2} \tag{11.4.5}$$

Comparison to Eq. 4.6.21 shows that such particles give rise to a *coupled resonance*. As emphasized in Sect. 4.6.3, the application of the SLA here requires some caution because the load may or may not be small in the sense of the SLA. The reader quickly convinces himself that the Sauerbrey limit and the elastic-coupling limit are obtained from Eq. 11.4.5 in the limits of $\omega_P' \gg \omega$ and $\omega_P' \ll \omega$, respectively (Both cases require $\omega_P'' \ll \omega_P'$).

A further algebraic simplification results if, firstly, $\omega_P'' \ll \omega_P'$ (small damping) and, secondly, $\omega \approx \omega_P'$. One can then write $\omega^2 - \tilde{\omega}_P^2 = (\omega + \tilde{\omega}_P)(\omega - \tilde{\omega}_P) \approx 2\tilde{\omega}_P(\omega - \tilde{\omega}_P)$, which leads to

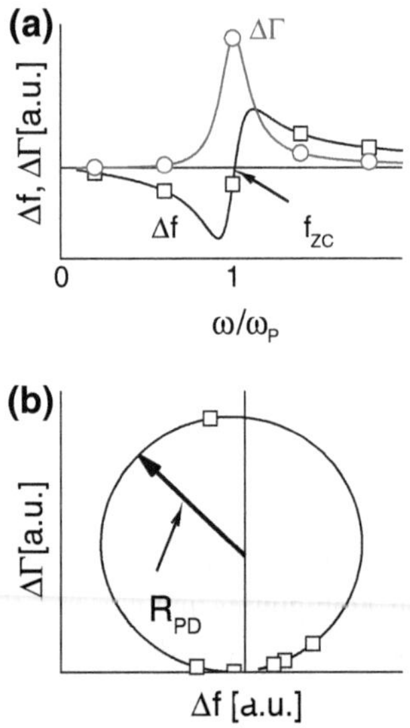

Fig. 11.8 When the sample gives rise to a coupled resonance, Δf and $\Delta\Gamma$ (shown in **a** and **b**, respectively) display some characteristic features: Δf crosses from negative to positive at the zero-crossing frequency, f_{ZC}. $2\pi f_{ZC}$ is similar to the undamped resonance frequency of the coupled resonance $\omega_{P,0} = (\kappa_P/M_P)^{1/2}$. At the zero-crossing frequency, $\Delta\Gamma$ goes through a maximum. Plotting $\Delta\Gamma$ versus Δf, one finds a circle (more generally: a spiral). The radius of this circle is a measure of the contact stiffness, κ_P. The open symbols were inserted to remind the reader that ω in Eq. 11.4.5 can only take discrete values equal to the overtones

$$\Delta\tilde{f} \approx -N_P \frac{f_0}{2\pi Z_q} \omega\, M_P \frac{\tilde{\omega}_P}{\tilde{\omega}_P - \omega} \tag{11.4.6}$$

Note that this simplification can be unrealistic when the coupled resonance is broad.

Figure 11.8a shows Δf and $\Delta\Gamma$ versus ω/ω_P according to Eq. 11.4.5. The frequency shift crosses from negative to positive at a frequency in the range of $\omega_{P,0}/(2\pi)$. This frequency is called "frequency of zero crossing", f_{ZC}, in Ref. [42] (Do not confuse the frequency of zero crossing with the damped particle resonance frequency. The two are different. The former increases with increased damping as shown in Eq. 11.4.7, while the latter decreases.) The open symbols in Fig. 11.8 are meant to remind the reader that experimental values of Δf and $\Delta\Gamma$ are only available for frequencies matching one of the overtones. Since f_{ZC} must be obtained by interpolation, there is some uncertainty. Setting the real part of Eq. 11.4.5 to zero, one finds the frequency of zero-crossing as

$$f_{ZC} = \frac{1}{2\pi} \frac{\omega_{P,0}}{\sqrt{1 - \frac{\gamma_P^2}{\omega_{P,0}^2}}} = \frac{1}{2\pi} \sqrt{\frac{\kappa_P}{M_P - \frac{\xi_P^2}{\kappa_P}}} \tag{11.4.7}$$

$\gamma_P = \xi_P/(2M_P)$ is the damping factor of the coupled resonator. A note on large damping: One might be tempted to define the parameter $\Gamma_P = \gamma_P/(2\pi)$ and view Γ_P as the half-bandwidth of the coupled resonance. However, this is misleading because many of the relations making life easier for the QCM rely on a sharp resonances. The coupled resonance, on the contrary, usually is broad. In liquids, it can easily be overdamped.

The target of investigation in such studies will usually be the bond stiffness. The bond stiffness should reflect the strength of the interaction between the particle and the surface. One might be tempted to invert Eq. 11.4.7, such that κ_P is on the left-hand side. There are three problems. Firstly, the right-hand side of Eq. 11.4.7 contains three unknowns (κ_P, M_P, and ξ_P). ξ_P can only be neglected if it is small, but this is not necessarily the case. Secondly, the parameter M_P is much less certain than one would hope. M_P is *not* equal to the mass of the particle. There is the possibility of rotation (in which case M_P represents a moment of inertia) and there may be some liquid mass coupled to the particle's motion. Thirdly, it is not even clear, what the meaning of κ_P should be. κ_P may be a shear stiffness or a bending stiffness. The virtue of Eq. 11.4.7 mostly is in the comparison between experiments. If f_{ZC} increases or decreases in a set of experiments, this probably goes back to an increase or decrease in contact stiffness (but, again, ξ_P and M_P also enter the equation).

The parameter M_P has relevance unrelated to the ambiguities in interpretation: By suitably choosing the particle size, one can move the particle resonance frequency into the frequency window of the QCM. The corresponding change in the particle resonance frequency is the analog of the isotope shift in IR-spectroscopy. As in IR–spectroscopy, it is not strictly certain that the properties of the bond are unchanged when the mass of the particle increases or decreases, but one can make that assumption. If the spheres of interest are available in different sizes, one can pick a size such that $\omega_P'/(2\pi)$ is between 5 and 65 MHz.

As always with coupled resonances, $\Delta\Gamma$ and Δf fall on a spiral when plotted against each other (Fig. 11.14, Ref. [42]). When the coupled resonance is weakly damped ($\omega_p'' \ll \omega_p'$), $\Delta\Gamma$ plotted versus Δf forms a circle. Such a polar diagram is shown in Fig. 11.8b. The parameter R_{PD} is the radius of the *circle*, where the index *PD* stands for *P*olar *D*iagram. Like the parameter f_{ZC}, the parameter R_{PD} can be read from the experimental data without fitting. R_{PD} is attractive in so far, as it can sometimes be determined in cases, where f_{ZC} is outside the accessible frequency range. An example is shown in Fig. 11.10. These data form a section of a circle, but the circle does not include the apex. Values for f_{ZC} are uncertain, but an estimate of R_{PD} can still be obtained from the graph.

If the damping is weak enough to let Eq. 11.4.5 be applicable, one can separate $\Delta\tilde{f}$ into real and imaginary parts as

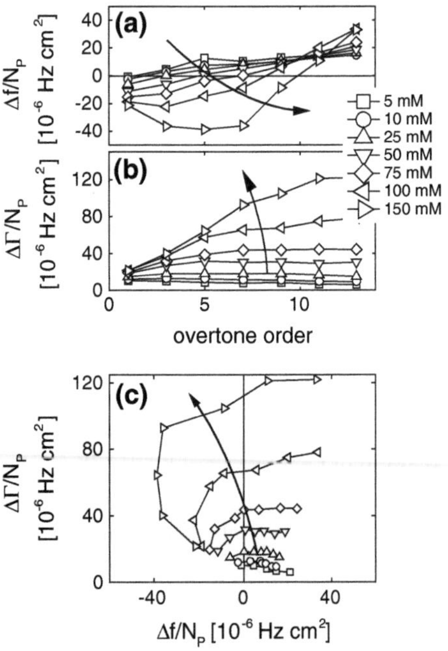

Fig. 11.9 $\Delta f/N_P$ and $\Delta\Gamma/N_P$ as a function of the overtone order n (**a**, **b**) and their relation as expressed in a polar diagram (**c**) for 0.5 μm radius silica particles adhering to sensor crystals coated with SiO_2 at various ionic strengths. Adapted from Ref. [42]

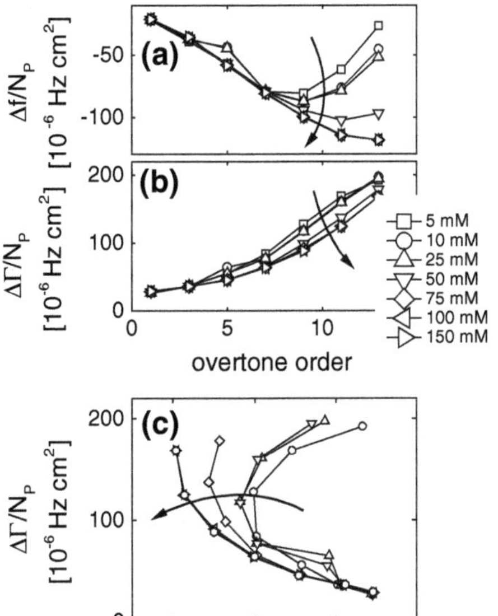

Fig. 11.10 Same as Fig. 11.9, but now with a soft layer immersed between the resonator surface and the sphere. The particle resonance frequency has moved up compared to Fig. 11.9. It is now beyond 65 MHz. Adapted from Ref. [42]

$$\frac{\Delta f}{f_0} = \frac{-N_P \omega\, M_P}{\pi\, Z_q} \frac{\omega_P\left(\omega_{P,0}^2 - \omega^2\right) + (\omega\, \gamma_P)^2}{\left(\omega_{P,0}^2 - \omega^2\right)^2 + (\omega\, \gamma_P)^2}$$
$$\frac{\Delta\Gamma}{f_0} = \frac{N_P \omega\, M_P}{\pi\, Z_q} \frac{\gamma_P \omega^3}{\left(\omega_{P,0}^2 - \omega^2\right)^2 + (\omega\, \gamma_P)^2} \quad (11.4.8)$$

R_{PD} can be approximated by half the value of $\Delta\Gamma$ at $\omega = \omega_{P,0}$:

$$R_{PD} \approx \frac{\Delta\Gamma\left(\omega = \omega_{P,0}\right)}{2} \approx \frac{N_P M_P f_0}{2\pi\, Z_q} \frac{\omega_{P,0}^2}{\gamma_P} = \frac{N_P f_0}{2\pi\, Z_q} \frac{M_P \kappa_P}{\xi_P} \quad (11.4.9)$$

Equation 11.4.9 is plagued by the same problems as Eq. 11.4.7. While one would certainly hope to be able to move κ_P to the left-hand side and derive it from R_{PD}, there are the uncertain parameters M_P and ξ_P on the right-hand side, as well. Assumptions are needed to derive κ_P.

At this point, we anticipate a result from the Sect. 11.5. Taking a distribution of particle resonance frequencies into account and, further, allowing for less-than-perfect coupling between the particle's vibration and the crystal's vibration, one finds:

$$R_{PD} \approx \frac{N_P f_0}{2\pi\, Z_q} f_{OS} \frac{\kappa_P}{\gamma_{het}} \quad (11.4.10)$$

The factor f_{OS} (the "oscillator strength") quantifies incomplete coupling. γ_P was replaced by the "heterogeneous linewidth", γ_{het} in order to account for $\tilde{\omega}_P$ varying between particles. Regardless of these modifications: If R_{PD} increases in a set of experiments, this increase can—with all caveats kept in mind—be attributed to an increase in contact stiffness. Clearly, this interpretation requires the parameters f_{OS} and γ_{het} to be constant between experiments.

Figure 11.9 shows an experimental example [42]. Glass spheres with a radius of 0.5 μm had been deposited on a SiO_2-coated resonator. The number of spheres per unit area had been determined independently with a camera. Δf and $\Delta\Gamma$ have been divided by particle number in Figs. 11.9 and 11.10. Note the difference to the experiment shown in Fig. 11.6. The spheres here were much smaller than in Ref. [43] and they did not densely cover the surface. Also, this experiment occurred in a liquid, while the experiment shown in Fig. 11.6 had occurred in air.

For glass spheres interacting with a silica surface across an electrolyte solution, the force between the sphere and the plate can be tuned by ion concentration because ions screen the electrostatic repulsion. The normal force has three contributions, which are gravity, van-der-Waals attraction, and electrostatic repulsion. The electrostatic repulsion opposes gravity and van-der-Waals attraction. The experiments shown in Figs. 11.9 and 11.10 rely on the tunability of the force. In pure water, the repulsion is so strong that the spheres can be easily washed away by pumping liquid through the cell. Upon adding salt, the electrostatic repulsion is screened and the overall attractive force increases correspondingly. The contact stiffness can be expected to follow this increase. The different symbols in Fig. 11.9 label the different ion strengths. The arrows indicate increasing attractive force.

Figure 11.10 shows a second data set from the same series of experiments. These data differ from Fig. 11.9 in that the resonator surface had been coated with a layer of biotin, linked to the surface across short chains of polyethylene glycol (PEG). While one might expect a soft interlayer to decrease the bond stiffness, the experiment evidences the opposite. The data still fall onto a section of a circle, but the frequency of zero crossing has moved up and now is outside the accessible range ($f_{ZC} > 65$ MHz).

From Eq. 11.4.7 one sees that an increase in f_{ZC} can have two sources, which are an increase in $\omega_{P,0}$ and an increase in γ_P. An increase in γ_P would be plausible because the surface has become softer after functionalization with the organic layer (Remember the flexible linker between the surface and the biotin groups). A more detailed analysis based on fits with Eq. 11.5.6 shows both effects contribute: γ_P has increased between the experiments shown in Figs. 11.10 and 11.9, but $\omega_{P,0}$ has also gone up. An increase in $\omega_{P,0}$ is less plausible than an increase in γ_P. After all, the surface has become *softer*. Possibly, an increased contact radius is the cause for $\omega_{P,0}$ going up (cf. Eq. 11.2.11). A similar conclusions is reached from the analysis of R_{PD} (Fig. 11.10c). R_{PD} is connected to the stiffness of the link by Eq. 11.4.10. R_{PD} is larger than in Fig. 11.9.

Figures 11.9 and 11.10 agree with the expectations in two regards:

- The frequency of zero crossing, f_{ZC}, increases with increasing attractive force. Following Eq. 11.4.7 and reasoning that the ionic strength has little effect on the parameters M_P and ξ_P, this implies that the contact stiffness increases with increasing normal force, which seems reasonable.
- The radius of the circle in the polar diagram, R_{PD}, increases with increasing attractive force. Assuming that the parameters f_{OS} and γ_{het} in Eq. 11.4.10 stay approximately constant, this again suggests increasing contact stiffness.

Given that the data from Figs. 11.9 and 11.10 together with Eq. 11.4.7 and Eq. 11.4.10 form a consistent picture, one might hope that quantitative fits of the data with Eq. 11.4.5 were possible and would make the above statements more reliable and exact. However, it turns out that convincing fits are not easily obtained. The poor fits reveal significant short-comings of the model. There are two separate problems: Firstly, the peaks in $\Delta\Gamma(u)$ are too broad to agree with Eq. 11.4.5. Secondly, the fitted parameter M_P is too small to be compatible with the known sphere size. We will come back to these problems after rephrasing the coupled-resonance-phenomenon, using the language of spectroscopy.

A historical note: Coupled resonances caused by adsorbed particles have been observed on one single overtone, before. In these experiments, $\omega_{P,0}$ was made to change with time by gradually stiffening the bond between the particle and the plate. In the first case, capillary aging was employed [43]; in the second case, the spheres were deposited onto to spin-cast polymer film [44]. Heating the resonator to the glass temperature of the polymer induced to particle embedding and an increased contact stiffness. If ramping $\omega_{P,0}$ in this way, one finds a peak in a plot of $\Delta\Gamma$ versus time and a corresponding asymmetric pattern in the Δf-trace. These features occur, when $\omega_{P,0}$ equals the resonance frequency of the QCM.

11.5 Coupled Resonances and Shear-Wave Spectroscopy

If one re-labels the axes in Fig. 11.8 with "$-n$" for Δf and "k" for $\Delta\Gamma$, Fig. 11.8 describes an optical absorption line. n and k here are the real and the imaginary part of the refractive index. An optical absorption does two different things to the electromagnetic wave: It attenuates the wave and it delays its propagation. Similarly, the coupled resonance removes energy from the vibration of the crystal and it adds a phase shift onto the amplitude of the reflected wave, thereby delaying its return to the crystal. A QCM performs shear-wave spectroscopy.

The analogy has aesthetic value, for one, but it also holds the clue to why the parameter M_P comes out as too small when fitting the data from Fig. 11.9 with Eq. 11.4.5. More precisely, M_P comes out as too small, when the data are fitted with a modified equation, which accounts for a distribution in $\omega_{P,0}$, (Eq. 11.5.6 below). M_P is a prefactor in Eq. 11.4.5 and it remains a prefactor in the modified equation. M_P is a robust fit parameter; its fitted value does not depend on which distribution is chosen for $\omega_{P,0}$. Stating that M_P is too small amounts to saying that the absorption line is weaker than expected. Weak absorption lines are known (very well) from optical spectroscopy. Weak lines are not related to a small modal mass, but rather to an inefficient coupling between the respective transition and the wave exciting it.

To illustrate the analogy between vibrational spectroscopy on molecules, on the one hand, and shear-wave spectroscopy on surface-attached colloids, on the other, consider the example of the CO_2 molecule in the gas phase. For this molecule, the "mode assignment problem" has been solved (The mode assignment problem will concern us later in the context of shear waves, as well). There are a total of 4 modes, which are the symmetric stretch mode at a wavenumber of about 1,340 cm^{-1}, the antisymmetric stretch mode at 2,350 cm^{-1}, and two degenerate bending modes at 666 cm^{-1}. For a sufficiently dilute gas, the symmetric stretch mode is so weakly coupled to the exciting electromagnetic wave, that it is actually not visible at all. More technically, the symmetric stretch mode has a vanishing transition dipole moment (TDM). The molecule's vibration does not involve an oscillating electric dipole because the oxygen atoms (which carry a small negative charge) move symmetrically. Because of the vanishing TDM, the vibrating molecule neither emits nor absorbs electric dipole radiation. In the language of spectroscopy, the symmetric stretch mode is "forbidden". "Forbidden" does not mean that the mode would not exist. It exists and is populated by thermal excitation like any other mode. "Forbidden" only implies that the mode does not couple to the IR electromagnetic wave. Which modes are forbidden, is the content of the "selection rules". Selection rules often are rooted in the symmetry of the molecule.

If one looks a little closer, one can find conditions, under which the symmetric stretch mode can still be observed [46]. The symmetric stretch mode becomes visible, when the density of the gas is large enough to let collisions between molecules play a role. Two colliding molecules form a transient dimer and the presence of a second molecule breaks the first molecule's inversion symmetry. The symmetric stretch mode now is visible (although the line is weak). In other words: Special conditions and a refinement of the theory have turned a "forbidden" transition into a "weakly allowed" transition. Weakly allowed transitions are very common in spectroscopy. There are numerous examples, where certain transitions appear to be forbidden, at first glance, but can nevertheless be observed because the model does not capture certain mechanisms of coupling or certain special conditions of the experiment. The "rocking mode" of a surface-adsorbed particle will turn out to be weakly allowed in shear-wave spectroscopy.

Clearly, there is no sharp line separating forbidden and allowed transitions. The coupling between a transition and its source of excitation is quantified by an "oscillator strength", f_{OS}, which is a number between 0 and 1. f_{OS} is close to zero for forbidden transitions. It will enter Eq. 11.4.5 as a prefactor; it is a further free parameter of the model.

Two more notes on the vibrational spectrum of CO_2:

- Whether a transition is forbidden or not depends on the type of excitation. For instance, the symmetric stretch mode of CO_2 *can* be observed in Raman spectroscopy (as opposed to IR spectroscopy), because the selection rules in Raman spectroscopy differ from the selection rules in IR spectroscopy. By analogy, the stretch mode of a surface-adsorbed colloidal particle is not excited by shear waves. It *is* excited by compressional waves, though.
- There is an interesting twist to the symmetric stretch mode of CO_2 (visible in Raman spectroscopy and in collision-induced IR absorption spectroscopy). There actually are *two* lines around 1,340 cm^{-1}, not one. That happens because there is a coupled resonance (cf. Sect. 4.6.3). The second overtone of the bending mode happens to have a frequency close to the frequency of the symmetric stretch mode. Because the two modes couple, the symmetric stretch mode splits up into a doublet. The phenomenon is called "Fermi resonance" in the context of molecular spectroscopy. What was called "activity dip" in Sect. 4.6.3 for macroscopic resonators, also happens on the molecular scale.

We now turn to heterogeneity and heterogeneous line broadening. Heterogeneous line broadening denotes a situation, where different molecules have slightly different transition frequencies. The absorption line observed from the outside then is a superposition of all individual lines. The linewidth does not reflect the damping of the resonance, but rather the width of the distribution of resonance frequencies. The damping would be reflected in the "natural linewidth", which often is smaller than the heterogeneous linewidth. The CO_2 molecule in the gas phase is not a good example for heterogeneous line broadening. Better examples are the electronic transition of dye molecules embedded in a solid matrix. The matrix affects the frequency of the electronic transition and broadens the line. At any rate: An experimentally observed linewidth cannot always be interpreted as the rate of damping.

The terms to be remembered from this discussion of optical spectroscopy and to be translated to shear-wave spectroscopy are "mode assignment", "oscillator strength", and "heterogeneous broadening". The mode assignment problem in shear-wave spectroscopy is approached with a finite element analysis in the following section.

11.5.1 Finite Element Simulations, Multiple Modes

The hypothesis, that an adsorbed sphere may resonate in more than one way and that these different modes couple to the motion of the resonator to different extents, can be supported with a finite element simulation. The simulation amounts to a clean experiment on the computer. The technicalities of the calculation are discussed in Chap. 12. The user prescribes a geometry, the viscoelastic constants, and the frequency of excitation; the software uses this information to calculate the displacement pattern. The software knows nothing about modes or selection rules. It just solves the partial differential equations given to it. From the displacement pattern, it calculates the area-averaged stress at the resonator surface, which leads to $\Delta\tilde{f}$ after insertion into the SLA. A short-coming of this particular code is that it only works in two dimensions. What is called a particle below, is a circle or, depending on the point of view, an infinite cylinder. Of course an adsorbed cylinder is different from an adsorbed sphere, but it turns out that the motion of the cylinder also displays two resonant modes.

As in most simulations, simplifying assumptions are needed with regard to the geometry. As other simulations, the finite element model can provide answers, which the real experiment cannot give. Of importance here is, that it predicts the displacement pattern. More on the technical side: The simulation is not limited to frequencies corresponding to one of the overtones. The frequency scale in Fig. 11.11 extends from 1 to 90 MHz and contains a few hundred data points. The scale might as well have extended from 1 Hz to a few GHz. There are no constraints on frequency.

Figure 11.12 shows the geometry and the flow field. The resonator surface is located to the left. A particle with a diameter of 4 μm touches the resonator across a link. The link is a rectangle 100 nm thick and 600 nm wide. Its stiffness is 1 GPa. The particle itself is rigid. The way this simulation is set up, the area of the contact remains unchanged if the particle moves. There is neither peeling not sliding (It is instructive to compare Figs. 11.12 to 12.3. In Fig. 12.3, the link was made widc enough to move the coupled resonance upwards out of the frequency window of the QCM. Also, the particle was much smaller).

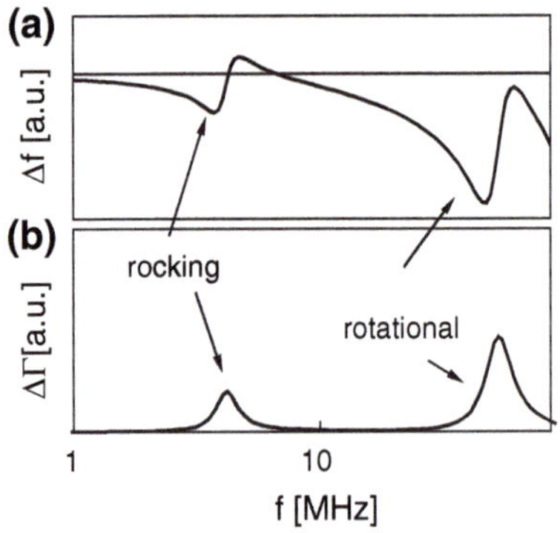

Fig. 11.11 Shifts of frequency (**a**) and half-bandwidth (**b**) as computed from the finite element model shown in Fig. 11.12. One observes *two* coupled resonances, which are the rocking mode and the rotational mode. The integral over a peak in $\Delta\Gamma(f)$ is proportional to the oscillator strength, f_{OS}

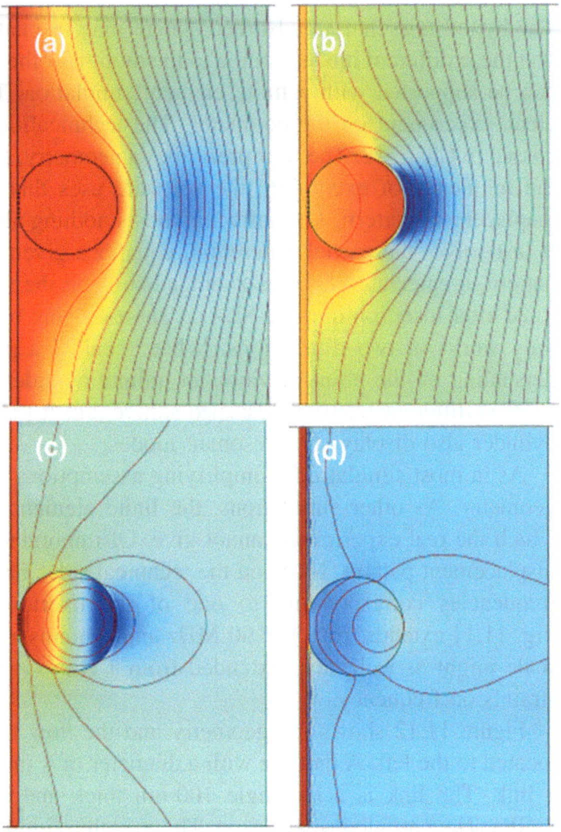

Fig. 11.12 Displacement pattern around an adsorbed sphere as calculated by the Finite Element Method. The quartz resonator is on the left-hand side. The sphere has a diameter of 4 μm and is rigid. It is linked to the substrate across a link with a thickness of 100 nm, a width of 600 nm, and a shear modulus of 1 GPa. Color encodes the local tangential velocity. For panels **a** and **b**, the dark lines are streamlines, where the velocity is measured relative to the substrate. Panels **a**, **b**, **c**, and **d** correspond to frequencies of 1, 4.2, 54, and 90 MHz, respectively

Figure 11.11 shows $\Delta f(f)$ and $\Delta\Gamma(f)$ for frequencies between 1 and 90 MHz. Clearly, there are *two* coupled resonances, not one. Moreover, one of the peaks in $\Delta\Gamma(f)$ is smaller than the other. Since the displacement patterns are known, one can look into the question of why the mode at 4.2 MHz is coupled to the resonator's tangential movement less efficiently than the mode at 54 MHz. The finite element simulation has solved the mode-assignment problem for us.

The different panels in Fig. 11.12 correspond to different frequencies, which are 1, 4.2, 54, and 90 MHz. At low frequencies (1 MHz, panel **a**), the particle essentially moves with the crystal. The sphere and the trapped liquid increase the resonator's mass, thereby decreasing the resonance frequency. The decrease is roughly proportional to the frequency itself, as shown on the left-hand side in Fig. 11.11a. Clearly, this is the Sauerbrey limit. Panel **d** shows the opposite limit of high frequencies (90 MHz). The particle almost rests in place. It still rotates, but only by a small amount. The link exerts a restoring force onto the resonator, which amounts to elastic coupling as discussed in Sect. 11.3. As the right-hand side in Fig. 11.11a shows, Δf is not positive in this limit, although the frequency shift decreases with increasing ω. Δf is negative because the viscous drag exerted by the liquid is superimposed onto the elastic force exerted by the sphere. The liquid causes a negative offset in Δf.

Panels **b** and **c** in Fig. 11.12 show the flow fields at the frequencies of the two resonances (4.2 and 54 MHz). The sphere rotates about the point of contact (or a point close to the contact) at 4.2 MHz, while it rotates about its center (or a point close to that) at 54 MHz. This explains why the mode at 4.2 MHz (the "rocking mode", see below) is the weaker mode than the mode at 54 MHz (the "rotational mode"). When the sphere rotates about the point of contact, it exerts a *torque* onto the substrate, rather than a tangential stress. If the axis of rotation were exactly at the point of contact and if there were no coupling across the liquid, the mode would be strictly forbidden. But there is coupling across the liquid and the axis of rotation is slightly displaced from the point of contact. For these reasons, the mode at 4.2 MHz is (weakly) coupled to the substrate's tangential motion. It is seen, but it is less strong than the rotational mode.

11.5.2 Vibrational Spectroscopy on Surface-Attached Colloids

At the beginning of this chapter, it was stated that Dybwad's model leaves open how the particle moves. Unrealistically, there is only one degree of freedom; the model is one-dimensional. The finite element model, on the other hand, treats a continuum, which amounts to infinitely many degrees of freedom (The number of degrees of freedom is finite in the actual calculation because the mesh-size is finite). In the following, the problem is discussed at an intermediate level of abstraction, where the number of degrees of freedom is large enough to capture the essential features of the problem, but not so large that these features would be hidden in the complexity of the model. This model consists of a rigid sphere and a deformable link. In a way, the model is close to a mechanical equivalent circuit; it

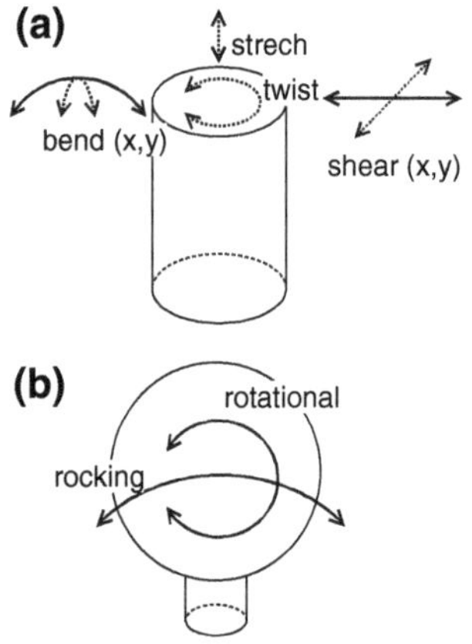

Fig. 11.13 A link between a particle and a substrate (here displayed as a cylinder) can be deformed from the two ends in 6 different ways (**a**). Only two of those deformations couple to a periodic translation of the substrate along *x*. These are termed the rotational mode and the rocking mode (**b**)

differs in that the link is not just a spring, but a rather a rod which can be distorted in different ways. Likewise, the particle is not just a point, but rather a sphere which can move in different ways (three translations, three rotations). The different ways to deform a link are depicted in Fig. 11.13a. Each way to deform the link corresponds to a mode of vibration of the substrate/link/sphere assembly. The modal stiffness and the modal mass differ between the types of deformation. Three of the modes translate the upper end of the link relative to the bottom, while the three other ones rotate it. The deformation modes of the link are:

- Vertical translation, stretching the link
- In-plane translation along *x* and *y*, shearing the link
- Rotation about *z*, twisting the link
- Rotation of the sphere about horizontal axes passing through the link, bending the link.

Symmetry enforces some of the modes to be degenerate. Because *x* and *y* are equivalent, the two modes shearing the bond (along *x* or *y*) and the two modes bending the link (along *x* or *y*) have the same frequency, respectively. Symmetry also implies certain selection rules: The substrate performs a translation along *x* and therefore cannot excite shear along *y*, bending along *y*, twist, or stretch. These four modes are forbidden by symmetry. Actually, the stretch mode would become weakly allowed, if there was a normal contribution to the motion of the substrate. In fact, there *is* such a contribution because of energy trapping (ignored in the following for the sake of simplicity). Similarly, shear along *y* and bending along *y* would become weakly allowed if the mirror symmetry about the *x*-axis were broken in some way. This would, for instance happen if there was some kind of interaction between the particle and one of its neighbors.

The two interesting modes are the one shearing the contact along *x* and the one bending it along *x*. These are again depicted in Fig. 11.13b, where the particle was included in the diagram. The modes have been renamed. The link is sheared if the particle performs a rotation about its center. This mode was termed "rotational".

The link is bent if the particle rotates about an axis passing through the link. This mode was named "rocking" mode. An oscillation of the substrate along x certainly couples to the rotational mode. This mode is allowed. With regard to the rocking mode, the situation is more complicated. One might think that bending of the link could only be induced by a torque, not by a tangential stress. Following this argument, the rocking mode would be forbidden. However, this is only true as long as the axis of rotation passes *exactly* through the bond. If the axis of rotation is slightly displaced from the bond, the rocking mode slightly shears the bond and the rocking mode becomes weakly allowed. For the displacement pattern shown in Fig. 11.12b, the axis of rotation is indeed not exactly at the point of the link and the rocking mode therefore is seen as a weak absorption line in Fig. 11.11. Also, there is some transfer of stress across the liquid phase.

The equivalent circuit from Fig. 11.1b (containing a mass, a spring, and a dashpot) is not invalidated by these insights; it only needs a reinterpretation. Consider the rotational mode, first. The mass must be replaced by a moment of inertia; the force must be replaced by a torque. The moment of inertia of a sphere rotating about its center is $I_P = (2/5)M_P R_P^2$. The torque exerted by the contact is given by $\hat{F}R_P$ (with $\hat{F}$ the force). The force is equal to $\tilde{\kappa}_P \hat{u}_S$ and $\hat{u}_S$ is related to the angle of rotation, $\hat{\theta}$, as $\hat{u}_S = R_P\hat{\theta}$. The dynamics of the system follows the equation

$$-\omega^2 I_P\hat{\theta} + \tilde{\kappa}_{shear}R_P^2\hat{\theta} = 0 \tag{11.5.1}$$

The spring constant was given an index "shear" because it must be distinguished from the bending stiffness, which governs the dynamics of the rocking mode. From Eq. 11.5.1 the resonance frequency of the rotational mode follows as

$$\omega_{P,rot,0} = \sqrt{\frac{\kappa_{shear}R_P^2}{I_P}} = \sqrt{\frac{5}{2}}\sqrt{\frac{\kappa_{shear}}{M_P}} \tag{11.5.2}$$

Rotation can simply be accounted for by a prefactor of $(5/2)^{1/2}$. Should the axis of rotation be slightly displaced from the center of the sphere, this shifts the values of I_P and also the calculation of the torque, but the effects are not dramatic.

Now consider the rocking mode. If the axis of rotation is at the point of contact, the moment of inertia is $I_P = 7/5\ M_P R_P^2$. The torque exerted at the point of contact is equal to the bending stiffness of the contact, $\tilde{\kappa}_{bend}$, times the angle. The bending stiffness needs to be calculated from contact mechanics. It has units of Nm. The ratio of the bending stiffness and the moment of inertia has units of s^{-2}. The particle resonance frequency is

$$\omega_{P,rock,0} = \sqrt{\frac{5}{7}}\sqrt{\frac{\kappa_{bend}}{M_P R_P^2}} \tag{11.5.3}$$

In Fig. 11.11 the rocking mode occurs at a frequency below the frequency of the rotational mode. This is always so, regardless of the details, as the following argument shows. For the Hertzian contact with $a \ll R_P$, the shear stiffness and the bending stiffness are

$$\begin{aligned}\kappa_{shear} &= b_{shear} G a \\ \kappa_{bend} &= b_{bend} G a^3\end{aligned} \tag{11.5.4}$$

b_{shear} and b_{bend} are numerical constants of the order of unity (see also Eq. 11.2.11 and Eq. 11.2.12). Equation 11.5.4 follows from dimensional arguments. The particle resonance frequencies of the rotational mode and the rocking mode are

$$\begin{aligned}\omega_{P,rot,0} &= \sqrt{\frac{\kappa_{shear} R_P^2}{2/5\, M_P R_P^2}} \approx \sqrt{\frac{5 b_{shear}}{2}} \sqrt{\frac{G a}{M_P}} \\ \omega_{P,rock,0} &= \sqrt{\frac{\kappa_{bend}}{7/5\, M_P R_P^2}} \approx \sqrt{\frac{5 b_{bend}}{7}} \sqrt{\frac{G a}{M_P} \frac{a}{R_p}}\end{aligned} \tag{11.5.5}$$

If b_{shear} and b_{bend} are both of order unity and if $a \ll R_P$, the rocking mode has the lower frequency.

11.5.3 Heterogeneous Broadening, Oscillator Strength

Eventually, the insights derived from the comparison with spectroscopy must result in a suitably modified scheme to analyze QCM experiments. Consider the width of the peak in the plot of $\Delta\Gamma(n)$ first. Broad peaks as seen in Fig. 11.9 cannot be explained with a large damping coefficient, γ_P. For a narrow coupled resonance, the half-width at half-height is given by $\gamma_P/(2\pi)$. However, when γ_P is comparable to $\omega_{P,0}$ or larger, this picture breaks down. Large values of γ_P turn the circle in the polar diagram into a spiral (Fig. 11.14, see also Fig. 10.3). Expressed differently, a large value of γ_P leads to Δf being negative at low n, but barely turning positive at high n. For $\gamma_P > \omega_{P,0}$, the frequency shift does not turn positive at all; there is no zero-crossing. In Fig. 11.9 the bandwidth of the coupled resonance clearly is larger than $\omega_{P,0}$ itself. The peak is wider than the accessible frequency range. However, there still is a zero crossing in $\Delta f(n)$ and the data can therefore not be fitted Eq. 11.4.5.

This inconsistency can be resolved by allowing different particles to have different particle resonance frequencies. The distribution in ω_P might go back to a distribution in stiffness, which might be a consequence of nano-scale roughness. Taking heterogeneity into account, the coupled-resonance picture explains the data well. Unfortunately, this extension of the model can be made to agree with experiment in a few different ways. Different distributions of ω_P will work equally well. Also: There might be distributions not only for ω_P, but also for γ_P, and possibly even for M_P. The parameter space for fitting is large.

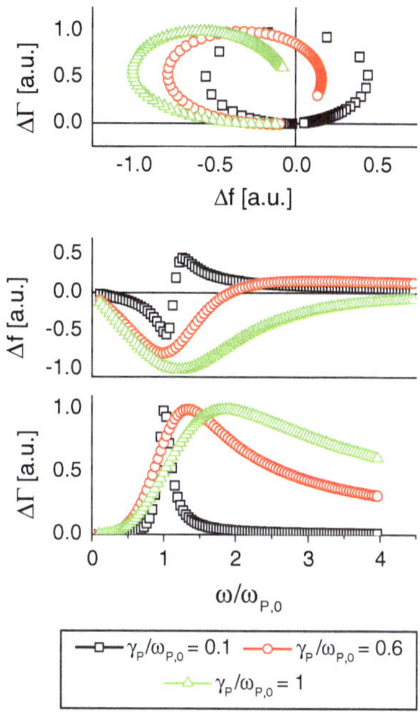

Fig. 11.14 Δf and $\Delta\Gamma$ as predicted by Eq. 11.4.5. A large damping of the coupled resonance (γ_P comparable to ω_P) turns the *circle* in the polar plot into a spiral. For clarity, Δf and $\Delta\Gamma$ were normalized to the peak value of $\Delta\Gamma$. For large damping, the plot of Δf versus ω becomes asymmetric. With $\gamma_P > \omega_{P,0}$, Δf does not even turn positive at high overtones. The frequency of zero crossing is lost

The second problem with quantitative fitting (the fitted M_P being unrealistically small) is fixed by introducing the oscillator strength, f_{OS}, as a prefactor into the equations.

Equation 11.4.5 turns into:

$$\tilde{\Delta f}(\omega) \approx -f_{OS} N_P \frac{f_0}{\pi Z_q} \omega M_P \int_0^\infty g\left(\omega_{P,0}\right) \frac{\omega_{p,0}^2 + i\omega\,\gamma_P}{\left(\omega_{p,0}^2 + i\omega\,\gamma_P\right) - \omega^2} d\omega_{P,0} \qquad (11.5.6)$$

$g(\omega_{P,0})$ is the distribution function of particle resonance frequencies. Again: The above equation only covers distributions in $\omega_{P,0}$. γ_P might vary between particles, as well.

The section concludes with an updated equation for the radius of the circle in the polar diagram, R_{PD}. Taking heterogeneous line broadening and the nontrivial oscillator strength into account, the radius becomes

$$R_{PD} \approx \frac{N_P f_0}{4\pi Z_q} f_{OS} \frac{\kappa_P}{\gamma_{het}} \qquad (11.5.7)$$

The right-hand side now contains two unknowns (γ_{het} and f_{OS}). Clearly, one must be careful with quantitative statements on κ_P.

There are difficulties and unknowns in this analysis, but these should not hide the central outcome of the discussion: The stiffness of the bond between a bio-colloid and a solid surface is accessible with the QCM.

11.6 Particles Coupled to the Resonator Surface Across a Dashpot

In Sect. 11.5 it has become clear that the 3-element network from Fig. 11.1b holds problems and that any simplification must be welcome. One simplification can be a large sphere, which makes the parameter M_P disappear from the equations. Another circuit bringing about a simplification is the particle coupled to the surface across a dashpot, only (Fig. 11.15). This circuit has been employed in the context of QCM-based nanotribology. Figure 11.15 assumes the link to behave like a liquid. The sphere slides on the surface. A remark on terminology: Do not confuse this type of motion with solid-like sliding in the Coulomb sense. In Coulomb friction, the tangential force is independent of the sliding velocity, while the tangential force here is proportional to molecule's velocity relative to the substrate. The molecule's motion has also been called "creep" to avoid this misunderstanding.

From Fig. 11.15 the load impedance is found to be:

$$\begin{aligned}\tilde{Z}_L &= N_P\left(\frac{1}{\mathrm{i}\omega M_P}+\frac{1}{\xi_P}\right)^{-1} = N_P\frac{\mathrm{i}\omega M_P\xi_P}{\mathrm{i}\omega M_P+\xi_P}\\ &= N_P\,\mathrm{i}\omega M_P\frac{1}{1+\mathrm{i}\omega\tau_{MR}} = N_P\,\mathrm{i}\omega M_P\frac{1-\mathrm{i}\omega\tau_{MR}}{1+\omega^2\tau_{MR}^2}\end{aligned} \tag{11.6.1}$$

$\tau_{MR} = M_P/\xi_P$ is the momentum relaxation time. The momentum relaxation time is called "slip time" in Refs. [47] and [48]. To understand the meaning of the slip time, consider a hypothetical experiment, where the particle and the substrate move at the same velocity, initially, and where the substrate at some time abruptly arrests. The velocity of the particle then also returns to zero, but it does so with a delay because of inertia. The particle velocity decreases exponentially, where the decay time is τ_{MR}.

Using the small load approximation, Eq. 11.6.1 leads to

$$\frac{\Delta\tilde{f}}{f_0} = \frac{-N_P\omega M_P}{\pi Z_q}\frac{1-\mathrm{i}\omega\tau_{MR}}{1+\omega^2\tau_{MR}^2} \tag{11.6.2}$$

The momentum relaxation time is inferred from the ratio of $\Delta\Gamma$ and $-\Delta f$ as

$$\tau_{MR} = \frac{1}{\omega}\frac{\Delta\Gamma}{(-\Delta f)} \tag{11.6.3}$$

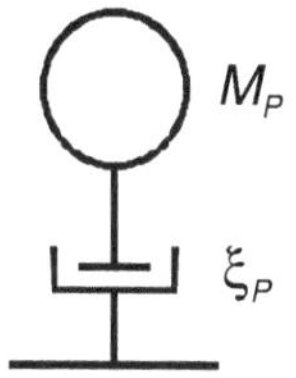

Fig. 11.15 A particle coupled to the resonator surface across a dashpot exerting liquid-like forces

Equation 11.6.3 has been used extensively by J. Krim and co-workers to analyze the behavior of adsorbed atoms and molecules. Similar studies have been reported by Mistura, who also studied the amplitude-dependence of the bandwidth [49]. A recent review is given in Ref. [50].

Two remarks:

- A check on whether or not τ_{MR} depends on overtone order is worthwhile. The author has occasionally found the momentum relaxation time as derived by Eq. 11.6.3 to differ between overtones.
- If there are clusters of some kind ("nanodroplets"), these might be more efficient in dissipating energy than adsorbed molecules. Even few clusters might have an effect which is large compared to the dissipation induced by a molecular monolayer.

Krim finds a significant increase in bandwidth even for monolayers of adsorbed noble gases. This is surprising because simple estimates of the drag coefficient lead to values which are much higher than what is found with the QCM using Eq. 11.6.3, the known mass, and the relation $\tau_{MR} = M_P/\xi_P$. Inertia should not be strong enough to make molecular monolayers slide on a QCM surface. The drag coefficient is related to the diffusivity, D, by the fluctuation-dissipation theorem [51]. One should have $D = k_BT/\xi_P$ with k_BT the thermal energy. If the shift in bandwidth is converted to diffusivity using Eq. 11.6.3 and the fluctuation-dissipation theorem, the derived diffusivities come out as unrealistically high. Also, there are estimates of how much the adsorption energy should vary across the surface. There are barriers between the preferred sites of adsorption and these should prevent sliding of single atoms. A number of suggestions have been made to remove this inconsistency. Naturally, some of them go beyond the simple model underlying Eq. 11.6.3. For a review, see Ref. [50].

11.7 Point Contacts as Centers of Acoustic Scattering

The discussion of point contacts above was based on the area-averaged stress, which was converted to $\Delta\tilde{f}$ using the SLA. Scattering was briefly alluded to. There is one particular aspect of the scattering picture, which merits discussion. Guided by the optical analog, one might think that even elastic scattering ($\xi_P = 0$ in Fig. 11.1, no dissipative processes at the link) would attenuate the shear wave and thereby damp the resonance because it deflects part of the acoustic energy sideways. This is what

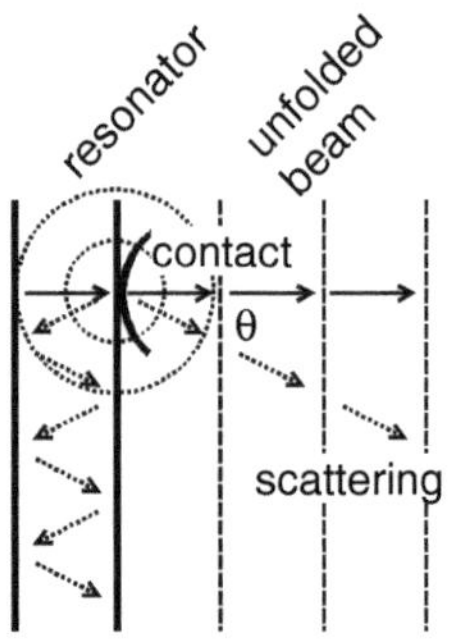

Fig. 11.16 Sketch of an unfolded beam path. The contact is a scattering center. The scattered wave is approximated as spherical. It superimposes itself onto the primary wave (the resonant mode). Forward scattering leads to a phase shift and to a frequency shift, in consequence. The energy scattered into directions other than the forward direction damps the resonance, in principle, but this effect is small compared to the shift in frequency

happens in turbid media. The optical extinction has two contributions, which are absorption and scattering. The same is true with shear waves, in principle. Elastic scattering does increase $\Delta\Gamma$. However, when applying the SLA to this problem, the contribution of scattering to $\Delta\Gamma$ is found to be negligible.

To see this more quantitatively, unfold the path of the acoustic beam as in Fig. 11.16. The resonant mode turns into a propagating wave. A localized contact scatters the wave. Because only an estimate is needed, simplifying assumptions can be made:

- The wave is represented as a scalar field; the direction of displacement is ignored.
- The material is approximated as isotropic.
- The contact is assumed to be so small that the scattered wave is spherical. In quantum mechanics, this limit is called s-wave scattering [52, 53].

With these simplifications the scattered wave takes the form

$$\hat{u}_{sc}(r) = \hat{u}_S \frac{\tilde{b}_{sc}}{r} \exp(-ikr) \tag{11.7.1}$$

r is the distance to the scattering center, $\hat{u}_S$ is the amplitude of the incoming wave, and $\tilde{b}_{sc}$ is the scattering length.

Scattering is isotropic, which includes the forward direction. Forward scattering shifts the phase (and possibly the amplitude) of the incident wave (the resonant mode). In the language of Sect. 4.4, scattering affects the reflectivity at the resonator surface.

The scattered field at large z contains contributions from many different contacts (Fig. 11.17). Replacing the sum over all scattering centers by an integral over

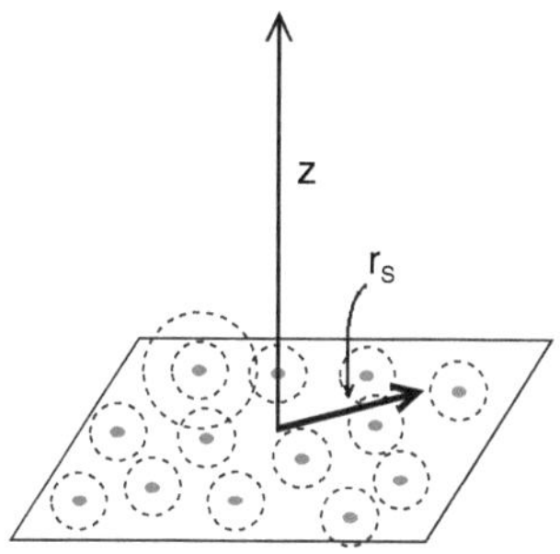

Fig. 11.17 The scattered field is a superposition of scattered waves originating from the different scattering centers. A large z, the sum over the different contributions can be replaced by an integral

area multiplied with the number density of particles and, further, using $r = (r_S^2 + z^2)^{1/2}$ with r_S the in-plane distance to the beam, one finds

$$\begin{aligned}\hat{u}_{sc}(z) &\approx \hat{u}_S N_P \tilde{b}_{sc} \int_0^\infty \frac{\exp\left(-\mathrm{i}k\sqrt{r_S^2+z^2}\right)}{\sqrt{r_S^2+z^2}} 2\pi r_S \mathrm{d}r_S \\ &= -\frac{2\pi\mathrm{i}}{k} N_P \tilde{b}_{sc} \hat{u}_S \exp(-\mathrm{i}kz)\end{aligned} \tag{11.7.2}$$

The term $-2\pi \mathrm{i} N_P \tilde{b}_{sc}/k$ should be identified with the quantity $(\tilde{r}_{q,S} - 1)$ in Eq. 4.4.12. Using Eq. 4.4.12, this amounts to

$$\frac{\Delta\tilde{f}}{f_0} \approx \frac{1}{2\pi\mathrm{i}}\left(\tilde{r}_{q,S} - 1\right) = \frac{1}{2\pi\mathrm{i}}\left(-2\pi\mathrm{i}\frac{1}{k}N_P\tilde{b}_{sc}\right) = -\frac{1}{k}N_P\tilde{b}_{sc} = -\frac{N_P}{k^2}k\tilde{b}_{sc} \tag{11.7.3}$$

As long as N_P/k^2 is not extremely large, one has $k\tilde{b}_{sc} \ll 1$ because $\left|\Delta\tilde{f}/f_0\right| \ll 1$. For elastic scattering, the scattering length is of the form [52]

$$\tilde{b}_{sc} = \frac{1}{2\mathrm{i}k}(\exp(2\mathrm{i}\delta_{sc}) - 1) \tag{11.7.4}$$

δ_{sc} is the scattering phase. Since $k\tilde{b}_{sc} \ll 1$, the scattering phase is small and one can Taylor-expand $\tilde{b}_{sc}$ to second order in δ_{sc} ($\exp(\varepsilon) \approx 1 + \varepsilon + \varepsilon^2/2$), leading to

$$\tilde{b}_{sc} \approx \frac{1}{k}\left(\delta_{sc} + \mathrm{i}\delta_{sc}^2\right) \tag{11.7.5}$$

The ratio of $\Delta\Gamma$ to Δf follows as

$$\frac{\Delta\Gamma}{\Delta f} \approx \delta_{sc} \tag{11.7.6}$$

Since δ_{sc} is small, elastic scattering has little influence on $\Delta\Gamma$.

11.8 Contact Stiffness of Rough Interfaces

The stiffness of rough interfaces is to be distinguished from the stiffness of discrete contacts. The geometry is shown in Fig. 11.18. Rough interfaces (multi-contact interfaces, MCIs) respond to tangential stress with an excess tangential displacement. The interface stiffness is the ratio of the tangential stress and the excess displacement. One might view the interface as a soft interlayer, but the thickness of this layer is not uniquely defined. The stiffness of the interface is normalized to area; it has dimensions of N/m^3. If the contact is established across asperities, which do not elastically interact, one can write $S = \Sigma\kappa_i/A$, where the index i labels the contacts and A is the total area (cf. Eq. 11.3.3 and Eq. 11.3.4). However, the contacts of MCIs usually do interact. The stiffness of MCIs is not easily calculated, but of considerable interest [54–56].

The prevalent method to determine the interfacial contact stiffness is ultrasonic reflectometry [57, 58]. One infers the stiffness, S, from the relation

$$|\tilde{r}| = \frac{1}{\sqrt{1 + \left(\frac{2S}{\omega Z}\right)^2}} \tag{11.8.1}$$

$\tilde{r}$ is the amplitude reflection coefficient and Z is the wave impedance of the adjacent media (assumed to be same on both sides of the interface). The above equation is derived in the following. It is a worthwhile exercise and builds on the discussion leading to Eq. 4.2.11. The interfacial contact stiffness can also be determined with a QCM, as shown later. Given that the QCM basically is a reflectometer, such a correspondence can be expected. Note: Eq. 11.8.1 formulates the problem in terms of the absolute value of the reflectivity. Other ultrasonic instruments determine the phase as well.

As usual, the analysis is based on the continuity of stress and displacement. Consider stress first. Stress is conserved, expressed by the relation

$$\hat{\sigma}_1^- + \hat{\sigma}_1^+ = \hat{\sigma}_{intf} = \hat{\sigma}_2^- \tag{11.8.2}$$

A shear wave with an associated stress of $\hat{\sigma}_1^-$ is assumed to enter from the left. The reflected wave has an amplitude of stress of $\hat{\sigma}_1^+$. $\hat{\sigma}_2^-$ is the stress of the transmitted wave; $\hat{\sigma}_{intf}$ is the stress at the interface.

Using that stress is the product of velocity and wave impedance (see Eq. 4.2.5), Eq. 11.8.2 turns into

$$i\omega\tilde{Z}_1\left(-\hat{u}_1^- + \hat{u}_1^+\right) = -i\omega\tilde{Z}_2\hat{u}_2^- \tag{11.8.3}$$

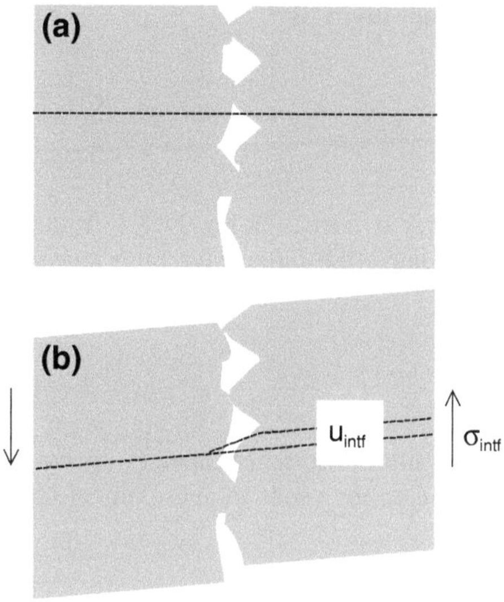

Fig. 11.18 The shear stiffness of rough interfaces is the ratio of the tangential stress, σ_{intf}, and the excess displacement, u_{intf}. One has $S = \sigma_{intf}/u_{intf}$. An analogous relation exists for the normal stiffness. Panels **a** and **b** show the undeformed and the deformed state, respectively

The amplitude of the transmitted wave is

$$\hat{u}_2^- = \frac{\tilde{Z}_1}{\tilde{Z}_2}\left(\hat{u}_1^- - \hat{u}_1^+\right) \tag{11.8.4}$$

Now turn to the displacement. Differing from Eq. 4.2.8, the continuity of displacement is expressed as

$$\hat{u}_1^- + \hat{u}_1^+ = \hat{u}_{intf} + \hat{u}_2^- \tag{11.8.5}$$

For the definition of the displacement at the interface, $\hat{u}_{intf}$, see Fig. 11.18. $\hat{u}_{intf}$ is related to the amplitudes of the different waves by

$$\hat{u}_{intf} = \frac{\hat{\sigma}_{intf}}{\tilde{S}} = \frac{\hat{\sigma}_2^-}{\tilde{S}} = \frac{-\mathrm{i}\omega\tilde{Z}_2}{\tilde{S}}\hat{u}_2^- = \frac{-\mathrm{i}\omega\tilde{Z}_1}{\tilde{S}}\left(\hat{u}_1^- - \hat{u}_1^+\right) \tag{11.8.6}$$

Equation 11.8.4 was used in the last step. Inserting Eq. 11.8.6 into Eq. 11.8.5, one finds

$$\hat{u}_1^- + \hat{u}_1^+ = \frac{-\mathrm{i}\omega\tilde{Z}_1}{\tilde{S}}\left(\hat{u}_1^- - \hat{u}_1^+\right) + \frac{\tilde{Z}_1}{\tilde{Z}_2}\left(\hat{u}_1^- - \hat{u}_1^+\right) \tag{11.8.7}$$

and

$$\ddot{u}_1^+\left(-\frac{\mathrm{i}\omega\tilde{Z}_1}{\tilde{S}} + \frac{\tilde{Z}_1}{\tilde{Z}_2} + 1\right) = \hat{u}_1^-\left(-\frac{\mathrm{i}\omega\tilde{Z}_1}{\tilde{S}} + \frac{\tilde{Z}_1}{\tilde{Z}_2} - 1\right) \tag{11.8.8}$$

The reflectivity is

$$\tilde{r} = \frac{\hat{u}_1^+}{\hat{u}_1^-} = \frac{-\frac{i\omega\tilde{Z}_1}{\tilde{S}} + \frac{\tilde{Z}_1}{\tilde{Z}_2} - 1}{-\frac{i\omega\tilde{Z}_1}{\tilde{S}} + \frac{\tilde{Z}_1}{\tilde{Z}_2} + 1} \tag{11.8.9}$$

In order to arrive at Eq. 11.8.1 from Eq. 11.8.9, assume that the two adjacent media have the same wave impedance ($\tilde{Z}_1 = \tilde{Z}_2 = \tilde{Z}$), which leads to

$$\tilde{r} = \frac{-\frac{i\omega\tilde{Z}}{\tilde{S}}}{2 - \frac{i\omega\tilde{Z}}{\tilde{S}}} = \frac{1}{1 + i\frac{2\tilde{S}}{\omega\tilde{Z}}} \tag{11.8.10}$$

Some ultrasonic reflectometers do not determine the phase. If the phase is not available, one needs to approximate both $\tilde{Z}$ and $\tilde{S}$ as real, which leads to

$$|\tilde{r}| = \frac{1}{\sqrt{1 + \left(\frac{2S}{\omega Z}\right)^2}} \tag{11.8.11}$$

Too see that the QCM functions in essentially the same way, start from Eq. 4.4.12 and insert $\tilde{r}$ from Eq. 11.8.9 for $\tilde{r}_{q,S}$.

$$\begin{aligned}\frac{\Delta\tilde{f}}{f_0} &= \frac{1}{2\pi i}\left(r_{q,S} - 1\right) = \frac{1}{2\pi i}\left(\frac{-\frac{i\omega\tilde{Z}_1}{\tilde{S}} + \frac{\tilde{Z}_1}{\tilde{Z}_2} - 1}{-\frac{i\omega\tilde{Z}_1}{\tilde{S}} + \frac{\tilde{Z}_1}{\tilde{Z}_2} + 1} - 1\right) = \frac{i}{\pi}\left(\frac{-1}{-\frac{i\omega Z_q}{\tilde{S}} + \frac{Z_q}{\tilde{Z}_2} + 1}\right) \\ &= \frac{i}{\pi}\left(\frac{\tilde{S}}{i\omega Z_q - \frac{\tilde{S}Z_q}{\tilde{Z}_2} - \tilde{S}}\right)\end{aligned} \tag{11.8.12}$$

$\tilde{Z}_1$ has been replaced by Z_q in step 3.

Equation 11.8.12 only holds if $|\Delta\tilde{f}| \ll f_0$. It can only be applied if the interfacial contact stiffness is small in the sense of $|\tilde{S}| \ll \omega Z_q$ (Equation 11.8.11 does not suffer from this limitation). If $|\tilde{S}| \ll \omega Z_q$ and if, further, $Z_q \approx \tilde{Z}_2$, the denominator in Eq. 11.8.12 can be replaced by $i\omega Z_q$, leading to

$$\frac{\Delta\tilde{f}}{f_0} \approx \frac{1}{\pi Z_q}\frac{\tilde{S}}{\omega} \tag{11.8.13}$$

If the contacts are all elastically independent (they usually are not), one has $\tilde{S} = N_P\tilde{\kappa}_P$, which results in

$$\frac{\Delta\tilde{f}}{f_0} \approx \frac{1}{\pi Z_q}\frac{\tilde{S}}{\omega} = \frac{1}{\pi Z_q}\frac{N_P\tilde{\kappa}_P}{\omega} \tag{11.8.14}$$

This is the familiar result known from the treatment of discrete contacts in Sect. 11.3.

If the small-load approximation is to be avoided, one can follow Eq. 4.6.8 and write

$$\tan\left(\pi\frac{\Delta\tilde{f}}{f_0}\right) \approx \mathrm{i}\frac{1 - r_{q,S}}{1 + r_{q,S}} \tag{11.8.15}$$

Inserting Eq. 11.8.9 into Eq. 11.8.15, an implicit equation in $\Delta\tilde{f}$ and $\tilde{S}$, is obtained, which can be solved numerically for $\tilde{S}$.

Two caveats:

- Using ultrasound, both the normal contact stiffness and the tangential contact stiffness can be determined, employing the suitably-polarized acoustic waves. With the QCM, one mostly determines the tangential contact stiffness, but there may be a contribution form the normal stiffness because of the flexural deformation.
- Both ultrasonic reflectometry and the QCM-based measurement neglect inertial effects. This is permissible as long as the contact size is much below the wavelength of sound. The assumption is called quasi-static approximation (QSA) [59] in the context of ultrasound.

Glossary

Variable	Definition (Comments)
0	As in index: undamped (Exception: f_0)
a	Contact radius
A	(Effective) area of the resonator plate (See Sect. 7.4)
$\tilde{b}_{sc}$	Scattering length (Sect. 11.7)
b_{shear}, b_{bend}	Numerical coefficients (Eq. 11.5.4)
D	Diffusivity (Sect. 11.6)
E	Young's modulus
E^*	Effective Young's modulus (Eq. 11.2.2)
f	Frequency

f_{OS}	Oscillator strength (a number, not a frequency)
f_{ZC}	Frequency of zero-crossing (Eq. 11.4.7)
f_0	Resonance frequency at the fundamental ($f_0 = Z_q/(2m_q) = Z_q/(2\rho_q d_q)$)
F_N	Normal force
F_x	Tangential force
g	Standard acceleration ($g = 9.81$ m/s^2)
G	Shear modulus
G^*	Effective shear modulus (Eq. 11.2.12)
het	As an index: *het*erogeneously broadened (Sect. 11.4)
intf	As in index: *interface* (Sect. 11.8)
I	Moment of area (Eq. 11.2.12)
k	Wavenumber
k_BT	Thermal energy
liq	As an index: *liq*uid
M	Mass
n	Overtone order
N_P	Number of particles per unit area
p	Normal stress
P	As an index: *P*article
PD	As an index: *P*olar *D*iagram
r	Distance from the center of a contact, distance from a scattering center
rock	As in index: *rock*ing mode
rot	As in index: *rotational* mode
R_P	Particle radius
R_{PD}	Radius of circle in polar diagram (Eq. 11.5.7)
$\tilde{r}_{q,S}$	Reflectivity evaluated at the resonator surface (Sect. 11.7)
$\mathbf{r_S}$	Location at the resonator surface
r_S	In-plane distance from the scattering center (Sect. 11.7)
S	As an index: *S*urface
$\tilde{S}$	Contact stiffness (Sect. 11.8)
sc	As an index: *sc*attered (Sect. 11.7)

$\hat{u}, u$	Tangential displacement
$\hat{v}$	Velocity
$\tilde{Z}_L$	Load impedance
Z_q	Acoustic wave impedance of AT-cut quartz ($Z_q = 8.8 \times 10^6$ kg m^{-2} s^{-1})
γ_P	Damping factor of a coupled resonance (Sect. 11.4)
γ_S	Surface energy (Sect. 11.2)
Γ	Imaginary part of a resonance frequency
δ_N	Normal compression (Fig. 11.2)
δ_{scm}	Scattering phase (Sect. 11.7)
Δ	As a prefix: A shift induced by the presence of the sample
κ	Spring constant
μ	Friction coefficient ($\mu = F_x/F_N$)
ν	Poisson number
θ	Angle of rotation
ρ	Density
σ	Tangential stress
τ_{MR}	Momentum relaxation time (also: slip time Sect. 11.6)
τ_S	A constant tangential stress in the sliding zone of a contact experiencing partial slip (Sect. 11.2)
ξ	Drag coefficient
ω	Angular frequency
ω_P	Particle resonance frequency

References

1. Luan, B.Q., Robbins, M.O.: The breakdown of continuum models for mechanical contacts. Nature **435**(7044), 929–932 (2005)
2. Johnson, K.L.: Contact Mechanics. Cambridge University Press, Cambridge (1985)
3. Popov, V.L.: Contact Mechanics and Friction: Physical Principles and Applications. Springer, Berlin (2010)
4. Wriggers, P.: Computational Contact Mechanics. Springer, Berlin (2006)
5. Hertz, H.R.: Üeber die Berüehrung elastischer Köerper. Gesammelte Werke, vol. 1, p.1895. J.A. Barth, Leipzig (1882)

6. Johnson, K.L., Kendall, K., Roberts, A.D.: Surface energy and contact of elastic solids. In: Proceedings of the Royal Society of London Series a-Mathematical and Physical Sciences, vol. 324, no. 1558, p. 301 (1971)
7. del Campo, A., Greiner, C., Arzt, E.: Contact shape controls adhesion of bioinspired fibrillar surfaces. Langmuir **23**(20), 10235–10243 (2007)
8. Hanke, S., Petri, J., Johannsmann, D.: Partial slip in mesoscale contacts: dependence on contact size. Phys. Rev. E **88**(3), 032408 (2013)
9. Cattaneo, C.: Sul contatto di due corpi elastici: distribuzione locale dei sforzi. Rendiconti dell' Academia Nationale dei Lincei **27**, 474–478 (1938)
10. Mindlin, R.D., Deresiewicz, H.: Elastic spheres in contact under varying oblique forces. J. Appl. Mech.-Trans. Asme **20**(3), 327–344 (1953)
11. Berthier, Y., Vincent, L., Godet, M.: Fretting fatigue and fretting wear. Tribol. Int. **22**(4), 235–242 (1989)
12. Varenberg, M., Etsion, I., Halperin, G.: Nanoscale fretting wear study by scanning probe microscopy. Tribol. Lett. **18**(4), 493–498 (2005)
13. Savkoor, A.R.: Dry Adhesive Contact of Elastomers. Tech. University Delft, The Netherlands (1987)
14. Johnson, K.L.: Adhesion and friction between a smooth elastic spherical asperity and a plane surface. Proc. R. Soc. Lond. A: Math. Phys. Eng. Sci. **1997**(453), 163–179 (1956)
15. Dybwad, G.L.: A sensitive new method for the determination of adhesive bonding between a particle and a substrate. J. Appl. Phys. **58**(7), 2789–2790 (1985)
16. Laschitsch, A., Johannsmann, D.: High frequency tribological investigations on quartz resonator surfaces. J. Appl. Phys. **85**(7), 3759–3765 (1999)
17. Vittorias, E., Kappl, M., Butt, H.J., Johannsmann, D.: Studying mechanical microcontacts of fine particles with the quartz crystal microbalance. Powder Technol. **203**(3), 489–502 (2010)
18. Akarapu, S., Sharp, T., Robbins, M.O.: Stiffness of contacts between rough surfaces. Phys. Rev. Lett. **106**(20), 204301 (2011)
19. Leopoldes, J., Jia, X.: Probing viscoelastic properties of a thin polymer film sheared between a beads layer and an ultrasonic resonator. EPL **88**(3), 34001 (2009)
20. Eriten, M., Lee, C.H., Polycarpou, A.A.: Measurements of tangential stiffness and damping of mechanical joints: direct versus indirect contact resonance methods. Tribol. Int. **50**, 35–44 (2012)
21. Rabe, U., Amelio, S., Kester, E., Scherer, V., Hirsekorn, S., Arnold, W.: Quantitative determination of contact stiffness using atomic force acoustic microscopy. Ultrasonics **38**(1–8), 430–437 (2000)
22. D'Amour, J.N., Stalgren, J.J.R., Kanazawa, K.K., Frank, C.W., Rodahl, M., Johannsmann, D.: Capillary aging of the contacts between glass spheres and a quartz resonator surface. Phys. Rev. Lett. **96**(5), 058301 (2006)
23. Hornbaker, D.J., Albert, R., Albert, I., Barabasi, A.L., Schiffer, P.: What keeps sandcastles standing? Nature **387**(6635), 765 (1997)
24. Mason, T.G., Levine, A.J., Ertas, D., Halsey, T.C.: Critical angle of wet sandpiles. Phys. Rev. E **60**(5), R5044–R5047 (1999)
25. Butt, H.J., Graf, K., Kappl, M.: Physics and Chemistry of Interfaces. Wiley-VCH, Berlin (2006)
26. Bocquet, L., Charlaix, E., Ciliberto, S., Crassous, J.: Moisture-induced ageing in granular media and the kinetics of capillary condensation. Nature **396**(6713), 735–737 (1998)
27. Persson, B.N.J.: Sliding Friction: Physical Principles and Applications. Springer, Berlin (2000)
28. Berthoud, P., Baumberger, T.: Role of asperity creep in time- and velocity-dependent friction of a polymer glass. Europhys. Lett. **41**(6), 617–622 (1998)
29. Dieterich, J.H., Kilgore, B.D.: Direct observation of frictional contacts—new insights for state-dependent properties. Pure. Appl. Geophys. **143**(1–3), 283–302 (1994)

30. Pomorska, A., Shchukin, D., Hammond, R., Cooper, M.A., Grundmeier, G., Johannsmann, D.: Positive frequency shifts observed upon adsorbing micron-sized solid objects to a quartz crystal microbalance from the liquid phase. Anal. Chem. **82**(6), 2237–2242 (2010)
31. Berglin, M., Olsson, A., Elwing, H.: The interaction between model biomaterial coatings and nylon microparticles as measured with a quartz crystal microbalance with dissipation monitoring. Macromol. Biosci. **8**(5), 410–416 (2008)
32. Fatisson, J., Domingos, R.F., Wilkinson, K.J., Tufenkji, N.: Deposition of TiO(2) nanoparticles onto silica measured using a quartz crystal microbalance with dissipation monitoring. Langmuir **25**(11), 6062–6069 (2009)
33. Zhang, Q.L., Lec, R.M., Pourrezaei, K.: The study of an interaction of solid particles with various surfaces using TSM sensors. IEEE Trans. Ultrason. Ferroelectr. Freq. Control **53**(1), 167–174 (2006)
34. Olofsson, A.C., Hermansson, M., Elwing, H.: Use of a quartz crystal microbalance to investigate the antiadhesive potential of N-acetyl-L-cysteine. Appl. Environ. Microbiol. **71**(5), 2705–2712 (2005)
35. Olsen, E.V., Pathirana, S.T., Samoylov, A.M., Barbaree, J.M., Chin, B.A., Neely, W.C., Vodyanoy, V.: Specific and selective biosensor for Salmonella and its detection in the environment. J. Microbiol. Methods **53**(2), 273–285 (2003)
36. Olsson, A.L.J., van der Mei, H.C., Busscher, H.J., Sharma, P.K.: Acoustic sensing of the bacterium-substratum interface using QCM-D and the influence of extracellular polymeric substances. J. Colloid Interface Sci. **357**(1), 135–138 (2011)
37. Poitras, C., Tufenkji, N.: A QCM-D-based biosensor for E.coli O157:H7 highlighting the relevance of the dissipation slope as a transduction signal. Biosens. Bioelectron. **24**(7), 2137–2142 (2009)
38. Su, X.L., Li, Y.B.: A QCM immunosensor for Salmonella detection with simultaneous measurements of resonant frequency and motional resistance. Biosens. Bioelectron. **21**(6), 840–848 (2005)
39. Vaughan, R.D., O'Sullivan, C.K., Guilbault, G.G.: Development of a quartz crystal microbalance (QCM) immunosensor for the detection of Listeria monocytogenes. Enzyme Microb. Technol. **29**(10), 635–638 (2001)
40. Molino, P.J., Hodson, O.A., Quinn, J.F., Wetherbee, R.: The quartz crystal microbalance: a new tool for the investigation of the bioadhesion of diatoms to surfaces of differing surface energies. Langmuir **24**(13), 6730–6737 (2008)
41. Poitras, C., Fatisson, J., Tufenkji, N.: Real-time microgravimetric quantification of Cryptosporidium parvum in the presence of potential interferents. Water Res. **43**(10), 2631–2638 (2009)
42. Olsson, A.L.J., van der Mei, H.C., Johannsmann, D., Busscher, H.J., Sharma, P.K.: Probing colloid-substratum contact stiffness by acoustic sensing in a liquid phase. Anal. Chem. **84**(10), 4504–4512 (2012)
43. D'Amour, J.N., Stalgren, J.J.R, Kanazawa, K.K., Frank, C.W., Rodahl, M., Johannsmann, D.: Capillary aging of the contacts between glass spheres and a quartz resonator surface. Phys. Rev. Lett. **96**(5), 058301 (2006)
44. Du, B.Y., Konig, A.M., Johannsmann, D.: On the role of capillary instabilities in the sandcastle effect. New J. Phys. **10** 053014 (2008)
45. Forrest, J.A., Mattsson, J., Borjesson, L.: Using adhesion to probe viscoelasticity of polymer film surfaces: a quartz crystal microbalance study. Eur. Phys. J. E **8**(2), 129–136 (2002)
46. Vigasin, A.A.: Intensity and bandshapes of collision-induced absorption by CO2 in the region of the Fermi doublet. J. Mol. Spectrosc. **200**(1), 89–95 (2000)
47. Krim, J., Solina, D.H., Chiarello, R.: Nanotribology of a Kr Monolayer—a quartz-crystal microbalance study of atomic-scale friction. Phys. Rev. Lett. **66**(2), 181–184 (1991)
48. Watts, E.T., Krim, J., Widom, A.: Experimental-observation of interfacial slippage at the boundary of molecularly thin-films with gold substrates. Phys. Rev. B **41**(6), 3466–3472 (1990)

49. Bruschi, L., Carlin, A., Mistura, G.: Depinning of atomically thin Kr films on gold. Phys. Rev. Lett. **88**(4), 046105 (2002)
50. Krim, J.: Friction and energy dissipation mechanisms in adsorbed molecules and molecularly thin films. Adv. Phys. **61**(3), 155–323 (2012)
51. Chandler, D.: Introduction to modern statistical mechanics. Oxford University Press, Oxford (1987)
52. Landau, L.D., Lifshitz, E.M.: Quantum Mechanics, Non-Relativistic Theory. Elsevier, Oxford (1997)
53. Rayleigh, L.: Theory of Sound. Macmillan, London (1894). Note: The theory of scattering underlying section 11.7 was developed by the 3rd Lord Rayleigh to explain acoustic scattering. Admittedly, Rayleigh was concerned with compressional waves, not shear waves
54. Baumberger, T., Caroli, C.: Solid friction from stick-slip down to pinning and aging. Adv. Phys. **55**(3–4), 279–348 (2006)
55. Akarapu, S., Sharp, T., Robbins, M.O.: Stiffness of Contacts between Rough Surfaces. Phys. Rev. Lett. **106**(20), 204301 (2011)
56. Persson, B.N.J.: Contact mechanics for randomly rough surfaces. Surf. Sci. Rep. **61**(4), 201–227 (2006)
57. Kendall, K., Tabor, D.: Ultrasonic study of area of contact between stationary and sliding surfaces. Proc. R. Soc. Lond. A: Math. Phys. Sci. **323**(1554), 321 (1971)
58. Krolikowski, J., Szczepek, J.: Assessment of tangential and normal stiffness of contact between rough surfaces using ultrasonic method. Wear **160**(2), 253–258 (1993)
59. Baik, J.-M., Thompson, R.B.: Ultrasonic scattering from imperfect interfaces: a quasi-static model. J. Nondestr. Eval. **4**, 177 (1984)

Chapter 12
Heterogeneous Samples

Abstract When the sample is structured in the plane of the resonator with a characteristic scale comparable to the wavelength of sound, analytical predictions of the displacement field and the frequency shift are difficult. Among the samples that are heterogeneous in this sense are nanobubbles, nanodroplets, nanoparticles, vesicles, and biological cells. In analyzing such samples, one can rely on common sense and empirical correlations. If one wants to go beyond those more qualitative pictures, one can calculate the area-averaged periodic stress at the resonator surface numerically. An example of a numerical method is discussed in detail. The finite element method (FEM) is employed to solve the incompressible Stokes problem and to predict the periodic interfacial stress. The frequency shift follows from the area-averaged stress and the SLA.

12.1 Laterally Heterogeneous Interfaces

Structured samples have been mentioned a few times already, and we therefore start with a list of problems, which this chapter is *not* about. It is, firstly, not about heterogeneity along the vertical, that is, layer systems. Stratified layers can be treated within the acoustic multilayer formalism, regardless of how thin the individual layers are. The chapter is, secondly, not about samples, the properties of which vary laterally on a scale much larger than λ (λ the wavelength of sound). Such samples can be treated as acoustic multilayers, locally, because the in-plane gradients can be neglected. Since the load impedance as derived by the multilayer formalism varies over the resonator surface, averaging with a suitable statistical weight (Eq. 6.1.31) must be applied. However, averaging does take care of heterogeneity; there is no need to abandon the multilayer formalism. Thirdly, materials with some random structure on a scale below λ are outside of this chapter. Colloidal dispersions (Sect. 9.3) are an example. Such materials should be treated within an effective medium theory. Finally, the advanced schemes discussed below are not needed for point contacts as described in chap. 11. Point contacts—by

D. Johannsmann, *The Quartz Crystal Microbalance in Soft Matter Research*,
Soft and Biological Matter, DOI: 10.1007/978-3-319-07836-6_12,

definition—are so small that one may integrate over the stress distribution and obtain a transverse force or a torque. These samples can be understood in the frame of a lumped element representation.

There is a large number of interesting samples falling outside these classes of heterogeneity. Vesicles, bacteria, other biological cells, nanobubbles, and many of the biocolloids display in-plane structure on the scale of λ. Neither the plane-wave picture (possibly adapted to a contact in the center only or to shallow roughness, see Sects. 9.4 and 9.5) nor the point-contact limit (Chap. 11) can be applied. Simplifying assumptions, which would allow for an analytical treatment, are not available. Apart from the complexity, there often is a second problem, which is insufficient knowledge of the geometry. The models described below need the geometry as an input. But for some cases, reasonable guesses can be made. Vesicles (Sect. 12.5), viruses, [1] proteins, [2] and other nanoparticles [3] have been characterized with other methods. Their size and shape is known or can be guessed with some confidence. If reasonable assumptions on the geometry and the viscoelastic parameters are available, one may try one's luck with a numerical model.

Typical questions that can be addressed by those computations are:

- What is the height of the adsorbed objects? (The height is different from the Sauerbrey thickness, if coverage is incomplete.)
- To what extent are the adsorbed objects clustered?
- What is the contribution of the trapped liquid to the frequency shift?
- If energy is dissipated in the adsorbed layer: Where is it dissipated? At the link? Inside the adsorbed object? Or rather in the liquid phase?
- Are there concentrations of stress at some locations? Can one expect rupture or slip at these places?
- If there is a membrane, what is the contribution of membrane bending to the dissipation of energy?
- If there is a liquid-liquid interface or a liquid-gas interface, to what extent does interfacial energy affect the flow profile and the frequency shift?

It will be emphasized below, that the existing code can only give preliminary answers. But the model as such exists. As the computational resources advance, such simulations will become more practical. In the following section, we give an overview of the essential steps.

12.2 Numerical Computation of Area-Averaged Stress

If the geometry and the viscoelastic constants of the object under study are known, the input to a numerical simulation is at hand. One computes the velocity field by solving the wave equation for the respective geometry. The software will often be a PDE-solver of some kind, where "PDE" stands for "partial differential equation". The PDE-solver might make use of the finite element method (FEM).

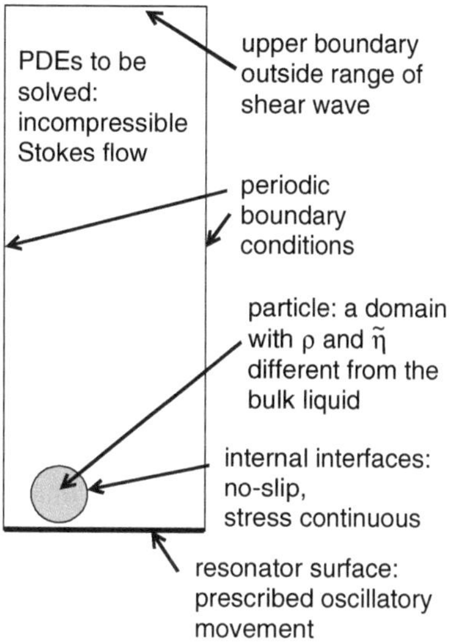

Fig. 12.1 In order to compute the periodic stress at the resonator surface using a PDE solver, the user has to specify the partial differential equation, the continuity conditions at the internal interfaces, and the boundary conditions at the border of the simulation cell. The existing code only works in two dimensions. What is called "particle", is a cross-section through an infinite cylinder

From the flow field, one calculates the stress field at the resonator surface and averages over the surface area. Note: Any numerical technique which predicts the stress at the resonator surface will work. The FEM-based methodology described below it nothing but a first try and a proof of principle. Computational fluid dynamics offers a wide variety of methods for treating problems of this kind. Even molecular dynamics calculations are conceivable [4].

The problem to be solved is a boundary value problem (BVP) as sketched in Fig. 12.1. A caveat: The existing code only works in two dimensions. The "particle" in Fig. 12.1 is a cross-section through an infinite cylinder, not a sphere. The velocity can be in the plane of the drawing (the xz-plane) or perpendicular to it (along y), but not at an angle. Evidently, these particular calculations are not truly realistic.

The elements of a BVP are the partial differential equation(s) (PDEs) to be solved and the boundary conditions. Consider the PDEs first. Soft matter at a QCM surface undergoes an incompressible Stokes flow. Incompressibility can be assumed because compressional sound has a wavelength much longer than the penetration depth of the shear wave. A Stokes flow (rather than a solution to the Navier-Stokes equation) can be assumed because the amplitude is small. The incompressible Stokes flow obeys the following set of equations:

$$\begin{aligned} \rho \frac{\partial \mathbf{v}}{\partial t} &= \eta \nabla^2 \mathbf{v} - \nabla p \\ \nabla \cdot \mathbf{v} &= 0 \end{aligned} \qquad (12.2.1)$$

For periodic motion, Eq. 12.2.1 can be written in the frequency domain as

$$\begin{aligned} i\omega\rho\hat{\mathbf{v}} &= \tilde{\eta}\nabla^2\hat{\mathbf{v}} - \nabla\hat{p} \\ \nabla\cdot\hat{\mathbf{v}} &= 0 \end{aligned} \tag{12.2.2}$$

Velocity and pressure have turned into complex amplitudes. (Note: Pressure is not truly a dynamic variable in these calculations. It takes whatever value is needed to keep the density constant.) ρ and $\tilde{\eta}$ are materials parameters depending on position, $\mathbf{r}$. They might be constant inside certain domains (where a domain would correspond to a particle, a film, or a liquid) or be continuous functions of $\mathbf{r}$ (which would be typical for a polymer brush). Since the viscosity is complex, the materials involved are allowed to be viscoelastic. A "rigid" sphere would be composed of a material with a large, imaginary viscosity. The complex shear modulus is given as $\tilde{G} = i\omega\tilde{\eta}$. Viscoelastic dispersion is not a problem, since $\tilde{\eta}$ can be made to depend on overtone order.

In Eq. 12.2.2 the velocity is assumed to be in the plane of the simulation. It is also possible to let all motion occur perpendicular to that plane. The velocity then only has a y–component, v_y. v_y depends on x and z, but not on y itself. Since $\partial v_y/\partial y = 0$, incompressibility is automatically fulfilled and there is no need for a hydrostatic pressure to ensure constant density. The equation to be solved simply is

$$i\omega v_y = \frac{\tilde{\eta}}{\rho}\nabla^2 v_y \tag{12.2.3}$$

This PDE is solved by COMSOL's steady state diffusion model. v_y must be renamed as c (concentration); $\tilde{\eta}/\rho$ (the kinematic viscosity) must be renamed as D (a diffusivity, which fortunately is allowed to be complex). The term on the left-hand-side is a source term (called "reaction rate", also allowed to be complex by COMSOL). Clearly, this computation is simpler than the one solving Eq. 12.2.2. The discussion below is phrased such that it applies to Eq. 12.2.2, but all remarks can be reformulated to apply to Eq. 12.2.3.

How can a model designed to solve the Navier-Stokes equation be applied to the Stokes equation and, further, how can a model designed to calculate steady flows be made to compute periodic flows? The Stokes equation differs from the Navier-Stokes equation (cf. Sect. 13.2) in that the advected-momentum term is absent. The advected momentum term can be effectively removed from the Navier-Stokes equation by making the amplitude small. The advected momentum term is not strictly absent at low amplitudes, but it is negligible for amplitudes lower than 10^{-10} m. The module then solves the Stokes problem (This problem does not arise with the steady state diffusion model solving Eq. 12.2.3 because the diffusion equation does not contain a nonlinear term.) A *periodic* flow pattern can be computed using a "steady state" module because both modules (Navier-Stokes and diffusion) allow for complex-valued solutions. Why this is so, is unclear to the author. He just (gratefully) accepts it. The complex fields may represent the complex amplitudes of periodic solutions. This leaves the problem of somehow accounting for inertia. Inertia is introduced by inserting the term $i\rho\omega\hat{\mathbf{v}}$ as a body force. A typical body force would be gravity, but if $i\rho\omega\hat{\mathbf{v}}$ is entered in the respective field, the module still works and it solves the periodic Stokes problem. (In the diffusion model, inertia—more precisely: acceleration—is entered in the form of a reaction rate.)

We now turn to the boundary conditions. Importantly, the simulation box as shown in Fig. 12.1 is only a few microns wide; it does not contain the resonator itself. The bottom of the cell is a small fraction of the resonator surface. That must be so because the sample has structure on the nanometer scale, while the resonator is a few millimeters wide. The finite element model cannot cover small length scales and large length scales at the same time. However, a multi-scale model is not actually needed. It suffices to let the software compute the stress at the bottom of a small simulation volume. As long as the small volume is representative of the entire sample, the stress averaged over the small surface suffices for the SLA to be applicable. The SLA provides for a link between the different spatial scales.

The boundary conditions of the problem are the following:

- The displacement is prescribed at the bottom of the cell (the resonator surface). It is constant over the bottom surface and given as $\hat{u}_S \cos(\omega t)$.
- Periodic boundary conditions apply at the external boundaries to the left and to the right.
- Stress and displacement are continuous at all internal interfaces.
- The boundary condition at the top is irrelevant because the height of the cell is made to exceed the penetration depth of the shear wave. Since the liquid is at rest at the top, the boundary condition at the top has no influence on the flow field at the bottom.

The model reproduces the Sauerbrey equation and Gordon-Kanazawa-Mason result to an excellent accuracy even with a coarse mesh. More interestingly, it reproduces the frequency shifts induced by a corrugated surface according to Ref. [5]. For details see the supporting information to Ref. [9]. The calculations from Ref. [5] lead to the shallow-roughness effects as discussed in Sect. 9.5.

There are technical complications, not discussed here in detail. For instance, there are stress singularities at corners, which require attention. Further limitations of the current code are the following:

- The simulation occurs in 2D.
- The model cannot deal with *random* distributions of particles [6, 7]. Remember the boundary conditions to the right and to the left: All geometries represent periodic arrays. In order to cover larger assemblies, some kind of coarse graining scheme would be needed. In a way, the work reported in Ref. [2] amounts to a coarse-graining approach. Assumptions were made on, firstly, the interactions between the molecules and the substrate and on, secondly, the interactions between neighboring molecules. In a second step, molecules were

deposited randomly on the resonator surface and the frequency shift was predicted from the interactions according to the assumption and the geometry. If the elementary interactions can be calibrated against a realistic calculation, such a procedure will link first principles to experiment, but such a coarse-graining scheme is not available at present.

- The boundaries are fixed. A particle can never slide, roll or detach. If it does in reality, the code will not catch the consequences [8].
- Boundaries do not have an interfacial energy themselves, they do not exert a capillary pressure.

Those limitations are not of a fundamental kind. They can be overcome by better code.

A detailed step-by-step account of the existing code is outside the scope of the book. Reference [9] contains details. The following paragraphs are intended to give the reader an understanding of the essential steps.

The adsorption of ferritin molecules to a gold surface was chosen as the toy system. Ferritin is an iron storage protein with a roughly spherical shape and a diameter of 12 nm [10]. It has been discussed as model system for bio- and nanotechnology by a different authors [11–13]. AFM images had been acquired in situ on the exact same samples. These showed that the ferritin molecules were distributed evenly on the surface. The surface coverage could be inferred from the AFM micrographs.

While the geometry of the adsorbate is known (from the literature) and the coverage is known (from the AFM images), it is still an open questions, how to represent contact to the substrate in the FEM simulation. At this point, the simulation was guided by the experiment. Experiment showed that, firstly, Δf was negative and that, secondly, Δf was larger in magnitude than $\Delta\Gamma$. Motivated by this knowledge, the link was modeled as a truncation of an otherwise spherical (strictly speaking: cylindrical) particle. The width of the link was 6 nm, which is half the diameter of the particle. Had the link been made much narrower, a coupled resonance would have resulted (Sect. 11.5.1).

Figure 12.2 shows the geometry and the mesh. The figure was cut off at the top for clarity. The cell extended to $z = 2$ μm, which is well beyond the penetration depth of the shear wave. The width of the cell was 20 nm. With a particle diameter of 12 nm, this amounts to a surface coverage of 60 %. The coverage was varied (see Fig. 12.5). A further noteworthy feature is the resolution of the mesh close to the edge of the contact. Because of the stress singularity (Fig. 12.4), the mesh must be fine there. The density of the liquid was chosen as 1 g/cm^3; its viscosity was 1 mPa s. The density of the ferritin molecules was chosen as either 1.6 g/cm^3 (corresponding to the state, where the protein is loaded with iron) or 1 g/cm^3 (corresponding to the unloaded state, in which case the protein is called apoferritin). The storage modulus was chosen as $G' = 1$ GPa. The loss angle was $\delta_L = 5°$. These latter choices are somewhat arbitrary. The molecules were modeled as "rigid". Choosing $G' = 0.1$ GPa rather than 1 GPa would have not made a noticeable difference (but choosing it as 1 MPa would).

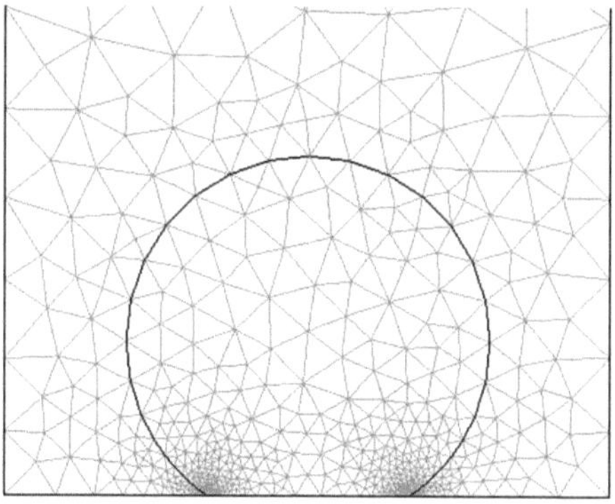

Fig. 12.2 An example of a geometry together with the mesh. The resonator surface is at the *bottom*. The truncated sphere represents an adsorbed ferritin molecule. Adapted from Ref. [9]

Figure 12.3 shows raw outputs. Panels a and b display the vertical and the horizontal components of the velocity as calculated from the Navier-Stokes equation. More precisely, what is shown are the real parts of these velocities, that is, the components of the velocity which are in-phase with the movement of the resonator surface. The velocities are complex amplitudes. Panel c shows the hydrostatic pressure. It is the imaginary part in this case. The hydrostatic pressure mostly is in-phase with the displacement, that is, out-of-phase relative to the velocity of the substrate. Note the peaks in pressure close to the edge of the contact.

The next step towards the frequency shift is the calculation of the stress at the resonator surface as shown in Fig. 12.4. The stress is supplied by COMSOL as part of the post-processing. There is a stress singularity at the edge, also discussed in Sect. 11.2. Most of the stress occurs close to the edge. Differing from the discussion in Sect. 13.3, there is no partial slip. There is no partial slip in the calculation, because all boundaries are fixed. Whether or not there is partial slip in the experiment, cannot be judged based on the simulation. However, experiment did not reveal an amplitude dependence of $\Delta\tilde{f}$, which is evidence against partial slip.

A side remark for the mathematically inclined readers: Since the singularity is an inverse square-root singularity, the integral is finite. Even though the stress is formally infinite at the edge, the integrated force is not. The stress concentration leads to technical difficulties, but it does does invalidate the calculation as such.

Hardly visible in Fig. 12.4 is a second feature of interest: The stress exerted by the liquid outside of the contact is not strictly constant; it depends on the distance to the edge of the contact. Some of the force exerted by the particle onto the surface is transmitted across the liquid. The effect is small here, but it is clearly significant for cases, where the link between the particle and the substrate is narrow. Remember the high frequency: The stress transmitted across a liquid scales as $\omega\eta$, which for water at 5 MHz amounts to about 4×10^4 Pa.

Integration over the stress profile shown in Fig. 12.4 produces the force. Dividing the force by the width of the cell produces the average stress, which—upon insertion into the SLA leads to the frequency shift, shown in Fig. 12.5. Note

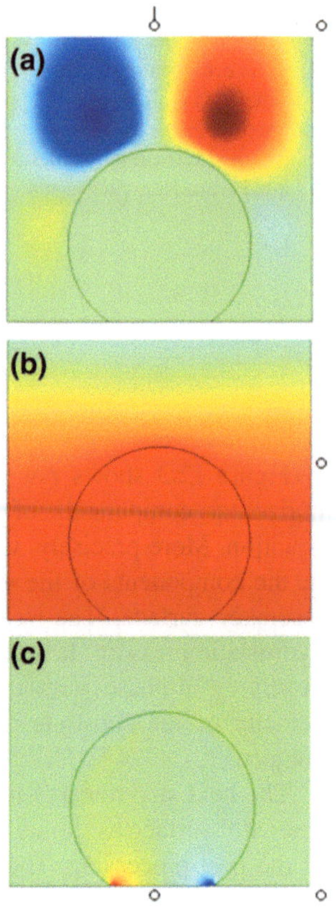

Fig. 12.3 Raw output of the finite element model shown in Fig. 12.2. The flow is in the plane of the paper; Eq. 12.2.2 is the PDE applied. Panels **a** and **b** and **c** show the vertical component of the velocity, the horizontal component of velocity, and the pressure, respectively. Adapted from Ref. [9]

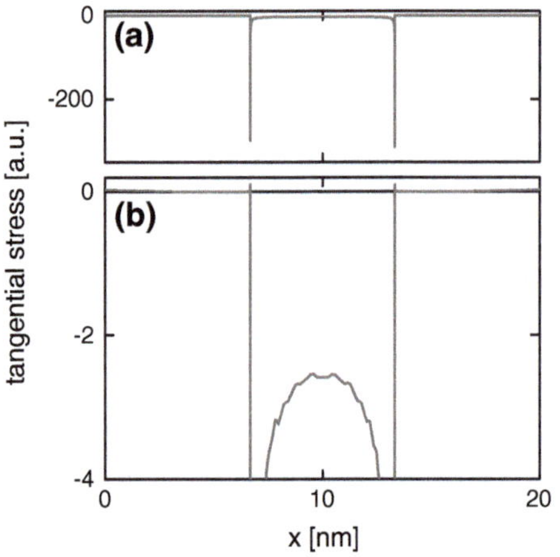

Fig. 12.4 Local tangential stress at the *bottom* of the simulation cell as shown in Fig. 12.2. Note the stress peaks close to the edge of the contact. Panel **b** shows the same data as panel **a** with the y-scale expanded. Adapted from Ref. [9]

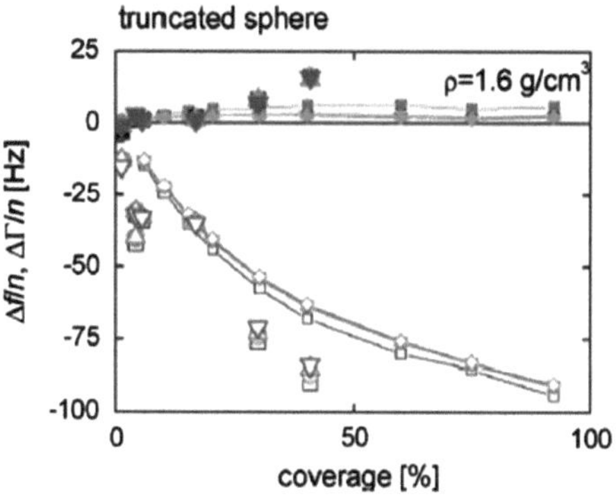

Fig. 12.5 Shifts of frequency and bandwidth versus coverage. *Large symbols* are experimental values. *Small symbols* connected with *lines* are the results from FEM modeling with truncated particles. The FEM results are arithmetic means of the results obtained with the flow direction perpendicular and along the cylinders (Eqs. 12.2.2 and 12.2.3). *Squares*, *circles*, and *upper triangles* denote overtone orders 1, 5, and 11, respectively. Reprinted with permission from Ref. [9]. Copyright 2006 American Chemical Society

the dependence of $\Delta f/n$ on overtone order. The n-dependence results from hydrodynamic effects, not from the softness of the adsorbate. The adsorbate had been given a storage modulus of $G' = 1$ GPa, which amounts to rigid object.

The agreement between the simulations and the experiment is as good as it can be expected. In the author's opinion, these tests can count as a validation—within the limits as discussed—and he has applied the model to a number of different geometries [14–19]. Clearly, the FEM model is a start. It is to be hoped that experts will pick up on the problem and make better code available.

The section concludes with a second case study (following Ref. [14]), which addresses the question of why certain samples dissipate more energy than others and where the dissipation occurs. The model here carries insight, which is independent from experiment and common sense. The example shall demonstrate how such computations can contribute to the understanding of experimental findings beyond what plain physical reasoning can do.

It is an experimental observation that the bandwidth (proportional to the dissipation factor) often goes through a maximum, when discrete objects adsorb to the resonator surface. A prominent example are vesicles (Sect. 12.5). A second example is streptavidin anchored to a lipid bilayer [20]. In the case of vesicles, there clearly is a transient state in the adsorption process, which is different from the final state, where a supported lipid bilayer has formed. (Actually, this final state is not always reached. Some types of liposomes stay intact.) In the case of streptavidin anchored to a lipid bilayer, the authors attribute the transient state to a crystallization process, which was observed in a separate study by atomic force microscopy [21]. For vesicles and streptavidin a peculiar transient state is plausible and certainly interesting. However, the same type of maximum in the dissipation factor was also observed when adsorbing cow pea mosaic virus (CPMV) particles and annexin proteins to a lipid bilayer [2]. In principle, one might hypothesize a transient state for these simple particles, as well. The particles might be loosely bound to the surface, but the nature of this soft bond would have to be explained on a molecular level. One might seek a more conventional explanation: A half-filled adsorbate layer might be more mobile than a layer at full coverage because of the interparticle interactions, which stiffen the layer, once it becomes dense.

The latter hypothesis can be supported with an FEM calculation. The experimentally acquired curves of $\Delta f(\theta)$ and $\Delta\Gamma(\theta)$ with θ the coverage can be reproduced with simulations based on the geometry shown in Fig. 12.5c. A single particle is attached to the resonator surface across a link. The diameter of the particle was 28 nm, matching the diameter of the cow pea mosaic virus, which had been used in the experiments (Fig. 12.6a, Refs. [2, 14]). The only parameter varied in the simulation was the lateral width of the simulation cell. Remember: What appears as a single particle in Fig. 12.6c is an array of particles because of the periodic boundary conditions. By varying the width of the cell, one in effect varies the coverage. Coverage in Fig. 12.6b is to be understood as the ratio of the particle diameter to the width of the cell. All other parameters of the model were kept fixed. Keeping them fixed excludes some kind of a transition state during adsorption.

Increasing the thus-defined coverage step-by-step, one finds that the bandwidth first increases with coverage and then later decreases as the layer becomes more dense. The agreement with the experimental results can be made quantitative (Fig. 12.5a, b) by choosing the stiffness of the link suitably. There is an interesting side result. One might suspect that the energy is mostly dissipated in the liquid between the particles because the particles themselves are rigid. For the simulation reported in Ref. [14], this is not the case. Figure 12.5c shows a map of the local rate of frictional heating. The pattern is complicated and not easily understood from simple arguments. For this particular geometry, the major part of the dissipation occurs at the link. The link periodically bends, thereby dissipating energy. Of course this only happens if the link has nonzero G'', which was the case here. When there is little hydrodynamic interaction between particles, the amplitude of this rocking motion is large and the energy dissipated at the link is correspondingly large (cf. Sect. 11.5). As the interparticle interaction starts to lower the amplitude of the rocking motion, it lowers the amount of dissipated energy. The hydrodynamic interaction is critical to the explanation of the peak in the dissipation factor, but the mechanism is indirect. The rate of entropy production is highest at the link, not in the space between particles.

Of course such a calculation is not strictly a proof for the second hypothesis (peak in bandwidth entirely caused by weak lateral interactions). It only says that a transient state is not strictly required to explain the experiments. More generally, there is an enormously wide space of parameters and geometries, which all might represent the sample. It cannot be ruled out that other choices would have reproduced the experiment equally well.

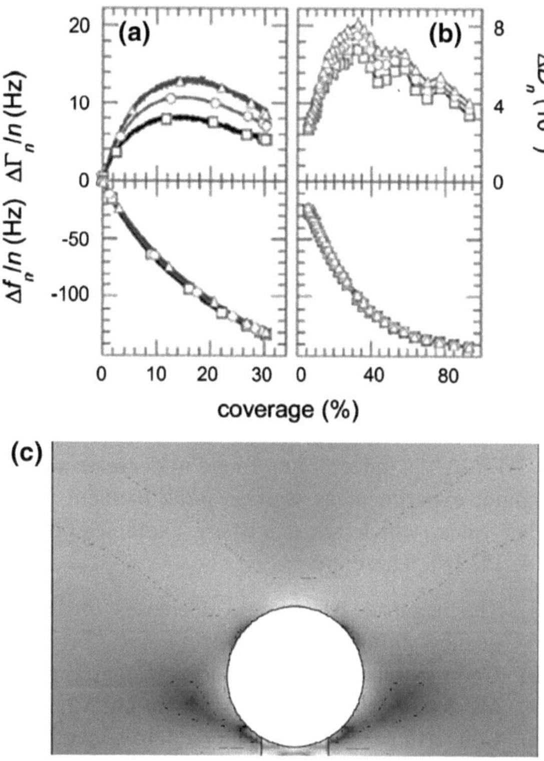

Fig. 12.6 The bandwidth often goes through a maximum during adsorption of small particulate objects like the cow pea mosaic virus (panel **a**). This has lead to the speculation, that there might be a loosely bound transient state. A maximum in ΔΓ versus coverage can be reproduced with a finite element simulation (panel **b**), where the only parameter varied is the number of particles per unit area (the coverage). Panel **c** shows the local entropy production rate. Bright corresponds to a large rate of frictional heating. The particle itself is rigid and does not dissipate energy. Integration shows that most of the dissipation occurs at the link. Panels **a** and **b** were reprinted with permission from Ref. [14]. Copyright 2005 American Chemical Society

12.3 ΔΓ/Δf-Extrapolation Scheme

The previous section hopefully brought across that there is much potential in numerical modeling, on the one hand, but that both physical reasoning and empirical input are still needed, on the other. The following section introduces a semi-empirical procedure for data evaluation (invented by Tellechea and Reviakine [18]) which is half-way between common sense and rigorous modeling. It amounts to what is sometimes called a "rule". The scheme has been shown to work a few times in cases, where independent information on the sample was available (more specifically, where the geometric height of the adsorbed discrete objects was known). The scheme has also been substantiated with FEM simulations. However, there is no simple physical reason for why this rule should hold. Also, there are deviations, both in experiment and in the FEM models. The (known) height of sample often agrees reasonably well (±20 %) with the prediction obtained from the $\Delta\Gamma/\Delta f$-extrapolation scheme, but there is scatter and there are outliers. The $\Delta\Gamma/\Delta f$-extrapolation scheme has to be used with some caution. Whether the rule applies to a given sample and what uncertainty should be assigned to the derived height remains in question. All caveats acknowledged, the rule is intriguing.

Equation 10.5.4 in Sect. 10.5 relates the $\Delta\Gamma/\Delta f$-ratio to the softness of adsorbate. Generally speaking, dilute adsorbates tend to be soft, while more dense adsorbates tend to be more stiff. Section 10.5 was concerned with continuous films. Dilute adsorbates might also be composed of discrete objects (vesicles, viruses, ...), which look like a film to the QCM, although they have an internal structure. "Soft" and "stiff" then are to be understood as effective properties. "Effective" is not necessarily meant to imply an effective medium theory; it is meant in the sense of "as seen by the QCM". The dynamics of such heterogeneous samples will contain a contribution from the stiffness of the link between the adsorbate and the substrate as well as a contribution from the interaction between the different objects. The interactions might be electrostatic, viscoelastic, or hydrodynamic in nature. The interactions become stronger with increasing coverage and one can therefore expect the effective stiffness of the layer to increase with coverage. In consequence, $\Delta\Gamma/(-\Delta f)$ should decrease during adsorption.

$\Delta\Gamma/(-\Delta f)$ is indeed often found to decrease with coverage. Beyond this general finding, experiment as well as finite element simulations have evidenced three other rules, which are not easily explained, but have re-surfaced a few times [14, 15, 18]. These are:

(a) The negative $\Delta\Gamma/\Delta f$-ratio decreases *linearly* with increasing negative frequency shift (increasing $-\Delta f/n$).
(b) Upon extrapolating the line to $\Delta\Gamma/(-\Delta f) = 0$ on the $\Delta f/n$-scale, the different overtones extrapolate to the same point (Fig. 12.8b).
(c) Converting $-\Delta f/n$ at the intercept to a length with the Sauerbrey equation, this length is close to the height of the adsorbate.

Note: The height of the adsorbate is larger than the Sauerbrey thickness because of incomplete coverage (Fig. 12.7). Of course rule c can only be formulated in cases where the height of the adsorbed objects is known. This was true in reference experiments and in the FEM calculations. Building on this experience, the last rule can be used to estimate the height of the adsorbate in those cases, where it is not known independently (Fig. 12.8a, b).

12.4 Droplets and Bubbles

The internal boundaries in the simulations discussed in Sect. 12.1 separate different domains. They do not exert any stress of their own, they only denote a line, where the materials abruptly change. This behavior is untypical of liquid-liquid or gas-liquid boundaries. Such phase boundaries have an interfacial energy themselves; they exert a capillary pressure. Since interfacial tension affects the flow field, it should also have an effect on $\Delta\tilde{f}$. One might be able to measure this effect when depositing droplets onto a QCM surface.

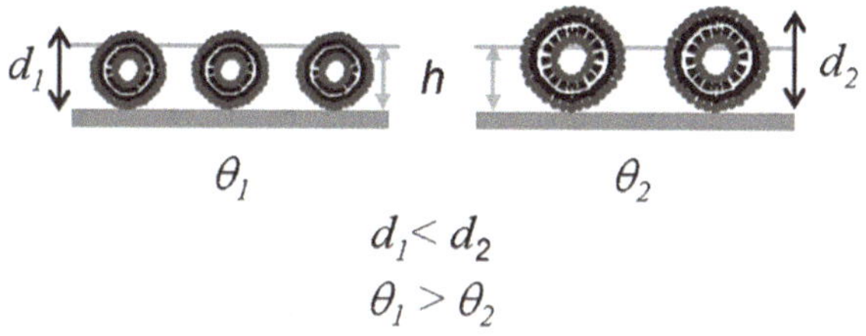

Fig. 12.7 The acoustic thickness of a layer as measured by the QCM is related to the size of the objects as well as the coverage. Different combinations of height and coverage therefore result in the same acoustic thicknesses. The unknown contribution of the surrounding liquid to the acoustic thickness makes it difficult to separate the contributions of size and coverage to the frequency and bandwidth shift. Reprinted with permission from Ref. [18]. Copyright 2009 American Chemical Society

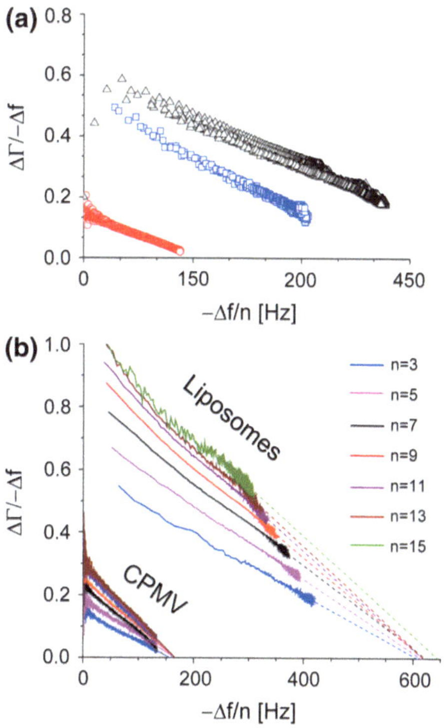

Fig. 12.8 **a** The negative $\Delta\Gamma/\Delta f$-ratio is often found to decrease linearly with $-\Delta f$. Only data from the third overtone are plotted. △ 114 nm liposomes. □ 86 nm liposomes. ○ Cow pea mosaic virus (CPMV) particles ($\approx$28 nm). **b** Plotting the negative $\Delta\Gamma/\Delta f$-ratio as a function of $-\Delta f/n$ for several overtones reveals a common intercept with the x-axis. Adapted from Ref. [18]

A quick estimate shows that surface tension only is of significance if the radius of curvature is below about a micron. Consider the geometry shown in Fig. 12.9. If the droplet radius is much larger than the penetration depth, the deformation pattern can be approximated by the deformation pattern in a semi-infinite medium, that is, $\hat{u}(z) \approx \hat{u}_S \exp(-ikz)$ with $k = (1 - i)/\delta$. Of course this is not strictly correct. Close to the edge, surface tension will affect the flow profile, but for the

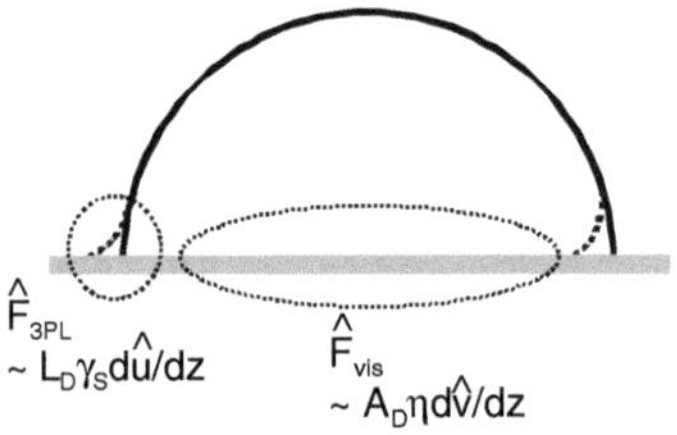

Fig. 12.9 For macroscopic droplets (with $r_D \gg \delta$), the deformation pattern at the resonator-liquid interface is about the same as for a semi-infinite liquid. The overall force at the substrate has two contributions: one from the interface (index *vis* for *vis*cous) and one from the 3-phase line (index $3PL$). The viscous contribution scales as the area, A_D. The force from the 3-phase line is proportional to the surface tension, γ_S, and scales as the length of the droplet's perimeter, L_D

sake of this estimate, ignore the difference. There are two separate forces onto the resonator. The first one originates from the viscosity of the liquid and is given as

$$\left|\hat{F}_{vis}\right| \approx \left|A_D\tilde{Z}_{liq}\hat{v}_S\right| = \left|A_D\tilde{Z}_{liq}i\omega\hat{u}_S\right| \tag{12.4.1}$$

A_D is the droplet area. As always, stress is equal to wave impedance × velocity. The second contribution to the stress at the interface originates from the 3-phase-line. Assume a hemispherical droplet as in Fig. 12.9. When the droplet is sheared, the contact angle changes periodically and there is a periodic tangential force roughly proportional the shear angle. This force is about

$$\left|\hat{F}_{3PL}\right| \approx \left|L_D\gamma_S \sin\left(\frac{\partial\hat{u}}{\partial z}\right)\right| \approx \left|L_D\gamma_S\frac{\hat{u}_S}{\delta}\right| \tag{12.4.2}$$

L_D is the perimeter of the droplet and γ_S is the surface energy. The index $3PL$ denotes the 3-phase-line. The ratio of the two forces is

$$\begin{aligned}\left|\frac{\hat{F}_{3PL}}{\hat{F}_{vis}}\right| &\approx \left|\frac{L_d\gamma_S}{A_d\tilde{Z}_{liq}i\omega\delta}\right| \\ &= \frac{2\pi r_D}{\pi r_D^2}\frac{\gamma_S}{\sqrt{\rho\omega\eta}\;\omega\;\sqrt{2\eta/(\rho\omega)}} = \frac{\sqrt{2}}{r_D}\frac{\gamma_S}{\omega\eta}\end{aligned} \tag{12.4.3}$$

In line 2 the droplet area, A_D, and the length of the perimeter, L_D, were expressed in terms of the droplet radius, r_D. Clearly, the influence of capillary forces increases with decreasing droplet size.

There is another way of putting this argument. If surface tension is insignificant compared to viscous stress, the droplet readily deforms in the flow field regardless of surface tension. Otherwise, the capillary pressure prevents deformation. The matter has been extensively studied in the context of droplet break-up [22].

To what extent capillary pressure and viscous stress influence droplet shape, is assessed by the capillary number, given as

$$Ca = \frac{\dot{\gamma}\eta r_D}{\gamma_S} \quad (12.4.4)$$

$\dot{\gamma}$ is the shear rate (more generally, the characteristic rate of the flow). Ca compares the viscous stress, $\dot{\gamma}\eta$ to the capillary pressure, $2\gamma_S/r_D$. For steady flows and $Ca \gg 1$, droplets get deformed and later break up. Surface tension delays this process, but does not prevent it. Smaller drops have a correspondingly smaller capillary number and the influence of capillary pressure increases. Large shear rates are needed to break these up.

Droplet deformation and bubble deformation have an associated time scale, termed "emulsion time" in the context of drops [23]. It is given as $\tau_r = \eta r_D/\gamma_S$. After cessation of a flow, the droplet returns to spherical shape within a time of about τ_r. The capillary number can also be written as $Ca = \dot{\gamma}\tau_r$. Written this way, Ca compares the flow rate to the inverse emulsion time.

The above remarks concerned steady flows. For oscillatory flow, shear rate is replaced by frequency. The capillary number becomes

$$Ca = \frac{\omega\eta r_D}{\gamma_S} \quad (12.4.5)$$

As before, Ca compares viscous stress to capillary pressure. It also compares the emulsion time to the period of oscillation. Following the first view, small Ca implies that the viscous stress does not suffice to deform the particle. Following the second view, small Ca implies that the period of oscillation is too short to allow for significant deformation. There is a difference to the case of steady flow: Bubbles subjected to a small-amplitude periodic shear flow never break up; they only deform. The capillary number separates the regimes of deformable and non-deformable droplets ($Ca \gg 1$ and $Ca \ll 1$).

Everything said about droplets above similarly applies to bubbles, at least in principle. There is a difference, though: If viscous forces dominate, bubbles attached to the resonator surface will *increase* the resonance frequency, because the viscous drag exerted by the gas phase is lower than the drag exerted by the liquid. For adsorbed bubbles, the effects of viscous drag and of surface tension have opposite sign with regard to their effect on Δf. At this point, the size dependence is critical. Large bubbles are soft objects. A resonator with large bubbles adhering to its surface has a frequency *higher* than the frequency of the same resonator in the bulk liquid. This contrasts to a surface covered with *hemispherical nanobubbles* (Fig. 12.10b). Nanobubbles are stiff objects. They trap liquid in the pockets between them.

Note: This argument only applies to hemispherical bubbles (Fig. 12.10a). Shallow nanobubbles also are stiff objects, but there is slip at the bubble surface with respect to *tangential* motion. Shallow nanobubbles will reduce the hydrodynamic drag; they will induce slip.

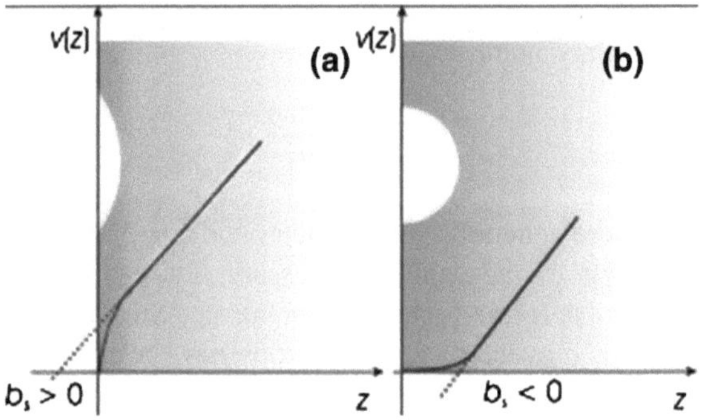

Fig. 12.10 Shallow nanobubbles induce slip because the liquid feels little hydrodynamic resistance when tangentially moving along at the bubble surface. This contrasts to hemispherical nanobubbles. Here, there is a significant normal component to the interfacial stress. Because the capillary pressure is high for nanobubbles, they look like solid objects. They trap water in the pockets between them, thereby *increasing* the hydrodynamic drag at the interface. The slip length is negative

The arguments above were made quantitative, making use of finite element analysis, in Ref. [19]. The results from the FEM simulations were corroborated by an experiment, where bubbles had been formed electrochemically. The frequency shifted *downward* under conditions, where nanobubbles were expected to form. The rigidity of the nanobubbles makes the bubble mattress look like a film. The bubbles *increase* the hydrodynamic drag at the interface. The slip length is negative, and so is the frequency shift. A similar result has been obtained for steady shear flows [24, 25]. A mattress of nanobubbles has a negative slip length in steady shear, as well.

At this point we slightly digress and discuss an unresolved problem. As discussed in Sect. 9.1, determining viscosities with a QCM has a few limitations. Among them is the fact that the scaling laws predicted by the Gordon-Kanazawa-Mason result ($\Delta f = -\Delta\Gamma$, $\Delta f \sim n^{1/2}$, $\Delta\Gamma \sim n^{1/2}$) are slightly but systematically different from what is observed in experiment. Du observed that these deviations can be captured by modifying the equations with one additional parameter, which he calls "*B*" [26]. He writes

$$\frac{\Delta\tilde{f}}{f_0} = \frac{\mathrm{i}}{\pi Z_q}\sqrt{\mathrm{i}\omega\rho_{liq}\eta_{liq}}(1 + \mathrm{i}Bn) \qquad (12.4.6)$$

Tentatively, this functional form was explained with laterally heterogeneous slip in the original publication. The calculation is analytical in nature. While laterally heterogeneous slip might occur in a number of different ways, the explanation of Eq. 12.4.6 with nanobubbles or nano-pancakes has not survived an experimental test performed by Zhang [27]. Zhang produced nanobubbles by the solvent exchange method [28]. That should have produced laterally homogeneous slip, but

Zhang could not explain her data with the model from Ref. [26]. Equation 12.4.6 as such has been confirmed on an experimental level in other laboratories, [29] but there is no satisfactory explanation.

12.5 Vesicles and Supported Lipid Bilayers

Among the numerous surface-sensitive techniques that work in liquids or in vacuum (SPR, ellipsometry, XPS, SIMS, ...) the QCM stands out because it is sensitive to how the material is organized at the interfaces, and not only to how much of it is there. A particularly impressive feature of the QCM is its ability to distinguish between surface-adsorbed liposomes (lipid vesicles) and planar lipid bilayers (also: "supported lipid bilayers", SLBs).

We slightly digress and comment on the relevance of liposome adsorption. Liposomes are spherical compartments with a size in the range of a few tens to a few hundred nanometers, bounded by a lipid bilayer. They form spontaneously; they even are a paradigmatic example of a self-organization. The vesicle size can to some extent be adjusted by the conditions of preparation. Otherwise, the structure and the properties are fully determined by the properties of the constituent molecules. In nature, the lipid membrane is host to numerous functional proteins. Liposomes can likewise contain proteins and other functional units.

Liposomes are rather stable, basically, but when brought into contact with a solid surface, they may rupture. If the conditions are chosen suitably, the lipid then spontaneously forms a new self-organized structure at the respective surface, which is the supported lipid bilayer (the SLB, inset in Fig. 12.11) [32, 33]. The SLB is attached to the surface by non-covalent interactions. It may contain functional molecules of a wide variety. Being attached to a surface, it is amenable to the array of surface analytical techniques. SLBs can be a platform for sensing, as well.

Given the importance of SLBs, the processes leading to their formation have attracted much interest, and the QCM has had a prominent role in this research. SLBs in their final state lead to a frequency shift of about –25 Hz, corresponding to a Sauerbrey thickness of about 5 nm. They do not induce an increase in bandwidth (blue curve in Fig. 12.11), while adsorbed liposomes induce large shifts in frequency and bandwidth, that depend on the size of the liposomes (red curve in Fig. 12.11) [34–36]. Based on the observation of the extrema in frequency and dissipation factor during the SLB formation, Keller and Kasemo [34] postulated that SLBs form via adsorbed liposome intermediates, which was later confirmed by atomic force microscopy [37]. There were further studies on the subject, which analyzed the kinetics of lipid rearrangements at the surface, building on the extrema in frequency and dissipation. The magnitude of these extrema is related to the amount of liposomes present at the surface during SLB formation. When the rate at which liposomes rupture is much faster than the rate at which they adsorb, the extrema vanish. This happens with negatively charged surfaces, i.e., when the

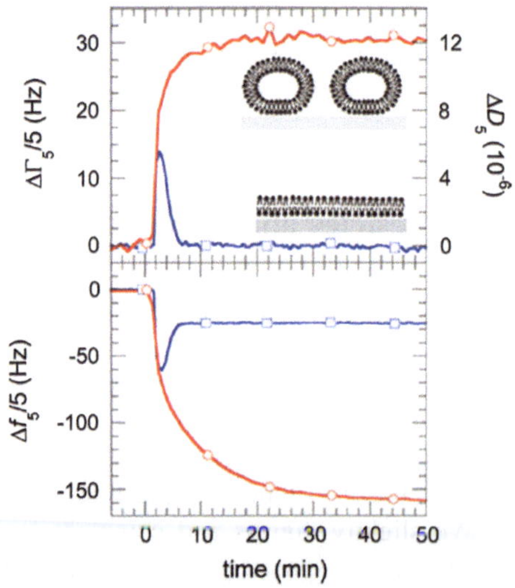

Fig. 12.11 The evolution of Δf and $\Delta\Gamma$ upon vesicle adsorption differs characteristically between liposomes forming bilayers (*blue*) and liposomes which remain intact upon adsorption (*red*) as reported in Ref. [34]. When a bilayer forms, there is a transient state with both $-\Delta f$ and $\Delta\Gamma$ being higher than in the final state. The behavior of liposomes at inorganic surfaces has been reviewed in Ref. [30]. Reprinted with permission from Ref. [31]. Copyright 2011 American Chemical Society

lipid-surface interactions are strong [38, 39]. These and related results have been reviewed in Ref. [30].

At this point it should be emphasized that there is no ab initio understanding of how adsorbed vesicles affect the frequency shift. The existing literature on SLB formation studied with the QCM makes use of the extrema as such, the comparison between samples, the kinetics, and supplementary techniques. There is no equation of any kind, which would relate the surface coverage and the liposome size to $\Delta\tilde{f}$. A quantitative model would indeed have to be rather advanced. In particular, the bending rigidity of the lipid bilayer at MHz frequencies would be needed as an input to a realistic representation.

There is a recent quantitative study on liposome *deformation* [15]. It goes beyond the kinetic analysis. Building on the $\Delta\Gamma/\Delta f$-extrapolation scheme and finite element analysis, values of the heights of adsorbed liposomes were derived from the QCM experiment. For a model system with well-known properties, a correlation was established between the liposome deformation and bending modulus, which was plausible and gives credibility to the technique. Building on the $\Delta\Gamma/\Delta f$-extrapolation scheme, one can move on to systems which are less well understood.

As sketched in Fig. 12.12, liposomes deform when they adsorb to a surface. The deformation is driven by the energy of adsorption to the surface; it is resisted by, firstly, the osmotic pressure of the medium inside the liposome and, secondly, by the membrane's bending rigidity [40, 41]. Liposome deformation has been studied previously by cryo-electron microscopy [42], which is a powerful, but time-consuming technique. Being able to study liposome deformation in situ with a QCM certainly is an attractive perspective.

Figure 12.13a, c shows adsorption kinetics acquired at temperatures of 10 and 32 °C. The lipid chosen was dimyristoyl phosphatidyl choline (DMPC) in

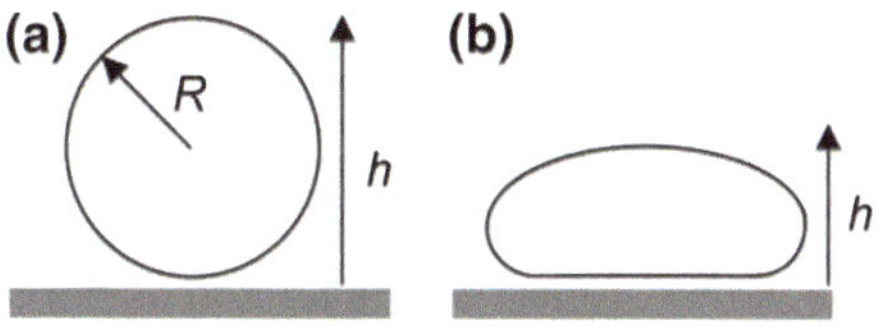

Fig. 12.12 Deformation of liposomes upon adsorption to surfaces. R is the radius of the liposomes in solution. h is the height of the surface-adsorbed liposomes. For non-deformed liposomes one has $h = 2R$ (**a**). Adsorption may lower the value of h (**b**). Reprinted with permission from Ref. [15]. Copyright 2012 American Institute of Physics

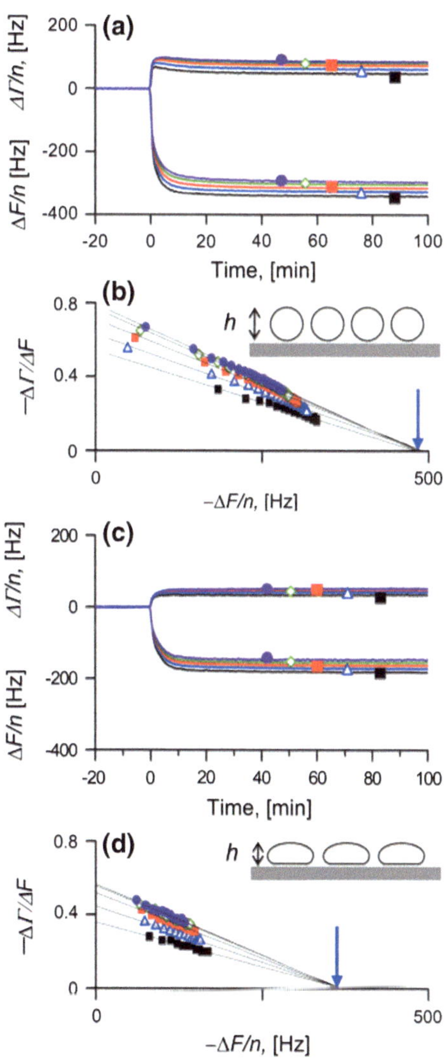

Fig. 12.13 Studies of adsorbed liposome deformation making use of the $\Delta\Gamma/\Delta f$-extrapolation scheme. Dimyristoyl phosphatidyl choline (DMPC) liposomes were adsorbed to TiO_2 at 10 °C (**a**, **b**), and 32 °C (**c**, **d**). Below the main transition temperature of ~24 °C, the liposomes are stiff and deform little, while they are soft and free to deform, above. This was confirmed by cryo-electron microscopy. The intercepts of the—$\Delta\Gamma/\Delta f$ versus $-\Delta f/n$ plots shown in **b** and **d**, pointed to by the *blue arrows*, indicate that the heights of the adsorbed liposomes are different at the two different temperatures (see insets), even though the same liposome solution was used in both cases. Reprinted with permission from Ref. [15]. Copyright 2012 American Institute of Physics

both cases. The bending rigidity of DMPC depends on temperature. At low temperatures, there is strong lateral interaction between the alkyl chains, which has numerous consequences, among them a bending stiffness of the membrane, which is much larger than the bending stiffness above the main transition at 24 °C. Adsorbing DMPC to a titania surface at temperatures above and below the main transition, one can study the influence of the bending rigidity on the liposome deformation with one and the same material.

Note: DMPC does not form a supported lipid bilayer under the conditions of this experiment. There are maxima in bandwidth in both panel (a) and panel (c), but these are of the kind discussed at the end of Sect. 12.5. The maxima do not reflect a transient state of the individual vesicle, but rather the fact that the hydrodynamic interaction between neighboring adsorbed liposomes is weak, initially, and grows stronger as the density of adsorbed liposomes increases. AFM images reveal that the final state consists of adsorbed liposomes, as opposed to an SLB. This is the situation, to which the $\Delta\Gamma/\Delta f$-extrapolation scheme can be applied. Panels (b) and (d) of Fig. 12.13 show $\Delta\Gamma/(-\Delta f)$ versus $-\Delta f/n$ for the first five overtones. The lines extrapolate to $-\Delta f/n \approx -480$ Hz and to $-\Delta f/n \approx -370$ Hz in panel (b) and (d), respectively. The lowered value in (d) can be interpreted as the consequence of deformation, as sketched in the insets.

The height of adsorbed DMPC liposomes is plotted as a function of temperature in Fig. 12.14. A correlation emerges between the liposome height and the bending modulus: The modulus changes abruptly across the transition, and so does the height. The correlation as such was expected, but Fig. 12.14 makes it quantitative. Building on this methodology, one can study rather easily and conveniently, how the admixture of other lipids, the addition of proteins, the pH, the salt concentration, and the presence of solutes inside the liposome affect liposome deformation. Liposome deformation is a central aspect of the formation of SLBs.

12.6 Biological Cells

Cell layers and bacteria attached to a QCM surface are a popular target of current research. As of March 2013, the Web of Science returned 254 publications matching "quartz crystal microbalance" and "bacteria*". It returned 274 entries for "quartz crystal microbalance" and "cell adhesion". The focus of the book is on shear waves more than on biological cells and the discussion therefore is limited to a few remarks:

- Viscoelastic modeling from first principles in the author's view is hopeless. Bioadhesion is the realm of empirical models [45]. An empirical model predicts trends in Δf and $\Delta\Gamma$ as a function of system parameters, based on experience and common sense. Such models exist, but reproducing them with a sophisticated ab-initio calculation is futile because the geometry contains too many unknowns.

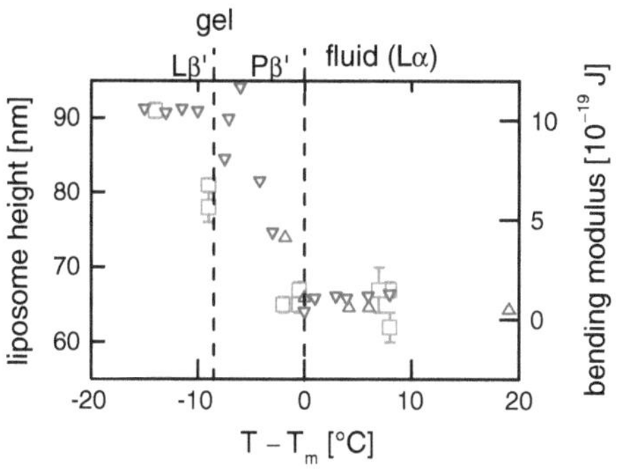

Fig. 12.14 The height of DMPC liposomes adsorbed to a TiO_2 surface at different temperatures (*open squares*) is overlaid with the bilayer bending moduli (*triangles*). The bilayer bending moduli were obtained from Refs. [43] (*lower triangles*) and [44] (*upper triangles*). *Vertical dashed lines* indicate the temperatures of two phase transitions. For details see Ref. [15]. Adapted from Ref. [15]

- Δf is often found to be positive. This can be explained with coupled resonances (Sect. 11.4).
- Biological cells are thick and soft to the extent that the shear wave does not reach to the top. The QCM mostly probes the viscoelastic properties of the region close to the resonator surface.
- With living objects, one can analyze the variability of the signal with time, which gives access to the cell's motility [46, 47].
- In Ref. [48], the half-bandwidth and its dependence on the system parameters (number of cells per unit area, in this case) was reproducible and made sense, while Δf fluctuated strongly and was not of much predictive use. This behavior is the consequence of the coupled resonance (Sect. 4.6.3). In the presence of a coupled resonance, $\Delta\Gamma$ is always positive. It might fluctuate when the experimentalist has poor control, but it never turns negative. Δf, on the other hand, may change sign and may be close to zero when the zero-crossing frequency of the coupled resonance is close to the resonance frequency. In such situations, $\Delta\Gamma$ is the more robust empirical parameter. Bandwidth in this case is not just an "added benefit" of the advanced QCM, it is the one bit of information left to be interpreted. Generally speaking, this situation (which challenges the conventional view of the QCM) can occur in other contexts as well (Sect. 17.9).
- There have been a number of attempts to monitor blood clotting with a QCM [49–51]. The matter is more complicated than one might think. A commercial instrument did not emerge.

More detailed accounts of the QCM applied to biological systems are available in Refs. [45, 52].

Glossary

Variable	Definition (Comments)
$3PL$	As an index: *3-Phase Line*
A_D	Area of a droplet
b_S	*S*lip length (see also Sect. 10.7)
Ca	Capillary number
D	Dissipation factor ($D = 2\Gamma/f_r$)
D	As an index: Droplet
f	Frequency
f_0	Resonance frequency at the fundamental ($f_0 = Z_q/(2m_q) = Z_q/(2\rho_q d_q)$)
G	Shear modulus
h	Height of an adsorbate layer
k	Wavenumber
liq	As an index: *liq*uid
n	Overtone order
L_D	Perimeter of a droplet
p	Pressure
r_D	Radius of a droplet
$\mathbf{r}$	Position (a vector)
R	Radius of a liposome (Fig. 12.12)
S	As an index: *S*urface
t	Time
T	Temperature
T_m	Melting temperature of a lipid membrane
$\mathbf{u}$, u	Tangential displacement (when bold: a vector)
v, $\mathbf{v}$, $\hat{\mathrm{v}}$, $\hat{\mathbf{v}}$	Velocity
vis	As an index: *vis*cous
x, y, z	Spatial coordinates, z: along the surface normal
$\tilde{Z}_{liq}$	Shear-wave impedance of a liquid ($\tilde{Z}_{liq} = (i\omega\rho_{liq}\eta_{liq})^{1/2}$)
$\dot{\gamma}$	Shear rate
γ_S	Surface energy

Γ	Imaginary part of a resonance frequency
δ	Penetration depth of a shear wave (Newtonian liquids: $\delta = (2\eta_{liq}/(\rho_{liq}\omega))^{1/2}$)
δ_L	Loss angle
Δ	As a prefix: A shift induced by the presence of the sample
$\eta, \tilde{\eta}$	Viscosity
θ	Coverage
ρ	Density
σ	Tangential stress
τ_r	Emulsion time (A relaxation time of emulsions and droplets)
ω	Angular frequency

References

1. Hayden, O., Lieberzeit, P.A., Blaas, D., Dickert, F.L.: Artificial antibodies for bioanalyte detection-sensing viruses and proteins. Adv. Funct. Mater. **16**(10), 1269–1278 (2006)
2. Bingen, P., Wang, G., Steinmetz, N.F., Rodahl, M., Richter, R.P.: Solvation effects in the quartz crystal microbalance with dissipation monitoring response to biomolecular adsorption. A phenomenological approach. Anal. Chem. **80**(23), 8880–8890 (2008)
3. Olsson, A.L.J., Quevedo, I.R., He, D., Basnet, M., Tufenkji, N.: Using the quartz crystal microbalance with dissipation monitoring to evaluate the size of nanoparticles deposited on surfaces. ACS Nano **7**(9), 7833–7843 (2013)
4. Grest, G.S.: Interfacial sliding of polymer brushes: a molecular dynamics simulation. Phys. Rev. Lett. **76**(26), 4979–4982 (1996)
5. Urbakh, M., Daikhin, L.: Roughness effect on the frequency of a quartz-crystal resonator in contact with a liquid. Phys. Rev. B **49**(7), 4866–4870 (1994)
6. Adamczyk, Z., Siwek, B., Zembala, M., Belouschek, P.: Kinetics of localized adsorption of colloid particles. Adv. Colloid Interface Sci. **48**, 151–280 (1994)
7. Kastl, K., Herrig, A., Luthgens, E., Janshoff, A., Steinem, C.: Scrutiny of annexin A1 mediated membrane—membrane interaction by means of a thickness shear mode resonator and computer simulations. Langmuir **20**(17), 7246–7253 (2004)
8. Wriggers, P.: Computational Contact Mechanics. Springer, Heidelberg (2006)
9. Johannsmann, D., Reviakine, I., Rojas, E., Gallego, M.: Effect of sample heterogeneity on the interpretation of QCM(-D) data: comparison of combined quartz crystal microbalance/atomic force microscopy measurements with finite element method modeling. Anal. Chem. **80**(23), 8891–8899 (2008)
10. Granick, S.: Ferritin—its properties and significance for iron metabolism. Chem. Rev. **38**(3), 379–403 (1946)
11. Hook, F., Rodahl, M., Brzezinski, P., Kasemo, B.: Measurements using the quartz crystal microbalance technique of ferritin monolayers on methyl-thiolated gold: dependence of energy dissipation and saturation coverage on salt concentration. J. Colloid Interface Sci. **208**(1), 63–67 (1998)
12. Johnson, C.A., Yuan, Y., Lenhoff, A.M.: Adsorbed layers of ferritin at solid and fluid interfaces studied by atomic force microscopy. J. Colloid Interface Sci. **223**(2), 261–272 (2000)

13. Hemmersam, A.G., Rechendorff, K., Besenbacher, F., Kasemo, B., Sutherland, D.S.: pH-dependent adsorption and conformational change of ferritin studied on metal oxide surfaces by a combination of QCM-D and AFM. J. Phys. Chem. C **112**(11), 4180–4186 (2008)
14. Johannsmann, D., Reviakine, I., Richter, R.P.: Dissipation in films of adsorbed nanospheres studied by quartz crystal microbalance (QCM). Anal. Chem. **81**(19), 8167–8176 (2009)
15. Reviakine, I., Gallego, M., Johannsmann, D., Tellechea, E.: Adsorbed liposome deformation studied with quartz crystal microbalance. J. Chem. Phys. **136**(8), 84702–84705 (2012)
16. Pomorska, A., Shchukin, D., Hammond, R., Cooper, M.A., Grundmeier, G., Johannsmann, D.: Positive frequency shifts observed upon adsorbing micron-sized solid objects to a quartz crystal microbalance from the liquid phase. Anal. Chem. **82**(6), 2237–2242 (2010)
17. Johannsmann, D.: Viscoelastic, mechanical, and dielectric measurements on complex samples with the quartz crystal microbalance. Phys. Chem. Chem. Phys. **10**(31), 4516–4534 (2008)
18. Tellechea, E., Johannsmann, D., Steinmetz, N.F., Richter, R.P., Reviakine, I.: Model-independent analysis of QCM data on colloidal particle adsorption. Langmuir **25**(9), 5177–5184 (2009)
19. Finger, A., Johannsmann, D.: Hemispherical nanobubbles reduce interfacial slippage in simple liquids. Phys. Chem. Chem. Phys. **13**(40), 18015–18022 (2011)
20. Hook, F., Ray, A., Norden, B., Kasemo, B.: Characterization of PNA and DNA immobilization and subsequent hybridization with DNA using acoustic-shear-wave attenuation measurements. Langmuir **17**(26), 8305–8312 (2001)
21. Reviakine, I., Brisson, A.: Streptavidin 2D crystals on supported phospholipid bilayers: toward constructing anchored phospholipid bilayers. Langmuir **17**(26), 8293–8299 (2001)
22. Stone, H.A.: Dynamics of drop deformation and breakup in viscous fluids. Annu. Rev. Fluid Mech. **26**, 65–102 (1994)
23. Minale, M.: Models for the deformation of a single ellipsoidal drop: a review. Rheol. Acta **49**(8), 789–806 (2010)
24. Hyväluoma, J., Kunert, C., Harting, J.: Simulations of slip flow on nanobubble-laden surfaces. J. Phys. Condens. Matter **23**(18), 184106 (2011)
25. Steinberger, A., Cottin-Bizonne, C., Kleimann, P., Charlaix, E.: High friction on a bubble mattress. Nat. Mater. **6**(9), 665–668 (2007)
26. Du, B.Y., Goubaidoulline, E., Johannsmann, D.: Effects of laterally heterogeneous slip on the resonance properties of quartz crystals immersed in liquids. Langmuir **20**, 10617–10624 (2004)
27. Zhang, X.H.: Quartz crystal microbalance study of the interfacial nanobubbles. Phys. Chem. Chem. Phys. **10**(45), 6842–6848 (2008)
28. Lou, S.T., Ouyang, Z.Q., Zhang, Y., Li, X.J., Hu, J., Li, M.Q., Yang, F.J.: Nanobubbles on solid surface imaged by atomic force microscopy. J. Vac. Sci. Technol. B **18**(5), 2573–2575 (2000)
29. Reviakine, I.: private communication
30. Richter, R.P., Berat, R., Brisson, A.R.: Formation of solid-supported lipid bilayers: an integrated view. Langmuir **22**(8), 3497–3505 (2006)
31. Reviakine, I., Johannsmann, D., Richter, R.P.: Hearing what you cannot see and visualizing what you hear: interpreting quartz crystal microbalance data from solvated interfaces. Anal. Chem. **83**(23), 8838–8848 (2011)
32. Tamm, L.K., McConnell, H.M.: Supported phospholipid-bilayers. Biophys. J. **47**(1), 105–113 (1985)
33. Sackmann, E.: Supported membranes: scientific and practical applications. Science **271**(5245), 43–48 (1996)
34. Keller, C.A., Kasemo, B.: Surface specific kinetics of lipid vesicle adsorption measured with a quartz crystal microbalance. Biophys. J. **75**(3), 1397–1402 (1998)
35. Reimhult, E., Hook, F., Kasemo, B.: Vesicle adsorption on SiO2 and TiO2: dependence on vesicle size. J. Chem. Phys. **117**(16), 7401–7404 (2002)

36. Reviakine, I., Rossetti, F.F., Morozov, A.N., Textor, M.: Investigating the properties of supported vesicular layers on titanium dioxide by quartz crystal microbalance with dissipation measurements. J. Chem. Phys. **122**(20), 204711 (2002)
37. Reviakine, I., Brisson, A.: Formation of supported phospholipid bilayers from unilamellar vesicles investigated by atomic force microscopy. Langmuir **16**(4), 1806–1815 (2000)
38. Richter, R., Mukhopadhyay, A., Brisson, A.: Pathways of lipid vesicle deposition on solid surfaces: a combined QCM-D and AFM study. Biophys. J. **85**(5), 3035–3047 (2003)
39. Edvardsson, M., Svedhem, S., Wang, G., Richter, R., Rodahl, M., Kasemo, B.: QCM-D and reflectometry instrument: applications to supported lipid structures and their biomolecular interactions. Anal. Chem. **81**(1), 349–361 (2009)
40. Seifert, U.: Adhesion of vesicles in 2 dimensions. Phys. Rev. A **43**(12), 6803–6814 (1991)
41. Seifert, U.: Configurations of fluid membranes and vesicles. Adv. Phys. **46**(1), 13–137 (1997)
42. Mornet, S., Lambert, O., Duguet, E., Brisson, A.: The formation of supported lipid bilayers on silica nanoparticles revealed by cryoelectron microscopy. Nano Lett. **5**(2), 281–285 (2005)
43. Lee, C.H., Lin, W.C., Wang, J.P.: All-optical measurements of the bending rigidity of lipid-vesicle membranes across structural phase transitions. Phys. Rev. E **64**(2), 020901 (2001)
44. Yi, Z., Nagao, M., Bossev, D.P.: Bending elasticity of saturated and monounsaturated phospholipid membranes studied by the neutron spin echo technique. J. Phys. Condens. Matter **21**(15), 15510 (2009)
45. Heitmann, V., Reiss, B., Wegener, J.: The quartz crystal microbalance in cell biology: basics and applications. In Steinem, C., Janshoff, A. (eds.) Piezoelectric Sensors. Springer, Berlin (2007)
46. Sapper, A., Wegener, J., Janshoff, A.: Cell motility probed by noise analysis of thickness shear mode resonators. Anal. Chem. **78**(14), 5184–5191 (2006)
47. Pax, M., Rieger, J., Eibl, R.H., Thielemann, C., Johannsmann, D.: Measurements of fast fluctuations of viscoelastic properties with the quartz crystal microbalance. Analyst **130**(11), 1474–1477 (2005)
48. Li, J., Thielemann, C., Reuning, U., Johannsmann, D.: Monitoring of integrin-mediated adhesion of human ovarian cancer cells to model protein surfaces by quartz crystal resonators: evaluation in the impedance analysis mode. Biosens. Bioelectron. **20**(7), 1333–1340 (2005)
49. Tessier, L., Patat, F., Schmitt, N., Lethiecq, M., Frangin, Y., Guilloteau, D.: Significance of mass and viscous loads discrimination for an at-quartz blood-group immunosensor. Sens. Actuators B-Chem. **19**(1–3), 698–703 (1994)
50. Bandey, H.L., Cernosek, R.W., Lee, W.E., Ondrovic, L.E.: Blood rheological characterization using the thickness-shear mode resonator. Biosens. Bioelectron. **19**(12), 1657–1665 (2004)
51. Muller, L., Sinn, S., Drechsel, H., Ziegler, C., Wendel, H.P., Northoff, H., Gehring, F.K.: Investigation of prothrombin time in human whole-blood samples with a quartz crystal biosensor. Anal. Chem. **82**(2), 658–663 (2010)
52. Saitakis, M., Gizeli, E.: Acoustic sensors as a biophysical tool for probing cell attachment and cell/surface interactions. Cell. Mol. Life Sci. **69**(3), 357–371 (2012)

Chapter 13
Nonlinear Interactions

Abstract If the stress at the resonator surface is not proportional to the displacement (if there is nonlinear response), the frequency shift depends on the amplitude of oscillation. Nonlinearities can also be detected in the form of steady forces, steady flows, second-harmonic signals, and third-harmonic signals. The chapter gives particular emphasis to the amplitude dependence of the frequency shift. It is analyzed in the frame of a time-averaged periodic stress inserted into the SLA. An application example is the study of partial slip.

13.1 General

The standard model of the QCM assumes linear stress-strain relations. Linearity is a prerequisite for the wave equation to hold. It also underlies the equivalent circuits and anything else covered in this book so far. Nonlinearities *do* occur in the physics of the QCM, but they are much weaker than the linear components of the periodic stress inside the resonator itself. For that reason, they can be treated as a small perturbation. Strong nonlinearities would lead to poorly predictable phenomena such as turbulence (in liquids) or fracture (in mechanics). Actually, quartz crystals do break when driven at high enough amplitudes, but damage is outside the scope here. Speaking of poorly predictable phenomena: Not only the crystals behave nonlinearly at high powers, the driving electronics does, as well. The problem needs attention.

The sources of nonlinear behavior in QCM experiments are twofold. A first field of application with ubiquitous nonlinearities is contact mechanics [1, 2]. In contact mechanics one deals with large stresses close to the contacts. Even if the stress averaged over the macroscopic apparent contact area is small, the local stress is high to the extent that the load-bearing asperities deform plastically or slide. Also, the true contact area often is a function of the load, which also entails nonlinear response. A second source of nonlinearities visible to the QCM is rooted in hydrodynamics. The nonlinearities observed in the liquid phase go back to the

D. Johannsmann, *The Quartz Crystal Microbalance in Soft Matter Research*,
Soft and Biological Matter, DOI: 10.1007/978-3-319-07836-6_13,

advected-momentum term in the Navier-Stokes equation. The importance of the nonlinear term relative to the viscous term is quantified by the Reynolds number, defined as $Re = \rho v L/\eta$ with ρ the density, v a characteristic velocity, L a characteristic length, and η the viscosity. Inserting numbers, one finds that $Re < 1$ in QCM experiments. One never observes turbulent flow, at least not close to the surface. Still, the Reynolds number can be large enough to let nonlinearities be visibly. The QCM can set off steady flows; it can act as a pump (Sect. 13.2).

If the nonlinear forces (or stresses) are small enough, they can be expressed as a power series in displacement (or strain). To simplify the algebra, we limit the scope and make a few assumptions. Firstly, we write the power series down for a force, F, and a displacement u. Doing so, we avoid tensors. Tensors would have been needed for stress-strain relations. (Ref. [3] uses tensors to treat the intrinsic nonlinearities of quartz crystals.) Secondly, we abandon the complex formulation. The coefficients κ_i in Eq. 13.1.1 are real. All interactions are approximated as elastic. Complex response functions might be used, but we avoid the complications connected to them. Thirdly, we exclude hysteresis. For the general case, a force at time t may not only depend on displacement and velocity at time t, but also on the state of the system at some earlier time. For instance, the force may depend on whether or not a contact has been sliding in the past and, if it has been sliding, where it has come to rest.

The force-displacement relation of a weakly nonlinear spring is expressed as

$$F(t) \approx \kappa_1 u(t) + \kappa_2 u^2(t) + \kappa_3 u^3(t) \tag{13.1.1}$$

$u(t)$ is the displacement. κ_1 is the conventional spring constant. κ_2 and κ_3 quantify the spring's nonlinear elasticity. Using $u(t) = u_0 \cos(\omega t)$, one finds

$$\begin{aligned} F(t) &\approx \kappa_1 u_0 \cos \omega t + \kappa_2 u_0^2 \cos^2 \omega t + \kappa_3 u_0^3 \cos^3 \omega t \\ &\approx \kappa_1 u_0 \cos \omega t + \kappa_2 u_0^2 \frac{1}{2}(1 + \cos 2\omega t) + \kappa_3 u_0^3 \frac{1}{4}(3 \cos \omega t + \cos 3\omega t) \end{aligned} \tag{13.1.2}$$

u_0 (a real parameter) is the amplitude of oscillation. Characteristic for weak nonlinearities, there are forces at frequencies other than the frequency of oscillatory displacement, which are 0, 2ω, and 3ω. Even higher frequencies occur if the power series is broken off at powers higher than 3. If the power series applies, the respective components have a characteristic scaling with u_0. For instance, the steady force at $\omega = 0$ scales as u_0^2. However, it is not guaranteed that the steady force is entirely caused by the quadratic term. Terms of 4th, 6th, and higher order also produce steady forces. Likewise, Eq. 13.1.1 predicts that the force at ω would only contain two contributions, one of them being linear in u_0 and the other one scaling as u_0^3. That does not say that other scaling laws were impossible. On the contrary, they are observed in experiment [4]. If the periodic force at ω depends on u_0 in a way, which differs from $\kappa_1 u_0 + 3/4\ \kappa_3 u_0^3$, this only says that Eq. 13.1.1 does not fully capture the force-displacement relation.

Importantly, symmetry can force the quadratic term in Eq. 13.1.2 to be zero. For instance, the quadratic term is forbidden if the system obeys inversion symmetry. u_0^2 does not change sign under inversion, while F does. If the relations $F(\omega = 0) = 1/2\ \kappa_2\ u_0^2$ and $-F(\omega = 0) = 1/2\ \kappa_2\ u_0^2$ are both supposed to be true, κ_2 must be zero. Surfaces always break the inversion symmetry. u might be a tangential displacement and $F(\omega = 0)$ might be a normal force. Steady normal forces induced by oscillatory tangential displacements *are* allowed because the surface does not have mirror symmetry along z. The exact same argument applies to second harmonic generation [5].

The following sections are concerned with nonlinear effects at frequencies of 0, ω, and 3ω. Steady (mostly repulsive) forces induced by tangential oscillatory motion can be observed with an AFM. The tangential motion can be achieved with a QCM [6], but any other piezoelectric transducer will do as well [7–9] (Sect. 16.3). We do not go into the details. A more relevant process is steady streaming of a liquid, which is also caused by a force at zero frequency. Steady streaming is discussed in Sect. 13.2. Forces at frequency ω change the resonance frequency, as expressed by the SLA. If such forces are of nonlinear origin, Δf and $\Delta\Gamma$ depend on amplitude. One finds nontrivial functions $\Delta f(u_0)$ and $\Delta\Gamma(u_0)$. An amplitude dependence of Δf and $\Delta\Gamma$ can go back to a 3rd-order nonlinearity, but also to nonlinearities of 5th, 7th, and higher orders. (More mundane processes such as frictional heating may also play a role.) The problem is treated in a general frame (not using the power series) in Sect. 13.3. Third-harmonic generation is briefly touched upon in Sect. 13.4.

13.2 Steady Streaming

In liquid environments, materials-based nonlinearities are not expected; the stress and the strain are too small. A typical "materials-based nonlinearity" would be a dependence of viscosity on shear rate. Such nonlinearities certainly exist [10], but they cannot be accessed with the QCM. To understand this quantitatively, remember that the displacement at the resonator surface, $|\hat{u}_S|$, is a few nanometers, at most (Sect. 7.4). The penetration depth, δ, is around 100 nm, depending on the frequency and the material. The strain is of the order of $|\hat{u}_S|/\delta$ and therefore amounts to a few percent, at most. A strain at that level does not suffice to reach the regime of nonlinear rheology. Harder materials can display nonlinear behavior at smaller strain, but in this case, the stress required to see them is correspondingly larger and cannot be exerted by the QCM surface. For more comments on that topic also see the introduction to slip in Sect. 10.7. (The above arguments apply to homogeneous materials, not to point contacts.)

Steady forces and steady flows *are* observed in simple liquids. These steady forces have little to do with material properties. They go back to the "advected momentum term" in the Navier-Stokes equation. Steady flows caused by high-frequency acoustic waves are a research field of their own. The subject was covered recently by a series of tutorial papers in Lab on a Chip [11]. These authors call the subject "acoustofluidics". Fluids can be pumped by acoustic means.

A few further comments:

- Researchers from acoustofluidics do not count shear waves as acoustic waves. They limit the use of the term "acoustic" to compressional waves. To them, an acoustic shear wave is a "transverse viscous wave", a "Stokes layer", or a "boundary layer". They have a point; the shear wave is a boundary layer.
- The steady flows described below are not discussed much in the context of acoustofluidics, but there is a related phenomenon, which is well-known and has been modeled in considerable detail [12]. When a liquid is permeated by conventional ultrasound, there is a layer close to the surface, inside which the velocity changes from zero (at the surface) to a finite value (in the bulk liquid). The thickness of this layer is δ. It is also called "Stokes layer" and it gives rise to steady forces and steady flows.
- Acoustofluidics covers fluids with finite compressibility. (Gases here count as fluids.) Finite compressibility can also give rise to nonlinearities. The discussion below is limited to incompressible fluids (to liquids). Approximating liquids at a QCM surface as incompressible is reasonable because the wavelength of compressional sound is much larger than the penetration depth of the shear wave.

The steady flow observed above a QCM surface is caused by the nonlinear term in the Navier-Stokes equation. The Navier-Stokes equation is:

$$\frac{\partial(\rho\mathbf{v})}{\partial t}+(\mathbf{v}\cdot\nabla)(\rho\mathbf{v})=\rho\left[\frac{\partial\mathbf{v}}{\partial t}+(\mathbf{v}\cdot\nabla)\mathbf{v}\right]=\eta\nabla^2\mathbf{v}-\nabla p+f_{body} \qquad (13.2.1)$$

ρ, **v**, t, and η have their usual meaning, p is the hydrostatic pressure and f_{body} is the density of a body force (such as gravity). A density constant in space was assumed in step 1. (The density is not constant at interfaces, which gives rise to steady surface force, see Ref. [13]). The term in square brackets is acceleration. Acceleration has two contributions, where the first goes back to unsteady motion and the second goes back to convection. A volume element can experience acceleration even if the flow is stationary (if $\partial\mathbf{v}/\partial t=0$), because it moves in the flow field. The latter contribution to the acceleration is given as $(\mathbf{v}\cdot\nabla)\mathbf{v}$. The term $\rho(\mathbf{v}\cdot\nabla)\mathbf{v}$ is also called the "advected momentum term". Since the velocity appears quadratically, the advected momentum gives rise to nonlinear behavior.

The influence of the nonlinear term onto the flow is assessed with the Reynolds number, $Re=\rho vL/\eta$. The Reynolds number compares the magnitudes of the advected momentum and the viscous drag. For $Re>1{,}000$, the flow usually is turbulent. For $Re\ll 1$, the flow obeys linear rheology. These flows are called "laminar". Above a QCM surface, Re is much below unity. To see that, insert the numbers $v\approx\omega u_0$, $u_0\approx 1$ nm, $\omega\approx 2\pi\times 5$ MHz, $L\approx\delta\approx 250$ nm, $\rho\approx 10^3$ kg/m^3, and $\eta\approx 10^{-3}$ Pa s. Re comes out to be in the range of 10^{-2}. Since $Re\ll 1$, there is no turbulence. The absence of turbulence does not imply that the nonlinear term in the Navier-Stokes equation was of no influence at all. It is of influence; it can drive a steady flow.

For the quantitative discussion of steady streaming, neglect the body forces and rearrange Eq. 13.2.1 as

$$\rho\frac{\partial \mathbf{v}}{\partial t} = -\rho(\mathbf{v}\cdot\nabla)\mathbf{v} + \eta\nabla^2\mathbf{v} - \nabla p \tag{13.2.2}$$

Written this way, the first term on the right-hand side can be viewed as an effective force density. Since it contains the velocity twice and since $\cos^2(\omega t) = 1/2(1 + \cos(2\omega t))$, part of this force occurs at $\omega = 0$.

Following perturbation theory, the flow field is decomposed into a steady part, $\mathbf{v}_{st}(\mathbf{r})$, and an unsteady part, $\mathbf{v}_{us}(\mathbf{r})$. The steady component is much smaller than the unsteady component. The Navier-Stokes-Equation applied to the (large) unsteady part at frequency ω is

$$\begin{aligned}-\omega\rho\mathbf{v}_{us,0}\sin(\omega t + \varphi_v) = &-\rho\left(\mathbf{v}_{us,0}\cdot\nabla\right)\mathbf{v}_{us,0}\cos^2(\omega t + \varphi_v)\\ &- \eta\nabla^2\mathbf{v}_{us,0}\cos(\omega t + \varphi_v) - \nabla p_{us,0}\cos\left(\omega t + \varphi_p\right)\end{aligned} \tag{13.2.3}$$

The index 0 denotes an amplitude, φ_v and φ_p are phases. Because of the $\cos^2$-term on the right-hand side, the unsteady flow generates a steady force density, which is to be inserted as a source term into the Navier-Stokes equation applied to the steady component. The 1st-order perturbation equation at $\omega = 0$ is

$$\begin{aligned}0 &= -\left\langle\rho\left(\mathbf{v}_{us,0}\cdot\nabla\right)\mathbf{v}_{us,0}\cos^2(\omega t + \varphi_v)\right\rangle_{time} - \eta\nabla^2\mathbf{v}_{st} - \nabla p_{st}\\ &= -\frac{1}{2}\rho\left(\mathbf{v}_{us,0}\cdot\nabla\right)\mathbf{v}_{us,0} - \eta\nabla^2\mathbf{v}_{st} - \nabla p_{st}\end{aligned} \tag{13.2.4}$$

The left-hand side is zero because the time-derivative of a steady flow vanishes by definition. The factor of 1/2 results from $\cos^2(\omega t) = ½\,(1 + \cos(2\omega t))$.

It was emphasized in the introduction that steady forces sometimes vanish for reasons of symmetry. Such a symmetry argument applies here. If the flow direction and the gradient direction are perpendicular, the steady force vanishes because the dot product in Eq. 13.2.4 returns a zero. As a consequence, the steady force vanishes for pure shear flows. In order for steady forces to exist above a QCM surface, the shear flow must be distorted somehow. The flexural modes resulting from energy trapping provide for such a distortion [14]. With flexural modes present, the surface moves into both the tangential and the normal direction. It is the combination of a transverse wave rapidly decaying into the z-direction and a compressional wave, which creates the steady force. A similar geometry was discussed by Wang and Drachman [15] and more recently by Sadhal [16].

Note: The type of streaming discussed above is different from Rayleigh streaming, which also occurs close to solid surfaces [12]. Rayleigh streaming involves strong in-plane gradients of amplitude. It is also different from the "quartz wind". For the quartz wind, see Ref. [12].

As discussed in more detail in Ref. [14], the flow occurs in the plane of the resonator, parallel to the displacement direction, and towards the center of the plate. Its absolute value can be estimated as

$$\mathbf{v}_{st} \approx \alpha \frac{\delta \rho \omega^2 u_0^2}{4\eta} \tag{13.2.5}$$

α is the ratio of the normal and the tangential component of the plate's motion. α depends on position (x and y); it may have positive or negative sign. As discussed in Sect. 6.1.3, α is below unity. With an amplitude of motion of around 40 nm on the fundamental, the velocity of the steady flow was found to be of the order of 1 mm/s. There is a side aspect to these experiments: From the fact that the flow is directed towards the center ($\alpha < 0$), one concludes that the sketch shown in Fig. 7. 6c (not the one in Fig. 7.6b) correctly depicts the deformation pattern of the plate in a liquid.

Steady streaming should be of importance in particle detachment events as described in Refs. [17, 18]. This type of measurement is termed Rupture Event Scanning (REVS) by its inventors. In REVS, the amplitude of oscillation is ramped up until the object of interest (often a virus) is shaken off the QCM surface. The interpretation is not primarily based on the frequency shift, but rather on the amplitude of oscillation, at which the particle detaches. A steady tangential flow may aid the debonding. Presumably, steady streaming was also implicitly observed by Edvardsson et al. in Ref. [19] and Heitmann et al. in Ref. [20]. These authors studied, whether particles or biological cells can be prevented from adsorbing by a large amplitude of oscillation. These experiments differ from REVS in that the high-amplitude vibration does not detach particles; it only prevents adsorption. The difference is emphasized in Ref. [19]. In the experiments by Edvardsson et al. the particles could not be shaken off. The authors conclude that the effect should be attributed to fluid dynamics rather than contact mechanics.

In the context of adsorption experiments, steady streaming can be used to stir the liquid. In a way, steady streaming is even more efficient in producing convective mass transport towards the sensor surface than conventional stirring. When stirring a liquid, convection does not reach to the resonator surface. There is a layer with a thickness of a few hundred microns (the so-called Nernst layer), inside which the transport occurs by diffusion only, no matter how fast the liquid is agitated. This is different if the resonator itself pumps the liquid. The steady force is generated inside the tail of the shear wave. The pump is located right at the resonator surface and convection therefore reaches closer to the surface than with stirring. (The same is true for flows driven by surface acoustic waves [21].)

13.3 Amplitude Dependence of Δf and ΔΓ

It was shown in Sect. 13.1 that nonlinear interactions produce forces at ω, which are not proportional to u_0. Since these periodic forces shift the resonance frequency, the resonance frequency depends on amplitude. The series expansion in Eq. 13.1.1 is rather restrictive. In general, the force will depend in some complicated way on displacement and velocity, and it does so hysteretically, that is, there is a dependence of the force on displacement and velocity at times in the past. In the context of QCM experiments, the displacement always is time-harmonic ($u = u_0\cos(\omega t)$). The QCM controls the displacement and measures the force. In that specific context it suffices to know how the force depends on u, u_0, and ω. This dependence will encompass memory effects, but only in so far as they occur with time-harmonic displacement.

For notational convenience, we define a normalized displacement as $u_N = u/u_0 = \cos(\omega t)$. The force can be written as $F_\pm(u_N, u_0, \omega)$, where −and + denote motion into direction of negative and positive u (Fig. 13.1). Since the displacement is time-harmonic, the functions $F_\pm(u_N, u_0, \omega)$ can be converted to a function of time, $F(t)$. It turns out that the SLA can be expanded, such that it covers arbitrary functions $F(t)$ and therefore arbitrary functions $F_\pm(u_N, u_0, \omega)$, as well. The theory of the weakly perturbed resonator leads to the relation [22–24]

$$\begin{aligned}\frac{\Delta\tilde{f}}{f_0} &= \frac{\mathrm{i}}{\pi Z_q}\frac{2\langle\sigma_S(t)\exp(\mathrm{i}\omega t)\rangle_{area,time}}{\hat{v}_S}\\ &= \frac{\mathrm{i}}{\pi Z_q}\frac{2}{\hat{v}_S}\langle\sigma_S(t)\cos(\omega t)+\mathrm{i}\sigma_S(t)\sin(\omega t)\rangle_{area,time}\\ &= \frac{N_P}{\pi Z_q}\frac{2}{\omega u_0}\langle F(t)\cos(\omega t)+\mathrm{i}F(t)\sin(\omega t)\rangle_{area,time}\end{aligned} \tag{13.3.1}$$

In line 3 it was assumed that the stress is caused by individual contacts, which are all identical and elastically independent. The stress is then equal to the force per contact, $F(t)$, multiplied by the number of contacts per unit area, N_P. Also, the velocity was expressed as $\hat{v}_S = \mathrm{i}\omega u_0$ in line 3. As Eq. 13.3.1 shows, the QCM essentially operates like a lock-in amplifier. It reports the first Fourier-component of $\sigma_S(t)$ at frequency ω. Δf and ΔΓ are proportional to the in-phase and the out-of-phase part of this component. The proof of Eq. 13.3.1 is sketched in the following. More detailed derivations are given in Refs. [22, 23].

Consider the unloaded resonator first (Sect. 4.1.3). Slightly rearranged, Eq. 4.1.8b reads as

$$\frac{\mathrm{d}^2x}{\mathrm{d}t^2} = -2\gamma_D\frac{\mathrm{d}x}{\mathrm{d}t} - \omega_0^2 x \tag{13.3.2}$$

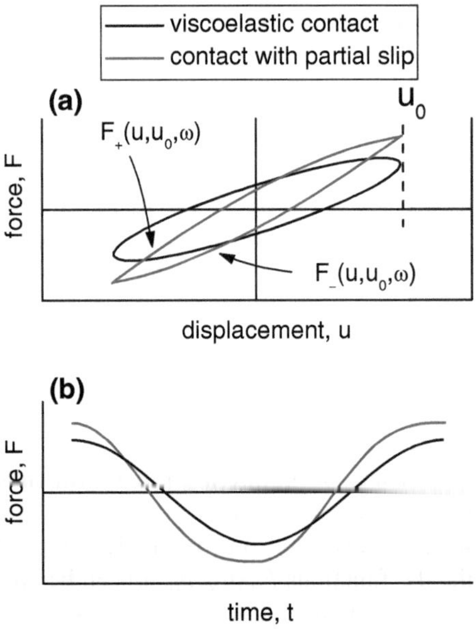

Fig. 13.1 **a** Force-displacement relations for the viscoelastic contact (*black, ellipse*) and the contact undergoing partial slip according to Cattaneo and Mindlin (*red, lens-shaped*). The force depends on the instantaneous displacement, u, on the amplitude of oscillation, u_0, and on the frequency, ω. **b** Since the displacement is approximately time-harmonic according to the SLA, the functions $F_-(u_N, u_0, \omega)$ and $F_+(u_N, u_0, \omega)$ can be converted to a function of time, $F(t)$. The QCM determines the first Fourier-component of $F(t)$

Define the variable y as

$$y = \frac{1}{\omega_d}\frac{\mathrm{d}x}{\mathrm{d}t} + \frac{\gamma_D}{\omega_d}x \tag{13.3.3}$$

$\omega_d = 2\pi f_d$ is the damped angular resonance frequency. From Eq. 4.1.18 one has

$$\omega_d^2 = \omega_0^2 - \gamma_D^2 \tag{13.3.4}$$

Using the variable y, the second order differential equation in Eq. 13.3.2 can be transformed to a system of two differential equations only containing first time-derivatives:

$$\begin{aligned}\frac{\mathrm{d}x}{\mathrm{d}t} &= -\gamma_D x + \omega_d y \\ \frac{\mathrm{d}y}{\mathrm{d}t} &= -\omega_d x - \gamma_D y\end{aligned} \tag{13.3.5}$$

A simpler equation system is obtained by introducing the absolute value, a, and the negative phase, ϕ, as

$$\begin{aligned} a &= \sqrt{x^2 + y^2} \\ \phi &= -\arctan\frac{y}{x} \end{aligned} \tag{13.3.6}$$

Repeated application of the chain rule and the product rule leads to:

$$\begin{aligned} \frac{\mathrm{d}a}{\mathrm{d}t} &= \frac{1}{a}\left(x\frac{\mathrm{d}x}{\mathrm{d}t} + y\frac{\mathrm{d}y}{\mathrm{d}t}\right) = -\gamma_D a \\ \frac{\mathrm{d}\phi}{\mathrm{d}t} &= \frac{1}{a^2}\left(y\frac{\mathrm{d}x}{\mathrm{d}t} - x\frac{\mathrm{d}y}{\mathrm{d}t}\right) = \omega_d \end{aligned} \tag{13.3.7}$$

The solution to this equation system with $a(t = 0) = x_0$ and $\phi(\mathrm{t} = 0) = 0$ is:

$$\begin{aligned} a(t) &= x_0 \exp(-\gamma_D t) \\ \phi(t) &= \omega_d t \end{aligned} \tag{13.3.8}$$

Clearly, the solution forms a spiral in the xy-plane as shown in Fig. 13.2. The damped resonance frequency and the half bandwidth are related to $\mathrm{d}a/\mathrm{d}t$ and $\mathrm{d}\varphi/\mathrm{d}t$ by

$$\begin{aligned} f_d &= \frac{1}{2\pi}\frac{\mathrm{d}\phi}{\mathrm{d}t} \\ \Gamma &= -\frac{1}{2\pi}\frac{1}{a}\frac{\mathrm{d}a}{\mathrm{d}t} \end{aligned} \tag{13.3.9}$$

In the next step, turn on the external force. There are two types of external forces, which are the driving force and the force exerted by the sample, F_{sam}. Only the force exerted by the sample is discussed in the following. With regard to the driving force, the discussion below Eq. 4.1.20 captures the situation well. There is little to be gained by redoing the calculation in terms of a and ϕ. Limiting the discussion to the force exerted by the sample can formally be justified by the equivalence between ring-down and impedance analysis. In ring-down, the driving force *is* zero. Arguably, this justification is on weak grounds because—strictly speaking—ring-down and impedance analysis only are equivalent within the frame of linear response. We still proceed with the force exerted by the sample.

With the force exerted by sample included, Eq. 13.3.2 turns into

$$\frac{\mathrm{d}^2 x}{\mathrm{d}t^2} = -2\gamma_D\frac{\mathrm{d}x}{\mathrm{d}t} - \omega_0^2 x + \frac{F_{sam}(t)}{M_R} \tag{13.3.10}$$

Defining y as before, one arrives at

$$\begin{aligned} \frac{\mathrm{d}x}{\mathrm{d}t} &= -\gamma_D x + \omega_d y \\ \frac{\mathrm{d}y}{\mathrm{d}t} &= -\omega_d x - \gamma_D y + \frac{1}{\omega_d}\frac{F_{sam}(t)}{M_R} \end{aligned} \tag{13.3.11}$$

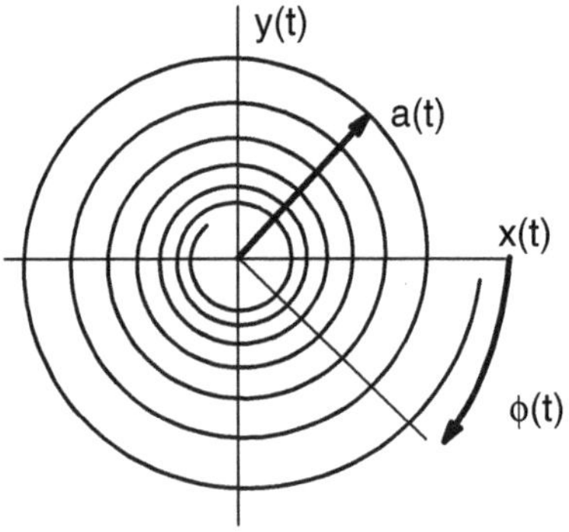

Fig. 13.2 a and ϕ are the absolute value and the negative phase of the complex variable $x + iy$. For the unperturbed resonance, $d\phi/dt$ and da/dt are equal to $2\pi f$ and $-2\pi\Gamma a$, respectively

The analog of Eq. 13.3.7 is

$$\begin{aligned}\frac{da}{dt} &= -\gamma_D a + \frac{y}{a}\frac{1}{\omega_d}\frac{F_{sam}(t)}{M_R}\\ \frac{d\phi}{dt} &= \omega_d - \frac{1}{a}\frac{x}{a}\frac{1}{\omega_d}\frac{F_{sam}(t)}{M_R}\end{aligned} \tag{13.3.12}$$

The next step exploits the SLA. Since the trajectory of the resonator, $x(t)$, is only marginally changed by application of the force, one may write $x/a \approx \cos(\phi)$ and $y/a \approx -\sin(\phi)$. Cutting a slightly more complicated argument short, we limit the discussion to long times. In the words of Ref. [22], we apply the "two-timing approximation". The complications arising from application of the SLA are irrelevant on long time scales. Replacing x/a and x/a in Eq. 13.3.12 by $\cos(\phi)$ and $-\sin(\phi)$ and using Eq. 13.3.8, one finds

$$\begin{aligned}\frac{da}{dt} &\approx -\gamma_D a + \frac{\sin(\omega_d t)}{\omega_d M_R} F_{sam}(t)\\ \frac{d\phi}{dt} &\approx \omega_d - \frac{\cos(\omega_d t)}{a\omega_d M_R} F_{sam}(t)\end{aligned} \tag{13.3.13}$$

The shift in resonance frequency is obtained by subtracting Eq. 13.3.7 from Eq. 13.3.13, inserting the difference into Eq. 13.3.9, and averaging over time. The following relations result:

$$\begin{aligned}\Delta f = f_d - f_{d,ref} &= \frac{1}{2\pi}\left\langle\frac{d\phi}{dt} - \frac{d\phi_{ref}}{dt}\right\rangle_{time} \approx -\frac{1}{2\pi M_R\omega_d a}\langle\cos(\omega_d t)F_{sam}(t)\rangle_{time}\\ \Delta\Gamma = \Gamma - \Gamma_{ref} &= -\frac{1}{2\pi a}\left\langle\frac{da}{dt} - \frac{da_{ref}}{dt}\right\rangle_{time} \approx -\frac{1}{2\pi M_R\omega_d a}\langle\sin(\omega_d t)F_{sam}(t)\rangle_{time}\end{aligned}$$

(13.3.14)

In complex notation, this reads as

$$\Delta\tilde{f} \approx \frac{1}{2\pi M_R \omega_d a} \langle -F_{sam}(t)\exp(i\omega_d t)\rangle_{time} \tag{13.3.15}$$

This result obtained for arbitrary resonators must now be mapped onto the model of the vibrating plate. The replacements leading from Eq. 13.3.15 to 13.3.1 are the following:

- Rename ω_d as ω.
- Replace M_R by $Am_q/2 = AZ_q/(4f_0)$ (Eqs. 4.5.27a and 4.3.17).
- Replace a by the amplitude of oscillation u_0. The dependence of $\Delta\tilde{f}$ on u_0 will be discussed in detail below.
- Replace F_{sam} by $-A\sigma_S$. The minus-sign occurs because the stress in the SLA is exerted by the resonator onto the environment, contrasting to the force in the simple harmonic resonator, which is exerted onto the resonator.

Again, Eq. 13.3.15 is valid *beyond* the power series introduced in the beginning of the chapter. It does not require the relation between σ_S and u_S to be weakly nonlinear. There is a subtlety here: The dynamics of the *entire* resonator *is* weakly nonlinear because the surface traction is small compared to the elastic stress inside the crystal itself. If the dynamics of the crystal as a whole was strongly nonlinear, it would behave chaotically. This would be the end of any QCM experiment. The QCM has a whole is weakly perturbed and therefore obeys a weakly nonlinear dynamical equation. Still, the sample may behave strongly nonlinear in the sense that the (small) surface stress is not well described by a power series in u_0. A strongly nonlinear stress-displacement relation is even plausible in experiments involving the rupture of contacts. When a contact starts to slide, the force transmitted across the contact abruptly drops. Such a sudden drop is not well covered by a series expansion of F_{sam} in u_0.

The matter can also be rephrased from an experimental point of view. If the stress as a function of surface displacement follows a cubic polynomial (Eq. 13.1.1), one can insert this law into Eq. 13.3.1 and arrive at a prediction for the function $\Delta f(u_0)$, which is

$$\begin{aligned} \frac{\Delta f}{f_0} &\approx \frac{2}{\pi Z_q \omega u_0}\left[\left\langle \kappa_1 u_0 \cos^2(\omega t)\right\rangle_{time} + \left\langle \kappa_3 u_0^3 \cos^4(\omega t)\right\rangle_{time}\right] \\ &= \frac{2}{\pi Z_q \omega u_0}\left[\frac{\kappa_1}{2} u_0 + \frac{3}{8}\kappa_3 u_0^3\right] \\ &= \frac{1}{\pi Z_q \omega}\left[\kappa_1 + \frac{3}{4}\kappa_3 u_0^2\right] \end{aligned} \tag{13.3.16}$$

Clearly, there is a constant term and a term proportional to u_0^2. $\Delta f(u_0)$ is predicted to be parabolic. Whether or not the series expansion from Eq. 13.1.1 holds, can be easily checked experimentally. Experiment shows that the power series terminated at third order does not usually hold.

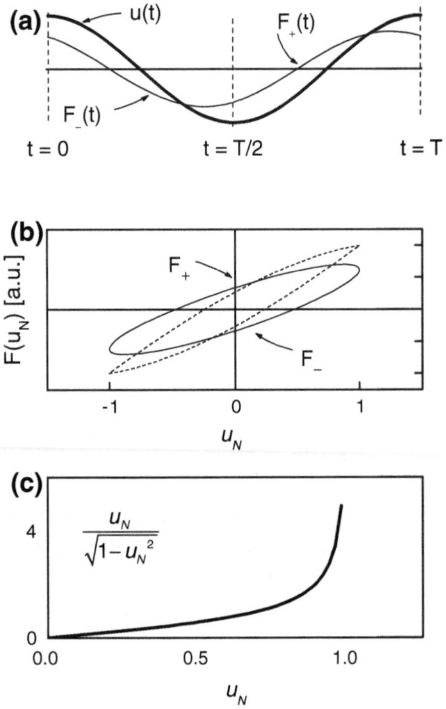

Fig. 13.3 **a** Force and displacement versus time for a viscoelastic contact. The displacement monotonously decreases and increases for the first half and the second half of the period of oscillation. Since $u(t)$ is piecewise monotonic, a time-average can be converted to an average over displacement. **b** Force-displacement relation. *Solid line* viscoelastic contact. *Dashed line* Partial slip according to Cattaneo [9] and Mindlin [10]. The shift in half-bandwidth is proportional to the area inside the force-displacement loop (that is, the difference between F_+ and F_-). The shift in frequency is a weighted integral of the sum of F_+ and F_-, where the weight function (panel **c**) has a sharp maximum at the turning point. *Note* F_+ and F_- are functions of u, u_0, and ω. Adapted from Ref. [31]

Equation 13.3.1 primarily concerns forces which are not time-harmonic. The displacement, $u(t)$, on the other hand, *is* almost time-harmonic. This is the consequence of the SLA. The force is so small that the motion of the resonator surface remains approximately sinusoidal. In the terminology of mechanics, the QCM controls the displacement and measures the force. (Other instruments in mechanical testing control the force and measure the displacement.) Since the function $u_N(t) = u(t)/u_0(t) = \cos(\omega t)$ is piecewise monotonic (Fig. 13.3a), the time-average in Eq. 13.3.1 can be converted to an average over displacement. The averaging process must occur separately for the two directions of motion, where the two forces are $F_-(u_N,u_0,\omega)$ and $F_+(u_N,u_0,\omega)$.

For the function $\Delta f(u_0)$, the transformation from $F(t)$ to $F_\pm(u_N,u_0,\omega)$ takes the form

$$\begin{aligned}\Delta f(u_0) &= \frac{N_P}{n\pi^2 Z_q}\frac{1}{u_0}\frac{1}{T}\left(\int_0^{T/2} F_-(t)\cos(\omega t)\mathrm{d}t + \int_{T/2}^{T} F_+(t)\cos(\omega t)\mathrm{d}t\right)\\
&= \frac{N_P}{n\pi^2 Z_q}\frac{1}{u_0}\frac{1}{\omega T}\left(\int_1^{-1} -F_-(u_N,u_0,\omega)\frac{u_N}{\sqrt{1-u_N^2}}\mathrm{d}u_N + \int_{-1}^{1} F_+(u_N,u_0,\omega)\frac{u_N}{\sqrt{1-u_N^2}}\mathrm{d}u_N\right)\\
&= \frac{N_P}{n\pi^2 Z_q}\frac{1}{u_0}\frac{1}{2\pi}\int_{-1}^{1}\left(F_-(u_N,u_0,\omega)+F_+(u_N,u_0,\omega)\right)\frac{u_N}{\sqrt{1-u_N^2}}\mathrm{d}u_N\\
&= \frac{N_P}{n\pi^2 Z_q}\frac{1}{u_0}\frac{1}{\pi}\int_{0}^{1}\left(F_-(u_N,u_0,\omega)+F_+(u_N,u_0,\omega)\right)\frac{u_N}{\sqrt{1-u_N^2}}\mathrm{d}u_N\end{aligned} \tag{13.3.17}$$

T is the period of oscillation. Line 2 made use of $\omega T = 2\pi$ and $\mathrm{d}t/\mathrm{d}u_N = \pm\,\omega^{-1}(1-u_N^2)^{-1/2}$. Since the integrand is symmetric in u_N, the integral can be limited to positive values of u_N (line 4). The term $u_N(1-u_N^2)^{-1/2}$ then is positive and can be viewed as a statistical weight. The frequency shift is proportional to a weighted average of $|F_- + F_+|$, where the weight function has a sharp peak at the turning point (Fig. 13.3c).

For the half-bandwidth, ΔΓ, the conversion reads as

$$\begin{aligned}\Delta\Gamma(u_0) &= \frac{N_P}{n\pi^2 Z_q}\frac{1}{u_0}\frac{1}{T}\left(\int_0^{T/2} F_-(t)\sin(\omega t)\mathrm{d}t + \int_{T/2}^{T} F_+(t)\sin(\omega t)\mathrm{d}t\right)\\
&= \frac{N_P}{n\pi^2 Z_q}\frac{1}{u_0}\frac{1}{\omega T}\left(\int_1^{-1} F_-(t)\mathrm{d}(\cos(\omega t)) + \int_{-1}^{1} F_+(t)\mathrm{d}(\cos(\omega t))\right)\\
&= \frac{N_P}{n\pi^2 Z_q}\frac{1}{u_0}\frac{1}{2\pi}\int_{-1}^{1}\left(-F_-(u_N,u_0,\omega)+F_+(u_N,u_0,\omega)\right)\mathrm{d}u_N\\
&= \frac{N_P}{n\pi^2 Z_q}\frac{1}{u_0^2}\frac{1}{2\pi}\int_{-u_0}^{u_0}\left(F_+(u_N,u_0,\omega)-F_-(u_N,u_0,\omega)\right)\mathrm{d}u\end{aligned} \tag{13.3.18}$$

Line 2 made use of $\sin(\omega t)\mathrm{d}t = -\mathrm{d}(\cos(\omega t))/\omega$. ΔΓ is proportional to the area inside the force-displacement loop. The integration variable u (rather than u_N) was used in the line 4 in order to give the integral the familiar dimension of an energy. ΔΓ is proportional to the energy dissipated per cycle divided by u_0^2.

If the dependence of F_+ and F_- on u_N, u_0, and ω is known, one can predict the functions $\Delta f(u_0)$ and $\Delta\Gamma(u_0)$ from Eqs. 13.3.17 and 13.3.18. Unfortunately, the inversion (the calculation of $F_+(u_N,u_{0,}\omega)$ from $\Delta\tilde{f}(u_0)$) is impossible unless $F_+(u_N,u_{0,}\omega)$ has some further properties not obeyed in friction experiments.

Inversion is possible if the force is a function of displacement only, that is, if there is no hysteresis. These conditions often are fulfilled in non-contact atomic force microscopy [25–27]. In frequency-modulation AFM, one varies the distance between the vibrating cantilever and the surface, rather than amplitude, but this does not affect the formal structure of the equations. The measurement of the resonance frequency versus distance to the surface is a rather powerful method in this context. One can explicitly derive the force-distance relation. In friction experiments, this methodology does not work because the force itself depends on the amplitude (if there is hysteresis). The formalism only runs forward; the analysis requires assumptions on $F_\pm(u_N,u_0,\omega)$ as an input. There is a second difference: The formalism applied to QCM-data starts from a viscoelastic contact, which becomes slightly nonlinear, as the amplitude increases. On the way, the contact becomes weaker and eventually breaks. In the AFM experiment, one starts from a non-contact situation, which is perturbed by a weak interaction with the sample. Both types of experiments deal with weak contacts, but the weak coupling is approached from opposite sides.

If a quantitative prediction on $F_\pm(u_N,u_0,\omega)$ is not available, one can still plot a guess of the force-displacement loops and understand its relation to the frequency-amplitude relation on a qualitative level. The essence of Eqs. 13.3.17 and 13.3.18 can be expressed as follows:

- The weight function $u_N(1 - u_N^2)^{-1/2}$ in Eq. 13.3.17 has a sharp peak at the turning point, and Δf therefore mostly measures the *restoring force at the turning point* (Fig. 13.3c). If the force at $u = u_0$ is proportional to u_0 (as in the viscoelastic case, Fig. 13.4a), $\Delta f(u_0)$ is independent of u_0. If the force at the turning point depends sub-linearly on amplitude (as in partial slip, Fig. 13.4b, or in gross slip, Fig. 13.4c), $\Delta f(u_0)$ has negative slope. If the frequency is experimentally found to increase with amplitude (as in Ref. [28]) this cannot be explained with partial slip.
- $\Delta\Gamma$ measures the *area inside the force-displacement loop divided by* u_0^2. If the dissipated energy scales as u_0^2 (as in the viscoelastic case, Fig. 13.4a), then $\Delta\Gamma$ is independent of amplitude. If the scaling with amplitude is stronger than quadratic (as in partial slip, Fig. 13.4b) or weaker than quadratic (as in gross slip, Fig. 13.4c), $\Delta\Gamma$ increases or decreases with amplitude, respectively.

In the following, the formalism is applied to the study of partial slip [29]. This discussion serves two purposes. It mainly concerns partial slip, but it also is a case study, demonstrating how the concepts developed above can be put to use.

Figure 13.5 shows a data set illustrating three characteristically distinct types of behavior. The resonator was touched from above with three glass spheres, which had been glued to a common support and all exerted the same vertical force. An acrylic polymer film with variable glass temperature, T_g, had been spin-cast onto the resonator. Polymer surfaces are advantageous for such studies because they

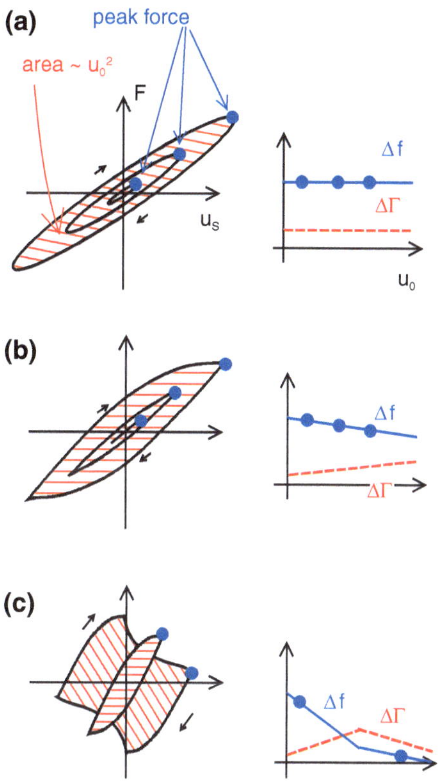

Fig. 13.4 Illustration of how Δf and $\Delta\Gamma$ depend on the shape of the force-displacement loop.
a Viscoelastic contact
b Partial slip according to Cattaneo and Mindlin
c Transition to gross slip. The frequency shift is proportional to the ratio of force and displacement at the turning point (*full dots*). $\Delta\Gamma$ is proportional to the area inside the loop divided by u_0^2 (*hatched*). Adapted from Ref. [31]

usually allow for some small amount of plastic flow, which makes the effects of nano-roughness slightly less pronounced than in the case of silica-silica contacts. Parameters varied in the experiment were the softness of the polymer surface, the sphere size, the normal force, and the humidity.

The following three cases can be distinguished:

- For small spheres ($R_P = 25$ µm) on a soft substrate, Δf and $\Delta\Gamma$ do not depend on oscillation amplitude. The contact obeys linear viscoelasticity (Fig. 13.5a, b).
- With a somewhat larger sphere radius ($R_P = 70$ µm), Δf decreases with amplitude, while $\Delta\Gamma$ increases. This behavior is characteristic of partial slip (Fig. 13.5c, d).
- For the largest spheres ($R_P = 137$ µm) on a more rigid substrate, one sees partial slip at low amplitudes, turning over into what is called "gross slip" above an amplitude of about 12 nm. At high amplitudes, the bandwidth decreases and the frequency levels off (Fig. 13.5e, f).

How the force-displacement relations and the frequency-amplitude relations are connected to each other, can be understood without quantitative modeling. One can read the frequency-amplitude relations from the force-displacement diagrams

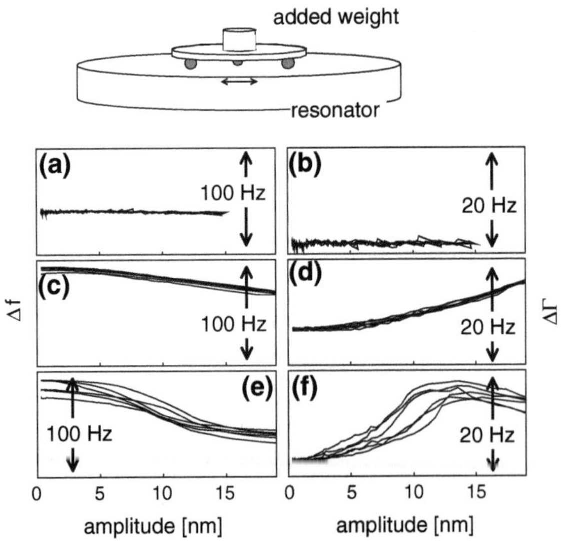

Fig. 13.5 A data set where all three regimes (linear viscoelasticity, partial slip, gross slip) are observed. Panels **a** and **b** Viscoelastic behavior (small contacts, soft substrate). Panels **c** and **d** Partial slip (medium-size contacts, soft substrate). Panels **e** and **f** Gross slip (large contacts, substrate with T_g much above room temperature). Adapted from Ref. [31]

as depicted on left-hand side in Fig. 13.4. Frequency and half-bandwidth are related to the peak force and to the area inside the loop, respectively. From Fig. 13.4, one understands that under conditions of gross slip, Δf is smaller than under conditions of stick. The ratio F_{max}/u_{max} is decreased relative to the sticking condition. Also, the area inside the force-displacement loop does no longer depend quadratically on u_0. In the limit of Coulomb-friction ($F_x = \mu_D F_N =$ const with μ_D the dynamic friction coefficient), the dissipated energy is proportional to u_0. In the transition range, the scaling will be somewhere between linear and quadratic. Since it is lower than quadratic, $\Delta\Gamma$ decreases with u_0.

The author is not aware of a quantitative model for the transition to gross slip under oscillatory load, but there are two such models for partial slip. The model by Cattaneo [25] and Mindlin [30] assumes the tangential stress to be proportional to the normal stress in the sliding zone, while the model devised by Savkoor assumes the tangential stress in the sliding zone to be constant [31, 32]. The stress distributions and the corresponding force-displacement relations are shown in Figs. 11.3 and 11.4. Presumably, both models are too simple, but one can still translate them to a prediction for $\Delta f(u_0)$ and $\Delta\Gamma(u_0)$. Based on experiment, one can make a statement on which of the two models is closer to reality.

For the mathematical details the reader is referred to Ref. [31]. The frequency-amplitude relation derived from the Cattaneo-Mindlin model in the low-amplitude limit is [31]:

$$\begin{aligned} \Delta f(u_0) &\approx \frac{N_P}{2n\pi^2 Z_q}\kappa\left(1-\frac{\kappa}{3\mu F_N}u_0\right) \\ \Delta\Gamma(u_0) &\approx \frac{N_P}{2n\pi^2 Z_q}\kappa\frac{4}{9\pi}\frac{\kappa}{\mu F_N}u_0 \end{aligned} \tag{13.3.19}$$

The Savkoor model differs from the Cattaneo-Mindlin model with regard to the stress in the sliding zone, which is constant (rather than being proportional to the normal stress). The resulting frequency-amplitude relation is [31]

$$\begin{aligned} \Delta f(u_0) &\approx \frac{N_P}{2n\pi^2 Z_q}\kappa\left(1-\frac{5}{8}\left(\frac{\kappa u_0}{4\tau_0 a^2}\right)^2\right) \\ \Delta\Gamma(u_0) &\approx \frac{N_P}{2n\pi^2 Z_q}\kappa\frac{8}{6\pi}\left(\frac{\kappa u_0}{2\tau_0 a^2}\right)^2 \end{aligned} \tag{13.3.20}$$

Clearly, Δf decreases with u_0 in both models, while $\Delta\Gamma$ increases. In this regard, the two models agree. However Savkoor predicts a parabolic relation, while Cattaneo and Mindlin predict a linear dependence (Fig. 13.6). Neither model is perfectly obeyed in Fig. 13.5c, but the data come closer to the model by Cattaneo and Mindlin. This conclusion is also reached from the experiments by Leopoldes and Jia [33].

Probing contact stiffness based on the frequency of a resonator is also practiced with the AFM [34] and with macroscopic instruments [35]. That includes non-linear interactions and the amplitude dependence of the resonance frequency. The procedure is called "contact resonance method" in that context. One may wonder why the contact resonance method is employed at all, given that the force-displacement relations cannot be explicitly obtained. The advantage of the contact resonance method lies in its precision. It is much easier to measure the frequency and the half-bandwidth of a resonance, than it is to detect small nonlinearities in a force-displacement diagram. The QCM arrives at this accuracy by averaging over millions of oscillation cycles. There is an inherent problem in this approach, though. When excited at elevated amplitudes, the properties of the contact might evolve *with time*. For instance, the contact might soften because of frictional heating. Translated to the diagrammatic scheme of Fig. 13.4, this would imply that the force-displacement loop is an ellipse at all amplitudes, but that the width and the orientation of the ellipse change with amplitude (that is, with temperature, see Fig. 13.7). Frictional heating is not necessarily the only reason for an evolution of a contact's properties upon prolonged cyclic loading. There also might be plastic deformation, known from mechanics. Elastic-plastic bodies under cyclic load tend to evolve towards a state with lower overall energy dissipated per cycle than in the initial state. The phenomenon is referred to as "shake down" [36]. The concept is not directly applicable to interfacial sliding [37], but one may nevertheless suspect that plastic deformation modifies a contact such that it becomes less dissipative. For instance, the two surfaces might become smoother with time, which would decrease the frictional forces.

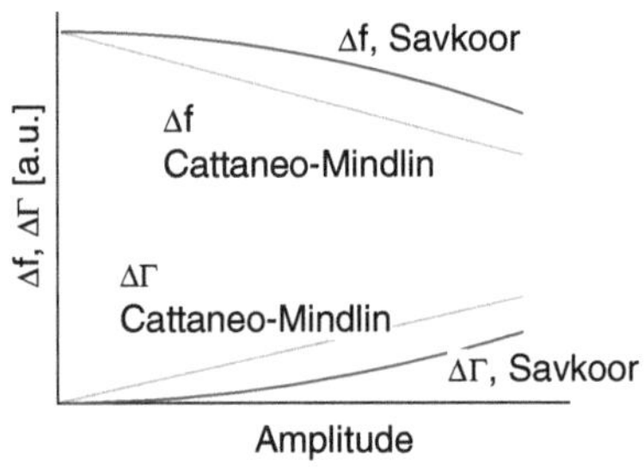

Fig. 13.6 Frequency-amplitude relations for conditions of partial slip according to Cattaneo-Mindlin [25] and Savkoor [33, 34]. Adapted from Ref. [31]

In impedance analysis, a few million oscillation cycles enter every single measurement of $\Delta\tilde{f}$; an evolution of a contact's stiffness, which would happen during the first frequency sweep (lasting for about a second), can therefore not be resolved in experiment. This problem is much alleviated in ring-down. Such experiments have been reported in Ref. [38]. In ring-down, the individual measurement lasts for about the inverse half-bandwidth, which is a few milliseconds. Every single decay is a fast amplitude sweep. Should the frequency decrease with amplitude, it would increase over the duration of the decay. The pulses would be "chirped". Birds chirp. They emit pulses of acoustic radiation, the frequency of which varies over the duration of the pulse. Frictional heating (or any other evolution of the contact's properties with time) would have to happen within a few milliseconds in order to produce chirp. That is not strictly impossible, but unlikely.

At any rate: Eq. 13.3.1 does not give access to the shape of the force-displacement curves itself; this caveat must be kept in mind. The alternative approach, which does not suffer from this ambiguity, is third-harmonic generation (THG, Sect. 13.4). The THG signal vanishes if the force-displacement loop is an ellipse, regardless of its width and orientation.

Two more comments:

- The resonator itself is slightly nonlinear (Sect. 17.3). Even the bare resonator shows an amplitude dependence of Δf; this intrinsic contribution must be subtracted from the raw data.
- When calculating the amplitude, the output resistance of the impedance analyzer, R_{out}, must be taken into account (Sect. 7.4). The amplitude of oscillation is proportional to the voltage at the electrode, as opposed to nominal voltage at the output of the analyzer, U_{out}. This has the interesting consequence, that the amplitude of oscillation does not necessarily increase, when the frequency of excitation is swept towards the resonance frequency. When the crystal's motional impedance, R_1, is much less than R_{out}, the resonator is under current control. The current through the circuit is given as $U_{out}/(R_{out} + R_1)$. On resonance, the larger part of the voltage drop occurs at R_{out}. The current into the resonator remains about constant, while the voltage across the electrodes goes through a minimum. This situation is beneficial for experiment because it implies that the resonance frequency does not change during a frequency sweep. If this was not the case, the resonance curve would be distorted. The

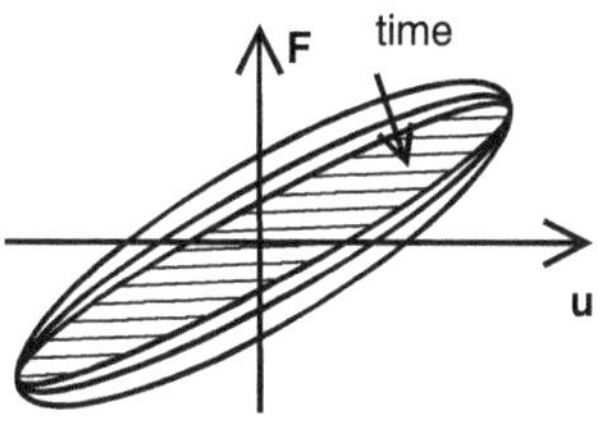

Fig. 13.7 The properties of a contact subjected to large-amplitude oscillatory shear may evolve with time. The contact might behave viscoelastically, but the width of the ellipse (and also its orientation) might differ from the low-amplitude state. Adapted from Ref. [31]

resonance frequency would change while the driving frequency approaches the peak of the conductance curve because the amplitude changes during the sweep. Such distorted resonance curves are known from the Duffing oscillator [39]. Fortunately, this effect is weak. Again, the above argument applies when $R_1 \ll R_{out}$, which is expected in air with large enough crystals.

13.4 Third-Harmonic Generation

Third-harmonic generation as a probe for nonlinear interactions was introduced to QCM-based sensing by the Cambridge group in the context of "Rupture Event Scanning" (REVS) [17]. The setup is simple. The signal returned from the crystal is fed to a lock-in amplifier, which monitors the third-harmonic component, referenced to the driving signal. In rupture event scanning, one is mostly concerned with sudden spikes in the third-harmonic signal, which occur when particles detach from the resonator surface. The corresponding oscillation amplitude is a measure of the contact strength and is characteristic of the respective bond. Later, the Cambridge group has also reported data with a smooth dependence of the third-harmonic on amplitude [40]. Gosh et al. plot the current at 3ω versus the amplitude of oscillation. Unfortunately, the conversion between the electric current at 3ω and the corresponding area-averaged stress is nontrivial. Still, such measurements should help to the unravel the properties of discrete contacts under large loads.

13.5 Appendix: Determination of the Shear Direction with Polarizing Microscopy

There are a few (although not many) instances, where one needs to know the direction of oscillation in order to properly interpret a QCM experiment. Examples are the study of anisotropic viscoelasticity [41], anisotropic viscosity in nematic liquid crystals (Sect. 9.2), and steady streaming (Sect. 13.2).

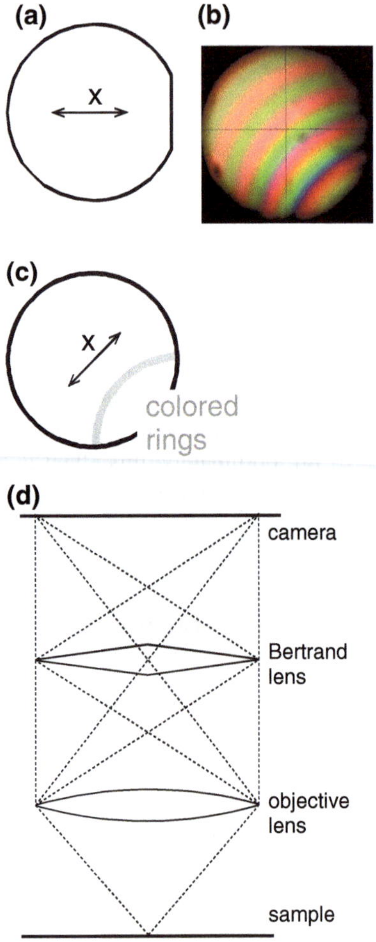

Fig. 13.8 On the determination of crystal orientation with conoscopy. **a** Some crystals have a flat cut at the edge, indicating the direction of the x-axis. **b** and **c** The crystallographic x-axis can be found with a polarizing microscope, using the conoscopic mode. One observes colored rings. The z-axis corresponds to the center of these rings. Panel **c** shows the direction of the x-axis. **d** In conoscopy, a second lens (the "Bertrand lens") maps on image of the objective lens onto the screen. The position of a pixel encodes angle, not position

AT-cut crystals oscillate along the crystallographic x-axis. Some crystals are slightly truncated on one side, where the flat denotes the x-axis (Fig. 13.8a). If the crystal does not have such a flat, a polarizing microscope equipped with a conoscopy-module (a Bertrand lens) can help. For an explanation of conoscopy see Refs. [42, 43]. In conoscopy, the pixels on the screen correspond to beams passing through the sample under different *angles* (Fig. 13.8d). One usually employs crossed polarizers and studies birefringent specimens. Quartz is uniaxially birefringent, where the uni-axis is the z-axis, also called the "optical axis". The conoscopic image of quartz blanks displays colored rings, the center of which is the optical axis. For the AT-cut, the z-axis is inclined away from the surface normal by 55°. Usually, the center of the rings is outside the field of view, but one can still identify the orientation of the x-axis based on the location of the rings (Fig. 13.8c). Depending on the orientation of the crystal relative to the crossed polarizers, the conoscopic image may display black stripes. These are the arms of a cross, centered on the optical axis. Again, the center typically is outside the field of view, but one can infer its location by rotating the sample and watching the movement of the stripes.

If one does not have access to a polarizing microscope, one can obtain equivalent information by placing the quartz blank between crossed polarizers. (These can be cheap sheet polarizers.) The sandwich becomes bright in transmission, if the crystallographic x-axis is at an angle of 45° relative to the polarizers. If one rotates the plate about the x-axis, one sees a sequence of colors (corresponding to the rings in conoscopy). If one rotates the plate about the other axis, there are color changes as well, but not as many as for the rotation about the x-axis.

Glossary

Variable	Definition (Comments)
$\langle \ldots \rangle$	An average
0	As an index: amplitude of a real-valued time-harmonic function
a	Instantaneous amplitude (Eq. 13.3.6)
A	(Effective) area of the resonator plate
f	Frequency
f_{body}	A body force (Eq. 13.2.1)
f_d	Damped resonance frequency
f_0	Resonance frequency at the fundamental ($f_0 = Z_q/(2m_q) = Z_q/(2\rho_q d_q)$)
F	Tangential force
F_+, F_-	Force at interface while u_S increases (+) and decreases (−)
F_N	Normal force
$F_{\max}$	Maximum force in a force-displacement loop
F_{sam}	Force exerted onto a resonator by a sample
F_x	Tangential force
m_q	Mass per unit area of the plate ($m_q = \rho_q d_q = Z_q/(2f_0)$)
M_R	Mass of a resonator
N_P	Number of particles per unit area
R_P	Particle radius
p	Pressure

ref	As an index: *ref*erence state
S	As an index: *S*urface
st	As an index: *st*eady
t	Time
T	Period of oscillation
T_g	Glass temperature
us	As an index: *uns*teady
$u(t)$	Displacement
$\hat{u}_S$, u_S	Tangential displacement
u_{max}	Maximum displacement in a force-displacement loop
u_N	Normalized displacement ($u_N = u_S/u_0$)
u_0	Amplitude of oscillation (equal to the absolute value of $\hat{u}_S$)
$\mathbf{v}$	Velocity
x, y, z	Spatial coordinates
x	Displacement of a simple harmonic resonator
y	Auxiliary variable (Eq. 13.3.3)
Z_q	Acoustic wave impedance of AT-cut quartz ($Z_q = 8.8 \times 10^6$ kg m^{-2} s^{-1})
α	Ratio of the normal and the tangential component of the plate's motion (depends on position, x and y; Sect. 13.2)
γ_D	Damping factor
Γ	Imaginary part of a resonance frequency
δ	Penetration depth of a shear wave (Newtonian liquids: $\delta = (2\eta_{liq}/(\rho_{liq}\omega))^{1/2})$)
Δ	As a prefix: A shift induced by the presence of the sample
φ_p, φ_v	Phases (Eq. 13.2.3)
η	Viscosity
κ	Spring constant
κ_1, κ_2, κ_3	Coefficients in a power series of force versus displacement (κ_1 is the conventional spring constant)
ϕ	Instantaneous phase (Eq. 13.3.6)
μ	Friction coefficient
μ_D	Dynamic friction coefficient

ρ	Density
ω	Angular frequency
ω_d	Damped angular resonance frequency
ω_0	Undamped angular resonance frequency

References

1. Johnson, K.L.: Contact Mechanics. Cambridge University Press, Cambridge (1985)
2. Popov, V.L.: Contact Mechanics and Friction: Physical Principles and Applications. Springer, Heidelberg (2010)
3. Patel, M.S., Yong, Y.K., Tanaka, M.: Drive level dependency in quartz resonators. Int. J. Solids Struct. **46**(9), 1856–1871 (2009)
4. Leopoldes, J., Jia, X.: Transverse shear oscillator investigation of boundary lubrication in weakly adhered films. Phys. Rev. Lett. **105**(26), 266101 (2010)
5. Wondraczek, K., Bund, A., Johannsmann, D.: Acoustic second harmonic generation from rough surfaces under shear excitation in liquids. Langmuir **20**(23), 10346–10350 (2004)
6. Heim, L.O., Johannsmann, D.: Oscillation-induced static deflection in scanning force microscopy. Rev. Sci. Instrum. **78**(1), 013902 (2007)
7. Krotil, H.U., Stifter, T., Marti, O.: Lock-in technique for concurrent measurement of adhesion and friction with the scanning force microscope. Rev. Sci. Instrum. **72**(1), 150–156 (2001)
8. Rabe, U., Amelio, S., Kopycinska, M., Hirsekorn, S., Kempf, M., Goken, M., Arnold, W.: Imaging and measurement of local mechanical material properties by atomic force acoustic microscopy. Surf. Interface Anal. **33**(2), 65–70 (2002)
9. Rabe, U., Janser, K., Arnold, W.: Vibrations of free and surface-coupled atomic force microscope cantilevers: theory and experiment. Rev. Sci. Instrum. **67**(9), 3281–3293 (1996)
10. Larson, R.G.: The Structure and Rheology of Complex Fluids. Oxford University Press, New York (1998)
11. Bruus, H., Dual, J., Hawkes, J., Hill, M., Laurell, T., Nilsson, J., Radel, S., Sadhal, S., Wiklund, M.: Forthcoming Lab on a Chip tutorial series on acoustofluidics: acoustofluidics-exploiting ultrasonic standing wave forces and acoustic streaming in microfluidic systems for cell and particle manipulation. Lab Chip **11**(21), 3579–3580 (2012)
12. Riley, N.: Steady streaming. Annu. Rev. Fluid Mech. **33**, 43–65 (2001)
13. Langhoff, A., Johannsmann, D.: Attractive forces on hard and soft colloidal objects located close to the surface of an acoustic-thickness shear resonator. Phys. Rev. E **88**(1), 013001 (2013)
14. König, R., Langhoff, A., Johannsmann, D.: Steady flows above a quartz crystal resonator driven at elevated amplitude. Phys.Rev. E **89**, 043016 (2014)
15. Wang, C.Y., Drachman, B.: The steady streaming generated by a vibrating plate parallel to a fixed plate. Appl. Sci. Res. **39**(1), 55–68 (1982)
16. Sadhal, S.S.: Acoustofluidics 13: analysis of acoustic streaming by perturbation methods Foreword. Lab Chip **12**(13), 2292–2300 (2012)
17. Cooper, M.A., Dultsev, F.N., Minson, T., Ostanin, V.P., Abell, C., Klenerman, D.: Direct and sensitive detection of a human virus by rupture event scanning. Nat. Biotechnol. **19**(9), 833–837 (2001)
18. Dultsev, F.N., Speight, R.E., Florini, M.T., Blackburn, J.M., Abell, C., Ostanin, V.P., Klenerman, D.: Direct and quantitative detection of bacteriophage by "Hearing" surface detachment using a quartz crystal microbalance. Anal. Chem. **73**(16), 3935–3939 (2001)

19. Edvardsson, M., Rodahl, M., Hook, F.: Investigation of binding event perturbations caused by elevated QCM-D oscillation amplitude. Analyst **131**(7), 822–828 (2006)
20. Heitmann, V., Wegener, J.: Monitoring cell adhesion by piezoresonators: impact of increasing oscillation amplitudes. Anal. Chem. **79**(9), 3392–3400 (2007)
21. Yeo, L.Y., Friend, J.R.: Ultrafast microfluidics using surface acoustic waves. Biomicrofluidics **3**(1), 012002 (2009)
22. Strogatz, S.: Nonlinear Dynamics and Chaos, Chapter 7.6. Addison-Wesley, Boston (1996)
23. Berg, S., Prellberg, T., Johannsmann, D.: Nonlinear contact mechanics based on ring-down experiments with quartz crystal resonators. Rev. Sci. Instrum. **74**(1), 118–126 (2003)
24. Johannsmann, D.: Viscoelastic, mechanical, and dielectric measurements on complex samples with the quartz crystal microbalance. Phys. Chem. Chem. Phys. **10**(31), 4516–4534 (2008)
25. Giessibl, F.J.: A direct method to calculate tip-sample forces from frequency shifts in frequency-modulation atomic force microscopy. Appl. Phys. Lett. **78**(1), 123–125 (2001)
26. Hölscher, H., Schwarz, U.D., Wiesendanger, R.: Calculation of the frequency shift in dynamic force microscopy. Appl. Surf. Sci. **140**(3–4), 344–351 (1999)
27. Sader, J.E., Jarvis, S.P.: Accurate formulas for interaction force and energy in frequency modulation force spectroscopy. Appl. Phys. Lett. **84**(10), 1801–1803 (2004)
28. Kunze, A., Zhao, F., Marel, A.K., Svedhem, S., Kasemo, B.: Ion-mediated changes of supported lipid bilayers and their coupling to the substrate. A case of bilayer slip? Soft Matter **7**(18), 8582–8591 (2011)
29. Hanke, S., Petri, J., Johannsmann, D.: Partial slip in mesoscale contacts: dependence on contact size. Phys. Rev. E **88**(3), 032408 (2013)
30. Mindlin, R.D., Deresiewicz, H.: Elastic spheres in contact under varying oblique forces. J. Appl. Mech. Trans Asme **20**(3), 327–344 (1953)
31. Savkoor, A.R.: Dry Adhesive Contact of Elastomers. University Delft, Tech (1987)
32. Johnson, K.L.: Adhesion and friction between a smooth elastic spherical asperity and a plane surface. Proc. R. Soc. Lond. Ser. A Math. Phys. Eng. Sci. **1997**(453), 163–179 (1956)
33. Leopoldes, J., Jia, X.: Transverse shear oscillator investigation of boundary lubrication in weakly adhered films. Phys. Rev. Lett. **105**(26), 266101–266104 (2010)
34. Rabe, U., Amelio, S., Kester, E., Scherer, V., Hirsekorn, S., Arnold, W.: Quantitative determination of contact stiffness using atomic force acoustic microscopy. Ultrasonics **38**(1–8), 430–437 (2000)
35. Eriten, M., Lee, C.H., Polycarpou, A.A.: Measurements of tangential stiffness and damping of mechanical joints: direct versus indirect contact resonance methods. Tribol. Int. **50**, 35–44 (2012)
36. Melan, E.: Zur Plastizität des räumlichen Kontinuums. Ing. Arch. **9**, 116 (1938)
37. Klarbring, A., Ciavarella, M., Barber, J.R.: Shakedown in elastic contact problems with Coulomb friction. Int. J. Solids Struct. **44**(25–26), 8355–8365 (2007)
38. Berg, S., Johannsmann, D.: High speed microtribology with quartz crystal resonators. Phys. Rev. Lett. **91**(14), 145505 (2003)
39. http://en.wikipedia.org/wiki/Duffing_equation. Accessed 14 Feb 2014
40. Ghosh, S.K., Ostanin, V.P., Seshia, A.A.: Anharmonic surface interactions for biomolecular screening and characterization. Anal. Chem. **83**(2), 549–554 (2010)
41. Johannsmann, D., Mathauer, K., Wegner, G., Knoll, W.: Viscoelastic properties of thin-films probed with a quartz-crystal resonator. Phys. Rev. B **46**(12), 7808–7815 (1992)
42. http://en.wikipedia.org/wiki/Conoscopy. Accessed 4 Feb 2013
43. Hartshorne, N.H., Stuart, A.: Crystals and the Polarizing Microscope. Hodder & Stoughton Educational, London (1970)

Chapter 14
Practical Consequences of Piezoelectric Stiffening

Abstract Piezoelectric stiffening can be exploited to probe the sample's electrical impedance. Conversely, the electrical impedance of the sample and the circuitry around the crystal can influence on the resonance frequency. Some precautions must be taken to avoid the corresponding artifacts.

14.1 Application of Piezoelectric Stiffening for Sensing

The use of piezoelectric stiffening for sensing was first proposed by Josse and Shana [1, 2]. These authors demonstrated a frequency-based measurement of the sample's electrical impedance. Josse and Shana focus on the measurement of the liquid's conductivity, but the liquid's dielectric permittivity can be sensed, as well [3]. Electrical influences on f_s have found renewed interest recently in the context of the Lateral Field Excitation (LFE) resonators [3]. These have both electrodes on one side of the crystal. There are strong electric fringe fields permeating the space around the crystal. The influence of the sample's conductivity and its dielectric permittivity on the resonance frequency is correspondingly strong. Zhang et al. varied the electrode shape between different crystals and exposed these crystals to the various liquids. They inferred the electrical permittivity from the differences in Δf between these resonators [3]. Other electrical parameters (like the capacitance of an adsorbed film) might be determined from such a measurement as well. The line of reasoning is that the distribution of the electric field inside the sample differs between the differently configured crystals (depending on the electrode shape) and that the electrical properties of the sample for this reason affect the resonance properties to a different extent. Superimposed onto the electrical contribution to Δf is the familiar acoustic contribution. The latter is assumed to be independent of electrode shape.

In more mathematical terms, sensing applications of piezoelectric stiffening are based on Eq. 5.3.24. In order to bring it closer to application, we transform the equation as

D. Johannsmann, *The Quartz Crystal Microbalance in Soft Matter Research*,
Soft and Biological Matter, DOI: 10.1007/978-3-319-07836-6_14,

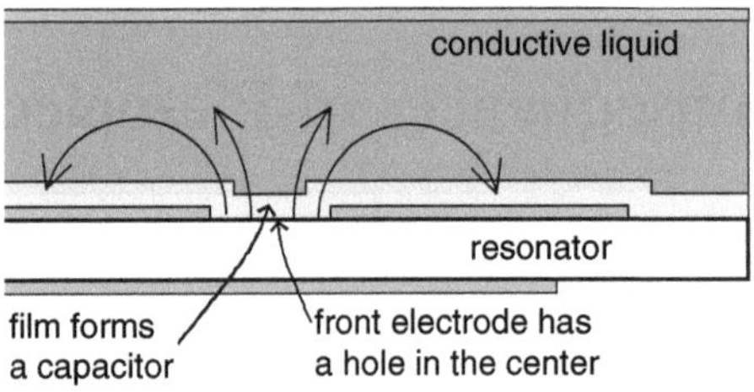

Fig. 14.1 A film's capacitance can be probed by using a conventional front electrode and grounding the back electrode. The film then contributes to the sample's electrical impedance. As an alternative, the front electrode may contain a *small hole* in the *center*. In this configuration, the front electrode stays grounded and a capacitor is formed by the uncoated and the coated parts of the resonator surface

$$\begin{aligned}\frac{\Delta\tilde{f}_{PE}}{f_0} &= \frac{\mathrm{i}}{A\pi Z_q}4\phi^2\left(\left(\mathrm{i}\omega C_0+\tilde{Z}_{ex}^{-1}\right)^{-1}-\left(\mathrm{i}\omega C_0+\tilde{Z}_{ex,ref}^{-1}\right)^{-1}\right)\\ &= \frac{4k_t^2}{\pi^2 n}\left(\left(1+\frac{\tilde{Y}_{ex}}{\mathrm{i}\omega C_0}\right)^{-1}-\left(1+\frac{\tilde{Y}_{ex,ref}}{\mathrm{i}\omega C_0}\right)^{-1}\right)\end{aligned} \tag{14.1.1}$$

The relations $\phi = Ae_{26}/d_q$, $k_t^2 = e_{26}^2/(G_q\varepsilon_q\varepsilon_0)$, $C_0 = A\varepsilon_q\varepsilon_0/d_q$, $Z_q^2 = \rho_q G_q$, and $d_q = Z_q/(2\rho_q f_0)$ were used. $\tilde{Y}_{ex} = \tilde{Z}_{ex}^{-1}$ is the external electrical admittance. The relation can be further simplified if $\tilde{Y}_{ex}$ is much larger than $\mathrm{i}\omega C_0$. This situation is to be expected in an electrochemical setting. In electrochemistry, one employs a supporting electrolyte, which lowers the bulk resistance to the extent that it does not influence the outcome of the measurement (Note: When employing a low bulk resistance and applying the signal to the front electrode, one at the same time short-circuits the crystal and the admittance trace becomes noisy. A compromise must be found). If $\tilde{Y}_{ex} \gg \mathrm{i}\omega C_0$, one arrives at

$$\Delta\tilde{f}_{PE} \approx k_t^2\frac{8\mathrm{i}}{\pi}f_0^2 C_0\left(\tilde{Z}_{ex}-\tilde{Z}_{ex,ref}\right) \tag{14.1.2}$$

With $f_0 = 5$ MHz and $A = 30$ mm^2, the prefactor outside the brackets evaluates to 2 Hz/Ω.

An easy way to modulate the electrical impedance between the electrodes is to switch between grounded front electrode and grounded back electrode. The electrical impedance of the sample is largely prevented from affecting $\Delta\tilde{f}$ for a grounded front electrode. It enters, if the front electrode is connected to signal.

A typical electrical property to be sensed by this scheme would be the capacitance of a film immersed in an electrolyte. This poses the problem that the corresponding impedance (equal to $1/\mathrm{i}\omega C$) might be much lower than the ohmic resistance of the bulk liquid, R_{bulk}. The capacitance of the interface and the resistance of the bulk are arranged in series (see the discussion of the Randles circuit in the textbooks of electrochemistry [4]). The resistance of the bulk liquid can be lowered by adding supporting electrolyte, as is common in electrochemistry. Interestingly, there is another option: One can employ resonators containing a small hole in the front electrode and ground the front electrode (which now is a ring surrounding the central hole) [5]. The resulting field distribution is shown in Fig. 14.1. This geometry lowers the influence of R_{bulk}.

14.2 Safeguards Against Electrical Artifacts

If the stray capacitances between the crystal and sample compartment are not under control, their variation induces frequency shifts, and these may amount to many tens of Hz. There are two remedies, which are grounding the front electrode well and inserting a pi-network. When grounding the front electrode, one reasons that, firstly, all stray capacitances originate from the back of the crystal and that, secondly, these stray capacitances are constant over the duration of the experiment. Of course grounding of the front electrode is also advisable to avoid electrical influences onto the process of adsorption. An oscillating electrical voltage may affect the behavior of the sample.

The "pi-network" (Fig. 14.2) implements another line of reasoning: The resonator is electrically blind to the environment if the electrodes are short-circuited. The environment here includes cables and stray capacitances. Short-circuiting avoids all electrical artifacts because all channels, by which the sample might lower the impedance between the electrodes, are by-passed. At the same time, however, there is no voltage across the crystal, which would drive it. The crystal is blind to its electrical environment, but the impedance analyzer is blind to the crystal, as well, which clearly is not a solution to the problem. The pi-network is a compromise. It is part of the international standard IEC 444. The pi-network is an arrangement of resistors around the crystal as shown in Fig. 14.2. With the pi-network implemented, the crystal sees an electrical impedance, which is much lower than $1/(i\omega C_0)$ (where the latter impedance is in the kΩ range). The electrodes are not strictly short-circuited, but almost. When shielding the crystal from the electrical environment, the pi-network at the same time attenuates the electrical signal by 15 dB on both sides. This is the price to be paid.

For liquid sensing, the pi-network poses a problem because the resonator is wired in the "series through" mode. Neither electrode is grounded. To fix this problem, one can place a transformer between the crystal and the pi-network (Fig. 14.2b). In this way, one has both safeguards in place: The active surface is grounded and the resonator does not see its electrical environment very well.

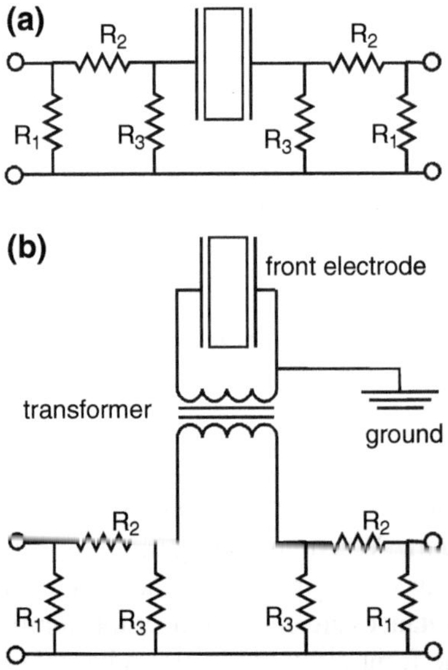

Fig. 14.2 **a** Resonator inserted into a pi-network. The pi-networks attenuates the signal by 15 dB on both sides. The resonator "looks" at a low impedance, which lowers its sensitivity to the electrical boundary conditions. According to IEC 444, the resistances are $R_1 = 159\ \Omega$, $R_2 = 66.2\ \Omega$, and $R_3 = 14.2\ \Omega$. **b** With regard to sensing, a disadvantage of the pi-network as shown in **a** is that neither electrode is grounded. This can be changed by inserting a transformer and grounding the leg connected to the front electrode. A suitable component is the unit ADT1-1 from Minicircuits

Glossary

Variable	Definition (Comments)
A	(Effective) area of the resonator plate
C_0	Parallel capacitance
d_q	Thickness of the resonator ($d_q = m_q/\rho_q = Z_q/(2\rho_q f_0)$)
e_{26}	Relevant component of the *e*-tensor (piezoelectric stress coefficient, $e_{26} = 9.65 \times 10^{-2}$ C/m^2 for AT-cut quartz)
f_0	Resonance frequency at the fundamental ($f_0 = Z_q/(2m_q) = Z_q/(2\rho_q d_q)$)
G_q	Shear modulus of AT-cut quartz ($G_q \approx 29 \times 10^9$ Pa for AT-cut quartz. G_q is the piezoelectrically stiffened modulus)
k_t	Piezoelectric coupling coefficient ($k_t = (e_{26}/(\varepsilon_q \varepsilon_0 G_q))^{1/2}$, $k_t^2 = 0.8\ \%$ for AT-cut quartz)
PE	As an index: *PiezoElectric* stiffening

q	As an index: *q*uartz resonator
ref	As an index: *ref*erence
R	Resistance
R_{bulk}	Resistance of a bulk liquid
R_1, R_2, R_3	Elements of the pi-network (Fig. 14.2)
$\tilde{Y}_{ex}$	External electrical admittance ($\tilde{Y}_{ex} = 1/\tilde{Z}_{ex}$)
$\tilde{Z}_{ex}$	External electrical impedance
Z_q	Acoustic wave impedance of AT-cut quartz ($Z_q = 8.8 \times 10^6$ kg m^{-2} s^{-1})
Δ	As a prefix: A shift induced by the presence of the sample
$\tilde{\varepsilon}_q$, ε_q	Dielectric constant of AT-cut quartz (ε_q is the clamped dielectric constant, $\varepsilon_q = 4.54$ for AT-cut quartz)
ε_0	Dielectric permittivity of vacuum ($\varepsilon_0 = 8.854 \times 10^{-12}$ C/(Vm))
ϕ	Factor converting between mechanical and electric quantities in the Mason circuit ($\phi = Ae_{26}/d_q$)
ρ_q	Density of crystalline quartz ($\rho_q = 2.65$ g/cm^3)
ω	Angular frequency

References

1. Shana, Z.A., Zong, H., Josse, F., Jeutter, D.C.: Analysis of electrical equivalent-circuit of quartz-crystal resonator loaded with viscous conductive liquids. J. Electroanal. Chem. **379**(1–2), 21–33 (1994)
2. Shana, Z.A., Josse, F.: Quartz-crystal resonators as sensors in liquids using the acoustoelectric effect. Anal. Chem. **66**(13), 1955–1964 (1994)
3. Zhang, C., Vetelino, J.F.: Chemical sensors based on electrically sensitive quartz resonators. Sens. Actuators B-Chemical **91**(1–3), 320–325 (2003)
4. Gileadi E.: Physical Electrochemistry: Fundamentals, Techniques and Applications: A Textbook for Students of Science and Engineering. Wiley (2011)
5. Johannsmann, D., Bucking, W., Bode, B., Petri, J.: Simple frequency-based sensing of viscosity and dielectric properties of a liquid using acoustic resonators. IEEE Transactions on Ultrasonics Ferroelectrics and Frequency Control **57**(3), 677–683

Chapter 15
Other Surface-Acoustic-Wave Based Instruments

Abstract There is a variety of sensing instruments making use of acoustic waves (often shear waves) near interfaces. Among the concepts shared between the QCM and these instruments are the mass sensitivity and the acoustic reflectivity as a central intermediate parameter. The kHz resonators measure viscoelastic parameters at frequencies more relevant to most applications than MHz frequencies. On the other hand, they are less sensitive. Smaller resonators operating at higher frequencies tend to have better mass-sensitivity.

15.1 Shear-Wave Reflectometers

The shear-wave reflectometer dates from the time when acoustic sensing started. It has seen numerous developments and extensions since then. Figure 15.1 shows the geometry as proposed by McSkimin in 1949 (Ref. [1]). More recent instrumental developments and experiments are discussed in Refs. [2, 3].

The design proposed by McSkimin separates excitation and detection from the sensor surface itself. The sensor is a trapezoidal bar. Two transducers are glued to the short faces. The first transducer launches a transverse acoustic wave, which is reflected at the interface with the sample. The second transducer is the detector. It measures amplitude and phase, which (together with a reference experiment) allows to calculate the complex amplitude reflection coefficient. If the transverse wave is polarized suitably, the displacement at the surface of the bar occurs into the tangential direction. The load impedance is calculated with line 3 of Eq. 4.4.13, where $\tilde{r}_{q,S}$ is replaced by the reflectivity at the upper surface of the bar and Z_q is replaced by the wave impedance of the bar.

Alig and co-workers have also built a reflectometer with vertical incidence of the acoustic wave [4]. They used both longitudinal and transverse waves, thereby gaining access to the Young's modulus and shear modulus of the sample (and to its Poisson ratio, in consequence). As with the QCM, the instrument can be combined

D. Johannsmann, *The Quartz Crystal Microbalance in Soft Matter Research*,
Soft and Biological Matter, DOI: 10.1007/978-3-319-07836-6_15,

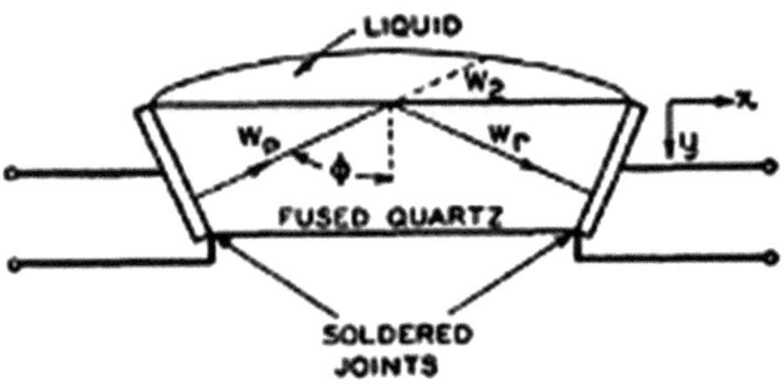

Fig. 15.1 An acoustic shear-wave reflectometer as sketched in one of the early publications. The material contacting the liquid is fused quartz. It is non-piezoelectric. Pulses of shear waves are launched by the transducers glued to sides of the bar. The shear-wave impedance of the liquid is inferred from the amplitude and the phase of reflected wave. Reprinted with permission from Ref. [1]. Copyright 1949 American Physical Society

with other types of surface analysis. Reference [4] reports on the combination with NIR spectroscopy.

The substrate often is fused quartz. Fused quartz is non-piezoelectric, which avoids an influence of the sample's conductivity. The transducers can be piezoelectric crystals, but other types have been explored, as well [5]. McSkimin investigated bulk liquids. More recently, the instrument has also been applied to monitor the drying and curing of polymer films [2, 6, 7]. Section 9.2 discusses experiments, where the technique was applied to nematic liquid crystals [8, 9]. Ultrasonic reflectometry is also used to measure interfacial stiffness Sect. 11.8.

On the down side, the shear-wave reflectometer is more complicated than the QCM. On up-side, there are no effects of piezoelectric stiffening and no flexural contributions to the displacement pattern. Also, being a non-resonant device, the shear-wave reflectometer is not limited in frequency to the overtones of a resonator. Temperature-frequency coupling is less of a problem for shear-wave reflectometers than for the QCM (although calibration *is* required). The shear-wave reflectometer has been applied to monitor temperature-driven phase transitions of various kinds [10, 11].

15.2 The High Overtone Bulk Acoustic Wave Resonator

The High-overtone Bulk Acoustic wave Resonator (HBAR) [12] combines the fragile FBAR architecture (characterized by a high operating frequency) with the robustness of a thick substrate. The reasoning underlying the original design was the use of a low-acoustic-loss substrate (often sapphire) to generate multiple echoes in a delay line application [13]. Recently, the focus has shifted towards high-Q resonators [14]. The acoustic wave is launched by a thin piezoelectric slab deposited on the low-loss (non-piezoelectric) substrate. The transfer function of the composite structure is the product of the thin film spectral transfer function

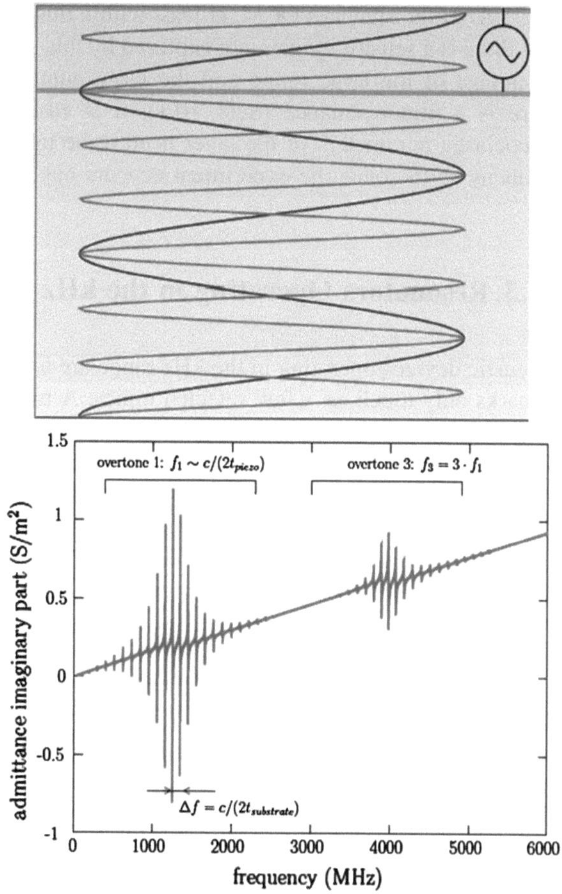

Fig. 15.2 *Top* HBAR architecture, with the piezoelectric layer on top, sandwiched between metal electrodes. There is a low-loss substrate below, in which the overtones propagate. Bottom: Simulated transfer function exhibiting overtones 1 and 3 of the thin piezoelectric film (envelopes at 1,300 and 4,000 MHz) and a comb of modes with a frequency gap governed by the thickness of the substrate. t_{piezo} is the thickness of the piezoelectric layer, $t_{substrate}$ is the thickness of the low-loss substrate, and c the acoustic velocity of the wave. Figure kindly provided by Jean-Michel Friedt

(similar to the transfer function of conventional BAW resonators, but at higher frequencies) and the thick substrate producing a comb of modes (Fig. 15.2).

Simple HBARs can be manufactured by deposition of a polycrystalline piezoelectric film on a low-loss plate, for instance using physical vapor deposition (PVD). However, these devices do not resonate in a pure thickness-shear mode. Thickness-shear modes (needed for operation in the liquid phase) can only be achieved with piezoelectric single crystals. Methods to bond thin single crystals (mostly $LiNbO_3$) to rigid substrates include the smartcut approach [15] and the room-temperature metal-metal diffusion approach [14].

Current work on HBARs in the context of sensing aims at exploiting the high quality factors of the high overtones. High Q-factors imply good baseline stability [16]. There is an intricacy with regard to the gravimetric limit of detection (LOD). Within a simple perturbation approximation the mass sensitivity constant scales as $f_0 f_r = n f_0^2$ (see the discussion below Eq. 8.1.6). Since the scatter in the frequency readings also scales with n, the gravimetric LOD of the HBAR is comparable to the LOD of the standard QCM, at least within this simple model. Mansfeld argues that there is a sensitivity gain not captured by this simple model [17]. One can take advantage of the large range and the large number of operating frequencies. If there is a film resonance (Sect. 10.1), it is easily recognized. As always, the viscoelastic parameters of the layer need to be taken into account in interpreting frequeny shifts, once the experiment occcurs outside the gravimetric regime.

15.3 Resonators Operating in the kHz Range

Acoustic devices operating in the kHz range are a field of their own; the following remarks only touch on a few selected topics. A major incentive to lower the base frequency of a resonator is an improved suitability for measurements of viscosity. The kHz–viscosity is more relevant to most technical applications than the MHz-viscosity. This argument would, for instance, apply to engine oils. Also, the lowered frequency lowers the sensitivity to adsorbed mass. Adsorbed mass (polymeric additives to a lubricant, debris,...) interferes with the measurement of the viscosity. Decreased mass-sensitivity in this context is helpful.

For colloidal dispersions (discussed in Sect. 9.3), there is a specific benefit of measuring the viscoelastic parameters at kHz frequencies. kHz rheology is better adapted to this particular problem than both quasi-static measurements and MHz rheology. For the details, see Sect. 9.3.

kHz resonators have also been applied to biological tissue [18, 19]. The resonator touches the object of interest. As an example, one may be able to recognize a tumor based on its viscoelastic properties being different from those of healthy tissue. Evidently, the problem of creating a well-defined contact area needs attention.

Using a thickness-shear resonator (that is, a plate) in the kHz range is unpractical because the plate must be much wider than it is thick. Otherwise, energy trapping will create flexural contributions of a magnitude, which is detrimental to the resonator's operation in liquid. When making the plate thick enough to support a resonance in the kHz range, its lateral dimensions are driven into the range of tens of centimeters. A more practical geometry is the torsional rod [20]. With torsional resonators, the frequency is set by the length of the rod. Viscometers based on torsional rods are commercially available [21].

Because of the large size of the kHz resonators, the simplest geometry is not necessarily the best geometry. Composite resonators with shapes adapted to the application have been devised [22–25]. Also, there is no need to let the entire resonator be piezoelectric. Composite resonators for the kHz range are more easily built than for the MHz range because of their larger size. The resonator can be driven by a transducer, which is part of the resonating structure. If the main body of the resonator is non-piezoelectric, this can alleviate problems related to piezoelectric stiffening. Complicated resonator shapes entail complicated modes of vibration and complicated flow patterns, but these problems can be solved with calibration and numerical modeling.

There is an interesting difference between kHz and MHz resonators with regard to compressional waves. At MHz frequencies, the wavelength of longitudinal sound is a few hundred microns and it is therefore difficult to construct resonant devices, which would be much smaller than this wavelength. This *is* possible for kHz resonators. Take the cantilever as an example. For devices smaller than λ, a non-tangential motion of the resonator surface can be made to be inefficient in radiating compressional waves into the far-field. When the cantilever moves upwards, the liquid can flow around the lever to the back side, thereby reducing the pressure and the amplitude of the compressional wave launched by the cantilever's movement. Because the flow is largely confined to the acoustic near field, the cantilever mostly probes the liquid's shear-viscosity, rather than its compressibility.

15.4 Instruments Based on Guided Waves

For the QCM and the other bulk acoustic wave (BAW) devices, the deformation pattern is a plane wave with the wave vector perpendicular to the plate. Allowing the wave to propagate *along* the surface has a few advantages (which, as always, come at the expense of complexity) [26]. There are numerous designs. Again, the discussion is limited to a few general remarks.

Most of the time, the wave is launched by interdigitated transducers (IDTs, also: "comb electrodes") as shown in Fig. 15.3b and c. The distance between the fingers of the comb sets the wavelength. Guided-wave devices can be configured as delay lines (Fig. 15.3b) or resonators (Fig. 15.3c). In both cases, the quantity to be sensed is contained in the speed of propagation of the wave. Why the speed of propagation depends on the properties of the adjacent medium can be understood from Fig. 15.3d, showing the Love wave device. The wave can be thought of as bouncing back and forth between the upper and the lower surface of the guiding layer. The reflectivity is large at both surfaces. At the lower surface, there is total internal reflection, meaning that the x-component of k-vector is smaller than ω/c_{bulk}, where c_{bulk} is the speed of sound in the substrate. At the upper surface, the reflectivity is large because the wave impedance of the sample is much lower than the wave impedance of the guiding layer. Without going into details, it is clear that the reflectivity at that upper surface must enter the speed of propagation somehow. For that reason, the reflectivity can be inferred from the speed of propagation. The reflectivity depends on the load impedance and from thereon, data analysis proceeds as for the QCM.

In Fig. 15.3d the guiding layer is a film with a low c_{GL} where the index *GL* stands for *Guiding Layer*. Since $c_{GL} < c_{bulk}$, one achieves total internal reflection at the lower interface. This design provides for a thin guiding layer (and a correspondingly good mass sensitivity), while the guiding layer at the same time is supported by a rigid substrate. Because of the rigid substrate the sensor is robust and does not suffer from bending.

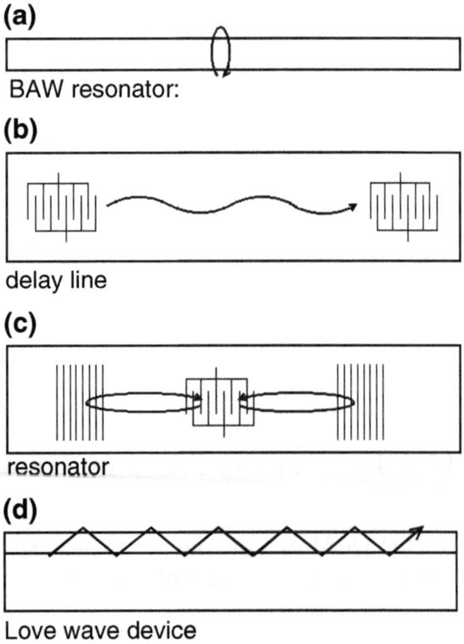

Fig. 15.3 Contrasting to bulk acoustic wave resonators (**a**), the wave propagates in the surface plane for guided-wave devices (**b**, **c**). The speed of propagation changes in the presence of a sample. Excitation usually occurs with interdigitated electrodes. Among the advantages of guided-wave devices is the rigid substrate (**d**). The substrate avoids bending. The guiding layer can be thinner than bulk acoustic wave resonators, which improves the mass-sensitivity

Of particular simplicity and importance are devices, where the wave is guided by the surface alone. These are the surface acoustic wave (SAW) devices [27, 28]. The wave here is a "surface mode". Surface modes exist in other fields of science, as well. Every-day examples are the waves on the water surface. In optics, surface modes are allowed at metal surfaces. These are the "surface plasmon polaritons" or "surface plasmons" (Sect. 16.2). Surface waves on elastic solids are familiar to the public from earthquakes. These modes are the "Rayleigh waves."

At the surfaces of isotropic solids and of the more common crystals, Rayleigh waves have one disadvantage, which is a component of the polarization normal to the surface. Rayleigh waves are corrugation waves. The more simple SAW devices are based on corrugation waves. These cannot be used for liquid sensing because they radiate acoustic energy into the bulk liquid. They can be (and are) employed in gas sensing. Their mass-sensitivity is better than the mass sensitivity of the conventional QCM. At some point, they were viewed as key components of the "electronic nose" [29, 30]. Today, metal-oxide detectors are more common in that role because their limit of detection is even better than what can be achieved with SAW devices.

Standard SAW devices are in use as electronic filters (competing with the FBAR). There is a third role of SAW devices, which is in pumping liquids on small scales [31, 32]. If the amplitude of oscillation is large enough, the nonlinear term in Navier-Stokes equation converts some of the acoustic energy into a steady flow (see also Sect. 13.2).

Certain types of crystals *do* support surface modes with a polarization in the surface plane. A prominent example is $LiTaO_3$. Acoustic devices based on such in-plane polarized surface modes are also called shear-horizontal surface acoustic wave devices (SH-SAW devices) [33–36]. For liquid sensing, SH–SAW devices should be superior in performance to the QCM, in principle, but there are a number of problems, which have prevented such devices from becoming a commercial success, so far. For a review see Ref. [37]. The author is aware of one start-up company marketing SH-SAW devices [38]. Among the difficulties are complexity and price. Also, the $LiTaO_3$-based devices do not show a quantum leap in the limit of detection (LOD) relative to the QCM. When the improvement in LOD is weighed against complexity and cost, practitioners often end up sticking to the QCM.

A few further remarks:

- While the higher frequency does not necessarily improve the LOD, it always is a benefit when it comes to wireless sensing [39]. Wireless sensing requires antennae and size of these decreases with increasing frequency. Wireless operation has also been reported for a QCM [40], but it is not routine.
- Because of the high frequency, soft adsorbates can easily have a thickness comparable to λ, which produces strong viscoelastic effects. These are a drawback when the prime interest is gravimetry.
- SH-SAW devices often need calibration.
- If the sensor surface is not grounded, there are strong electric fringe fields and effects of piezoelectric stiffening are correspondingly strong [41]. (Evidently, the same problem exists for the QCM). The term "electromechanical coupling" is used in that context more often than "piezoelectric stiffening".
- There are pressure sensors (including wireless sensors) based on SAW devices [42].

15.5 MEMS and NEMS Resonators

The debate on whether and why small resonators (MEMS resonators) can be, should be, or in fact are better than the QCM has somewhat of a history. The question certainly is an interesting one, but the author still prefers to stay away from a general statement. MEMS-based resonators are less mature than the QCM, at this point, but that might change. Making resonators smaller has the evident advantage that they can fit into small instruments more easily. At the same time, small resonators tend to be fragile, which may prevent their use in liquids.

There are many different configurations of MEMS and NEMS resonators. Examples are the cantilever [43], the paddlever [44], the trampoline resonator [45], the doubly clamped beam [46], the FBAR [47, 48], the contour-mode resonator [49], the capacitive micromachined ultrasonic transducer (CMUT) [50], and the suspended microchannel resonator (SMR) [51]. For a review, see Ref. [52].

Cantilevers have been extensively studied in the context of atomic force microscopy [53]. The resonance frequency of cantilevers is routinely analyzed in dynamic AFM to determine the interaction potential between the tip and the substrate [54–56]. This methodology does not usually count as sensing, though. Material deposited on the surface of a cantilever also affects the resonance frequency. With regard to conventional cantilevers, static bending is analyzed more often than a shift in resonance frequency (the latter being induced by the deposition of a mass). The doubly clamped beam, which in many respects is similar to the cantilever, displays higher resonance frequencies for comparable size and therefore has the better mass-sensitivity. LODs in the zeptogram range have been demonstrated [46]. Both the cantilever resonator and the doubly clamped beam cannot be easily employed in liquids because of the large damping. Most MEMS resonators suffer from this problem. See Ref. [57] for a proposed design, which would avoid large damping.

A few further remarks:

- To the best of the author's knowledge, there is no fundamental difference between MEMS resonators and NEMS resonators (where the *N* stands for *nano*). NEMS resonators are supposed to be smaller and lighter. Also, their resonance frequency tends to be higher, which implies improved sensitivity (Sect. 15.8).
- The smallest resonators are the nanowire resonators. These include carbon nanotubes. Reference [58] reports a limit of detection in the yoctogram range (where "yocto" denotes 10^{-24}).
- Actuation and read out can occur piezoelectrically, but other option exist. Among them are electrostatic actuation and piezoresistive read-out. An overview is given in Ref. [59].
- If the active area is small, this improves the limit of detection when calculated in terms of mass. With regard to mass, the sensitivity of MEMS devices is much better than what can be done with a QCM. When calculated in terms of mass *per unit area*, the performance is less spectacular because of the small area. Consequently, the application of MEMS-based sensors can be anticipated to be in mass-spectroscopy rather than in the determination of a concentration in an ambient medium.
- Given the small electrodes and the correspondingly small currents, MEMS resonators require advanced electronics for read-out.
- With regard to noise sources, see the tutorial by Rubiola at http://rubiola.org/pdf-slides/2008I-fsm-phase-noise.pdf. There is a book, as well [60].

15.6 The FBAR as a Sensor

The thin film bulk acoustic wave resonator (FBAR) can be viewed as a MEMS resonator. It is treated separately here because of the conceptual similarities with the QCM. The idea of using FBARs for sensing is old [61]. The renewed interest in recent years follows from the FBAR's success in the telecom industry. For a recent review on the FBAR in general see Ref. [62].

The standard FBAR (employing a sputtered piezoelectric layer) does not support a thickness shear mode. The thickness shear mode requires a tilted orientation of the crystallographic *c*-axis, which can be achieved with oblique evaporation. A review on shear-mode FBARs is given in Ref. [47]. Reference [47] reports a limit of detection comparable to what can be achieved with a QCM.

Currently there are the following problems:

- Neither the standard FBAR nor the shear mode FBAR as such are temperature compensated. They have negative temperature coefficients of frequency (TCFs). The problem can be overcome with coatings out of materials with a positive TCF, at least in principle.
- The shear mode FBAR often has longitudinal admixtures to the movement of the surface. These may be reduced with improved control over the growth process.
- The standard FBAR is a thin membrane, which means that it is fragile. Alternatively, one can employ an acoustic Bragg mirror between the resonator and a rigid support [63].

As always, the limit of detection is not the only performance parameter of a sensor. Because the production process is CMOS-compatible, FBARs can be monolithically integrated with the driving electronics and—given their small size—might be produced as arrays on a single chip. The situation is open.

15.7 Novel Resonator Materials

Generally speaking, novel resonator materials are not a prime demand in the field because quartz has a few properties, which are difficult to surpass. It has low intrinsic damping, it is hard but not brittle, chemically inert, and temperature-stable up to 200 °C. Its piezoelectric coupling coefficient is on the low side (compared to $LiNbO_3$ or AlN), but this is not necessarily a disadvantage. On the contrary, low k_t^2 implies small electric disturbances to the measurement of f_r (caused by piezoelectric stiffening).

The two main reasons for searching alternatives to quartz are high-temperature stability and CMOS compatibility. At a temperature of 573 °C, quartz undergoes a transition from the α- to the β-phase. Even below this temperature, defects are spontaneously generated, which reduce the piezoelectric coupling and increase the damping [64]. Prevalent materials for use beyond 500 °C are gallium-orthophosphate ($GaPO_4$, Ref. [65]), Langasite ($La_3Ga_5SiO_{14}$, Refs. [66] and [67]), and certain members of the rare earth calcium oxoborate family [68]. $GaPO_4$ is widely used as a pressure sensor [69], but is difficult to grow to the size required for BAW resonators. At temperatures beyond 800 °C, Langasite has a noticeable ionic conductivity, which increases the bandwidth. The consequences have been studied in detail by Fritze [70]. Finding a suitably material is one thing and being able to grow crystals of sufficient size and quality is another.

The principal material competing with quartz for reasons of CMOS-compatibility is aluminum nitride (AlN). These films form the FBAR (Sect. 15.6).

15.8 Smaller and Better?

Contradicting a popular paradigm, making devices smaller does not always improve their performance. Scale effects enter the physics of the QCM in the form of thickness, area, and frequency. Below is a list of arguments, which have to be weighed against each other when designing a sensor:

- On the side of practicalities: If the sensor shall be part of a larger system (such as a microfluidic system), it must fit in. If many sensing elements are involved (as in an array [71, 72]), the individual elements must be small. Fabrication may be a problem for small devices.
- Thin sensor plates tend to be fragile. Also, the dependence of frequency on hydrostatic pressure applied to one side of a plate increases with decreasing plate thickness. Thin plates bend more easily than thick ones. This argument does not apply to most guided-wave devices (Sect. 15.4) because these have a rigid substrate.
- A very important scale effect in QCM-based sensing is embodied in Eq. 6.1.18. The numerator and the denominator are a surface and a volume integral, respectively. If one shrinks the resonator, the surface-to-volume ratio (and thereby the sensitivity) increases.
- With decreasing plate thickness, the fundamental frequency increases, with numerous consequences. Note: The fundamental frequency, f_0, is to be distinguished from the frequency at the respective overtone. The arguments below apply to the fundamental frequency. Going to high overtone orders is less effective in increasing the mass sensitivity than increasing f_0. The primary benefit in operation at overtones is the comparison between overtones.
- For the gravimetric QCM in the vapor phase, the mass-sensitivity constant increases as f_0^{-2}. However, one should not confuse the mass-sensitivity constant with the limit of detection (LOD). The LOD scales as f_0^{-1} because the scatter of the frequency readings must be taken into consideration (Sect. 8.1).
- A gravimetric QCM in a liquid feels both the adsorbed mass and the viscous drag. The mass-induced frequency shift scales as f_0^2, while the increase in bandwidth due to viscous drag scales as $f_0^{3/2}$ (Eq. 9.1.1). Since the precision of a frequency reading scales as the bandwidth, the noise-equivalent mass scales as $f_0^{-1/2}$. See also the discussion around Eq. 8.1.6.
- The time of data acquisition scales as Γ^{-1}. In ring-down, each decay must last for a few multiples of the decay time. In impedance analysis, the sweep rate can be increased in proportion to Γ.

  Note: The crystal *itself* responds to changes at its surface within a few periods of oscillation. In more mathematical terms, a resonator responds quickly to changes in M_R, κ_R, and ξ_R, while it responds slowly (on the time scale of Γ^{-1}) to an external force (see Eqs. 4.1.7 and 4.1.20). Processes occurring on a time scale below Γ^{-1} *can* be resolved in ring-down [73]. In ring-down, one directly sees the current trace versus time on the screen of the digital oscilloscope. The resonator obeys the resonator equation at all times. If the spring constant rapidly changes, this will immediately affect the current trace, regardless of the resonator's damping constant. Such sudden changes will usually be triggered in one way or another. They might be triggered by the process of ring-down itself [73].

- When measuring the stiffness of a single contact, the sensitivity scales as the inverse of the active area, A^{-1} (cf. Eq. 11.3.4).
- The current into the electrode scales as electrode area. Small electrodes require sensitive amplifiers. The argument applies in frequency control as well. Low-cost oscillator circuits run into difficulties with small resonators because of the small currents involved.
- When an analyte concentration is to be inferred from the adsorbed mass per unit area, one cannot gain in accuracy by making the area smaller. However, if the total mass is the target of the measurement, one will try to decrease the total mass of the resonator (Eq. 6.1.18). For mass spectrometry (a potential future application of acoustic resonators), NEMS resonators are the devices of choice [45, 46].
- There is a scale effect in the study of an adsorption kinetics because the efficiency of mass transport to a sensor surface depends on the geometry of the sensor [74, 75].
- For wireless interrogation, higher frequency translates to smaller antennas [76].
- If quantum effects are to be observed, a high frequency makes life easier because of $\Delta E = hf$. (ΔE the energy quantum and h the Planck constant). Smaller resonators have higher frequencies and larger quanta of vibrational energy. In consequence, the requirements on cooling are the less stringent. As reported in Refs. [77, 78], quantization of the vibrational energy has been seen with MEMS devices. Here are the words of Pierre Meystre, commenting on this work in the Science Magazine [79]: *…The successful cooling of macroscopic mechanical systems to their motional ground state is an exciting and essential step toward that goal, but it is "beyond ground state" physics that promises to be most exciting in bringing the emerging field of quantum acoustics to the level of sophistication of quantum optics. The demonstration of phonon lasers, squeezed and other nonclassical phonon states, entangled states, state transfer between phonon and photon fields, and much more, are already on the horizon, with an explosion of new results expected soon.*

Glossary

Variable	Definition (Comments)
A	Effective area of the resonator plate
c	Speed of propagation
E	Energy
f	As an index: *f*ilm

f_r	Resonance frequency
f_0	Resonance frequency at the fundamental ($f_0 = Z_q/(2m_q) = Z_q/(2\rho_q d_q)$)
GL	As an index: *G*uiding *L*ayer
h	Planck constant
M_R	Mass of a resonator
$\tilde{r}_{q,S}$	Reflectivity evaluated at the resonator surface
Z_q	Acoustic wave impedance of AT-cut quartz ($Z_q = 8.8 \times 10^6$ kg m^{-2} s^{-1})
λ	Wavelength of sound
Γ	Imaginary part of a resonance frequency
κ_R	Spring constant of a resonator
ξ_R	Drag coefficient of a resonator
ω	Angular frequency

References

1. Mason, W.P., Baker, W.O., McSkimin, H.J., Heiss, J.H.: Measurement of shear elasticity and viscosity of liquids at ultrasonic frequencies. Phys. Rev. **75**(6), 936–946 (1949)
2. Alig, I., Lellinger, D., Sulimma, J., Tadjbakhsch, S.: Ultrasonic shear wave reflection method for measurements of the viscoelastic properties of polymer films. Rev. Sci. Instrum. **68**(3), 1536–1542 (1997)
3. Wang, X.J., Subramaniam, K.V., Lin, F.B.: Ultrasonic measurement of viscoelastic shear modulus development in hydrating cement paste. Ultrasonics **50**(7), 726–738 (2010)
4. Alig, I., Steeman, P.A.M., Lellinger, D., Dias, A.A., Wienke, D.: Polymerization and network formation of UV curable materials monitored by hyphenated real-time ultrasound reflectometry and near-infrared spectroscopy (RT US/NIRS). Prog. Org. Coat. **55**(2), 88–96 (2006)
5. Wegener, M., Oehler, H., Lellinger, D., Alig, I.: Note: piezoelectric polymers as transducers for the ultrasonic-reflection method and the application in mechanical property-screening of coatings. Rev. Sci. Instrum. **83**(1), 3 (2012)
6. Alig, I., Oehler, H., Lellinger, D., Tadjbach, S.: Monitoring of film formation, curing and ageing of coatings by an ultrasonic reflection method. Prog. Org. Coat. **58**(2–3), 200–208 (2007)
7. Lellinger, D., Tadjbach, S., Alig, I.: Determination of the elastic moduli of polymer films by a new ultrasonic reflection method. Macromol. Symp. **184**, 203–213 (2002)
8. Kiry, F., Martinoty, P.: Ultrasonic investigation of anisotropic viscosities in a nematic liquid-crystal. J. De Phys. **38**(2), 153–157 (1977)
9. Martinoty, P., Candau, S.: Determination of viscosity coefficients of a nematic liquid crystal using a shear waves reflectance technique. Mol. Cryst. Liq. Cryst. **14**(3–4), 243 (1971)
10. Alig, I., Tadjbakhsch, S., Floudas, G., Tsitsilianis, C.: Viscoelastic contrast and kinetic frustration during poly(ethylene oxide) crystallization in a homopolymer and a triblock

copolymer. Comparison of ultrasonic and low-frequency rheology. Macromolecules **31**(20), 6917–6925 (1998)

11. Alig, I., Tadjbakhsch, S., Hadjichristidis, N., Floudas, G.: Order-to-disorder transition in a diblock copolymer studied at ultrasonic frequencies with a shear wave reflection technique. Europhys. Lett. **52**(3), 291–296 (2000)
12. Baron, T., Lebrasseur, E., Bassignot, F., Martin, G., Pétrini, V., Ballandras, S.: High-overtone bulk acoustic resonator. In: Beghi, M.G. (ed.) Modeling and Measurement Methods for Acoustic Waves and for Acoustic Microdevices. Intech, Rijeka (2013). www.intechopen.com/books/modeling-and-measurement-methods-for-acoustic-waves-and-for-acoustic-microdevices/high-overtone-bulk-acoustic-resonator
13. Huang, H.C., Knox, J.D., Turski, Z., Wargo, R., Hanak, J.J.: Fabrication of submicron $LiNbO_3$ transducers for microwave acoustic (bulk) delay-lines. Appl. Phys. Lett. **24**(3), 109–111 (1974)
14. Gachon, D., Courjon, E., Martin, G., Gauthier-Manuel, L., Jeannot, J.-C., Daniau, W., Ballandras, S.: Fabrication of high frequency bulk acoustic wave resonator using thinned single-crystal lithium niobate. Ferroelectrics **362**(1), 30–40 (2010) www.tandfonline.com/doi/abs/10.1080/00150190801997872
15. Rabus, D., Martin, G., Carry, E., Ballandras, S.: Eight channel embedded electronic open loop interrogation for multi sensor measurements. Proc. Eur. Freq. Time Forum (EFTF) 436–442 (2012)
16. Pijolat, M., Reinhardt, A., Defay, E., Deguet, C., Mercier, D., Aid, M., Moulet, J., Ghyselen, B., Gachon, D., Ballandras, S.: Large Qf product for HBAR using smart cut transfer of $LiNbO_3$ thin layers onto $LiNbO_3$ substrate. Proc. IEEE Ultrason. Symp. 201–204 (2008)
17. Mansfeld, G.D.: Theory of high overtone bulk acoustic wave resonator as a gas sensor. In: Proceedings of 13th International Conference on Microwaves, Radar and Wireless Communications (MIKON) (2000)
18. Valtorta, D., Mazza, E.: Measurement of rheological properties of soft biological tissue with a novel torsional resonator device. Rheol. Acta **45**(5), 677–692 (2006)
19. Stroop, R., Uribe, D.O., Martinez, M.O., Brokelmann, M., Hemsel, T., Wallaschek, J.: Tactile tissue characterisation by piezoelectric systems. J. Electroceram. **20**(3–4), 237–241 (2008)
20. Stokich, T.M., Radtke, D.R., White, C.C., Schrag, J.L.: An instrument for precise measurement of viscoelastic properties of low-viscosity dilute macromolecular solutions at Frequencies from 20 to 500 khz. J. Rheol. **38**(4), 1195–1210 (1994)
21. http://www.flucon.de/. Accessed 28 Feb 2013
22. Kirschenmann, L., Pechhold, W.: Piezoelectric rotary vibrator (PRV)—a new oscillating rheometer for linear viscoelasticity. Rheol. Acta **41**(4), 362–368 (2002)
23. Pechhold, W., Mayer, U., Raju, G.B., Guillon, O.: Piezo rotary and axial vibrator (PRAV) characterization of a fresh coating during its drying. Rheol. Acta **50**(3), 221–229 (2011)
24. Crassous, J.J., Regisser, R., Ballauff, M., Willenbacher, N.: Characterization of the viscoelastic behavior of complex fluids using the piezoelastic axial vibrator. J. Rheol. **49**(4), 851–863 (2005)
25. Vadillo, D.C., Tuladhar, T.R., Mulji, A.C., Mackley, M.R.: The rheological characterization of linear viscoelasticity for ink jet fluids using piezo axial vibrator and torsion resonator rheometers. J. Rheol. **54**(4), 781–795 (2010)
26. Grate, J.W., Martin, S.J., White, R.M.: Acoustic-wave microsensors. 1. Anal. Chem. **65**(21), A940–A948 (1993)
27. Martin, S.J., Frye, G.C., Senturia, S.D.: Dynamics and response of polymer-coated surface-acoustic-wave devices—effect of viscoelastic properties and film resonance. Anal. Chem. **66**(14), 2201–2219 (1994)
28. Reindl, L., Scholl, G., Ostertag, T., Scherr, H., Wolff, U., Schmidt, F.: Theory and application of passive SAW radio transponders as sensors. IEEE Trans. Ultrason. Ferroelectr. Freq. Control **45**(5), 1281–1292 (1998)

29. Rock, F., Barsan, N., Weimar, U.: Electronic nose: current status and future trends. Chem. Rev. **108**(2), 705–725 (2008)
30. http://us.msasafety.com/CBRNE-Detectors/CBRNE-Detectors/HAZMATCAD%26reg%3B-and-HAZMATCAD%26reg%3B-Plus/p/000400000200001000. Accessed 11 May 2013
31. Yeo, L.Y., Friend, J.R.: Ultrafast microfluidics using surface acoustic waves. Biomicrofluidics **3**(1), 012002 (2009)
32. Friend, J., Yeo, L.Y.: Microscale acoustofluidics: microfluidics driven via acoustics and ultrasonics. Rep. Prog. Phys. **83**(2), 647–704 (2011)
33. Josse, F., Bender, F., Cernosek, R.W.: Guided shear horizontal surface acoustic wave sensors for chemical and biochemical detection in liquids. Anal. Chem. **73**(24), 5937–5944 (2001)
34. Saha, K., Bender, F., Gizeli, E.: Comparative study of IgG binding to proteins G and A: nonequilibrium kinetic and binding constant determination with the acoustic waveguide device. Anal. Chem. **75**(4), 835–842 (2003)
35. Fu, Y.Q., Luo, J.K., Du, X.Y., Flewitt, A.J., Li, Y., Markx, G.H., Walton, A.J., Milne, W.I.: Recent developments on ZnO films for acoustic wave based bio-sensing and microfluidic applications: a review. Sens. Actuators B Chem. **143**(2), 606–619 (2010)
36. Martin, F., Newton, M.I., McHale, G., Melzak, K.A., Gizeli, E.: Pulse mode shear horizontal-surface acoustic wave (SH-SAW) system for liquid based sensing applications. Biosens. Bioelectron. **19**(6), 627–632 (2004)
37. Lange, K., Rapp, B.E., Rapp, M.: Surface acoustic wave biosensors: a review. Anal. Bioanal. Chem. **391**(5), 1509–1519 (2008)
38. http://saw-instruments.com/index.php. Accessed 20 Sept 2013
39. Pohl, A.: A review of wireless SAW sensors. IEEE Trans. Ultrason. Ferroelectr. Freq. Control **47**(2), 317–332 (2000)
40. Zhou, X.F., Zhang, J., Jiang, T., Wang, X.H., Zhu, Z.Q.: Humidity detection by nanostructured ZnO: a wireless quartz crystal microbalance investigation. Sens. Actuators A Phys. **135**(1), 209–214 (2007)
41. Ricco, A.J., Martin, S.J., Zipperian, T.E.: Surface acoustic-wave gas sensor based on film conductivity changes. Sens. Actuators **8**(4), 319–333 (1985)
42. Jungwirth, M., Scherr, H., Weigel, R.: Micromechanical precision pressure sensor incorporating SAW delay lines. Acta Mech. **158**(3–4), 227–252 (2002)
43. Berger, R., Gerber, C., Lang, H.P., Gimzewski, J.K.: Micromechanics: a toolbox for femtoscale science: "towards a laboratory on a tip"'. Microelectron. Eng. **35**(1–4), 373–379 (1997)
44. Varshney, M., Waggoner, P.S., Tan, C.P., Aubin, K., Montagna, R.A., Craighead, H.G.: Prion protein detection using nanomechanical resonator arrays and secondary mass labeling. Anal. Chem. **80**(6), 2141–2148 (2008)
45. Waggoner, P.S., Varshney, M., Craighead, H.G.: Detection of prostate specific antigen with nanomechanical resonators. Lab Chip **9**(21), 3095–3099 (2009)
46. Yang, Y.T., Callegari, C., Feng, X.L., Ekinci, K.L., Roukes, M.L.: Zeptogram-scale nanomechanical mass sensing. Nano Lett. **6**(4), 583–586 (2006)
47. Wingqvist, G.: AlN-based sputter-deposited shear mode thin film bulk acoustic resonator (FBAR) for biosensor applications—a review. Surf. Coat. Technol. **205**(5), 1279–1286 (2010)
48. Wingqvist, G., Bjurstrom, J., Liljeholm, L., Yantchev, V., Katardjiev, I.: Shear mode AlN thin film electro-acoustic resonant sensor operation in viscous media. Sens. Actuators B Chem. **123**(1), 466–473 (2007)
49. Zuniga, C., Rinaldi, M., Khamis, S.M., Johnson, A.T., Piazza, G.: Nanoenabled microelectromechanical sensor for volatile organic chemical detection. Appl. Phys. Lett. **94**(22), 223122 (2009)
50. Tang, P.K., Wang, P.H., Li, M.L., Lu, M.S.C.: Design and characterization of the immersion-type capacitive ultrasonic sensors fabricated in a CMOS process. J. Micromech. Microeng. **21**(2), 129901 (2011)

51. Arlett, J.L., Roukes, M.L.: Ultimate and practical limits of fluid-based mass detection with suspended microchannel resonators. J. Appl. Phys. **108**(8) , 084701 (2010)
52. Fanget, S., Hentz, S., Puget, P., Arcamone, J., Matheron, M., Colinet, E., Andreucci, P., Duraffourg, L., Myers, E., Roukes, M.L.: Gas sensors based on gravimetric detection—a review. Sens. Actuators B Chem. **160**(1), 804–821 (2011)
53. Lange, D., Brand, O., Baltes, H.: CMOS Cantilever Sensor Systems: Atomic Force Microscopy and Gas Sensing Applications. Springer (2002)
54. Hölscher, H., Schwarz, U.D., Wiesendanger, R.: Calculation of the frequency shift in dynamic force microscopy. Appl. Surf. Sci. **140**(3–4), 344–351 (1999)
55. Sader, J.E., Jarvis, S.P.: Accurate formulas for interaction force and energy in frequency modulation force spectroscopy. Appl. Phys. Lett. **84**(10), 1801–1803 (2004)
56. Giessibl, F.J.: A direct method to calculate tip-sample forces from frequency shifts in frequency-modulation atomic force microscopy. Appl. Phys. Lett. **78**(1), 123–125 (2001)
57. Dufour, I., Josse, F., Heinrich, S.M., Lucat, C., Ayela, C., Menil, F., Brand, O.: Unconventional uses of microcantilevers as chemical sensors in gas and liquid media. Sens. Actuators B Chem. **170**, 115–121 (2012)
58. Chaste, J., Eichler, A., Moser, J., Ceballos, G., Rurali, R., Bachtold, A.: A nanomechanical mass sensor with yoctogram resolution. Nat. Nanotechnol. **7**(5), 300–303 (2012)
59. http://www.ifcs-eftf2011.org/sites/ifcs-eftf2011.org/files/editor-files/Slides_Piazza.pdf. Accessed 18 June 2014
60. Rubiola, E.: Phase Noise and Frequency Stability in Oscillators. Cambridge University Press, New York (2010)
61. Lakin, K., Wang, J., Kline, G., Landin, A., Chen, Y., Hunt, J. Thin film resonators and filters. In: Ultrasonics Symposium Proceedings, pp. 466–475 (1982)
62. Wingqvist, G.: Thin-film electro-acoustic sensors based on AlN and its alloys: possibilities and limitations. Microsyst. Technol. Micro Nanosyst. Inf. Storage Process. Syst. **18**(7–8), 1213–1223 (2012)
63. Ruby, R.: Review and comparison of bulk acoustic wave FBAR, SMR technology. In: 2007 IEEE Ultrasonics Symposium Proceedings, vols. 1–6, pp. 1029–1040, New York (2007)
64. Haines, J., Cambon, O., Keen, D.A., Tucker, M.G., Dove, M.T.: Structural disorder and loss of piezoelectric properties in alpha-quartz at high temperature. Appl. Phys. Lett. **81**(16), 2968–2970 (2002)
65. Krempl, P., Schleinzer, G., Wallnofer, W.: Gallium phosphate, $GaPO_4$: a new piezoelectric crystal material for high-temperature sensorics. Sens. Actuators A Phys. **61**(1–3), 361–363 (1997)
66. Fritze, H., Tuller, H.L.: Langasite for high-temperature bulk acoustic wave applications. Appl. Phys. Lett. **78**(7), 976–977 (2001)
67. http://newpiezo.com/langasite.html. Accessed 28 Mar 2013
68. Yu, F.P., Zhang, S.J., Zhao, X.A., Yuan, D.R., Wang, Q.M., Shrout, T.R.: High temperature piezoelectric properties of yttrium calcium oxyborate single crystals. Phys. Status Solidi Rapid Res. Lett. **4**(5–6), 103–105 (2010)
69. https://www.avl.com/pressure-sensors-for-combustion-analysis. Accessed 28 Mar 2013
70. Fritze, H.: High-temperature piezoelectric crystals and devices. J. Electroceram. **26**(1–4), 122–161 (2011)
71. Jin, X.X., Huang, Y., Mason, A., Zeng, X.Q.: Multichannel monolithic quartz crystal microbalance gas sensor array. Anal. Chem. **81**(2), 595–603 (2009)
72. Hung, V.N., Abe, T., Minh, P.N., Esashi, M.: Miniaturized, highly sensitive single-chip multichannel quartz-crystal microbalance. Appl. Phys. Lett. **81**(26), 5069–5071 (2002)
73. Berg, S., Johannsmann, D.: High speed microtribology with quartz crystal resonators. Phys. Rev. Lett. **91**(14), 145505 (2003)
74. Arlett, J.L., Myers, E.B., Roukes, M.L.: Comparative advantages of mechanical biosensors. Nat. Nanotechnol. **6**(4), 203–215 (2011)
75. Squires, T.M., Messinger, R.J., Manalis, S.R.: Making it stick: convection, reaction and diffusion in surface-based biosensors. Nat. Biotechnol. **26**(4), 417–426 (2008)

76. Rabe, J., Seidemann, V., Buettgenbach, S.: Monolithic fabrication of wireless miniaturized quartz crystal microbalance (QCM-R) arrays and their application for biochemical sensors. Sens. Mater. **15**(7), 381–391 (2003)
77. O'Connell, A.D., Hofheinz, M., Ansmann, M., Bialczak, R.C., Lenander, M., Lucero, E., Neeley, M., Sank, D., Wang, H., Weides, M., Wenner, J., Martinis, J.M., Cleland, A.N.: Quantum ground state and single-phonon control of a mechanical resonator. Nature **464**(7289), 697–703 (2010)
78. Teufel, J.D., Donner, T., Li, D.L., Harlow, J.W., Allman, M.S., Cicak, K., Sirois, A.J., Whittaker, J.D., Lehnert, K.W., Simmonds, R.W.: Sideband cooling of micromechanical motion to the quantum ground state. Nature **475**(7356), 359–363 (2011)
79. Meystre, P.: Cool vibrations. Science **333**(6044), 832–833 (2011)

Chapter 16
Combined Instruments

Abstract The QCM is often combined with other techniques of interface analysis. In some cases, doing that in situ is straight-forward. An example is the electrochemical QCM (EQCM). The combination with optical reflectometry is particularly interesting because the data analysis proceeds along similar lines, but still often leads to an effective optical thickness, which is lower than the Sauerbrey thickness.

16.1 Electrochemical QCM

One of the goals when immersing the QCM into liquids in the 1980s was its combination with electrochemical instrumentation [2, 3]. An electrochemical QCM (EQCM) converts Δf to the mass transfer at the electrode (Fig. 16.2). The EQCM is widely used on a routine basis. Early reviews were provided by Buttry and Ward [4] and by Schumacher [5]. More recent accounts can be found in the book chapters by Marx [6] and by Dhaikin et al. [7] or in the reviews by Doblhofer and Weil [8] and by Hillman [9]. Also, see the work by the Kutner group (Ref. [10] and others).

When interrogated by impedance analysis, most EQCMs are configured as depicted in Fig. 16.1. The front electrode of the resonator at the same time is the working electrode (WE) of an electrochemical three-electrode setup. In principle, there is a problem, because two electronic instruments talk to the same electrode. The impedance analyzer (operating at MHz frequencies) and the potentiostat (operating at low frequencies) might interfere with each other. A similar problem occurs when the resonator is driven by ring-down or by an oscillator circuit. It can be solved by inserting a high-pass filter (a capacitor, typically 0.1 μF) and a low pass filter (an inductor, typically 1 mH) between the resonator and the driving circuitry as shown in Fig. 16.1.

Here are issues to keep an eye on:

- Mass transfer is only one out of a few different causes for a frequency shift. A thorough discussion of the matter can be found in Ref. [7]. Δf is also affected

D. Johannsmann, *The Quartz Crystal Microbalance in Soft Matter Research*, Soft and Biological Matter, DOI: 10.1007/978-3-319-07836-6_16,

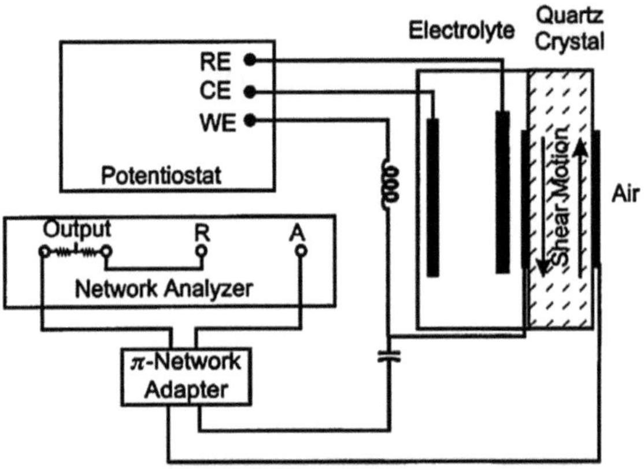

Fig. 16.1 Sketch of an EQCM-setup employing impedance analysis. The front electrode is connected to both the impedance analyzer ("network analyzer") and the potentiostat. An inductor and a capacitor act as a high-pass and a low-pass filter, respectively. Reprinted with permission from Ref. [1]. Copyright 2000 Elsevier

by roughness, slip, the viscosity of the coupled layer [11], piezoelectric stiffening, and the softness of the adsorbate.

- At positive voltages, the front electrode may be damaged because metals tend to dissolve under anodic potentials. Often, the chromium adhesion layer is the weakest element of the chain. According to Ref. [12], titanium adhesion layers are more robust than chromium layers in this regard. At any rate, positive potentials can pose problems.
- The EQCM requires calibration because the charge is proportional to the electrochemically active area, while the frequency shift is not. Calibration can occur with a well-understood electroplating process (such as copper deposition [13]). Do not confuse the electrochemically active area with the acoustically active area according to Eq. 7.4.7. Faraday processes occur on the metal surface with equal efficiency everywhere, while the amplitude distribution varies and may extend to beyond the electrode.
- The speed of data acquisition can be a problem. Electrochemical measurements often rely on transients. An example is cyclovoltammetry. A typical cyclovoltammogram is acquired over a few seconds and contains a few hundred readings of voltage and current. Impedance analysis cannot go any faster than 100 ms per frequency sweep.

As in all other applications, the advanced EQCM has more to offer than the measurement of mass transfer [15–17]. Marx reports at length on changes of the motional resistance (proportional to the half-bandwidth) with the progress of a reaction [6]. Making use of frequency and half-bandwidth on the different overtones, one can infer the softness of the deposit or assess the roughness of the layer. Hillman extensively studied the viscoelastic properties of poly(3,4-ethylene dioxythiophene) films (PEDOT films) [15]. There is an interesting twist to this work. When fitting Δf and $\Delta\Gamma$ on the different overtones with the parameters of a layer system, the thickness is one of the free parameters. Errors in thickness and softness are often cross-correlated. In an EQCM experiment, one has independent

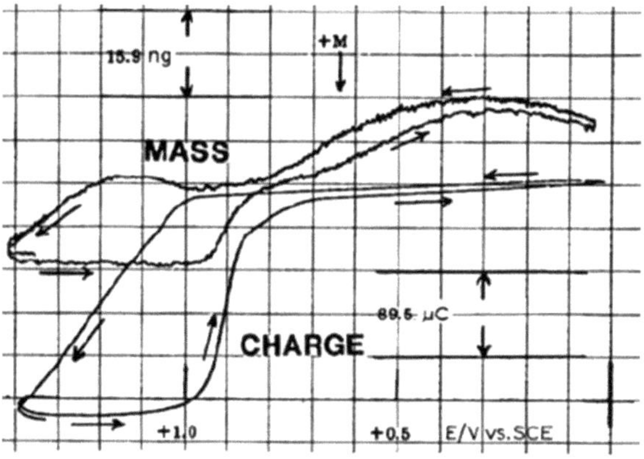

Fig. 16.2 Results from an EQCM experiment, where the shift of frequency and the current have been converted to mass transfer and charge. A gold electrode was employed. The resonator was immersed into an aqueous $HClO_4$ solution. Under anodic conditions (to the *left*) a monolayer of adsorbed oxygen was formed. The layer was re-dissolved when the voltage returned to less positive values. Reprinted with permission from Ref. [14]. Copyright 1985 Elsevier

knowledge on the layer thickness from the charge passed through the electrode. Using this knowledge helps in the determination of G' and G''.

While not immediately evident, there seems to be no influence of the oscillation amplitude onto the speed of an electrochemical reaction. To the best of the author's knowledge, no such effect has ever been reported. Düwel searched for them [18], but what he found could be explained with effects of temperature.

A final remark: Running an electro-chemical experiment on an EQCM is different from "sono-electrochemistry" [19, 20]. The transverse oscillation of the QCM surface is a gentle disturbance compared to the large-amplitude compressional waves applied in sono-electrochemistry. In particular, there is no cavitation above an EQCM surface, while cavitation is important in sono-electrochemistry.

16.2 Optical Reflectometry

The combination of shear-wave reflectometry (which is what a QCM measurements amounts to) with optical reflectometry has two separate aspects. Firstly, the experimental realization is less easy than one might think. Secondly, the apparent optical thickness (defined below) is often lower than the apparent acoustic thickness (the Sauerbrey thickness) [21–24], and there is a deeper reason for this being so.

As in acoustics, an optical reflectivity can mean reflected intensity (power per unit area) or reflected amplitude. In the latter case, the reflectivity is complex. For thickness measurements, the amplitude reflection coefficient is more useful than the power reflection coefficient. The information on film thickness is mostly contained in the phase of the amplitude reflection coefficient. At least this is so for films on metal surfaces. The reflected power can be analyzed [25, 26], but the dependence of the power reflectance coefficient on the thickness of a dielectric layer on top of a metal surface is weak. The phase of a reflected wave varies with film thickness much more than the intensity. Two techniques giving access to the phase are ellipsometry [27, 28] and surface plasmon resonance (SPR) spectroscopy [29]. In ellipsometry one relies on the interference between the two waves polarized such that the electric field vector is perpendicular and parallel to the plane of incidence ("s" and "p"). Ellipsometry on a QCM surface is straight-forward [21, 30], but it is still not strictly trivial because windows are needed for measurements in a liquid and windows are a reason to worry because of stress birefringence.

The following discussion is mostly based on SPR spectroscopy. SPR spectroscopy relies on a guided optical wave. (So does the so-called "waveguide spectroscopy", [31, 32] which is related to SPR spectrocsopy.) An SPR spectrometer is the optical analog of the surface acoustic wave (SAW) device (Sect. 15.4). There are solutions to Maxwell's equations, which only exist at metal surfaces. They are surface modes, called surface plasmons (also: surface plasmon polaritons). The surface plasmon is the optical analog of the Rayleigh wave. Its speed of propagation depends on the dielectric properties of the ambient medium. One usually leaves the frequency fixed (that is, one employs a laser) and infers the speed of propagation from a measurement of the wavenumber and the relation $k_{pl} = \omega/c_{pl}$. The index *pl* denotes a surface *pl*asmon. The wavenumber of the surface plasmon, k_{pl}, is determined with little effort from the coupling angle and the latter needs an explanation.

In order to excite a surface plasmon (or any other type of guided wave) from the outside, the x-component of the two wave vectors (guided wave and external beam) must be the same. This is the content of Snell's law. The x-component of the incoming beam is given as $k_x = \sin(\theta)\ n_{amb}\omega/c$ with n_{amb} the refractive index of the ambient medium, c the speed of light in vacuum, and θ the angle of incidence (measured from the surface normal). Clearly, k_x is always smaller than $n_{amb}\omega/c$. The plasmon's wavenumber, on the other hand, is always larger than $n_{amb}\omega/c$. Unless some trick is employed, Snell's law prohibits the excitation of surface plasmons with external beams. Some mechanism for coupling has to be found [29].

The easiest way to achieve coupling (prism coupling [38]) is not possible on a QCM because the prism disturbs the acoustic resonance. Grating coupling, on the other hand, is a possibility [33, 34]. One creates a shallow corrugation grating on the resonator surface. The grating diffracts the surface plasmon to the outside and it can also diffract an external beam such that the beam excites a surface plasmon. For excitation to be efficient, the diffracted beam and the surface plasmon must have the same k_x, which amounts to the condition:

$$k_{pl} = \frac{n_{amb}\omega}{c}\sin\theta \pm \frac{2\pi}{\Lambda} \qquad (16.2.1)$$

Λ is the grating constant of the corrugation grating. The surface plasmon is only excited if the angle of incidence, θ, has the correct value. The respective angle is

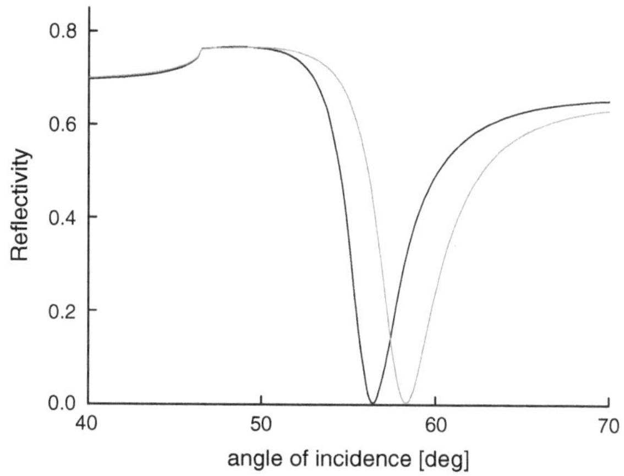

Fig. 16.3 Reflectivity of p-light as obtained with SPR spectroscopy. This numerical example was calculated with prism coupling. The dip in reflectivity takes the role of the resonance curve in a QCM experiment. It shifts to higher angles of incidence upon deposition of a film. "Reflectivity" here refers to reflected power, not to amplitude

the *coupling angle*. In SPR spectroscopy, one measures the coupling angle as precisely as one can.[1] One does so by acquiring the reflectivity of the sample versus incidence angle, $R(\theta)$ (Fig. 16.3). Note: $R(\theta)$ is the power reflectance coefficient, not the amplitude reflectance coefficient. The latter is at the core of SPR spectroscopy, but the amplitude reflectance coefficient is *not* what the raw data show. The function $R(\theta)$ has a characteristic dip at the coupling angle; SPR spectroscopy relies on this dip. The sine of the coupling angle is proportional to k_{pl}, and k_{pl}, in turn, shifts in response to the deposition of a film. All numerical factors aside, the coupling angle takes the role, which the frequency has in a QCM measurement. As in acoustics, a shift in k_{pl} can be translated to a shift of an amplitude reflection coefficient. In SPR spectroscopy, one—in effect—determines the optical amplitude reflection coefficient at the interface between the metal and the sample. The mathematical details are outside the scope of this chapter.

In principle, the analogy between the QCM and the SPR-spectrometer extends to the width of the resonance and the width of the dip of the reflectivity curve. One can analyze this width; it is the analog of Γ. This type of analysis is not routine, though, among the reasons being the possibility of scattering (see Sect. 11.7) and in-plane heterogeneity (see Sect. 11.5.3).

[1] A mantra of the book is "Measure a frequency if you can". There is a corresponding wisdom in astronomy (and even optics) which says: "Measure an angle if you can". The positions of the stars count as angles and so does their color because color turns into an angle after a diffraction grating. The apparent brightness of a star cannot be turned into an angle and the measurement of a star's distance therefore is a difficult matter. It is about as difficult has measuring a resonator's effective area (proportional to the peak conductance). Of course the apparent brightness can be measured, but it cannot be measured with the same ease and precision as the star's position in the sky and its color. Now to SPR spectroscopy: Film thickness turns into an angle (the coupling angle) when probed with an SPR spectrometer. SPR spectroscopy is a scheme to turn the parameter of interest into an angle.

A few remarks on the technical realization of combined QCM and SPR measurements:

- The presence of a grating actually shifts the plasmon's wavenumber, k_{pl}, by a small amount. The details are unessential here.
- The gratings needed for coupling have in the past been etched into the quartz blank. [33–35] In the author's opinion, this process does not have much of a future because of cost. The process makes use of reactive ion etching (RIE), which is laborious and needs expensive equipment. The effort would be well spent if the blanks could be used numerous times, but unfortunately they have to be replaced after a few tens of experiments, at best. Also, the quality of the quartz blank suffers from the etching process.
- Corrugation gratings can also be produced on polymer surfaces by hot embossing. [36] One might coat the crystal with a polymer film and imprint the grating into that film.
- SPR spectroscopy has also been combined with guided acoustic waves. [37] The remarks on optical and acoustic thickness below apply to this scheme, as well. The benefit in using a guided-wave device is that coupling can be achieved with a prism [38] rather than a grating. Prism coupling is easier. The prism can be mounted on the back of the substrate without damping the acoustic wave. There is no need for a corrugation grating.
- The in situ combination of optical reflectometry and piezo-based microgravimetry allows to study one and the same sample with both techniques. However, there still is a difference, which is the area of averaging. The optical technique (which ever it is) averages over the beam size. This area is different from the acoustically active area.
- In view of the various difficulties, the comparison between Sauerbrey thickness and apparent optical thickness may be easier and even more reliable when based on measurements with two separate instruments on two samples (which hopefully have the same properties).

Given that Δk_{pl} and Δf play a similar role in their respective contexts, one can expect the analysis to occur along similar lines. Indeed, the equations linking film thickness to Δk_{pl} in SPR spectroscopy and to Δf in a QCM measurement are similar. For the shift in Δk_{pl}, there is the following approximate relation [38, 39]:

$$\Delta k_{pl} \approx \frac{\omega}{c}\left(\frac{\varepsilon'_{sub}\varepsilon'_{liq}}{\varepsilon'_{sub}+\varepsilon'_{liq}}\right)^{2}\frac{1}{\sqrt{-\varepsilon'_{sub}\varepsilon'_{liq}}}\frac{1}{\varepsilon'_{sub}-\varepsilon'_{liq}}\left[\frac{\left(\varepsilon'_{sub}-\varepsilon'_{f}\right)\left(\varepsilon'_{f}-\varepsilon'_{liq}\right)}{\varepsilon'_{f}}\right]\frac{2\pi d_{f}}{\lambda} \tag{16.2.2}$$

The indices *sub* and *liq* denote the substrate (the metal layer) and the liquid. $\tilde{\varepsilon} = \tilde{n}^2$ is the dielectric constant at optical frequencies. ε'_f can be converted to the volume fraction of the adsorbate if the refractive index as a function of coverage is known. However, this is not usually the case and one may even debate, whether a refractive index is a meaningful parameter for a sparse layer of adsorbed molecules.

The term in square brackets in Eq. 16.2.2 collects all parameters depending on the properties of the film. Since Eq. 16.2.2 is linear in d_f, it can be written in integral form:

$$\begin{aligned}\Delta k_{pl} &\approx \frac{\omega}{c}\left(\frac{\varepsilon'_{sub}\varepsilon'_{liq}}{\varepsilon'_{sub}+\varepsilon'_{liq}}\right)^2 \frac{1}{\sqrt{-\varepsilon'_{sub}\varepsilon'_{liq}}}\frac{1}{\varepsilon'_{sub}-\varepsilon'_{liq}}\frac{2\pi}{\lambda}\int_0^\infty \frac{\left(\varepsilon'_{sub}-\varepsilon'(z)\right)\left(\varepsilon'(z)-\varepsilon'_{liq}\right)}{\varepsilon'(z)}dz \\ &= \frac{\omega}{c}\left(\frac{\varepsilon'_{sub}\varepsilon'_{liq}}{\varepsilon'_{sub}+\varepsilon'_{liq}}\right)^2 \frac{1}{\sqrt{-\varepsilon'_{sub}\varepsilon'_{liq}}}\frac{1}{\varepsilon'_{sub}-\varepsilon'_{liq}} I_1\end{aligned} \tag{16.2.3}$$

In the last line, the integral was renamed as I_1. I_1 is the "first integral invariant" [40], more generally defined as

$$I_1 = \int_{-\infty}^{\infty} \frac{(\varepsilon_1 - \varepsilon(z))(\varepsilon(z)-\varepsilon_2)}{\varepsilon(z)}dz \tag{16.2.4}$$

The indices 1 and 2 denote the two adjacent media (here: the metal and the bulk liquid). The first invariant plays a central role in both SPR spectroscopy and ellipsometry. When the wavelength is much larger than the thickness of the layer of interest, both techniques determine I_1. In the long-wavelength limit, neither SPR spectroscopy nor ellipsometry resolve the inner structure of the film.

The following argument needs one more approximation, which is that the properties of the metal are much different from the properties of both the adsorbate and the ambient medium. This approximation is reasonable because the adsorbate is an organic medium with a refractive index different but not vastly different from the refractive index of the solvent. The refractive index of the metal, on the contrary, is large and predominantly imaginary. This approximation allows to replace $\varepsilon'_{sub}-\varepsilon'(z)$ by $\varepsilon'_{sub}-\varepsilon'_{liq}$. Applied to Eq. 16.2.3, this leads to

$$\Delta k_{pl} \approx \frac{\omega}{c}\left(\frac{\varepsilon'_{sub}\varepsilon_{liq}}{\varepsilon'_{sub}+\varepsilon'_{liq}}\right)^2 \frac{1}{\sqrt{-\varepsilon'_{sub}\varepsilon_{liq}}}\frac{2\pi}{\lambda}\int_0^\infty \frac{\varepsilon'(z)-\varepsilon'_{liq}}{\varepsilon'(z)}dz \tag{16.2.5}$$

Equation 16.2.5 has the same structure as the corresponding equation from shear-wave reflectometry. Above Eq. 10.2.7, the long-wavelength limit had been applied, which had resulted in the relation

$$\frac{\Delta\tilde{f}}{f_0} \approx -\frac{-\omega}{\pi Z_q}\int_0^\infty \frac{\tilde{Z}^2(z)-\tilde{Z}^2_{liq}}{\tilde{Z}^2(z)}\rho(z)dz \approx -\frac{\rho\omega}{\pi Z_q}\int_0^\infty \frac{\tilde{G}(z)-\tilde{G}_{liq}}{\tilde{G}(z)}dz \tag{16.2.6}$$

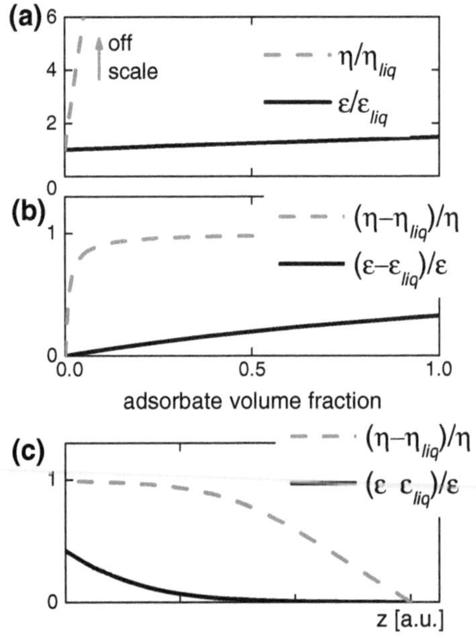

Fig. 16.4 In a model calculation, the dependence of the dielectric constant and the viscosity on adsorbate volume fraction are both approximated as linear. (**a**) However, the slope is much larger for the viscosity. The viscosity is assumed to increase by a factor 10 for a 10 % solution, while ε increases by only a factor of 1.05 for the same concentration. The contrast function (panel **b**, Eqs. 16.2.5 and 16.2.6) following from these assumptions saturates for the viscosity, while it is close to linear for the dielectric constant. Panel **c** shows the contrast function versus distance from the surface, where the underlying profile of the adsorbate volume fraction was assumed to be a decaying exponential. The shape of the optical contrast function closely follows the shape of the concentration. Its integral is proportional to the adsorbed amount. The shear-wave contrast function, on the contrary, is close to unity for all volume fractions larger than 0.1. It only decays to the zero in the tail of the profile. Its integral is not a good measure of the adsorbed amount; it increases with swelling

(A short-cut was made here in not discussing the role of the first invariant in the context of the QCM. The first invariant is different from the right-hand side in Eq. 16.2.6.) Comparing Eqs. 16.2.5 and 16.2.6, one might conclude that the information contained in surface plasmon resonances and quartz crystal resonances should be similar. The structure of the underlying equations is the same. In both cases, there is a response function (ε or G), which is converted to a "contrast function" (integrands in Eqs. 16.2.5 and 16.2.6). The outcome of the experiment is proportional to the z-integral of the contrast function.

The apparent thickness of the sample determined optically is often found to be less than the Sauerbrey thickness. The apparent optical thickness is meant to be the thickness derived from Eq. 16.2.2, assuming the ε'_f is equal to the dielectric constant of the unswollen film. Of course the derivation of this thickness needs an assumption on ε'_f. It is not always clear what the appropriate refractive index of the equivalent dry film would have to be and the conversion to optical thickness therefore is often not even attempted. The results of an SPR measurements are usually quoted in terms of "refractive index units" rather than equivalent thickness. This practice avoids the ambiguities resulting from having to make assumptions on n. Still, one can make such assumptions and vary them within reasonable bounds. When doing that, one usually finds that the derived equivalent optical thickness is lower than the Sauerbrey thickness, regardless of what exactly the assumption on n is.

The difference goes back to swelling of the sample in the liquid phase. Swelling affects the acoustic thickness to a larger extent than the optical thickness. In simple words, the liquid is dragged along by the polymer chains and therefore contributes to the acoustic thickness, while the liquid's polarizability (closely connected to ε) is not affected by whether or not there are polymer chains near-by. This simple explanation can be put onto quantitative physical grounds, which is done below.

The matter is illustrated in Fig. 16.4, which displays a model calculation. The crucial difference between optical waves and shear waves is that the response function varies by a few percent in optics, while it can easily vary by more than a decade in shear-wave reflectometry. The calculation underlying Fig. 16.4 is based on viscosity, rather than shear modulus. The two are related by $\tilde{G} = \mathrm{i}\omega\eta$. Somewhat simplifying the matter, it was assumed that both response functions (η and ε) depend linearly on adsorbate volume fraction. However, η was made to increase by a factor of 10 for a volume fraction of 10 %, while ε was made to only increase by a factor 1.05 for the same concentration. These choices reflect the argument made above. Upon adding a polymer to a solvent, the viscosity increases much stronger than the volume-averaged polarizability.

Figure 16.4b shows the resulting contrast functions. The contrast function cannot exceed unity. It comes close to unity for shear waves, while it stays well below unity and is closely proportional to the adsorbate volume fraction for the optical beam. Panel c shows how the contrast function depends on z if the underlying adsorbate volume fraction is a decaying exponential. Since the optical contrast function is roughly proportional to volume fraction, it also is close to a decaying exponential. The integral over the contrast function is roughly proportional to the adsorbed amount. This is different for shear waves. Because the contrast function saturates, the profile is closer to a box profile than to a decaying exponential. Its integral is not proportional to the adsorbed amount, not even roughly [21]. If a compact film swells by some limited amount, this will increase the thickness, but will not much decrease the acoustic contrast function. The latter is close to unity with and without solvent inside the film [21]. As a consequence, the Sauerbrey thickness increases with swelling even for rather dilute polymer layers. Swelling increases the acoustic thickness, while it affects the optical thickness to a lesser extent.

The above discussion focused on swellable polymer films. There is another situation with a similar background, which is the clustering of proteins (or other nanoparticles) upon adsorption. That topic has been addressed by Richter and co-workers at length [23, 41]. A "swollen" layer in this context is a layer with much liquid volume between the adsorbed objects. When the objects deposit as dimers (or as larger clusters), the amount of trapped liquid decreases relative to the state with random positions. The clustered and the random state can be distinguished based on the comparison of optical and acoustic thickness.

The consequence of this situation is that SPR spectroscopy is the more appropriate sensing method if the primary interest is the determination of adsorbed amount. QCM data are affected by swelling. Trapped liquid can of course increase the sensitivity. It amounts to an amplification scheme. The disadvantage lies in the difficulties with interpretation.

16.3 QCM/AFM Combination

One may want to place a QCM under an AFM tip for two reasons. Firstly, the AFM can image the sample, which aids the interpretation of Δf and $\Delta\Gamma$. [42–44] For instance, one might count the number particles per unit area and divide Δf by this number in order to infer the frequency shift per particle (see Sect. 12.1 for an example). A second target of the combination might be the behavior of the tip-sample contact under MHz shear excitation. The principal drawback of this type of studies is the poor sensitivity. The resonance frequency does not shift noticeably upon contact with an AFM tip. To see this, use the relation $\Delta f = 1/(2n\pi^2 Z_q)\kappa_P$ (Eq. 11.3.5) and estimate the constant stiffness as $\kappa_P = 2G^*a$ (Eq. 11.2.11) with G^* the relevant modulus (Eq. 11.2.12, use a value of $G^* \approx 10$ GPa for this estimate) and a the contact radius (use $a \approx 5$ nm for this estimate). Inserting this contact stiffness into Eq. 11.3.3 (use $A = 30\ \text{mm}^2$), one arrives at a frequency shift of

$$\Delta f = \frac{1}{2n\pi^2 Z_q A} 2G^* a \approx 0.02\ \text{Hz}. \tag{16.3.1}$$

One might be able to observe such a small frequency shift with a stable system, but the approximations leading to this estimate have already excluded all soft samples because G^* was assumed as 10 GPa. Also, when inserting a contact radius of 5 nm, the resolution, which can be achieved with an AFM, was substantially compromised. Increasing the fundamental frequency does not help: f_0 does not enter Eq. 16.3.1. The only knob to turn is the effective area. The effective area can be brought down by employing inverted MESA resonators or MEMS resonators.

There is a second implication of Eq. 16.3.1. A few small contacts do not change the resonant frequency. For that reason, one can place non-electroded quartz blanks on a slightly rough metal plate and use the metal plate as the lower electrode (Fig. 6.1). The resonance frequency is hardly changed by the few asperities holding the blank.

There are reports in the literature, where a QCM has been seen to respond to contact with an AFM tip [45–48]. One explanation would be tip blunting. Tip blunting may have occurred accidentally by the contact with the QCM. Actually, blunted tips (or small spheres glued to the AFM tip) are an option for contact mechanics experiments based on the QCM/AFM combination. Using an indenter rather than a tip also is an option, as demonstrated by Krim and Wahl [49]. These modifications avoid the problem of the contact area being too small. A second explanation for why an AFM tip can appear to induce a large frequency shift is an interaction with the tip holder, mediated by compressional waves [50]. The latter argument applies to experiments in the liquid phase.

While the QCM cannot be used for detection, one can still observe the response of the AFM tip to the vibration of the substrate (the quartz resonator). This is not at all difficult. One might look for a vibration of the tip at MHz frequencies [51], but the vibration might also affect the forces of detachment [51] or the friction force [52]. If one looks for an easy way to oscillate the substrate of an AFM, the QCM is one solution. This being said: There are other solutions. Any kind of ultrasonic transducer will do [53, 54], there is no need to employ a resonant device. Also, one might vibrate the tip, rather than the substrate. The tip's vibration might occur at MHz frequencies and one might employ quartz as the resonator material [55–57]. The cantilever *is* small, it will therefore respond to a contact with a frequency shift. (Of course a small cantilever is also overdamped more easily than a QCM). Not coincidentally, the analysis of frequency shifts of the cantilever's resonance induced by the interaction with a substrate is well established in dynamic atomic force microscopy [58, 59].

16.4 Combination with the Surface Forces Apparatus

The "surface forces apparatus" (SFA, also "surface forces balance, SFB") was designed to study forces between molecularly flat surfaces (mostly mica), where the distance between the two surfaces is measured by an interferometric scheme [60]. The same scheme also provides the area of contact. From the perspective of QCM-based contact mechanics experiments, there are two separate issues, which are the use of mica to provide for an atomically flat surface and the interferometric scheme to measure distance and contact area. The author has participated in experiments, where the QCM was integrated into an SFA. For the details see Refs. [61, 62]. It is fair to say that the technical constraints are demanding. Somewhat less ambitious is it, to glue mica sheets onto a QCM [63]. However, in the author's laboratory, the mica-coated resonators had poor mode shapes on the overtones (large anharmonic side bands, cf. Sect. 7.2). This meant that the principal benefit of the advanced QCM (the comparison of frequency and bandwidth from different overtones) could not be exploited. That is not meant to say that the use of mica on a QCM was impossible. A better procedure for preparation might succeed.

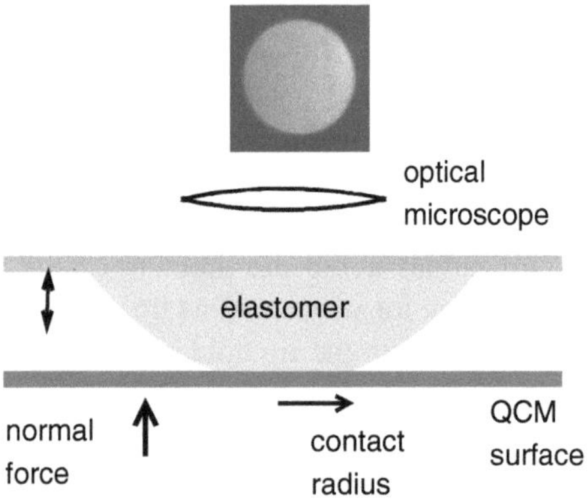

Fig. 16.5 A setup combining a QCM with a JKR apparatus. The JKR apparatus applies a vertical force, thereby pushing a soft polymer lens against a flat substrate. It measures the vertical displacement and the radius of contact. Adapted from Ref. [65]

16.5 Combination with the JKR Apparatus

Most polymer melts are too viscous to be studied as a bulk, semi-infinite medium with a QCM. They overdamp the resonance. They can still be investigated, if they are made to touch the resonator in the center only. The small contact area makes up for the polymer's large wave impedance. There are two ways of running this experiment. In the simpler approach, one does not measure the contact area and limits the analysis to the ratio of $\Delta\Gamma$ and $-\Delta f$. In simple models, the ratio is independent of contact area and can be converted to the material's loss tangent, $\tan(\delta_L) = G''/G'$. There is an approximation in these models though, elaborated on in Sect. 9.4.

The contact area can also be measured, making use of the so-called JKR apparatus (Fig. 16.5). The abbreviation "JKR" stands for Johnson, Kendall, and Roberts, who formulated the model underlying this instrument (Ref. [64], see also Sect. 11.2). A JKR apparatus pushes a hemisphere of the material under test against a flat substrate and images the contact area, using a microscope. The material must be soft enough to form a sufficiently extended contact. If JKR theory applies, the dependence of the contact area on normal force can be used to derive the material's (static) elastic modulus and the work of adhesion (Eq. 11.2.8). Flanigan et al. have replaced the flat substrate by a quartz resonator and used the frequency shift as an additional source of information [65]. With known contact area, Δf and $\Delta\Gamma$ can be converted to G' and G'' (that is, the near-surface value of the MHz shear modulus) because the displacement pattern is almost a plane wave.

16.6 Combination with Optical Instrumentation

Since the resonator surface is optically accessible, optical investigations on the sample are easy and are being done at many places. In particular, a camera is often placed over the resonator. (Optical reflectometry is discussed separately in Sect. 16.2.) The discussion is limited to two remarks:

- Transparent electrodes made of indium-tin-oxide (ITO) are available [66, 67]. The low conductivity of ITO is somewhat of a problem. It can broaden the resonance via piezoelectric stiffening and it can also tilt the resonance curve, as discussed in the context of Fig. 4.14. The problems are manageable, but need attention.
- A semi-quantitative optical measure of film thickness can be obtained with localized surface plasmons. See Ref. [68] for more details.

16.7 Combination with Calorimetry

Smith and co-workers have combined a QCM with calorimetry. They call the instrument QCM/HCC, where HCC stands for "heat conduction calorimetry" [69]. The resonator is mounted on small stacks of thermopiles. When heat flows through the thermopile, the Seebeck effect generates a voltage, which is recorded and analyzed. The source of heat usually is some kind of a functional film deposited on the resonator surface. The film might simply be a polymer layer exposed to vapor, in which case the instrument measures the heat of sorption [70]. The active layer might also contain a catalyst layer [71]. For more details see Ref. [70].

16.8 A Magnetic QCM

The Janata group has coated a QCM with a magnetic material and finds a dependence of the resonance frequency on the magnetic field [72]. The mechanism by which the change comes about is not entirely clear. In Ref. [73], Δf is not proportional to the magnetic field. The mechanism may be indirect (and there may be more than one mechanism).

16.9 Combination with Various Forms of Electrical Spectroscopy

Given that a front electrode is in place anyway, one can use it for purposes other than driving the crystal. It can be used to do electrochemistry (Sect. 16.1), electrochemical impedance spectroscopy (EIS, Refs. [74, 75, 76]), dielectric spectroscopy (Ref. [77]), or electric cell-substrate impedance sensing (ECIS, Ref. [78]). Presumably, the list is incomplete. Note that other electrical elements (such as thermocouples) or more than one electrode can be placed onto the resonator surface if the plate is large enough. Quartz plates with diameters of up to three inches are available.

Glossary

Variable	Definition (Comments)
a	Contact radius
amb	As an index: *amb*ient medium
A	Effective area of the resonator plate
c	Speed of propagation, speed of light
d_f	Film thickness
f	Frequency
f	As an index: *f*ilm
f_0	Resonance frequency at the fundamental ($f_0 = Z_q/(2m_q) = Z_q/(2\rho_q d_q)$)
$\tilde{G}$	Shear modulus ($\tilde{G} = G' + iG''$)
$G*$	Effective shear modulus (Eq. 11.2.12)
k_{pl}	Wavenumber of surface plasmon
k_x	x-component of the wavenumber of the incident beam
liq	As an index: *liq*uid (mostly the ambient medium)
n	Overtone order
$\tilde{n}$	Refractive index
pl	As an index: *pl*asmon
Q	Quality factor
$R(\theta)$	A reflectivity curve. (In SPR spectroscopy, R is the reflected intensity; θ is the angle of incidence.)
sub	As an index: *sub*strate (mostly a metal film)
z	Spatial coordinate along the surface normal
$\tilde{Z}$	Acoustic wave impedance
Z_q	Acoustic wave impedance of AT-cut quartz ($Z_q = 8.8 \times 10^6$ kg m^{-2} s^{-1})
Γ	Imaginary part of a resonance frequency
Δ	As a prefix: A shift induced by the presence of the sample
$\tilde{\varepsilon}$	Dielectric constant optical frequencies ($\tilde{\varepsilon} = \tilde{n}^2$)
η	Viscosity
κ_P	Spring constant of a contact with a particle

λ	Wavelength
Λ	Wavelength of a corrugation grating (Eq. 16.2.1)
θ	Angle of incidence
ρ	Density
ω	Angular frequency

References

1. Bund, A., Schwitzgebel, G.: Investigations on metal depositions and dissolutions with an improved EQCMB based on quartz crystal impedance measurements. Electrochim. Acta **45**(22–23), 3703–3710 (2000)
2. Nomura, T., Okuhara, M.: Frequency shifts of piezoelectric quartz crystals immersed in organic liquids. Anal. Chim. Acta **142**, 281–284 (1982)
3. Nomura, T., Hattori, O.: Determination of micromolar concentrations of cyanide in solution with a piezoelectric detector. Anal. Chim. Acta **115**, 323–326 (1980)
4. Buttry, D.A., Ward, M.D.: Measurement of interfacial processes at electrode surfaces with the electrochemical quartz crystal microbalance. Chem. Rev. **92**(6), 1355–1379 (1992)
5. Schumacher, R.: The quartz microbalance—a novel-approach to the insitu investigation of interfacial phenomena at the solid liquid junction. Angew. Chem. Int Ed. Engl. **29**(4), 329–343 (1990)
6. Marx, KA.: The quartz crystal microbalance and the electrochemical QCM: applications to studies of thin polymer films, electron transfer systems, biological macromolecules, biosensors, and cells. In: Janshoff, A., Steinem, C. (eds.) Piezolelectric Sensors, pp. 371–424. Springer, Berlin (2007)
7. Daikhin, L.; Tsionsky, V.; Gileadi, E.; Urbakh, M.: Looking at the metal/solution interface with the electrochemical quartz crystal microbalance: theory and experiment. In: Bard, A. J., Rubinstein, I., (eds.) Electroanalytical Chemistry: A Series of Advances, pp. 1–99. Marcel Dekker Inc, New York (2003)
8. Doblhofer, K., Weil, K.G.: Application of the quartz microbalance in electrochemistry. Bunsen Mag. **9**, 162 (2007)
9. Hillman, A.R.: The EQCM: electrogravimetry with a light touch. J. Solid State Electrochem. **15**(7–8), 1647–1660
10. Kochman, A., Krupka, A., Grissbach, J., Kutner, W., Gniewinska, B., Nafalski, L.: Design and performance of a new thin-layer radial-flow holder for a quartz crystal resonator of an electrochemical quartz crystal microbalance. Electroanalysis **18**(22), 2168–2173 (2006)
11. Tsionsky, V., Daikhin, L., Gileadi, E.: Response of the electrochemical quartz crystal microbalance for gold electrodes in the double-layer region. J. Electrochem. Soc. **143**(7), 2240–2245 (1996)
12. http://www.thinksrs.com/downloads/PDFs/Manuals/QCM200m.pdf. Accessed 14 Feb 2013
13. http://www.gamry.com/assets/Uploads/eQCM-Calibration-of-an-Au-coated-Quartz-Crystal-AppNote.pdf. Accessed 6 Apr 2103
14. Bruckenstein, S., Shay, M.: Experimental aspects of use of the quartz crystal microbalance in solution. Electrochim. Acta **30**(10), 1295–1300 (1985)
15. Hillman, A.R., Efimov, I., Ryder, K.S.: Time-scale- and temperature-dependent mechanical properties of viscoelastic poly(3,4-ethylenedioxythlophene) films. J. Am. Chem. Soc. **127**(47), 16611–16620 (2005)

16. Wudy, F., Multerer, M., Stock, C., Schmeer, G., Gores, H.J.: Rapid impedance scanning QCM for electrochemical applications based on miniaturized hardware and high-performance curve fitting. Electrochim. Acta **53**(22), 6568–6574 (2008)
17. Wickman, B., Gronbeck, H., Hanarp, P., Kasemo, B.: Corrosion induced degradation of Pt/C model electrodes measured with electrochemical quartz crystal microbalance. J. Electrochem. Soc. **157**(4), B592–B598 (2010)
18. Düwel, M.: Diploma thesis, Clausthal University of Technology (2007)
19. Compton, R.G., Eklund, J.C., Marken, F.: Sonoelectrochemical processes: a review. Electroanalysis **9**(7), 509–522 (1997)
20. Schneider, O., Matic, S., Argirusis, C.: Application of the electrochemical quartz crystal microbalance technique to copper sonoelectrochemistry—part 1. Sulfate-based electrolytes. Electrochim. Acta **53**(17), 5485–5495 (2008)
21. Domack, A., Prucker, O., Ruhe, J., Johannsmann, D.: Swelling of a polymer brush probed with a quartz crystal resonator. Phys. Rev. E **56**(1), 680–689 (1997)
22. Plunkett, M.A., Wang, Z.H., Rutland, M.W., Johannsmann, D.: Adsorption of pNIPAM layers on hydrophobic gold surfaces, measured in situ by QCM and SPR. Langmuir **19**(17), 6837–6844 (2003)
23. Bingen, P., Wang, G., Steinmetz, N.F., Rodahl, M., Richter, R.P.: Solvation effects in the quartz crystal microbalance with dissipation monitoring response to biomolecular adsorption. A phenomenological approach. Anal. Chem. **80**(23), 8880–8890 (2008)
24. Wang, Z.H., Kuckling, D., Johannsmann, D.: Temperature-induced swelling and de-swelling of thin poly(N-isopropylacrylamide) gels in water: combined acoustic and optical measurements. Soft Mater. **1**(3), 353–364 (2003)
25. Wang, G., Rodahl, M., Edvardsson, M., Svedhem, S., Ohlsson, G., Hook, F., Kasemo, B.: A combined reflectometry and quartz crystal microbalance with dissipation setup for surface interaction studies. Rev. Sci. Instrum. **79**(7), 075107 (2008)
26. Edvardsson, M., Svedhem, S., Wang, G., Richter, R., Rodahl, M., Kasemo, B.: QCM-D and reflectometry instrument: applications to supported lipid structures and their biomolecular interactions. Anal. Chem. **81**(1), 349–361 (2009)
27. Azzam, R.M.A., Bashara, N.M.: Ellipsometry and Polarized Light. Springer, New York (1987)
28. Lekner, J.: Theory of Reflection of Electromagnetic and Particle Waves. Springer, New York (1987)
29. Homola, J.: Present and future of surface plasmon resonance biosensors. Anal. Bioanal. Chem. **377**(3), 528–539 (2003)
30. Bodvik, R., Macakova, L., Karlson, L., Thormann, E., Claesson, P.: Temperature-dependent competition between adsorption and aggregation of a cellulose ether-simultaneous use of optical and acoustical techniques for investigating surface properties. Langmuir **28**(25), 9515–9525 (2012)
31. Tiefenthaler, K., Lukosz, W.: Sensitivity of grating couplers as integrated-optical chemical sensors. J. Opt. Soc. Am. B Opt. Phys. **6**(2), 209–220 (1989)
32. http://www.owls-sensors.com/. Accessed 13 June 2013
33. Laschitsch, A., Menges, B., Johannsmann, D.: Simultaneous determination of optical and acoustic thicknesses of protein layers using surface plasmon resonance spectroscopy and quartz crystal microweighing. Appl. Phys. Lett. **77**(14), 2252–2254 (2000)
34. Bailey, L.E., Kambhampati, D., Kanazawa, K.K., Knoll, W., Frank, C.W.: Using surface plasmon resonance and the quartz crystal microbalance to monitor in situ the interfacial behavior of thin organic films. Langmuir **18**(2), 479–489 (2002)
35. Zong, Y., Xu, F., Su, X.D., Knoll, W.: Quartz crystal microbalance with integrated surface plasmon grating coupler. Anal. Chem. **80**(13), 5246–5250 (2008)
36. Worgull, M.: Hot Embossing: Theory and Technology of Microreplication. Elsevier, Amsterdam (2008)

37. Francis, L.A., Friedt, J.M., Zhou, C., Bertrand, P.: In situ evaluation of density, viscosity, and thickness of adsorbed soft layers by combined surface acoustic wave and surface plasmon resonance. Anal. Chem. **78**(12), 4200–4209 (2006)
38. Kretschmann, E.: Determination of optical constants of metals by excitation of surface plasmons. Zeitschrift Fur Physik **241**(4), 313 (1971)
39. Pockrand, I.: Surface plasma-oscillations at silver surfaces with thin transparent and absorbing coatings. Surf. Sci. **72**(3), 577–588 (1978)
40. Lekner, J.: Invariant formulation of the reflection of long waves by interfaces. Phys. A **128**(1–2), 229–252 (1984)
41. Carton, I., Brisson, A.R., Richter, R.P.: Label-free detection of clustering of membrane-bound proteins. Anal. Chem. **82**(22), 9275–9281 (2010)
42. Friedt, J.M., Choi, K.H., Frederix, F., Campitelli, A.: Simultaneous AFM and QCM measurements—methodology validation using electrodeposition. J. Electrochem. Soc. **150**(10), H229–H234 (2003)
43. Bund, A., Schneider, O., Dehnke, V.: Combining AFM and EQCM for the in situ investigation of surface roughness effects during electrochemical metal depositions. Phys. Chem. Chem. Phys. **4**(15), 3552–3554 (2002)
44. Hayden, O., Bindeus, R., Dickert, F.L.: Combining atomic force microscope and quartz crystal microbalance studies for cell detection. Meas. Sci. Technol. **14**(11), 1876–1881 (2003)
45. Kim, J.M., Chang, S.M., Muramatsu, H.: Scanning localized viscoelastic image using a quartz crystal resonator combined with an atomic force microscopy. Appl. Phys. Lett. **74**(3), 466–468 (1999)
46. Sasaki, A., Katsumata, A., Iwata, F., Aoyama, H.: Scanning shearing-stress microscope. Appl. Phys. Lett. **64**(1), 124–125 (1994)
47. Sasaki, A., Katsumata, A., Iwata, F., Aoyama, H.: Scanning shearing-stress microscopy of gold thin-films. Jpn. J. Appl. Phys. Part 2 Lett. **33**(4A), L547–L549 (1994)
48. Yamada, R., Ye, S., Uosaki, K.: Novel scanning probe microscope for local elasticity measurement. Jpn. J. Appl. Phys. Part 2 Lett. **35**(7A), L846–L848 (1996)
49. Borovsky, B., Krim, J., Syed Asif, S.A., Wahl, K.J.: Measuring nanomechanical properties of a dynamic contact using an indenter probe and quartz crystal microbalance. J. Appl. Phys. **90**(12), 6391–6396 (2001)
50. Friedt, J.M., Choi, K.H., Francis, L., Campitelli, A.: Simultaneous atomic force microscope and quartz crystal microbalance measurements: Interactions and displacement field of a quartz crystal microbalance. Jpn. J. Appl. Phys. Part 1 Regul. Pap. Short Notes Rev. Pap. **41**(6A), 3974–3977 (2002)
51. Lubben, J.F., Johannsmann, D.: Nanoscale high-frequency contact mechanics using an AFM tip and a quartz crystal resonator. Langmuir **20**(9), 3698–3703 (2004)
52. Inoue, D., Machida, S., Taniguchi, J., Suzuki, M., Ishikawa, M., Miura, K.: Dynamical frictional force of nanoscale sliding. Phys. Rev. B **86**(11), 4 (2012)
53. Scherer, V., Arnold, W., Bhushan, B.: Lateral force microscopy using acoustic friction force microscopy. Surf. Interface Anal. **27**(5–6), 578–587 (1999)
54. Krotil, H.U., Weilandt, E., Stifter, T., Marti, O., Hild, S.: Dynamic friction force measurement with the scanning force microscope. Surf. Interface Anal. **27**(5–6), 341–347 (1999)
55. Jersch, J., Maletzky, T., Fuchs, H.: Interface circuits for quartz crystal sensors in scanning probe microscopy applications. Rev. Sci. Instrum. **77**(8), 083701 (2006)
56. Günther, P., Fischer, U., Dransfeld, K.: Scanning near-field acoustic microscopy. Appl. Phys. B Photophys. Laser Chem. **48**(1), 89–92 (1989)
57. http://www.specs.de/cms/front_content.php?idcat=246. Accessed 3 Mar 2014
58. Giessibl, F.J.: A direct method to calculate tip-sample forces from frequency shifts in frequency-modulation atomic force microscopy. Appl. Phys. Lett. **78**(1), 123–125 (2001)
59. Hölscher, H., Schwarz, U.D., Wiesendanger, R.: Calculation of the frequency shift in dynamic force microscopy. Appl. Surf. Sci. **140**(3–4), 344–351 (1999)

60. Israelachvili, J.N.: Intermolecular and Surface Forces. Academic Press, London (2011)
61. Berg, S., Ruths, M., Johannsmann, D.: Quartz crystal resonators with atomically smooth surfaces for use in contact mechanics. Rev. Sci. Instrum. **74**(8), 3845–3852 (2003)
62. Berg, S., Ruths, M., Johannsmann, D.: High-frequency measurements of interfacial friction using quartz crystal resonators integrated into a surface forces apparatus. Phys. Rev. E **65**(2), 026119 (2002)
63. Xu, B., Wang, H.D., Wang, Y., Zhu, G.Y., Li, Z., Wang, E.K.: A mica-modified quartz resonator for a quartz crystal microbalance study. Anal. Sci. **16**(10), 1061–1063 (2000)
64. Johnson, K.L., Kendall, K., Roberts, A.D.: Surface energy and contact of elastic solids. Proc. Roy. Soc. Lond. Ser. A Math. Phys. Sci. **324**(1558), 301 (1971)
65. Flanigan, C.M., Desai, M., Shull, K.R.: Contact mechanics studies with the quartz crystal microbalance. Langmuir **16**(25), 9825–9829 (2000)
66. Zhang, J., Hu, J.Q., Zhu, F.R., Gong, H., O'Shea, S.J.: ITO thin films coated quartz crystal microbalance as gas sensor for NO detection. Sens. Actuators B Chem. **87**(1), 159–167 (2002)
67. ITO coated resonator crystals are available from microvacuum: http://www.microvacuum.com/
68. Larsson, E.M., Edvardsson, M.E.M., Langhammer, C., Zoric, I., Kasemo, B.: A combined nanoplasmonic and electrodeless quartz crystal microbalance setup. Rev. Sci. Instrum. **80**(12), 10 (2009)
69. http://www.masscal.com/library/EarlyQCMHCC.pdf. Accessed 28 Mar 2013
70. Smith, A.L., Mulligan, R.B., Shirazi, H.M.: Determining the effects of vapor sorption in polymers with the quartz crystal microbalance/heat conduction calorimeter. J. Polym. Sci. Part B Polym. Phys. **42**(21), 3893–3906 (2004)
71. Smith, A.L., Shirazi, H.M., Smith, F.C.: Real-time monitoring of catalytic surfaces using a mass/heat flow sensor: hydrogenation of ethylene on platinum and palladium. Catal. Lett. **104**(3–4), 199–204 (2005)
72. Yu, G.Y., Hunt, W.D., Josowicz, M., Janata, J.: Development of a magnetic quartz crystal microbalance. Rev. Sci. Instrum. **78**(6) (2007)
73. Vavra, K.C., Yu, G., Josowicz, M., Janata, J.: Magnetic quartz crystal microbalance: magneto-acoustic parameters. J. Appl. Phys. **110**(1), 013905-1 (2011)
74. Sabot, A., Krause, S.: Simultaneous quartz crystal microbalance impedance and electrochemical impedance measurements.Investigation into the degradation of thin polymer films. Anal. Chem. **74**(14), 3304–3311 (2002)
75. Briand, E., Zach, M., Svedhem, S., Kasemo, B., Petronis, S.: Combined QCM-D and EIS study of supported lipid bilayer formation and interaction with pore-forming peptides. Analyst **135**(2), 343–350 (2010)
76. Paul, D.W., Clark, S.R., Beeler, T.L.: Instrumentation for simultaneous measurement of double-layer capacitance and solution resistance at a QCM electrode. Sens. Actuators B Chem. **17**(3), 247–255 (1994)
77. Roth, M., Dera, T., Drost, A., Hartinger, R., Wendler, F., Endres, H.E., Hillerich, B.: Directly heated quartz crystal microbalance with an integrated dielectric sensor. Sens. Actuators A Phys. **68**(1–3), 399–403 (1998)
78. Heitmann, V., Reiss, B., Wegener, J.; The quartz crystal microbalance in cell biology: basics and applications. In: Steinem, C., Janshoff, A. (eds.) Piezoelectric Sensors. Springer, Berlin (2007)

Chapter 17
Considerations for Well-Controlled QCM Experiments

Abstract Tips and tricks for successful and meaningful experiments are mentioned at various places in the book. They are collected here. The list was expanded with advice of entirely practical nature.

17.1 Temperature-Frequency Coupling

The stiffness of any material depends on temperature. As a rule of thumb, the stiffness decreases with increasing temperature because the material expands. As the atoms move apart, the forces resisting deformation decrease. The matter is more complicated for anisotropic crystals, though. Certain components of the stiffness tensor of α-quartz actually *increase* with temperature. As a consequence, there are the so-called temperature-compensated cuts, which lead to a resonance frequency almost independent of temperature. One of these is the famous AT-cut. The AT-cut has the crystallographic x-axis in the resonator plane. The y-axis is rotated away from the surface normal by about 35°. The existence of a temperature-compensated cut, per se, is an advantage of quartz. Not all crystals have such a cut. Silicon, for instance, does not. (There are other ways to achieve temperature compensation.) Even if a suitable cut exists, there are limits to temperature compensation. A vanishing temperature coefficient of frequency, TCF, can only be achieved for one temperature and for one overtone at a time.

For the AT-cut, the T-f characteristic follows a cubic polynomial (Fig. 17.1):

$$\frac{\delta f}{f} = a(T - T_0) + b(T - T_0)^2 + c(T - T_0)^3 \tag{17.1.1}$$

δf is the deviation from the frequency at the reference temperature, T_0. The parameters of the polynomial are [2]

D. Johannsmann, *The Quartz Crystal Microbalance in Soft Matter Research*, Soft and Biological Matter, DOI: 10.1007/978-3-319-07836-6_17,

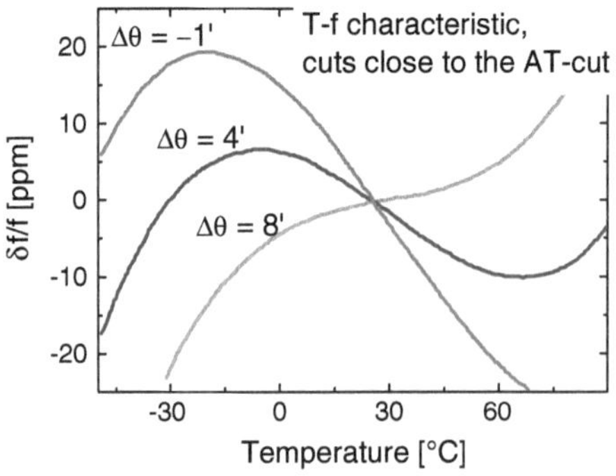

Fig. 17.1 For AT-cut quartz, the temperature dependence of f_r follows a cubic polynomial. The parameters of the polynomial slightly vary between overtones. They can be changed to some extent by adjusting the angle of the cut, θ (Eq. 17.1.2). Δθ is the deviation from 35°12.5′. Adapted from Ref. [1]

$$
\begin{aligned}
T_0 &= 25\,^\circ\mathrm{C} \\
a &= (-1.32 - 0.085\,\Delta 0)10^{-6} \\
b &= (0.15 - 0.093\,\Delta\theta)10^{-9} \\
c &= (101.3 + 0.200\,\Delta\theta)10^{-12} \\
\Delta\theta &= \theta - 35^\circ 30'
\end{aligned}
\qquad (17.1.2)
$$

One can adjust the T-f characteristic to some extent a slightly adjusting the angle of the crystal cut.

While Fig. 17.1 and Eq. 17.1.1 suggest that temperature effects could be accounted for by measuring temperature and correcting Δf suitably, such a correction is always incomplete because of hysteresis. Upon heating the crystal and returning to room temperature, the resonator does not return to the exact same frequency. The reason are crystal defects, which slowly migrate and rearrange in response to changes of temperature and stress.

Since the different overtones respond to changes of temperature differently, one can infer temperature from the comparison of frequency shifts on different overtones. There is no need to separately measure temperature. This is one of the rare cases, where anharmonic side bands can be of use. The comparison of Δf on a few anharmonic sidebands can provide the temperature. Inferring temperature from the comparison between overtones is a routine procedure in time and frequency control. In sensing, this correction scheme only works if temperature effects can be reliable subtracted from the effects of the sample, which also differ between the overtones.

Advice:

- Control temperature. You hope for constant temperature within ±0.01 °C.
- Be aware of the fact that crystals in air have poor thermal contact to the holder. It can take time until the temperature equilibrates.
- When doing sorption experiments in vapor, remember that solvent evaporation and solvent condensation are accompanied by a latent heat.
- When shining light onto a crystal, check whether absorption heats the crystal.

17.2 Effects of Pressure and Static Stress

The dependence of frequency on stress has similar roots as the dependence on temperature. The stiffness of a crystal depends on the distance between the atoms, which changes when the crystal is deformed. As with temperature effects, defects play a role. The stress-induced migration of defects produces memory effects.

A crystal is rarely stress-free. Even if suspended on an air-cushion, there is unrelaxed stress remaining from the evaporation of the metal electrodes and the mismatch of the thermal expansion coefficients. Stress relaxation is a major source of aging. The most stable resonators available for frequency control (the BVA resonators [3]) have the electrodes separated from the resonator.

The dependence of f_r on bending can be put to use to measure surface stresses [4–6]. Such surface stresses might originate from the sample. Also, there are commercial quartz resonators operated as pressure sensors [7].

There is a crystal cut less sensitive to stress than the AT-cut, which is the so-called "SC-cut". "SC" stands for "stress-compensated". Unfortunately, the SC-cut does not resonate in the thickness-shear mode. It cannot be used in liquids because of compressional waves.

Advice:

- Avoid static stress.
- Give the crystal time to adjust to changes in stress.
- Do not touch the crystal holder (and the cables, as well) during an experiment.
- Be careful with pumping liquids; consider measurements in quiescent solution.
- Be aware of the fact that hydrostatic pressure has an effect on frequency. The respective coefficient is about 1.26 ppm/bar [8]. This can be of relevance for experiments in vacuum or in high-pressure cells.
- When the crystal experiences a rapid change in environmental conditions, always keep an eye on whether these involve transients in temperature or stress.
- If using thin crystals ($f_0 > 10$ MHz), consider a pressure compensation scheme following Ref. [9].
- Because of the effects of stress, the resonance frequency usually changes by a few Hertz, when a crystal is taken out of the holder and reinserted back into it. The crystal must be removed from the holder for some kinds of surface treatment (such as spin-casting). If such effects are to be avoided, the crystal must stay in the holder at all times.

17.3 Drive Level Dependence

At low enough oscillation amplitude, the QCM is a linear device and the resonance frequency therefore does not depend on amplitude. The fact that the QCM obeys linear acoustics is a consequence of the deformation being small. In the context of oscillations, there is a simple test for linear response, which is whether or not f_r

depends on amplitude. Readers with a physics education may recall this topic from the discussion of the pendulum. The restoring force onto the pendulum is given as $mg\ \sin(\theta)$ with m the mass, $g = 9.81\ \mathrm{m/s^2}$ standard gravity, and θ the angle of excursion. For small θ, one uses $\sin(\theta) \approx \theta$ and finds that the oscillation period is independent of θ in this limit. If $\sin(\theta)$ is noticeably different from θ, the pendulum swings slower at larger amplitudes. Precise pendulum clocks therefore employ long arms with a correspondingly small angle.

Interestingly, the exact same problems creeps back in through the back door with quartz clocks. With quartz clocks, there is no such thing as $\sin(\theta)$, but the resonance frequency nevertheless slightly depends on amplitude because of a slightly nonlinear intrinsic stress-strain relation of AT-cut quartz. The larger the amplitude, the stronger these effects. From an electrical point of view, one wants the amplitude to be high because the correspondingly high current facilitates read-out. On the other hand, nonlinear elasticity becomes noticeable at high amplitudes. The amplitude eventually chosen is a compromise. The problem is less severe than with pendulum clocks, but it is of the same nature, in principle.

The dependence of resonance frequency on driving voltage is routinely determined and therefore has acquired an abbreviation, which is the "DLD" (for "drive level dependence") [10, 11]. One should make sure to keep the drive level low enough to not observe the crystal's intrinsic elastic nonlinearity.

Advice:

- Check for the drive level dependence of your crystal, that is, determine $\Delta f(u_0)$ on the bare resonator and subtract these frequency shifts later, if necessary.
- For experiments in air a drive level of −5 dBm (~180 mV) is reasonable. (This assumes an output resistance of the impedance analyzer of 50 Ω.)
- For experiments in liquids, a drive level of 15 dBm (~1.8 V) is a good start.
- Remember that the nominal voltage at the output of an impedance analyzer can be different from the voltage at the electrode because of the impedance analyzer's finite output resistance (Sect. 7.4).

A drive level of 0 dBm corresponds to a power of 1 mW. For a 50 ohm cable, 0 dBm translates to 317 mV (zero-to-peak). The voltage is $317 \times 10^{DL/20}$ mV with DL the drive level in units of dBm.

17.4 Compressional Waves

The origin of compressional waves and their consequences are discussed in Sect. 7.6. We proceed to the recommendations:

- Always remain aware of compressional waves when doing experiments in liquids.
- In air, compressional waves are not much of a problem because air has a low compressional-wave impedance. (Effects of compressional waves are not strictly absent, though.)

- Compressional waves are particularly problematic for measurements of viscosity because the reference state is the crystal in air. The effects of compressional waves differ between air and liquid.
- When doing an adsorption experiment in a liquid (aiming at adsorbed mass), one can hope that the effects of compressional waves are the same before and after absorption.
- Compressional waves are particular detrimental if they are reflected from the back of the cell. That can be avoided by a cell design, where the opposing wall is inclined with respect to the crystal surface or has a shape other than being a flat wall. A "spoiler" (an irregularly shaped piece of Teflon mounted in front of the resonator surface) has also been used [12].
- Effects of compressional waves are often variable during temperature sweeps in a liquid cell. The observed temperature dependence might go back to a coupled resonance, rather than the sample's intrinsic temperature dependence (Fig. 7.7). Keep an eye on coupled resonances when ramping temperature.

17.5 Front Electrode Grounded

Piezoelectric stiffening was discussed in Chap. 5 and in Sect. 14.2. The recommendations are:

- Ground the front electrode well.
- If possible, use a pi-network and a transformer as discussed in Sect. 14.2. The transformer can be used to ground the front electrode.
- If you have long cables between the driving electronics and the crystals, fix these well and do not move them during the experiment.
- If in doubt, compare the frequency readings from pure water and salt solutions. If the addition of salt lowers the frequency with no corresponding change in bandwidth, you are looking at an effect of piezoelectric stiffening. For experiments in air, approach a grounded metal plate to the resonator surface from the top (thereby increasing the capacitance between the front electrode and the environment). If the frequency decreases upon bringing the metal piece close the resonator, your measurements may be affected by piezoelectric stiffening.

17.6 Baseline Stability

Baseline drifts can never be completely avoided. The drift mostly goes back to slow stress relaxation. In liquids, you can hope for a drift of less than 1 Hz/h.

Advice:

- Give the resonator some time after mounting; allow for a relaxation of the stress induced by mounting.
- If possible, keep the crystal running while no measurement is performed. This helps to maintain a stationary state.
- Allow for a run-in time after a measurement has started.
- Be careful with pumping liquids because the concurrent stress may affect the frequency.
- If possible, go back to the initial state after an experiment and check if the frequency returns to where it was before.
- If working in liquids, consider degassing the liquid in order to avoid bubbles and presoak the crystal in the liquid before an experiment.

17.7 The 01-Mode Must Be Well Separated from the Anharmonic Side Bands

Among the benefits of impedance analysis is that one can judge the quality of a crystal from the shape of the conductance trace. The 01-mode should be well separated from the anharmonic side bands (Fig. 7.3). Otherwise, there is coupling between the two modes, which interferes with the measurement.

17.8 Check for Electrical Contact by Calculating the Effective Area

If in doubt on whether the crystal is well connected, calculated the effective area with Eq. 7.4.7. If the computed area is much less than the real area, the contacts at the back to the crystal may have suffered from corrosion. (An effective area much different from the true electrode area can also result from calibration errors.)

17.9 Bandwidth Can Be More Robust than Frequency

The resonance frequency responds to changes in temperature, stress, and drive level more than the bandwidth. In harsh environments, the frequency is sometimes useless (that is, uncorrelated with the state of the sample), while the bandwidth still carries information (is correlated with the state of the sample). In such situations, one will be careful to not overinterpret the sensor output, but one might still accept the bandwidth as some kind of an indicator.

17.10 Do not Use Frequency Shifts from the Fundamental in an Analysis of Overtone Dependence

For reasons, which are poorly understood, Δf on the fundamental often deviates from what is expected, based on Δf as determined on the other overtones and the available models. Potential sources are poor energy trapping and electric fringe fields. Since there is no trust-worthy model accounting for these effects, many researchers only measure on overtones 3 and higher.

17.11 When Doing Impedance Analysis, Be Aware of the Fact that the Sweep Rate Affects the Values on the Admittance Trace

If sweeping quickly in impedance analysis, the measured admittance at any given frequency is affected by the previous reading. This happens if the time elapsed between two measurement of G_{el} and B_{el} on the frequency scale is less than the inverse bandwidth of the resonance. (That usually is the case.) The fitted resonance frequency slightly shifts when the sweep rate is changed. Importantly, the apparent resonance frequency can also shift slightly, when the bandwidth changes by large amounts. The effect can be accounted for by performing two sweeps in opposite directions and taking the averages of the corresponding fit parameters.

17.12 When Comparing Overtones, Be Aware of the Fact that the Data at the Different Overtones Are Acquired at Different Times

The comparison between overtones is an essential element of the modeling process. When the sample's properties change with time, an overtone dependence of $\Delta f/n$ may reflect a time-dependence of $\Delta f/n$, rather than an n-dependence. The effects can be checked for by alternating the sequence, in which overtones are accessed (low $n \rightarrow$ high n followed by high $n \rightarrow$ low n).

17.13 When Analyzing the Overtone Dependence of Frequency Shifts Acquired on Samples in the Dry State, Be Aware of the Electrode Effects

Although not immediately apparent, the acoustic impedance of the electrode has an influence on the frequency shift induced by coating the electrode with some other layer. Electrode effects are particularly pronounced for films in the dry (Sect. 6.2). Also, remember that the modal mass has a small dependence on overtone order (Sect. 6.1.4).

17.14 Re-use of Crystals, Cleaning, Maintenance of Holder

Unfortunately, the lifetime of the crystals is rather limited. Depending one the details, crystals can only be re-used in a few (up to a few ten) experiments.

Advice:

- Check for crystal quality by monitoring the conductance trace of the anharmonic side bands. If the crystal has suffered, this spectrum changes. The band with the lowest frequency in a group should also be the one with the largest conductance (Sect. 7.2).
- As crystals age, their intrinsic bandwidth increases.
- Do not clean crystals with Piranha solution if you can avoid it. Piranha solution etches SiO_2. More generally, avoid alkaline solutions.
- UV-Ozone cleaning is effective in removing organic contaminants.
- If working in liquids, keep an eye on chemical compatibility of the liquid with the O-rings of the holder. Also, check the electrical contacts (the soft springs contacting the back of the crystal) for corrosion.
- See Ref. [13] for more suggestions on how to clean crystals after contact with various types of samples.

Glossary

Variable	Definition (Comments)
a, b, c	Coefficients of the 3rd-order polynomial describing the T-f characteristic of AT-cut crystals
f	Frequency
T	Temperature
T_0	Reference temperature of the T-f characteristic
δf	Shift in resonance frequency caused by temperature drifts
$\Delta f(u_0)$	Dependence of the frequency shift on amplitude, u_0
θ	Angle of a crystal cut (Angle between the surface normal and the crystallographic y-axis)

References

1. http://www.amus/Local_Papers/U11625%20VIG-TUTORIAL.pdf. Accessed 18 June 2014
2. Neubig, B., Briese, W.: Das große Quarzkochbuch. Franzis Verlag GmbH (2000)
3. http://tf.boulder.nist.gov/general/pdf/214.pdf. Accessed 15 Feb 2013
4. Heusler, K.E., Grzegorzewski, A., Jackel, L., Pietrucha, J.: Measurement of mass and surface stress at one electrode of a quartz oscillator. Ber. Der Bunsen Ges. Phys. Chem. Chem. Phys. **92**(11), 1218–1225 (1988)
5. Wagner, H.J., Hartig, W., Büttgenbach, S.: Design and fabrication of resonating at-quartz diaphragms as pressure transducers. Sens. Actuators A Phys. **42**(1–3), 389–393 (1994)
6. EerNisse, E.P.: Simultaneous thin-film stress and mass-change measurements using quartz resonators. J. Appl. Phys. **43**(4), 1330 (1972)
7. http://www.quartzdyne.com/quartz-pressure-sensor/quartz.html. Accessed 21 Sept 2013
8. Stockbridge, C.D.: In Behrndt, K.H. (ed.) Vacuum Microbalance Techniques, vol. 5, 4th edn. Plenum Press, New York (1966)
9. Sota, H., Yoshimine, H., Whittier, R.F., Gotoh, M., Shinohara, Y., Hasegawa, Y., Okahata, Y.: A versatile planar QCM-based sensor design for nonlabeling biomolecule detection. Anal. Chem. **74**(15), 3592–3598 (2002)
10. Tiersten, H.F., Stevens, D.S.: An analysis of nonlinear resonance in contoured-quartz crystal resonators. J. Acoust. Soc. Am. **80**(4), 1122–1132 (1986)
11. Nosek, J.: Drive level dependence of the resonant frequency in BAW quartz resonators and his modeling. IEEE Trans. Ultrason. Ferroelectr. Freq. Control **46**(4), 823–829 (1999)
12. Eggers, F., Funck, T.: Method for measurement of shear-wave impedance in the MHZ region for liquid samples of approximately 1 Ml. J. Phys. E Sci. Instrum. **20**(5), 523–530 (1987)
13. http://www.thinksrs.com/downloads/PDFs/Manuals/QCM200m.pdf. Accessed 14 Feb 2013

Collection of Essential Equations

Small load approximation (SLA)	$\frac{\Delta\tilde{f}}{f_0} = \frac{i}{\pi Z_q}\tilde{Z}_L = \frac{i}{\pi Z_q}\frac{\langle\hat{\sigma}_S\rangle_{area}}{\hat{v}_S}$
Gordon-Kanazawa-Mason result	$\frac{\Delta\tilde{f}}{f_0} = \frac{-1+i}{\sqrt{2}\pi Z_q}\sqrt{\omega}\sqrt{\rho_{liq}\eta_{liq}}$
Viscoelastic medium	$\tilde{G}(\omega)\rho = -\left(\pi Z_q \frac{\Delta\tilde{f}}{f_0}\right)^2$
Thin rigid film (Sauerbrey Equation)	$\frac{\Delta\tilde{f}}{f_0} \approx -\frac{2nf_0}{Z_q}m_f = -n\frac{\langle m_f\rangle_{area}}{m_q}$
Viscoelastic film in air	$\frac{\Delta\tilde{f}}{f_0} = \frac{-1}{\pi Z_q}\tilde{Z}_f \tan\left(\tilde{k}_f d_f\right)$
Thin viscoelastic film in air	$\frac{\Delta\tilde{f}}{f_0} = \frac{-\omega m_f}{\pi Z_q}\left[1 + \frac{(n\pi)^2}{3}\left(\frac{\tilde{J}_f}{\rho_f}Z_q^2 - 1\right)\left(\frac{m_f}{m_q}\right)^2\right]$
Viscoelastic film in liquid	$\frac{\Delta\tilde{f}}{f_0} = \frac{-\tilde{Z}_f}{\pi Z_q}\frac{\tilde{Z}_f tan\left(\tilde{k}_f d_f\right) - i\tilde{Z}_{liq}}{\tilde{Z}_f + i\tilde{Z}_{liq} tan\left(\tilde{k}_f d_f\right)}$
Thin film in liquid	$\frac{\Delta\tilde{f}}{f_0} \approx \frac{-\omega m_f}{\pi Z_q}\left[1 - 2\pi i n\frac{\tilde{J}_f(\omega)}{\rho_f}f_0\rho_{liq}\eta_{liq}\right]$
Thin adsorbate in liquid	$\frac{\Delta\tilde{f}}{f_0} \approx -\frac{\rho\omega}{\pi Z_q}\int_0^\infty \left[\frac{\tilde{G}(z) - \tilde{G}_{liq}}{\tilde{G}(z)}\right]dz$
Contact stiffness	$\Delta f = \frac{1}{n}\frac{N_P}{2\pi^2 Z_q}\kappa_P$
Time averaging	$\frac{\Delta\tilde{f}}{f_0} = \frac{i}{\pi Z_q}\frac{2\langle\sigma_S(t)\exp(i\omega t)\rangle_{area,time}}{\hat{v}_S}$
SLA in tensor form	$\Delta\tilde{f} \approx \frac{i}{4\pi\rho_q}\frac{\sum_{ijk}\int_{Surface}\hat{u}_{0,S,i}n_j\tilde{Z}_{L,ijk}(\mathbf{r}_S)\hat{u}_{0,S,k}d^2\mathbf{r}_S}{\sum_k\int_{Volume}\hat{u}_{0,k}\hat{u}_{0,k}d^3\mathbf{r}}$
Piezoelectric stiffening	$\frac{\Delta\tilde{f}_{PE}}{f_0} = \frac{i}{A\pi Z_q}4\phi^2\left(\left(i\omega C_0 + \tilde{Z}_{ex}^{-1}\right)^{-1} - \left(i\omega C_0 + \tilde{Z}_{ex,ref}^{-1}\right)^{-1}\right)$

D. Johannsmann, *The Quartz Crystal Microbalance in Soft Matter Research*, Soft and Biological Matter, DOI: 10.1007/978-3-319-07836-6,

MIX
Papier aus verantwortungsvollen Quellen
Paper from responsible sources
FSC® C105338

If you have any concerns about our products,
you can contact us on
ProductSafety@springernature.com

In case Publisher is established outside the EU,
the EU authorized representative is:
Springer Nature Customer Service Center GmbH
Europaplatz 3, 69115 Heidelberg, Germany

Printed by Libri Plureos GmbH
in Hamburg, Germany